The Road Map of China's Steel Industry

Xinchuang Li

# The Road Map of China's Steel Industry

## Reduction, Innovation and Transformation

冶金工业出版社
Metallurgical Industry Press

Springer

Xinchuang Li
China Metallurgical Industry Planning
and Research Institute
Beijing, China

ISBN 978-981-15-2073-0 ISBN 978-981-15-2074-7 (eBook)
https://doi.org/10.1007/978-981-15-2074-7

Jointly published with Metallurgical Industry Press
The print edition is not for sale in China (Mainland). Customers from China (Mainland) please order the print book from: Metallurgical Industry Press.
ISBN of the Co-Publisher's edition: 978-7-5024-8251-0



This Springer imprint is published by the registered company Springer Nature Singapore Pte Ltd.
The registered company address is: 152 Beach Road, #21-01/04 Gateway East, Singapore 189721, Singapore

# Foreword

The iron and steel industry is an important basic industry of the national economy. There are numerous writings on iron and steel, but most on metallurgical principles, engineering design, process technology, etc. There are only a few ones on the macro-developments of the industry, and *The Road Map of China's Steel Industry* is the epitome. As a monograph focusing on the analysis of the development history, law and direction of the iron and steel industry, the book written by Mr. Xinchuang Li is based on his experiences of more than 30 years on the planning and research of the iron and steel industry. It is my great honor to prepare the preface for this book.

Since the founding of the People's Republic of China and through the efforts by several generations of the people working in the iron and steel industry in several decades, China's iron and steel industry has grown from small to large and from weak to strong, and has embarked on a unique development road that conforms to the law of both global iron and steel industry and China's national conditions. At the new historical starting point, how can the iron and steel industry review the path it has taken in the perspective of internationalization? How should it adapt, lead and grasp the new normal? How should it thoroughly implement the supply-side structural reform? In the book, Mr. Xinchuang Li makes in-depth and meticulous analysis, exploration and research on the above issues.

In this manuscript, the logical veins are very clear. "Reduction" is a strategic judgment on the development of the iron and steel industry, "innovation" is the drive for the development of the iron and steel industry under the new normal, and "transformation" is the implementation path of supply-side structural reform in the iron and steel industry. "Reduction, innovation and transformation" were proposed by Mr. Xinchuang Li in an impromptu speech when he went to a steel company to give lecture a few years ago. At that time, the problem of overcapacity in the iron and steel industry was not fully recognized and paid with enough attention. The development of the iron and steel industry later confirmed Mr. Li's judgment, which well indicates his insight and predictability to the industry. In the first part, the book not only reviews the long-term development history of both China's and global iron and steel industry, but also summarizes the laws of the transfer and industrial layout of the global iron and steel industry, conducts an in-depth analysis of China's resolute decision on cutting overcapacity of the iron and steel industry from 2016 to the present, points out the huge role played by the supply-side structural reform in improving the competitiveness of China's iron and steel industry, and provides a useful Chinese solution to solve the global problem of overcapacity in the iron and steel industry. What is the future direction of the iron and steel industry? How shall we do? In the second part, a clear answer is given, i.e. promoting coordinated development in "Nine Aspects" (greenness, coordination, quality focus, standardization, differentiation, service-oriented, intelligence, diversification, and internationalization) to reshape the industrial value chain and continuously improve the competitiveness of the iron and steel industry. Each part is independent and inseparable and organically related yet distinct from each other. When reading the book, I was deeply aware of the clear viewpoints, objective statements, rich cases, and detailed data therein, showing the author's deep feelings and ingenious insights to the iron and steel industry.

Mr. Xinchuang Li has formed an indissoluble bond with the iron and steel industry. He had studied "iron and steel" in the Central South University and University of Science and Technology Beijing, "designed" iron and steel in Shanxi Metallurgical Design Institute, "planned" iron and steel in the Ministry of Metallurgical Industry, and "researched" iron and steel in China Metallurgical Industry Planning and Research Institute. Mr. Li once studied in an international MBA program at Peking University. He has long been making researches, has profound knowledge about the development trend of the iron and steel industry, policy research, corporate strategic positioning, project investment, etc., and has published many heavyweight papers and two monographs, which provides a more systematic research concept for promoting the healthy development of the iron and steel industry. I believe that the book plays an important guiding role for the readers to understand the development history of the iron and steel industry and master its future development.

Beijing, China
October 2017

Yong Gan

# Preface

Under the background that China's economic development has entered into a new normal, with the transformation of economic growth into medium and high speed, the move of the industrial structure towards mid-to-high end and the shift to new economic growth drives, the consumption intensity of steel products per unit of GDP tends to decline, and the total steel consumption has entered a declining zone after the peak arc zone. Despite the fluctuations and repetitions, the general trend of reduction-based development of China's iron and steel industry has become clear, which will lead to gradual declines in total energy consumption, total pollutant emissions and total resource consumption, have a far-reaching effect on the process restructuring, technical equipment structure, product structure, energy structure, resource structure, organizational structure and industrial layout of China's iron and steel industry, and put forward new requirements for industrial development mode and development drives.

With the 19th National Congress of the Communist Party of China as an important mark, a new era of socialism with Chinese characteristics has been opened. The principal contradiction faced by Chinese society has evolved into the one between unbalanced and inadequate development and the people's ever-growing needs for a better life, and the economic growth has turned from a high-speed stage to a high-quality stage. The iron and steel industry is an important basic industry of the national economy. Looking back at history, the Chinese nation needs iron and steel to stand up; the Chinese nation needs more iron and steel to become rich. Looking ahead, the Chinese nation needs better iron and steel to become more powerful. This puts higher demands on not only steel products but also iron and steel enterprises. How to do better requires the iron and steel industry to further implement the development concept of "innovation, coordination, greenness, openness, and sharing" and deepen the supply-side structural reform, which is the key to leading the development of the iron and steel industry in the new era.

China Metallurgical Industry Planning and Research Institute (MPI) is the advisory body for government agencies, the leader of industrial development, and the think tank for corporate planning. MPI has undertaken many important research projects on the policies for iron and steel industry, environment, safety, energy and quality, hundreds of special analysis related to raw materials, production and consumption of the iron and steel industry, and consultation reports for thousands of Chinese and foreign companies. As the President of China Metallurgical Industry Planning and Research Institute, I have taken charge of and participated in the above-mentioned major consultation projects. I am grateful for the great new era we are in and the iron and steel industry which enjoys an important position as a basic industry, and wish to express sincere thanks to the unique platform of China Metallurgical Industry Planning and Research Institute and to our energetic and united colleagues, which have given me many opportunities, special perspectives and personal feelings, thus forming an organic and unified understanding of the external situation, strategic position, development drives and transformation path of the iron and steel industry in the new era.

Following contents are mainly elaborated in this book—*The Road Map of China's Steel Industry—Reduction, Innovation and Transformation*: the development history and the industrial center shift of the global and China's iron and steel industry are reviewed, the industrial laws behind it is analyzed, and new understandings and judgments on the development stage of the global and China's iron and steel industry are made; from the perspective of the characteristics of the iron and steel industry and population economics, the competitive advantages of China's iron and steel industry is deeply analyzed, and a proposal that the Chinese iron and steel industry will lead the global iron and steel industry for a long time is made; the international and domestic situations faced by China's iron and steel industry under the new normal are analyzed, and the requirements for supply-side structural reform of China's iron and steel industry are put forward; relevant measures of "cutting overcapacity" for implementing the supply-side structural reform in the iron and steel industry are illustrated from comprehensive perspective, the major historical role of supply-side structural reform to promote the restructuring and upgrading of China's iron and steel industry and China's contributions to resolving the overcapacity of global iron and steel industry driven by supply-side structural reform are put forward through study; as socialism with Chinese characteristics enters a new era, the coordinated and innovative development of the iron and steel industry in "nine aspects" will be promoted in the future to reshape the industrial value chain and continuously improve industrial competitiveness.

I am humble in writing, and also felt that my knowledge is inadequate and the opinions are limited during the compilation process, but I do hope that the book can induce valuable opinions and make contributions to the society. If the book can give some references and guidance to the development of the iron and steel industry in the new era, the care, help, support and encourage given by many leaders and experts to me will not be in vain. As an ordinary practicer working in the iron and steel industry, I will be grateful, continue to work hard, and live up to expectations.

I sincerely thank the leaders, colleagues and all the friends who have provided support for the book! I express special thanks to Mr. Yong Gan, Academician of the Chinese Academy of Engineering for taking the time to write a preface to the book and provide guidance for the development of the iron and steel industry, and give thanks to the hard work made by my colleagues including Tiejun Fan, Longqiang Zhang, Xiaodong Jiang, Chuang Li, Jijun Cheng, Tao Liu, Xue Gao, Bing Li, Qi Liu, Cheng Chen, Ning Lu, Dongmei Huo, Sheng Gao, Jia Zhao, Zhijie Guan, Weiwei Zhang, Cantao Shi, Xiuting Wu, Feng Peng, Songbo Zhang, Deng Pan, Xiao Li, Xiang Zhou, Mei Li, Yifan Wang, etc. I sincerely hope that readers will provide criticism and guiding opinions so that I can continue to make improvement.

Beijing, China
December 2017

Xinchuang Li
President of China Metallurgical
Industry Planning and Research Institute

# Introduction

With the socialism with Chinese characteristics entering a new era, how to achieve high-quality development of the iron and steel industry has become a major proposition in front of the Chinese working in the iron and steel industry. This book focuses on the general requirements for high-quality development, combines the main line of supply-side structural reform with the reality of China's iron and steel industry, and takes the reduction, innovation and transformation as the veins. The book is divided into two parts and thirteen chapters, in which the development history of the global and China's iron and steel industry are reviewed, the industry laws are extracted from industry transfer footprints, a large number of enterprise practice cases are analyzed, the development situations faced by the iron and steel industry in the new era are systematically analyzed, and the strategic direction and implementation path for the future development of the iron and steel industry are proposed. Through the analysis in the book, the author believes that in order to achieve high-quality development of the iron and steel industry, it must strengthen innovation drive, promote coordinated development of "Nine Aspects", namely, greenness, coordination, quality focus, standardization, differentiation, service-oriented, intelligence, diversification and internationalization, and reshape the value chain of the iron and steel industry, continuously improving the competitiveness of the industry.

This book can be used by the leaders of metallurgical enterprises and institutions, the management and scientific research technicians engaged in planning research, engineering consulting, finance and other business as well as the teachers and students of metallurgy-related professions in colleges and universities.

Book Cataloging-in-Publication (CIP) Data

*The Road Map of China's Steel Industry*: *Reduction, Innovation and Transformation*/Xinchuang Li.–Beijing: Metallurgical Industry Press, March, 2018
ISBN 978-7-5024-7736-3

I. ①... II. ① Li... III. ① Iron and Steel Industry-Industrial Development-China IV.1 F426.31

No. 035972 (2018) of CIP data checking by Archival Library of Chinese Publications

Publisher: Tan Xueyu

Address: 39 Songzhuyuan North Lane, Dongcheng District, Beijing 100009, China, Telephone: 0086 (10) 64027926

Website: www.cnmip.com.cn; E-mail: yjcbs@cnmip.com.cn

Responsible Editors: Li Peilu, Xia Xiaoxue; Art Editor: Lu Xintong; Layout Design: Sun Yuehong

Responsible Proofreader: Qing Wenchun; Responsible Printer: Li Yushan

ISBN 978-7-5024-7736-3

Published by Metallurgical Industry Press; Distributed by Xinhua Bookstores around the country; Printed by Sanhe Shuangfeng Printing & Binding Co., Ltd.

First edition in March 2018, and first printing in March 2018

169*239 mm; 24.25 printed sheets; 1 color page; 398 thousand words; 374 pages 85.00 yuan

Metallurgical Industry Press Submission Telephone Submission Box

Marketing Center of Metallurgical Industry Press Telephone Fax

T-mall Flagship store of Metallurgical Industry Press

(If the book has any printing quality problem, we are responsible for the refund or replacement).

# Contents

## Part I Introduction

# About the Author

**Mr. Xinchuang Li** was born in November 1964 in Yuncheng City of Shanxi Province, China. He is currently the Chairman and Chief Engineer of China Metallurgical Industry Planning and Research Institute (MPI), Vice Chairman of China Iron and Steel Association (CISA), Vice Director General of China Energy Conservation Association (CECA) as well as Director of the Technical Economics Branch of the Chinese Society for Metals, enjoying special allowance approved by the State Council, Professorate Senior Engineer and registered consulting engineer (investment). He received his bachelor's degree from Central South University in 1984 and master's degree from the University of Science and Technology Beijing in 1989. He was awarded an M.B.A. degree by Fordham University, USA, in 2001. Mr. Li has been employed by Shanxi Metallurgical Design Institute and the Department of Development Planning of the former Ministry of Metallurgical Industry.

Mr. Li has participated in and undertaken development planning for China's iron and steel industry from the "Eighth Five-Year Plan" to the "Thirteenth Five-Year Plan" and has been involved in development planning for several hundred domestic and overseas iron & steel enterprises, as well as early stage debates on key projects. He has participated in a number of international planning/consulting meetings and cooperation negotiations, which promoted China's iron & steel enterprises to accelerate the pace of globalization.

Mr. Li has been an expert advisor to the National Development and Reform Commission (NDRC), the Ministry of Industry and Information Technology of the PRC (MIIT), the Ministry of Environmental Protection of the PRC, and the Ministry of Science and Technology of the PRC. He has also been an advisor to the China Development Bank, an independent director of the State-Owned Assets Supervision and Administration Commission of Tianjin Municipal People's Government, a member of the expert committee for the China Overseas Development Association, and a member of the first strategic advisory committee of HBIS Group Co. Ltd. As a result, he has extensive experience in the management of major projects, as well as a comprehensive grasp of industrial development and market dynamics. He has published about 150 articles on the development of iron & steel industry, and published several books entitled *Sustainable Development of Iron and Steel Industry*, *How to Transform and Upgrade China's Steel Industry* and *The Road Map of China's Steel Industry*, etc.

# Part I
# Introduction

# Chapter 1
# Development Road of China's Modern Iron and Steel Industry

The development of the modern iron and steel industry in China began in 1949 when the People's Republic of China was founded. In the past 70 years, along with China's growth and development, the modern iron and steel industry in China has experienced a great course of recovery, growth, and rise. In general, it is divided into four stages (see Fig. 1.1): the first stage begun since the founding of the People's Republic of China in 1949 and ended during the "Cultural Revolution", and it was in a stage of exploration, showing a trend of fluctuating development; the second stage begun from the beginning of reform and opening-up to the end of the twentieth century, and it was in its start-up stage, showing a trend of stable development; the third stage begun from the beginning of the twenty-first century to 2014, and it was in an accelerated stage, showing a trend of leapfrog development; the fourth stage begun from 2015 to the present, and it is in a stage of reduction, showing a trend of innovative development.

From the dimension of history and global perspective, the development, rise, and adjustment of China's iron and steel industry is a necessity, and it is the result of the combined effect by the development rule of global iron and steel industry and the actual characteristics of China's iron and steel industry. Different from the iron and steel industry in developed countries such as Europe and the USA, China's iron and steel industry started in the period of planned economy and emerged in the process of reform and opening-up. It has taken a unique and unprecedented development road and strongly supported the development of China's national economy. The practice has proved that this road is successful. No other road can meet the demand for iron and steel by China's rapid development. At the same time, in the process of rising development and meeting demand, China's iron and steel industry has also taken over the baton for the transfer of the global iron and steel industry, expanded the application filed of iron and steel materials, and adopted a large number of new technologies and equipment, which has pushed the development of the global iron and steel industry to a new high level.



X. C. Li, *The Road Map of China's Steel Industry*,
https://doi.org/10.1007/978-981-15-2074-7_1

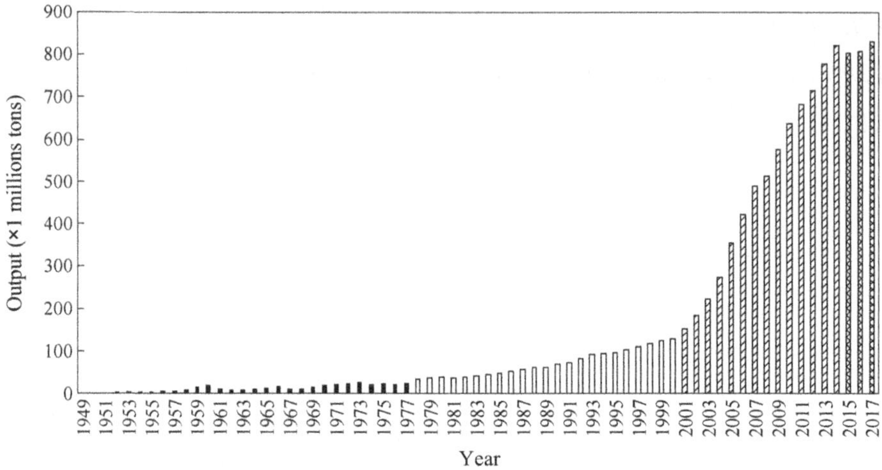


**Fig. 1.1** Development stages of the modern iron and steel industry in China

## 1.1 Exploration Stage of Fluctuating Development

In 1949, the national steel output of China was only 158,000 tons, ranking 26th in the world. After three years of production recovery, it experienced a rapid development during the 1st Five-Year Plan period, the three years of nationwide ironmaking and steelmaking during "Great Leap Forward", five-year adjustment and the follow-up decade of "Cultural Revolution", during which the output of crude steel had gone through three times of dramatic fall and rise, with both experiences of success and lessons of failure accumulated. During that period, the older generation of proletarian revolutionaries constantly explored the path to build socialism in China, and the modern iron and steel industry in China was also in its infancy, showing a very obvious trend of exploration and development. At that stage, the cumulative steel output of China was more than 300 million tons, and three large-scale national steel bases including Anshan Steel, Wuhan Steel, and Baotou Steel as well as Taiyuan Steel, Ma'anshan Steel, Panzhihua Steel, Bayi Steel, Hangzhou Steel, Liuzhou Steel, Tonghua Steel, Xinyu Steel and other key local steel plants were built.

### *1.1.1 Restoration and Transformation of Iron and Steel Production Facilities*

Before 1949, the iron and steel production capacity in China was small, the number of technological equipment was less, and the technology was backward. Due to years of war, most of them were in a state of production suspension, and some even were demolished and destroyed. After 1949, the modern iron and steel industry in

China was developed based on the restoration, transformation, and expansion of the iron and steel production facilities built before 1949. From 1949 to 1952, 34 blast furnaces and 26 open hearth furnaces were restored and expanded. In 1952, 1.929 million tons of iron were produced in China, 7.1% greater than the highest annual iron output before 1949; 1.349 million tons of steel were produced, 91.2% greater than the highest annual steel output before 1949. In 1953, the electric arc furnaces had resumed production in Daye Steel, Chongqing Steel, Taiyuan Steel, and other steel plants [1]. By 1957, after the repair and transformation, all the iron and steel production facilities left before 1949 basically resumed production.

### *1.1.2 Opening-Up of Iron and Steel Capital Construction to Increase Production Capacity*

1. Construction During the 1st Five-Year Plan Period

After 1949, the capital construction of the iron and steel industry began with the expansion and reconstruction of Anshan Steel from 1950 to 1952. Since the 1st Five-Year Plan period, China had started a large-scale capital construction based on new plants. From 1953 to 1957, China carried out the construction of eight iron and steel projects among the 156 large-scale projects aided by the Soviet Union, including the newly built Wuhan Steel Plant, Beiman Steel Plant, Baotou Steel Plant, Jilin Ferroalloy Plant, Jilin Carbon Plant, and Rehe Ferrovanadium Plant (the predecessor of Chengde Steel Plant). Anshan Steel and Benxi Steel were rebuilt and expanded as well during that period. At the same time, the reconstruction and expansion projects for more than 20 enterprises, including Daye Steel Plant, Taiyuan Steel Plant, Tangshan Steel Plant, Chongqing Special Steel Plant, Beijing Shijingshan Steel Plant, Shanghai Steel Plant No. 1, Shanghai Steel Plant No. 3, Shanghai Steel Plant No. 5, Fushun Steel Plant, and Dalian Steel Plant, were completed.

2. "Three Big, Five Medium and Eighteen Small" Iron and Steel Projects

In 1956, China began to plan and build "three big, five medium and eighteen small" iron and steel projects. The "three big ones" refer to the continued construction of three large iron and steel bases with a steel production capacity of more than 1 million tons, i.e., Anshan Steel, Wuhan Steel, and Baotou Steel. The "five medium ones" refer to the construction of medium-sized steel plants with an annual steel production capacity of 300,000–1,000,000 tons, including the expansion of Taiyuan Steel, Chongqing Steel, Ma'anshan Steel, and Shijingshan Steel Plant, and the construction of Xiangtan Steel Plant. The "eighteen small ones" refer to the construction of eighteen small steel plants with an annual steel production capacity of 100,000–300,000 tons in eighteen provinces and/or autonomous regions, including Handan Steel Plant in Hebei Province, Jinan Steel Plant in Shandong Province, Linfen Steel Plant in Shanxi Province, Xinyu Steel Plant in Jiangxi Province, Nanjing Steel Plant in Jiangsu Province, Liuzhou Steel Plant in Guangxi Zhuang Autonomous

Region, Guangzhou Steel Plant in Guangdong Province, Sanming Steel Plant in Fujian Province, Hefei Steel Plant in Anhui Province, Jiangyou Steel Plant in Sichuan Province, Bayi Steel Plant in Xinjiang Uyghur Autonomous Region, Hangzhou Steel Plant in Zhejiang Province, Echeng Steel Plant in Hubei Province, Lianyuan Steel Plant in Hunan Province, Anyang Steel Plant in Henan Province, Lanzhou Steel Plant in Gansu Province, Guiyang Steel Plant in Guizhou Province, and Tonghua Steel Plant in Jilin Province.

3. Iron and Steel Projects during "Third-Front Construction"

In 1964, under the direction of the Central Committee of the Communist Party of China and Chairman Mao, China's iron and steel industry began to promote the construction of the third-front iron and steel bases, with Panzhihua Iron and Steel Company as the landmark project. At the end of 1964, Deng Xiaoping was commissioned by the Central Committee of the Communist Party of China to Panzhihua to approve the construction plan of Panzhihua Steel. In 1967, a 1,000 $m^3$ blast furnace was completed and put into operation in Panzhihua Steel. In the construction of the third-front iron and steel bases, Great Wall Steel Plant, Emei Ferroalloy Plant, Shuicheng Steel Plant, Zunyi Metal Products Factory, etc. were newly built; Jiuquan Steel, Lanzhou Steel, and Lueyang Steel were restored; Chengdu Seamless Steel Pipe Plant, Chongqing Steel and Chongqing Special Steel, Kunming Steel, Guiyang Steel, and Zunyi Ferroalloy Plant were expanded; Anshan Steel assisted the construction of Shizuishan Wire Rope Plant in Ningxia Hui Autonomous Region, part of Benxi Steel was separated and moved to build Xining Special Steel, and Dalian Steel Plant assisted the construction of Shaanxi Steel Plant.

In addition, under the positive guidance of both the central and local governments, local medium and small iron and steel enterprises were also built in that period, such as Nanchang Steel Plant, Suzhou Steel Plant, Shijiazhuang Steel Plant, Xingtai Steel Plant, Wuhu Steel Plant, Chengdu Steel Plant, Qingdao Steel Plant, Beitai Steel Plant, Shaoguan Steel Plant, Lingyuan Steel Plant, Wuxi Steel Plant, Wuyang Special Heavy Plate Plant, and Laiwu Steel Plant.

Since 1949, China's iron and steel industry got off the ground from rubbles. Till the beginning of reform and opening-up, China had 982 blast furnaces with a total volume of 87,204 $m^3$, 98 open hearth furnaces with a total nominal capacity of 17,188 tons, 276 converters with a total nominal capacity of 3,034 tons, and 1,678 electric furnaces with a total nominal capacity of 4,156 tons. Furthermore, at that stage, China had established Northeastern Institute of Technology, Beijing Institute of Iron and Steel Technology, Central South Institute of Mining and Metallurgy, Kunming Institute of Technology, Xi'an Institute of Metallurgy and Architecture, Anshan Institute of Iron and Steel Technology, Wuhan Institute of Iron and Steel Technology, Baotou Institute of Iron and Steel Technology, Ma'anshan Institute of Iron and Steel Technology, and other metallurgical institutes; set up scientific research and construction institutions such as iron and steel research institutes, planning institutes, design institutes, mining and metallurgical institutes, building institutes, and metallurgical construction teams; developed and applied such technologies as oxygen top-blown converters, continuous casting, secondary refining, and tandem steel rolling; thus, a relatively complete development system for the iron and steel industry was initially formed.

## 1.2 Initial Stage of Stable Development

Since the reform and opening-up, on the one hand, with the determination of various tasks centered on economic development, China's economy developed rapidly leading to constantly increased demand on iron and steel. On the other hand, the reform had removed some of the previous institutional and systematic obstacles, the planning system gradually moved to the market system, and the productivity was gradually released. That period was the start-up stage of China's iron and steel industry, which showed a stable development situation for more than 20 years. At that stage, China's crude steel output was increased from 31.78 million tons in 1978 to 128.5 million tons in 2000, an increase of 4.04 times, marking an average annual growth rate of 6.56%. Meanwhile, China's crude steel output broke through 100 million tons in 1996, reaching 10.124 million tons, and China became the largest steel producer in the world. Under the care of the Party and the state, the nationwide resources and power were mobilized to build Baosteel, the most modern and competitive steel enterprise in China, which also was truly the first modern iron and steel base in China.

### *1.2.1 Technical Transformation of Existing Iron and Steel Enterprises*

At the beginning of reform and opening-up, the development of China's iron and steel industry lagged far behind the global advanced level, and the production efficiency was very low. There was no modern iron and steel enterprise in the true sense. For example, the steel output by open hearth furnaces still accounted for about one-third, the output of continuous-cast billet accounted for less than 10%, a considerable number of iron and steel enterprises were subject to unmatched process, and there were many phenomena of making steel by melting iron and rolling steel with purchased steel billets. In addition, most of the investment given by the state to the iron and steel industry was used for the construction of Baosteel. The contradiction between the lack of funds and the construction needs was more prominent for other iron and steel enterprises. In order to solve that problem, the iron and steel industry adopted a development road to implement technological transformation of existing iron and steel enterprises and carry out connotation-oriented expanded reproduction. In the early-to-mid 1980s, under the guidance of the policy of "mining potential, transforming, supporting, and expanding", the iron and steel enterprises such as Anshan Steel, Handan Steel, Baotou Steel, Tangshan Steel, Anyang Steel, and Hangzhou Steel continued to strengthen management and mobilized movable properties to transform out-of-date process equipment, complete and replenish supporting process facilities, and fill the gap of unmatched processed, which made their technical level and production efficiency improved to a certain level.

In the mid-to-late 1980s, the development of the iron and steel industry focused on solving the problem of "three incompatibilities" in terms of product quantity, variety structure, and product quality, and the technological transformation of old enterprises was continuously increased. During that period, the ironmaking facilities including one 3,200 $m^3$ blast furnace of Wuhan Steel, one 2,500 $m^3$ blast furnace of Ma'anshan Steel, one 2,200 $m^3$ blast furnace of Baotou Steel, two 1,200 $m^3$ blast furnaces of Tangshan Steel, one 1,200 $m^3$ blast furnace of Chongqing Steel, and one 1,260 $m^3$ blast furnace of Handan Steel were built; the construction of the 210 t converter steelmaking plant of Capital Steel, the 50 t converter steelmaking plant of Taiyuan Steel, the 50 t converter steelmaking plant of Chongqing Steel, and the 50 t electric furnace plant of Xinyegang Steel was commenced; more than 70 casting machines were also built, making the total continuous casting capacity of China reach more than 30 million tons. In addition, the construction of the 4,300 mm heavy plate mill of Anshan Steel, the 4,200 mm/3,500 mm mill of Shanghai Steel Plant No. 3 and the 2,800 mm medium plate mill of Handan Steel greatly improved the production capacity of medium and heavy plates. Ma'anshan Steel and Laiwu Steel introduced H-beam rolling mills, which filled the domestic gap in that steel product variety.

In the 1990s, the effect to eliminate backward technical equipment for China's iron and steel industry was increased, and advanced and applicable common technologies focusing on continuous casting and extended to both upstream and downstream processes were adopted to transform existing process flows and technologies, for example, adding hot metal pretreatment and secondary refining, elimination of open hearth furnace and break-down rolling, development of top and bottom combined blown converter and ultra-high power electric furnace, and application of ironmaking concentrate technology for blast furnaces, pulverized coal injection technology, hot transport and hot charging technology of continuous-cast billets, and splashing slag protection technology of oxygen converter. Through the elimination, transformation, and new construction during the 8th Five-Year Plan and 9th Five-Year Plan in the 1990s, the adjustment and optimization of technological equipment for China's iron and steel industry was promoted, and the production efficiency and the product quality were improved. The domestic market share of the steel products such as automobile panels, pipeline steel, petroleum pipes, boilerplates, and heavy rails continued to increase.

### *1.2.2 Newly-Built Baosteel had Greatly Accelerated the Modernization Pace of China's Iron and Steel Industry*

Baosteel was the product of reform and opening-up. In order to solve the iron shortage problem that had been plagued by the iron and steel industry of Shanghai for a long time, in early 1977, the Shanghai Government and the Ministry of Metallurgical Industry planned to build an ironmaking plant in Shanghai, which was the original idea of Baosteel project. From the overall situation of realizing the modernization of China's iron and steel industry and promoting the further development of the national economy, the leaders of the Central Committee of the Communist Party of China and the State Council upgraded the intention of building an ironmaking plant to the construction of a modern large-scale steel plant. On December 23, 1978, on the second day following the closing of the Third Plenary Session of the 11th CPC Central Committee, the construction of Baosteel project was commenced.

Baosteel project was the largest project in China since 1949. At that time, China had no experience in building large-scale modern iron and steel enterprise. At the initial stage, it was completely imported from Japan, and the investment was huge. Therefore, there was always controversy in the early and initial stages of construction. Shortly after the official commencement of the Baosteel project, it coincided with the adjustment of the national economy, during which shortening the basic construction front and controlling the investment scale became a top priority. As a giant project invested by the state, Baosteel project was listed as the first among the projects to be reapproved. Under the care of the leading comrades of the central government, after full justification, the opinion of building the Baosteel project in two phases was adopted. In 1981, the Phase-I project of Baosteel was reclassified as a continued project. In 1983, the continued Phase-II project of Baosteel was approved. In 1985, the Phase-I project of Baosteel was completed and put into operation. In 1992, both Phase-I and Phase-II projects of Baosteel were completed and put into operation. The total investment of the Phase-I and Phase-II projects of Baosteel was 30.12 billion yuan, and a total of 670,000 tons of equipment were purchased, including 466,000 tons of imported equipment and 204,000 tons of domestically manufactured equipment. The localization rate of the Phase-I project was about 12%, and that of the Phase-II project was increased to about 57%; in 1993, the Phase-III project of Baosteel was built, and the localization rate was increased to 80%.

The construction, operation, and innovative development of Baosteel had greatly shortened the gap between the technological equipment of China's iron and steel industry and the advanced level in the world. Most of the advanced technology and equipment in the Phase-I project of Baosteel were imported, the Phase-II project was mainly based on cooperative design and manufacturing, while the Phase-III project was mainly based on independent design and manufacturing. The completion of Baosteel had effectively compensated the shortage of iron and steel varieties and quality in China and satisfied the urgent demand for high-end steel products by downstream industries such as automobile, petroleum, and shipbuilding in China. For example, in 1991, Baosteel produced seven varieties of steel products out of 14, which were in shortage in China accounting for 50%; Baosteel produced 1.46 million tons of steel products which were in shortage, accounting for 23% of the total in shortage over the country. In addition, the enterprise management of Baosteel with a centralized and consistent system as its core had greatly promoted the transformation of Chinese iron and steel enterprises from traditional management to modern style.

### *1.2.3 Summarizing and Promoting the Experience of Handan Steel to Explore Market-Oriented Operation*

In the 1990s, the former Handan General Iron and Steel Plant implemented a cost management system of "simulating market accounting and implementing cost-based veto" after exploratory trail,optimization, and adjustment in order to stop their heavy losses. The essence is to apply the market operation mechanism into the enterprise to back-count the cost based on the market price and decompose the target cost to branch plants, workshops, shifts, and individuals, and hereby to make business accounting. The completion of the target cost indicator was linked to the bonus and salary; if it cannot be completed, the bonus or even the salary would be deducted according to the extent; in the case of completion or over-completion, certain reward would be given. That market-oriented assessment mechanism had greatly mobilized the enthusiasm of the leaders and the masses of the enterprise, and the result was quite remarkable. From 1991 to 1993, they reduced the cost by 6.4%, 4.8%, and 6.1%, respectively, compared with the previous year and realized a cumulative profit of more than 700 million yuan for three consecutive years. Even in the difficult period of the Asian financial crisis, they still maintained a profit of more than 500 million yuan. In 1993, the Ministry of Metallurgical Industry summarized the management experience of "simulating market accounting and implementing cost-based veto" of Handan Steel and promoted it to the whole industry. In 1996, the State Council approved the *Investigation Report on the Management Experience of Handan General Iron and Steel Plant* submitted by the State Economic and Trade Commission and the Ministry of Metallurgical Industry.

## 1.3 Accelerated Stage of Leapfrog Development

The third stage for the development of the modern iron and steel industry in China begun from the beginning of the twenty-first century to 2014. After the impact on the Asian financial crisis had gradually subsided, along with the upgrading of domestic consumption structure and China's accession to the WTO, a new round of economic growth period brought a golden decade for the development of the iron and steel industry. Despite the impact of the international financial crisis during that period, the crude steel output of China had generally maintained a rapid growth, from 128.5 million tons in 2000 to 822.7 million tons in 2014. The variety and quality of the products had improved significantly, realizing a historic transformation from a net importer into a net exporter, and the era of iron and steel shortages that had long plagued China's economic development was gone forever. That period can be regarded as the accelerated stage of China's iron and steel industry, achieving a leapfrog development.

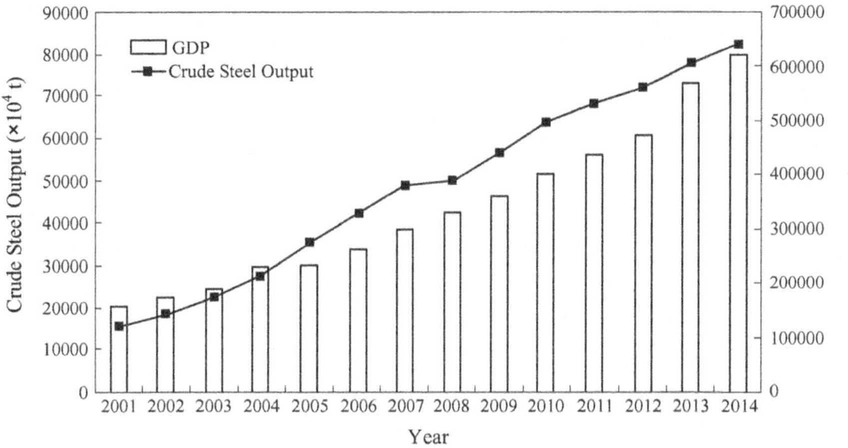


**Fig. 1.2** National economic growth and crude steel output increase from 2001 to 2014

## *1.3.1 Expansion of Scale of Iron and Steel Driven by the Market*

During that period, China's economy grew at a high speed, industrialization and urbanization were accelerated, foreign trade developed rapidly, and the steel demand increased dramatically, which drove the expansion of China's steel output and capacity. For example, during the 10th Five-Year Plan Period, China's economy grew at an average annual rate of 9.8%, and the urbanization rate was increased by 1.356% per year. The crude steel output increased by 2.75 times during the same period, with an average annual growth rate of 22%. During the 11th Five-Year Plan Period, the average annual growth rate of China's economy was close to 11.4%, the urbanization rate was increased by 1.39% per year, and the crude steel output was increased by 1.81 times during the same period, with an average annual growth rate of 12.6%. The growth of national economy and the increase of crude steel output [2] from 2001 to 2014 are shown in Fig. 1.2. The urbanization rate and the increase of crude steel output from 2001 to 2014 are shown in Fig. 1.3.

## *1.3.2 The End of an Era of Iron and Steel Shortage*

In the period of more than 50 years since the founding of the People's Republic of China, China had always faced the problem that the quantity of steel products was difficult to meet the needs of national economic development, and China was a net importer of steel products for a long time. Therefore, solving the contradiction of an insufficient supply of steel products had become the unshakable goal of generations of the Chinese engaged in the iron and steel industry. In 2006, China exported 43.01

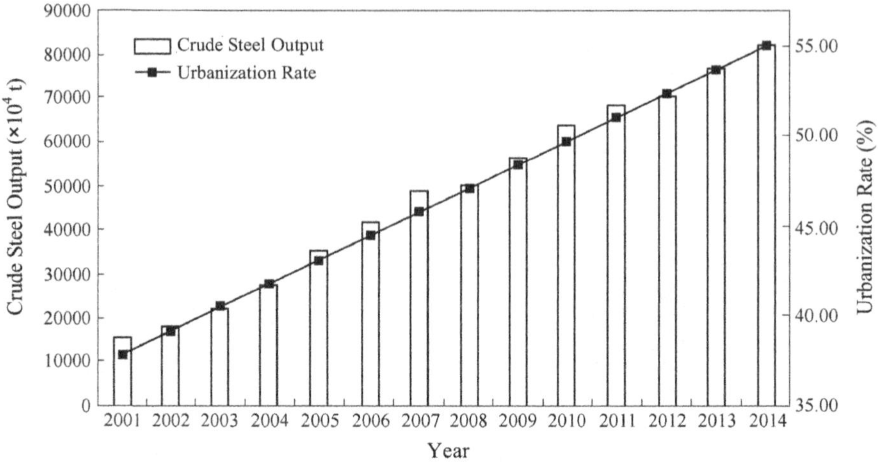


**Fig. 1.3** Urbanization rate and crude steel output increase from 2001 to 2014

million tons of steel products and imported 18.51 million tons of steel products, achieving 24.5 million tons of net export in the whole year, ending the history of a net importer of steel products for 57 consecutive years since 1949. China realized a historic transformation from a net steel importer to a net steel exporter, and the era of iron and steel shortage that had long plagued China's economic development was gone forever. Figure 1.4 shows the comparative evolution of China's steel import and export [3] from 2001 to 2014.

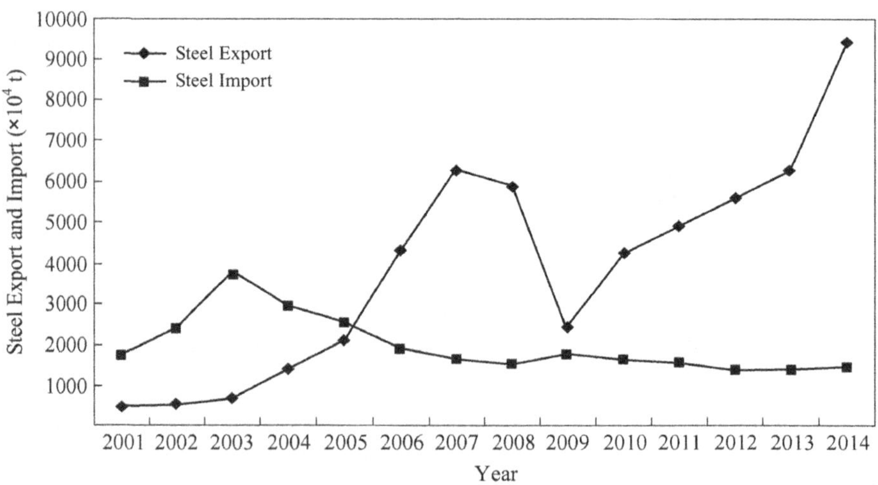


**Fig. 1.4** Comparative evolution of China's steel import and export from 2001 to 2014

### *1.3.3 Further Improvement of Technical Equipment Level*

During that period, with a large number of new ironmaking, steelmaking, and steel rolling facilities put into operation and the continuous elimination of outdated process equipment, the technical level of China's iron and steel industry had been further enhanced, and persistent efforts were made to promote large-sized, automated, and efficient equipment.

Major production projects under the national productivity layout such as Shougang Jingtang Company and Ansteel Bayuquan Company were completed and put into production, marking that China has the ability to independently integrate and construct modern coastal iron and steel bases. The technical equipment level of such enterprises as Baosteel, Anshan Steel, Wuhan Steel, Capital Steel, Ma'anshan Steel, Taiyuan Steel, Baotou Steel, Shagang, Tangshan Steel, Handan Steel, Xingcheng Special Steel, and the Dalian Base of Northeast Special Steel had reached the international advanced level. Tangshan Steel and Taiyuan Steel explored the green development, transformation, and upgrading of urban steel plants to develop the cleanest steel plants in the world, making them becoming models for learning by the global iron and steel industry. During the 11th Five-Year Plan Period, China eliminated 170 million tons of backward ironmaking capacity and 69 million tons of backward steelmaking capacity. During the 12th Five-Year Plan Period, China eliminated 91 million tons of backward ironmaking capacity and 94.8 million tons of backward steelmaking capacity.

As of the end of 2014, among the enterprises meeting the standards, the blast furnaces of 1,000 $m^3$ and above accounted for about 65% of the total ironmaking capacity, and the converters (electric furnaces) of 100 tons and above accounted for more than 56% of the total steelmaking capacity. Eighty-nine sets of hot wide strip rolling mills, 84 sets of medium and heavy plate mills, and 58 sets of cold tandem rolling mills had been built in the whole industry. The level of metallurgical automation and informatization had been greatly improved. The new business model represented by e-commerce had taken root in some iron and steel enterprises and has shown its vitality.

## 1.4 Reduction Stage of Innovative Development

Under the new normal, China's economic growth has shifted from high speed to medium-high speed. With the transformation of development mode, economic restructuring, and growth momentum transformation, the steel consumption intensity by specific GDP has dropped significantly. In 2015, the crude steel output in China was 804 million tons, a decrease of 2.3% compared to the previous year, marking the first decline since 1982. In the same year, the actual steel consumption in China was 664 million tons, a decrease of 5.4% compared to the previous year, marking the first decline since 1996. The "both declines" of steel consumption and crude steel output indicates that China's iron and steel industry has entered a development stage of reduction, showing a trend of innovative development. Generally, the reduction-oriented development will experience a long-term fluctuating bottoming process, during which the industry development will show the characteristics of survival of the fittest, process flow adjustment, multiple-policy implementation, integration, and restructuring, and innovation will become the strongest driving force and a key factor for success for enterprise development in this period.

### *1.4.1 Development Situation of China's Iron and Steel Industry Under the New Normal*

The iron and steel industry in China has already possessed strong international competitiveness and will play an important role in building a strong manufacturing country and the international capacity cooperation, and will lead the global iron and steel development for a long time. It should also be noted that under the new normal, the iron and steel industry is facing serious challenges of severe overcapacity and weak innovation capability. It is necessary to grasp the law of development, accelerate innovative development, resolutely implement the requirements of supply-side structural reform, and effectively promote transformation and upgrading in order to achieve sustainable development.

From an international perspective, the global economy has undergone a tortuous recovery from the deep adjustment. The global steel demand has entered a plateau period, showing a trend of stable but fluctuating development. The overcapacity of iron and steel has become a global problem. The trade protectionism in the international steel market is spreading, the competition will be more intense, the rules system for international investment and trade will be accelerated, and the opportunities and challenges for international iron and steel capacity cooperation and international trade will coexist. A new round of scientific and technological revolution and industrial transformation is in the ascendant. The industrial form, production management, and development mode of the global iron and steel industry are undergoing an unprecedented profound transformation. The steel materials and other materials show a general trend of mutual competition and collaborative integration. The financial attributes of staple commodities such as iron ore and coking coal have increased, their price fluctuations have intensified, the risk of investment in the global mining industry has increased, and uncertainties have increased significantly. Relying on the strategy of re-industrialization, the advanced steel powers strengthen the scientific and technological innovation and the strategic layout in the frontier domains to occupy the commanding heights in the middle and high-end steel markets; some emerging economies take shares from the ordinary steel market by relying on their advantages of low-cost factors such as labor; thus, the risks and challenges faced by China's iron and steel industry are increasing.

From the domestic perspective, the economic growth under the new normal is at a medium-to-high speed, and periodical contradictions and structural contradictions coexist, of which the structural problems are the main contradiction, and the total steel consumption and consumption intensity tend to decline. In addition, China's development mode is extensive, the unbalanced, uncoordinated, and unsustainable problems are still outstanding, resource constraints are tightening, and the deterioration of the ecological environment has not been fundamentally reversed. The iron and steel industry urgently needs to improve its innovation capability and accelerate adjustment and upgrading. However, the fundamentals that China's economy will sustain long-term growth remain unchanged, and the forward trend of economic restructuring and optimization has not changed. The iron and steel industry in China enjoys the largest and most active domestic demand market, the most complete industrial system, and the most abundant human resources, the latest and most advanced technology and equipment, and the fastest and most timely service system in the world; thus, it will continue to maintain strong competitiveness in the long run. In the process of achieving a jump from low level supply–demand balance to a high level supply–demand balance, improving the steel product supply quality and the steel product supply structure will be a new engine for China's iron and steel industry to move toward the mid-to-high end.

### *1.4.2 Requirements for China's Iron and Steel Industry by the Supply-Side Structural Reform*

To solve the problem of China's medium- and long-term economic development, the fundamental way is to promote supply-side structural reform. This is a path and barrier that is forced to go through. It is also the trend of the times and the situation. How can the supply-side structural reform be practically promoted in the iron and steel industry? In general, it shall resolutely implement the five development concepts of "innovation, coordination, greenness, openness, and sharing", and focus on the five key tasks of "cutting overcapacity, reducing excess inventory, deleveraging, reducing cost, and strengthening areas of weakness" to intensively analyze the problems existing in the industry development and then explore solutions based on that.

According to the requirements of those five development concepts, the iron and steel industry has a certain extent of disparity in several aspects. For example, in terms of innovation, China's iron and steel industry still has such problems as insufficient long-term investment in independent innovation, lack of original technology, weak collaborative innovation in production, education and research and application as well as a repeated allocation of innovative resources. In addition, there are still other uncoordinated problems such as unreasonable industrial layout and poor connection between steel manufacturing, service, and market demand. The above-mentioned "cutting overcapacity, reducing excess inventory, deleveraging, reducing cost, and strengthening areas of weakness" are the five key tasks of supply-side structural reform. Among them, cutting overcapacity is in the first place, and the iron and steel industry is a typical representative industry with severe overcapacity, and it is also the focus of cutting overcapacity. Therefore, cutting overcapacity is given the first priority for promoting supply-side structural reform in the iron and steel industry. However, while focusing on cutting overcapacity in the iron and steel industry, it could not neglect the requirements by other key tasks on the iron and steel industry. The objective analysis indicates that, as an important part of the real economy, a lot of achievements still can be achieved in the iron and steel industry in terms of deleveraging, reducing cost, and strengthening areas of weakness. The debt problem of some enterprises is outstanding; thus, the risk of inducing regional financial problem is increasing dramatically. In recent years, the iron and steel enterprises are subject to increasing and overwhelming rigid expenditure pressure in terms of financial expenses, which has become a critical factor influencing their market competitiveness. At the same time, the weaknesses of China's iron and steel industry in terms of technological innovation, standard system, quality manufacturing, intelligent manufacturing, service-oriented manufacturing, green manufacturing, and international development also need to be strengthened.

At present, the promotion of supply-side structural reform has clear concepts and clear ideas as well as specific tasks. In contrast, the iron and steel industry still has some of the above problems and gaps, but the problem means opportunities and gap implies potentials. The iron and steel industry should take the initiative to adapt, grasp and lead the new normal, and make good use of the macro-policy, industrial policy, micro-policy, reform policy, and social policy of the supply-side structural reform to promote the coordinated and innovative development of "nine aspects" (namely greenness, coordination, quality focus, standardization, differentiation, service-oriented, intelligence, diversification, and internationalization). Reshaping the value chain and continuously improving the competitiveness of the iron and steel industry are the fundamental way for the iron and steel industry to achieve sustainable development.

### *1.4.3 The Iron and Steel Industry in China has Competitive Advantages and will Lead the Global Iron and Steel Development for a Long Time*

China is the largest steel producer and consumer in the world. As an important fundamental industry for the national economy, China's iron and steel industry has satisfied the development of various industries such as construction, machinery, energy resources, and automobiles, its industry quality has improved remarkably, and its international status has been greatly improved as well. It has formed the comparative advantage of "good products, good prices, good scale, good services, and good brands". It is one of the most globally competitive industries among China's manufacturing industry category and will maintain its competitive advantages for a long time to come. The details are as follows:

**First**, China's iron and steel industry enjoys the largest and most active domestic demand market in the world. In 2016, the actual steel consumption (excluding inventory changes) in China was about 728 million tons, accounting for about 45% of global steel consumption. The situation that China has the largest and most active steel market in the world will not change for a long time, which is the strongest foundation for China's iron and steel industry to continue to maintain and enhance its competitiveness.

**Second**, China's iron and steel industry has the most comprehensive and complete industrial system in the world, including planning and design, equipment manufacturing, construction, production and operation, marketing services, technology research and development, material supply, supporting industries, etc., which promote each other and develop coordinately, making the overall competitiveness extremely strong.

**Third,** China's iron and steel industry has the richest and most promising human resources in the world. Many metallurgical and related universities, institutes, organizations, and enterprises have cultivated and gathered a large number of talents in the long-term development process, especially during the golden period of rapid development in the new century, a large number of backbone talents have been trained; thus, world-class top metallurgical talents will surely appear in China.

**Fourth**, China's iron and steel industry has the latest and most advanced technological equipment in the world. After the intensive investment during the last round of golden development period, a large number of advanced iron and steel production technologies, energy-saving and environmental protection technologies, and automation control technologies have been applied, generally the large-scale, automated, continuous, green, systematic, and precise production have been realized, and more than half of the production process equipment has reached an internationally advanced level.

**Fifth**, China's iron and steel industry has the fastest and most timely service system in the world. In the past decade, China's iron and steel industry has strengthened its service system to continuously improve its service capabilities. Not only complete warehousing, shearing and processing, logistics and distribution services systems have been established in the major cities over the country, but also e-commerce is applied to provide customers with timely and considerate services. In addition, the warehousing, processing, and distribution service systems have also been established in the major steel consuming countries in the world to provide timely and efficient services to the customers around the world.

Throughout the development history of the global iron and steel industry, three countries have occupied half of the global crude steel output and have long maintained the leading position in the global iron and steel industry. They are the UK in Western Europe, the USA in North America and China in East Asia. That is also the main center shifting axis of the global iron and steel industry. The time span from the crude steel output of the UK accounting for half of the global total to being surpassed by the

USA was about 80 years, while that from the crude steel output of the USA accounting for half of the global total to being surpassed by Japan was about 70 years. Iron and steel is an important symbol of humankind's entry into an industrialized society and industrial economic development. The analysis of population economics indicates that: the UK becoming the center of global iron and steel industry was a result that the UK was the first country to enter the industrial revolution and rapidly developed industrialization and urbanization among the countries with a population of more than ten million. The USA becoming the center of global iron and steel industry was a result that the USA rapidly developed industrialization and urbanization among the countries with a population of more than one hundred million. China becoming the center of global iron and steel industry is a result that China rapidly developed industrialization and urbanization among the countries with a population of over one billion. Looking into the future, there will be no countries or regions with a higher-scale (scale of 10 billion) population, and even it is hard for another country with a population of over one billion to rapidly develop industrialization and urbanization. In general, the time space that the iron and steel industry of China will lead the world exceeding those experienced by the UK and the USA, and it may last a hundred years.

### *1.4.4 The Direction of the Coordinated Development in "Nine Aspects" for China's Iron and Steel Industry*

The iron and steel industry in China should actively adapt to recognize and grasp the new normal in the development stage of reduction, adhere to the five new development concepts, take innovation as the first driving force, take generally improving the comprehensive competitiveness as the fundamental target, give full play to the decisive role of the market in resources allocation, and give better play to the role of the government, so as to vigorously promote the supply-side structural reform of the iron and steel industry while appropriately expanding market demand, and intensive efforts shall be made to promote the coordinated development in "nine aspects" (namely greenness, coordination, quality focus, standardization, differentiation, service-oriented, intelligence, diversification, and internationalization) and reshape the value chain.

1. Greenness

Green development is an important prerequisite for the survival of the iron and steel industry and a fundamental guarantee for achieving sustainable development. The narrow-minded view on pollution control shall be abandoned for the iron and steel industry; the green development shall be comprehensively strengthened from enterprise's all-round aspects, full manufacturing process, and total product cycle, that is, implementing the "six-in-one" concept of green development (green mine, green procurement, green logistics, green manufacturing, green products, and green industry) of the iron and steel industry, so as to establish profitable and sustainable modern green plants and a green industrial system.

2. Coordination

Enhancing the competitiveness shall be taken as the core to promote the reorganization of iron and steel enterprises. Major supports shall be given to the combination of powerful iron and steel enterprises and the reorganization of the enterprises with featured products on the market, so as to build world-class steel groups with international competitiveness. It is required to promote the reorganized enterprises to strengthen their substantive integration and re-design their business processes in order to give practical play to the coordination effect. One-stop integrated service platforms shall be established by relying on specialized consulting organizations for merger and reorganization of iron and steel enterprises to provide services in respects of information, coordination, policies, laws, and financing. Supports shall be given to the establishment of coordination mechanisms for regional enterprises and those producing similar products, so as to reduce disorderly competition. Laws and regulations shall be strictly observed, and law enforcement shall be strengthened to create a fair competitive market.

3. Quality Focus

Quality focus, that is, attaching great importance to brand building and product quality improvement, is an important guarantee for achieving sustainable development of enterprises. Quality is the cornerstone of brand building, service is the guarantee of brand improvement, innovation is the source of brand continuation, and culture is the essence of brand promotion. A "four-in-one" quality system integrated with standardization, inspection, traceability, and informatization shall be established through technological innovation and cultural development to comprehensively improve product quality and service quality, fully build a brand system integrated with quality, service, innovation, and culture, and build a brand enterprise with strong comprehensive competitiveness.

4. Standardization

In accordance with the requirements of relevant documents such as *Made in China 2025*, *Reform Plan for Deepening Standardization Work*, and *Development Plan for National Standardization System* (2016–2020), comprehensive and in-depth development of the standardization for China's iron and steel industry shall be promoted. Cooperative effects shall be made from the four levels of state, industry, local government, and organizations to promote standardization in areas such as steel logistics, green transformation, integration of informatization and industrialization, electricity demand-side management, and brand building. Enterprises shall be guided to establish and improve the enterprise standard systems that meet the needs of market competition and technological progress, and key technologies with independent intellectual property rights shall be incorporated into enterprise standards.

5. Differentiation

Differentiation is the only way leading to the restructuring, transformation, and upgrading of the iron and steel industry, and it is also an inevitable choice for iron and steel enterprises to survive and develop. The first is strategic differentiation, i.e., strengthening research on market demand and competitors, identifying the positioning of enterprises, and clearly defining development strategies and implementation paths to realize industrial chain differentiation and regional differentiation; the second is product differentiation, i.e., optimizing product structure, increasing the share of knock-out products, and focusing on product serialization as well as improving "one-stop" supply capacity; the third is production line differentiation, i.e., achieving specialized production through equipment technological modification and upgrade to significantly improve production efficiency and reduce production costs; the final one is service differentiation, i.e., applying different service modes for different customers and different regions to improve focalization and recognition.

6. Service-Oriented

The development of service-oriented manufacturing is a realistic requirement for China to enter the middle and late stages of industrialization, and it is also an important measure to seize the new round of opportunities for global scientific and technological revolution and industrial transformation. Iron and steel enterprises shall shift value adding way from simple manufacturing to coordination of manufacturing and service. More importantly, through the organic combination of manufacturing and service, the business model of iron and steel enterprises will be innovated to promote intelligent manufacturing, capital operation, "Internet+", technical services, further processing of products, etc. then to improve the integrated profit-making capability from aspects of industry and informatization combination, industry and finance interaction as well as product and service extension.

7. Intelligence

The iron and steel enterprises where conditions permit shall be encouraged to improve the establishment of five-level informatization systems consisting of basic automation, production process control, manufacturing execution, enterprise management, and decision support. Advanced iron and steel enterprises shall be guided to establish big data platforms and promote digitalization and networking in the whole process of iron making and steel making to obtain knowledge and experience. Supports shall be given to iron and steel enterprises to implement robot replacement projects in work posts with harsh environments, high safety risks, and high operational consistency. Efforts shall be made to promote the comprehensive integrated application of Industrial Internet, cloud computing, and big data in the full process flow including research and development, design, manufacturing, business management, and sales services of iron and steel enterprises as well as the entire industrial chain. Iron and steel enterprises shall be encouraged to build intelligent detection systems for key equipments and carry out new remote operation and maintenance services such as fault prediction and automatic system diagnosis.

8. Diversification

A full play shall be given to the role of the iron and steel industry in terms of long industrial chain, big influence, and high driving effect to make it deeply integrated into national and local economic and social development, the industrial development, and the needs of the social public, in order to develop market-oriented diversified industries. The coordination from varied aspects of strategy, investment, business, channel, and management between iron and steel industry and diversification shall be strengthened to focus on the development of the business with superior products, advanced technologies, and market competitiveness, and commit to sustainable improvement of performance and high level improvement, so as to form a diversified management system characterized by longitudinal extension, horizontal coupling, or territorial combination along the iron and steel industry chain. In conjunction with the reform of state-owned assets and enterprises, iron and steel enterprises shall be encouraged to actively explore various forms of property rights diversification to release development vitality.

9. Internationalization

The development opportunities under the Belt and Road Initiative shall be seized to accelerate the internationalization development of iron and steel enterprises. First, with major development opportunities such as international capacity cooperation, carrying out engineering project cooperation, and providing supporting services; second, exploring the way to establish strategic cooperation and reorganization over the whole industrial chain of enterprises, and joining hands with international and domestic strategic partners to deepen industrial cooperation, and improve international cooperation level and ability; third, actively integrating and utilizing global innovation resources to achieve breakthroughs in overseas talent training, international project cooperation, and overseas high-end talent introduction; fourth, relying on the advantages of own trading companies or strategic alliances with domestic trading companies to provide worldwide metallurgical enterprises and industrial enterprises with value-added supply chain services and actively develop and cultivate overseas strategic end customers.

## References

1. Zhang, X. Y. (2012). China's iron and steel. Beijing: Metallurgical Industry Press. June.
2. China steel statistics. Beijing: China Iron and Steel Association, 2001–2016.
3. Statistics of the General Administration of Customs, P. R. China.

# Chapter 2
# International Experiences

## 2.1 Iron and Steel Industry in Western Europe

### *2.1.1 Post-War Reconstruction Driving the Rapid Development of the Iron and Steel Industry*

After the end of the World War II, with the promotion of the Marshall Plan, Western European countries introduced funds and advanced technologies from the USA, which, together with the accelerated process of European integration, boosted rapid economic recovery and growth in those countries.

In 1951, six countries including France and the Federal Republic of Germany signed the *European Coal and Steel Community Treaty* for a period of 50 years (1952–2002) in Paris, which determined that its basic task is to create a single common market for coal and steel in order to eliminate relevant tariff restrictions and make intervention in production, circulation, and distribution [1]. At the beginning of the establishment of the European Coal and Steel Community (ECSC), it was responsible for coordinating the coal and steel production, investment, price, and raw material distribution within ECSC, so as to ensure effective internal competition. At the same time, ECSC had a say in the development or shutdown of certain enterprises and was in charge of the relationship of ECSC with the third countries and relevant international organizations.

Western Europe is the cradle land of the modern iron and steel industry. In the two decades after the World War II, the two most important strategic materials of steel and coal at that time had achieved planned complementary advantages in production and circulation among ECSC members, and effective coordination of resources and capacity allocation was realized. Together with the research and development and application of technical equipment such as oxygen top-blown converter, continuous casting machine, hot tandem strip rolling mill, and cold tandem rolling mill, the steel production capacity in Western Europe was greatly improved, which had effectively



X. C. Li, *The Road Map of China's Steel Industry*,
https://doi.org/10.1007/978-981-15-2074-7_2

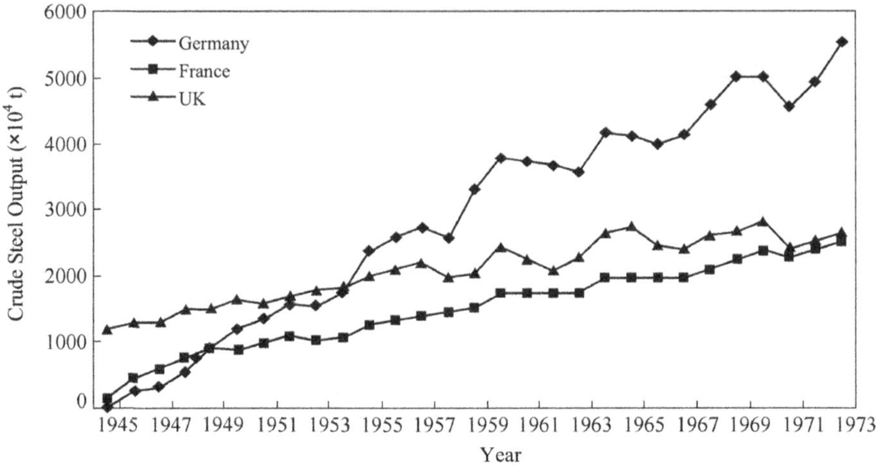


**Fig. 2.1** Growth of crude steel output in Germany, France, and the UK during the reconstruction period after World War II

supported the basic needs of those countries in capital construction and economic development.

For example, the crude steel output in the Federal Republic of Germany was increased from 2.55 million tons in 1946 to 41.19 million tons in 1965 and 55.41 million tons in 1973; that in France was increased from 4.41 million tons in 1946 to 19.6 million tons in 1965 and 25.26 million tons in 1973; that in the UK was increased from 16.55 million tons in 1950 to 27.44 million tons in 1965 and 26.59 million tons in 1973. The growth of crude steel output in Germany, France, and the UK during the reconstruction period after World War II is shown in Fig. 2.1.

### *2.1.2 Structural Adjustment of the Iron and Steel Industry in Western Europe*

After the 1970s, the iron and steel industry in Western Europe entered an adjustment period. In 1973, the first oil crisis directly led to a declining demand in the steel market. Due to the rapid growth of the production capacity of the iron and steel industry during the expansion period and the increasing competitiveness of the iron and steel industry in Japan, the original export market of the iron and steel enterprises in Western Europe was impacted and squeezed, and the iron and steel industry of the European Community was caught in a crisis of severe overcapacity. From 1974 to 1985, the steel output of the European Community fell by 30%, and the operating rate was once only 60%. Most iron and steel enterprises in Western Europe were subject to stubbornly high production costs, and their products were not competitive, leading to very serious losses. In the process of restructuring of the iron and steel industry, the governments in Western Europe had played a huge role and adopted various measures such as increasing subsidies, limiting production, and nationalization.

Although the *European Coal and Steel Community Treaty* explicitly prohibits governments from granting financial subsidies to their coal and iron and steel industries, in the face of the development difficulties of the iron and steel industry, most countries in Western Europe had given large-scale subsidies and even adopted measures such as price limits to their iron and steel industry in the 10 years from 1975 to 1985. From 1980 to 1985, the governments in the European Community subsidized the iron and steel industry by as much as 83 billion West German marks. Part of the government subsidy funds was used to adjust the industrial structure, resettle the unemployed persons, strengthen environmental protection, etc., and a considerable part of the funds was used for price subsidies, which to some extent concealed the problems of backward technology and poor management in some enterprises, thus artificially extending the life of uncompetitive enterprises.

In 1980, in order to coordinate the interests of member states, EC implemented a crude steel output quota system, which required the government to gradually reduce their financial subsidies to iron and steel enterprises and stop subsidies by the end of 1985. However, they were allowed to provide subsidies to enterprises that reduce production capacity. Moreover, the maximum total production quota and the maximum trade volume quota of most varieties of steel products among the EC member states were stipulated. The quotas were allocated according to the actual production capacity of each iron and steel enterprise, and the indicators can be sold among them after the approval by the EC Commission. That measure had somewhat eased the contradiction between supply and demand in the internal market of the European Community. However, since the quota allocation was made based on the actual production capacity of enterprises, all iron and steel enterprises were encouraged to maintain their existing production capacity in an effort to obtain as many quotas as possible, making the production capacity reduction more difficult.

At the beginning of the crisis, the countermeasures taken by Western European countries were to put the iron and steel industry under state control. By the mid-1980s, the crude steel output of the state-owned companies in Europe had accounted for half of Europe's total output. With the end of the steel production quota system applied by EC in 1988, the governments' control measures for the iron and steel industry in Western Europe had been gradually reduced [2], and the iron and steel industry could receive government funding only in a few special cases. For example, in 1993, EC approved the rationalization programs of six state-owned enterprises in Germany, Spain, Italy, and Portugal. The governments provided a subsidy of 6 billion European currency units (7.7 billion US dollars) to reduce their production capacity by 5.5 million tons. But overall, the subsidies received by the iron and steel enterprises in Western European countries from their governments had been greatly reduced.

With the relaxation of government regulation, the privatization of iron and steel enterprises in Western Europe began to rise. For instance, British Steel Corporation was privatized in the late 1980s. France, Italy, and Spain carried out privatization of state-owned iron and steel enterprises in the mid-to-late 1990s. By 1998, the steel output of the state-owned iron and steel enterprises in European Union (EU) accounted for less than 5% of the total output. The privatization and the enhanced market competition had led to the merger and reorganization of iron and steel enterprises. For example, in the late 1980s, two French steel companies—Usinor and Sacilor—were reorganized, and two German steel companies—Thyssen and Krupp—were reorganized in the late 1990s. It should be noted that the merger and reorganization of the iron and steel industry in Western Europe at that time were not carried out under the influence of administrative forces but rather the autonomous behavior of enterprises.

### *2.1.3 Iron and Steel Action Plan of Europe in the Post-International Financial Crisis Era*

After the international financial crisis, the steel demand in Europe fell sharply, enterprises had to reduce production due to operation difficulties, and even, production capacity was reduced to reduce job posts. In order to improve the competitiveness and sustainable development capability of the iron and steel industry in Europe, the European Commission issued the "Steel Action Plan" in June 2013, which is different from the tough measures taken during the adjustment period in the 1980s and 1990s. The "Steel Action Plan" is relatively mild. The main contents include:

(1) Rebuilding the regulatory framework. That is to say, to reform the management system in order to optimize and improve the existing systems through the reassessment of relevant laws and regulations, policies, and control frameworks, so as to make them more suitable to the actual development of the iron and steel industry in Europe.
(2) Promoting the upgrading of steel products and carrying out research and promotion of Sust Steel. The purpose was to formulate and establish new standards and norms by upgrading and updating steel products, aiming at forming standard barriers and safeguard their own interests while promoting market demand.
(3) Creating a fair market environment. This includes the EU internal market and the international market. In the internal market, EU focused on combating tax evasion to safeguard the legitimate interests of taxpaying enterprises. In the international market, the essence of this measure was to implement trade protection for EU's iron and steel enterprises and enhance the competitiveness of EU's steel products in the international market.
(4) Promoting the reduction of production costs. It includes adjusting the structure of raw materials, increasing the proportion of scrap application to reduce the demand for expensive iron ore, expanding the recycling, improving the quality of scrap by optimizing the design of end products, setting standards for final steel wastes, and combating illegal export of scrap. In addition, reducing the cost of electricity in steel production was also taken into account.

(5) Supporting technological innovation. It includes the research and development and application of low-carbon technology, new processes, new technical equipment, etc., in order to build the cornerstone for the competitiveness of the iron and steel industry of Europe in the future.

## 2.2 Iron and Steel Industry in the USA

### *2.2.1 Development History*

The development of the modern iron and steel industry in the USA can be dated back to as early as 1868. During the industrial revolution in the US from 1868 to 1880, the steel output in the US was increased at an average annual rate of about 40%. While the production capacity was expanding rapidly, complete varieties of steel products in good quality had been developed, and its production technology was at a relatively advanced level in the world at that time.

During the initial industrialization process from 1881 to 1920, the steel output in the USA was increased at an average annual rate of 10%, faster than that in the European countries such as the UK, Germany, and France. In 1899, the annual crude steel output in the USA reached 10.81 million tons.

In the middle stage of industrialization from 1920 to 1955, the steel output in the USA grew at an average annual rate of 7%. In 1953, the annual steel output in the USA broke through the mark of 100 million tons for the first time, far higher than that in other major steel-producing countries in the world.

In the late stage of industrialization from 1956 to 1975, the steel output in the USA was increased slightly at an average annual rate of 0.5% only. During that period, the steel output in the USA reached 136.8 million tons in 1973, making a record in the history of the iron and steel industry in the USA. Since then, the crude steel output in the USA has begun to fall, and it has never recovered to that level.

### *2.2.2 Restructuring of the Iron and Steel Industry in the USA*

In the long period from the end of the nineteenth century to the 1970s, the USA had been the largest steel producer in the world. Due to impact of the oil crisis, the competitiveness of the iron and steel industry of the US gradually declined in the 1970s. In the 1980s, large-scale technological transformation and restructuring began in the USA. From 1980 to 1989, the crude steel production capacity in the USA was reduced from 153.7 million tons to 105.1 million tons. At the same time, the production capacity utilization rate was increased from 66% to 85%, and the number of employees decreased from 500,000 to 210,000. The US government's control over the iron and steel industry during the industrial restructuring period mainly includes the following aspects.

(1) Laying emphasis on trade protection and creating a good environment for the restructuring of the domestic iron and steel industry. The iron and steel industry in the USA has powerful trade unions. In order to smoothly promote industrial restructuring and reduce the impact on employment, the US government often implemented steel trade protection through tax adjustment, quotas, and anti-dumping litigations.
(2) Revising taxation, depreciation, and financial policies to accelerate the accumulation of funds in iron and steel enterprises improve their financing capacity, and promote the transformation and upgrading of the iron and steel industry. For instance, the depreciation period of iron and steel production facilities was shortened from 12 to 5 years, tax reduction and exemption were implemented for iron and steel enterprises, and the duration for tax reduction and exemption could be extended according to the situations.
(3) Emphasizing support given by advanced technologies to industrial restructuring, transformation and upgrading; considering varied factors comprehensively such as the market, resources, industrial base etc; encouraging the enterprises to strive to achieve the leading level in terms of process flow and technical equipments; promoting the enterprises to carry out modernization in order to improve productivity.
(4) Supporting and subsidizing enterprises and scientific and technological institutions to carry out major fundamental research, application of advanced technical equipments, and training of researchers.
(5) No longer implementing stricter standards and regulations on iron and steel enterprises in terms of environmental protection.
(6) In the late 1990s, the USA once again adjusted the structure of the iron and steel industry, with a focus on promoting mergers and acquisitions to make the output of top three iron and steel enterprises accounting for 60% of the national total.

### *2.2.3 Development of Short Process Steel Making by Electric Arc Furnace*

From the perspective of the production process, the iron and steel industry in the USA is dominated by short process of electric arc furnace. In recent years, the proportion of electric furnace steel in the USA has been maintained at around 60%. The proportion and the development speed of the steel output of electric furnaces are closely related to the adequacy of scrap resources. This feature is quite obvious in the development history of electric furnace steel in the USA.

In the 1950s, the scrap adequacy in the USA remained at around 0.3, which was at a low level. At that time, the steel output of electric furnaces was less than

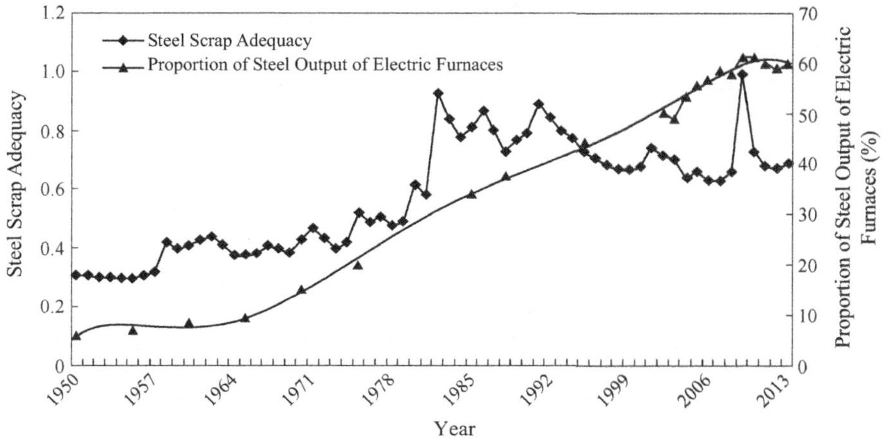


**Fig. 2.2** Scrap adequacy and the proportion of steel output of electric furnaces in the US

10 million tons, and the proportion of electric furnace steel was less than 10%. From the late 1950s to the early 1960s, the scrap adequacy in the USA reached 0.4 or more, and it remained at that level for about 15 years. During that period, the steel output of electric furnaces in the USA gradually increased to nearly 20 million tons, and the proportion of the steel output of electric furnaces was also over 15%. The scrap adequacy and the proportion of the steel output of electric furnaces in the USA are shown in Fig. 2.2.

In the mid-to-late 1970s and early 1980s, the scrap adequacy in the USA showed a "W" trend, rising fluctuated up to 0.5. During that period, the steel output of electric furnaces in the USA continued to grow steadily, reaching 25 million tons. The proportion of the steel output of electric furnaces was close to 30%. In the 1980s, due to the decline in crude steel output in the USA, the scrap adequacy surged to 0.9 and has been remaining at a high level of 0.6 or above till now. During that period, the scrap resources were sufficient in the USA, the steel output of electric furnace climbed to nearly 60 million tons, and the proportion of the steel output of electric furnaces was 60%.

Looking back at the development of the steel production by electric furnaces in the USA, it can be roughly divided into three stages according to the scrap adequacy, as shown in Table 2.1.

In summary, scrap resources are an important external condition for the development of steel production by electric furnaces. When the scrap resources of a country or region are inadequate (scrap adequacy <0.3), the development of steel production by electric furnaces is slow with a low proportion. When the scrap adequacy rises above 0.3, the development of steel production by electric furnaces begins to accelerate (there is a lag of 5–7 years, mainly due to market response, technological development, and engineering construction). When the scrap adequacy rises above 0.6, a further rapid development of steel production by electric furnaces will come.

**Table 2.1** Relationship between the scrap adequacy and the development of steel production by electric furnaces

| No. | Scrap adequacy $X$ | Scrap resources | Development of steel production by electric furnaces | Proportion of the steel output of electric furnaces |
|---|---|---|---|---|
| 1 | $X < 0.3$ | Inadequate | Slow | Low |
| 2 | $0.3 < X < 0.6$ | Adequate | Fast | High |
| 3 | $X < 0.6$ | More adequate | Faster | Higher |

The development of steel production by electric furnaces in a country is closely related to the development stage of its iron and steel industry. Taking the USA as an example, the development of the crude steel output and the proportion of steel output of electric furnaces are shown in Fig. 2.3.

The overall situations of the crude steel output and the development of steel production by electric furnaces shown in the above figure indicate that, during the crude steel output growth stage from the World War II to the early to middle 1950s, the proportion of steel production by electric furnaces was very low; with the crude steel output entering the peak arc zone in the middle to late 1950s, the proportion of steel production by electric furnaces began to gradually increase to 20–30%; after the peak zone of crude steel output, the steel production by electric furnaces was developed rapidly, and its steel output was increased to about 60% in more than 20 years.

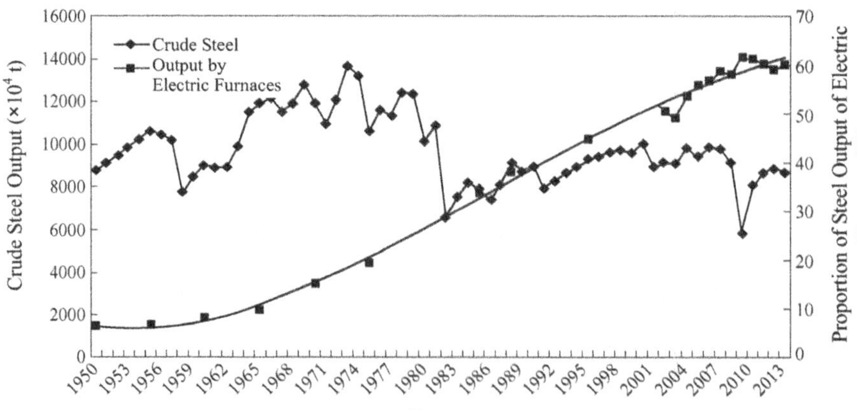


**Fig. 2.3** Crude steel output and proportion of steel output of electric furnaces in the US

### 2.2.4 Layout Evolution

1. Resource-Based Layout Features in the Early Stage

The early development of the iron and steel industry in the USA was characterized by typical resource-based layout. There were charcoal and small iron ore resources distributed along the Atlantic coast in the northeastern USA [3], and the early iron and steel enterprises in the USA was developed using these resources. With the discovery of anthracite resources in Bethlehem and Scranton and its gradual replacement of charcoal in iron and steel smelting, the focus of the iron and steel industry began to move westward to get close to anthracite-producing regions. Later, the coking coal-based ironmaking technology was successfully developed, and the focus of the iron and steel industry in the USA continued to move westward to the Pittsburgh region which was rich in coking coal resources, building up Pittsburgh's position as the "Steel Capital" in the development history of the iron and steel industry in the USA. At the beginning of the twentieth century, after the development of iron ore mines in the Lake Superior region, some iron and steel enterprises were established in the cities along the lake, such as Chicago, Detroit, Cleveland, and Buffalo. Meanwhile, some steel plants had also been built in the largest steel consumption area in the USA, i.e., the Atlantic coast in the north.

2. Transformation from Resource-Based Layout to Consumer-Oriented Layout

During and after the World War II, the iron and steel industry in USA showed a trend of concentration along lakes and coastal consumption regions. Most of the newly built steel plants were located in Chicago and along the Atlantic coast. During that period, the position of the Pittsburgh's iron and steel industry was clearly weakened, and some iron and steel enterprises in the region even closed down. The main reason was that the development of transportation infrastructures, especially cheaper water transportation means, allowed long-distance, low-cost, and large-scale transportation of the raw materials like coal and iron ore required for steel production. However, the price of steel product transportation by railway was very high. In addition, after the 1950s, the production of high-grade iron ore in the Lake Superior region continued to decrease, and the quantity of high-quality and cheaper iron ore imported from Canada, Venezuela, and other countries was increased. To facilitate the use of imported iron ore, the USA built the Burns Harbor Steel Plant and the Fairis Steel Plant in the coastal region and expanded the Sper Roth Steel Plant near Baltimore. In addition, with the increase in the proportion of steel production by electric furnaces in the USA, most of its main raw material—scrap—came from the main consumption areas of steel products. After the 1950s, some mini-steels that emerged in the USA were mainly built in important steel consumption areas.

## 2.3 Iron and Steel Industry in Japan

### *2.3.1 Development History*

Japan is a traditional steel power. Before China, it was ranked as the largest steel producer in the world. Its steel complexes like Nippon Steel and Sumitomo Metal and JFE and special steel companies like Aichi Steel, Sanyo Steel, and Daido Steel have strong market competitiveness, enjoying a good reputation in the global iron and steel industry. Together with several pillar industries such as the automobile industry and the shipbuilding industry, the iron and steel industry in Japan plays an important role in Japan's economy.

As early as 1901, Japan had established steel complexes integrated with iron-making, steelmaking, and steel rolling. After the World War II, the demand for steel products due to Japan's reconstruction and the outbreak of the Korean War was very strong. The Japanese government, by taking the advantage of the opportunities, adopted the policies for supporting the prior development of the iron and steel industry, which made its iron and steel industry develop rapidly in the several decades after the war. The iron and steel development in Japan highly focused on the technology introduction and re-innovation, such as liquid steel refining technology, which, together with the implementation of delicacy management, had promoted rapidly increasing of the competitiveness of Japan's iron and steel industry. In 1973, the crude steel output in Japan reached 100 million tons. After the dissolution of the Soviet Union, Japan's crude steel output ranked first in the world in the early and mid-1990s. There are hundreds of iron and steel enterprises in Japan, which can be roughly divided into three categories: the first one is large steel complex groups, such as Nippon Steel and Sumitomo Metal and JFE; the second one is independent steel companies, such as Tokyo Steel Manufacturing Co., Ltd., Sanyo Special Steel Products Co., Ltd., Daido Steel Co., Ltd., and Kobe Steel, Ltd.; the third one is small professional production companies, which generally only produce or process single- and special-purpose steel products.

### *2.3.2 Government Control Measures*

The Japanese government has always attached great importance to the development of the iron and steel industry, and its guidance and intervention in the iron and steel industry are highly targeted and highly efficient. The competent authority of the iron and steel industry in Japan is the Ministry of Economy, Trade, and Industry (former Ministry of Commerce and Industry), the specific responsible institutions are the Manufacturing Industry Bureau and the Trade Policy Bureau, of which the former is a professional institution (bureau), and its Steel Division is responsible for the specific matters concerning the iron and steel industry; the latter is a comprehensive institution (bureau) responsible for macroeconomic regulation and control, as well as policy formulation in various industries in terms of restructuring, development environment, and enterprise reform. In addition, the Japan Iron and Steel Federation also plays an important role as a corporate legal person in the management of the iron and steel industry in Japan.

The Japanese government's control over the iron and steel industry during the industrial restructuring period mainly includes the following contents.

(1) Promoting and approving enterprises to form a coordinated consortium, proposing the guiding output for the iron and steel industry on a quarterly basis, supervising its implementation and making coordination, and controlling the price adjustment of some varieties when necessary. Coordinating steel exports, including adjusting tax rates, and coordinating relevant iron and steel enterprises to carry out concerted actions on certain steel product varieties (or certain target markets) in terms of price and quantity.
(2) Conducting temporary interventions and adjustment on the varieties of steel produced by the enterprises or controlling the trade of certain steel product varieties according to market changes if necessary.
(3) Adjusting the prices of raw materials and energy such as scrap and electricity and conducting trade control on the quantity of imported scrap.
(4) Controlling the new projects and investments in the iron and steel industry; issuing the "Provisional Measures for Stabilizing Specific Depressed Industries" for the elimination of excess facilities in the sluggish industries such as iron and steel industry.
(5) Coordinating financial institutions such as banks to provide financial support for enterprises to reduce or transfer equipment capacity and support enterprises that join the coordination consortiums to implement production reduction; increasing preferential policies for enterprises in terms of the relief funds for employment adjustment.
(6) By relying on the intermediary organization, coordinating the iron and steel industry and its downstream industries to formulate and revise standards and norms for related products, building a steel product quality assurance system, and implementing it in parallel with the plant certification system.
(7) Promoting the concentration of production capacity to large enterprises in combination with the compression of excess facilities.
(8) Giving tax preferences for enterprises to invest in environmental protection equipment and allowing accelerated depreciation; encouraging the banks to provide strong financing supports for the frontier technologies and basic researches in the iron and steel industry; allocating government subsidies to major projects.

### 2.3.3 Development Experiences

Looking at the development of global iron and steel industry, the Japan's iron and steel industry is a model of both "big" (scale) and "strong" (competitiveness). Its development experiences are as follows:

The first is to pool the funds to introduce foreign advanced technologies and equipments, make continuous improvement and innovation in addition to steel production to further optimize their technologies and processes, and then promote the secondarily developed technologies, equipments, and processes as commodities to the market;
The second is to actively carry out laboratory research and industrialization pilot tests of various new technologies through steel alliances, so as to accumulate a large number of practical basic technologies to continuously enhance the competitiveness of the iron and steel industry;
The third is to strengthen the coordinated development with the steel product consuming industries and improve the technical level and production level to meet the ever-changing requirements of customers;
The fourth is to enhance product quality and attach importance to establishing the brand image of enterprises and products, boosting the improvement of product grades and strengthening the premium effect;
The fifth is to speed up the adaptation of the company's shareholding structure, organization, and personnel composition to the market and technology and adjust and optimize them in a timely manner.

### 2.3.4 Industrial Layout

The layout and distribution trend of the iron and steel industry in Japan has undergone a transformation from raw material-oriented to consumption-oriented [4]. The use of raw materials was mainly considered in the establishment of the earliest steel plants in Japan such as Kamaishi Steel Works and Muroran Steel Works. For instance, the first "imported blast furnace" in Japan was built in Kamaishi in order to make use of the local iron ore, fuel charcoal, clay raw materials, and stone materials; Muroran Steel Works, which was built in 1907, was an iron and steel enterprise relying on coking coal from Shikun coalfield in Hokkaido. The general layout of Kashima Steel Works of Sumitomo Metal is shown in Fig. 2.4.

With the use of raw materials and fuels outside Japan, the steel plants in Nagoya, Kimitsu, Oita, and Fukuyama, had considered to get close to consumption areas to facilitate the use of imported iron ore and coke and the export of steel products. Large-scale steel complexes in Japan generally make use of port conditions to build coastal plants on the land reclaimed from the sea. They are generally concentrated in the metropolitan circles and surrounding areas, namely the belt region in the Pacific

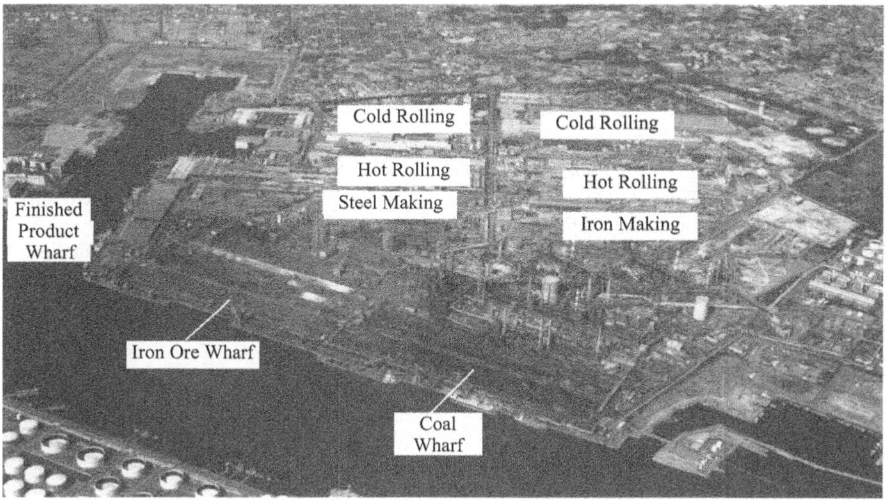


**Fig. 2.4** General layout of Kashima Steel Works of Sumitomo Metal

Ocean formed by the five major industrial areas of Keihin, Hanshin, Chukyo, Seto Inland Sea, and Kitakyushu, and the steel output in that region once accounted for more than 80% of the total in Japan. The general layout of Oita Steel Works of Nippon Steel is shown in Fig. 2.5.

**Fig. 2.5** General layout of Oita Steel Works of Nippon Steel

## References

1. (2005, September) Development Policy Guideline for iron and steel industry. Beijing: Economic Science Press.
2. Karl-UCh Koehler. Lessons from Mergers and Acquisitions and Globalization in the Iron and Steel Industry of Europe. China Iron and Steel Industry, May, 2006.
3. Analysis of the Transfer Characteristics of Iron and Steel Industry in Typical Countries. Metallurgical Management, September, 2013.
4. A summary report made after the delegation of the Chinese Society for Metals attending the 12th Japan-China Bilateral Steel Academic Conference visited the Japanese and Korean companies, April, 2011.

# Chapter 3
# Development Trend

## 3.1 Shift of the Center of Global Iron and Steel Industry

Since the nineteenth century, the center of the global iron and steel industry has experienced four major shifts: The UK was the first center of the global iron and steel industry; after that, the USA and Germany become the center of the global iron and steel industry at the turn of the nineteenth and twentieth centuries; the third shift was made to the Soviet Union and Japan after the World War II; at the end of the twentieth century, China successfully surpassed the other countries and became the new center of the global iron and steel industry. Refer to Fig. 3.1 for details.

After the first Industrial Revolution, the UK became the global factory. In the early nineteenth century, the steel output in the UK accounted for more than half of the global total [1]. In the late nineteenth century, with the rise of the USA and the

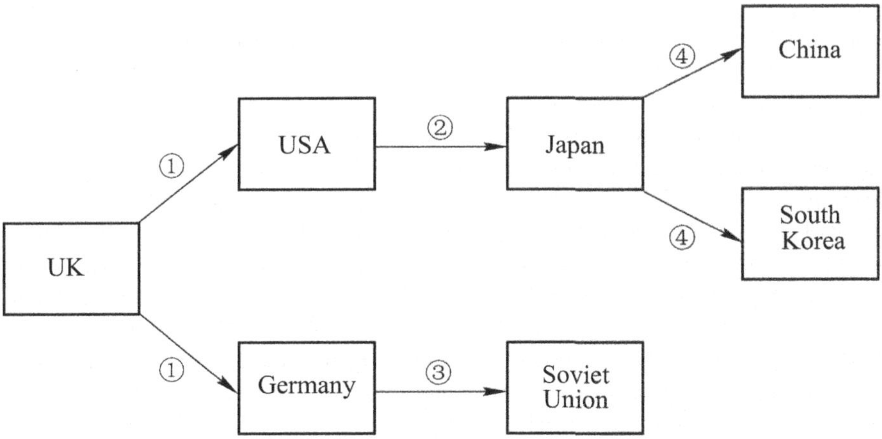


**Fig. 3.1** Shift of the center of global iron and steel industry: ①—late nineteenth century; ②—around 1970; ③—around 1945; ④—around 1995



X. C. Li, *The Road Map of China's Steel Industry*,
https://doi.org/10.1007/978-981-15-2074-7_3

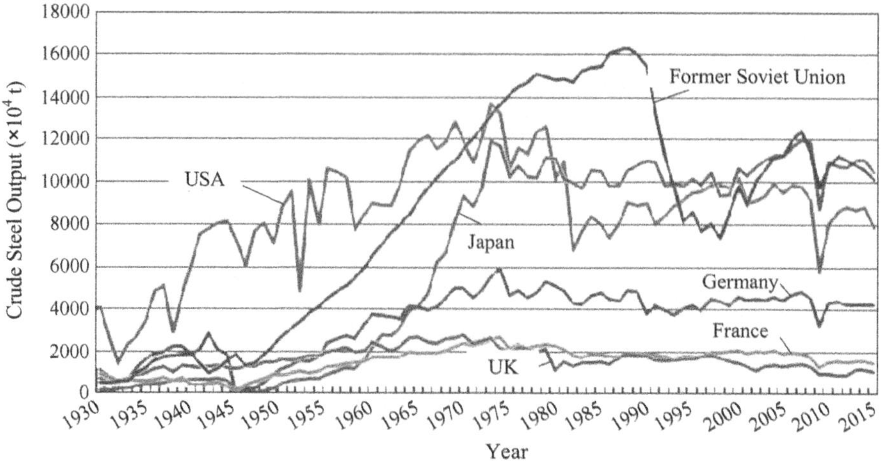


**Fig. 3.2** Changes in production of some major steel-producing countries

acceleration of industrialization, the scale of the iron and steel industry in the USA increased rapidly. By 1880, the crude steel output of the USA climbed to 30% of the global total, basically the same as that of the UK. By the end of the first decade of the twentieth century, the crude steel output of USA had further increased to 26.5 million tons, accounting for about half of the global total. While the iron and steel industry in the USA rapidly developed, making the USA become a new center of the global iron and steel industry, the iron and steel industry in Germany had also kept pace and caught up. In 1880, the crude steel output of Germany accounted for 15% of the global total, equivalent to half of that of the USA and the UK. On the eve of the outbreak of the World War I, the crude steel output of Germany accounted for more than 20% of the global total, making Germany become the second largest steel producer after the USA in the world. Refer to Fig. 3.2 for details.

After being impacted by two world wars, the iron and steel industry in Europe, which was the former center of the global iron and steel industry, was in a great depression. Although the postwar reconstruction and economic growth also brought rapid development, the iron and steel industry of the former Soviet Union and Japan developed faster. The market share of the steel products of Germany and the USA dropped significantly. In 1953, the crude steel output of the Soviet Union accounted for more than 20% of the global total, surpassing Germany to become the second largest steel producer in the world. By the 1970s, the Soviet Union had further leapt to the largest steel producer in the world, and Japan became the second largest steel producer. Nippon Steel, which was reorganized in the early 1970s, was the largest steel company in the world. After the dissolution of the Soviet Union, Japan becomes the largest steel power in the world. After the third shift of the center of global iron and steel industry dominated by the Soviet Union and Japan, China's iron and steel industry and South Korea began to rise. Since China's crude steel output exceeded 100 million tons for the first time in 1996, making it become the new largest steel

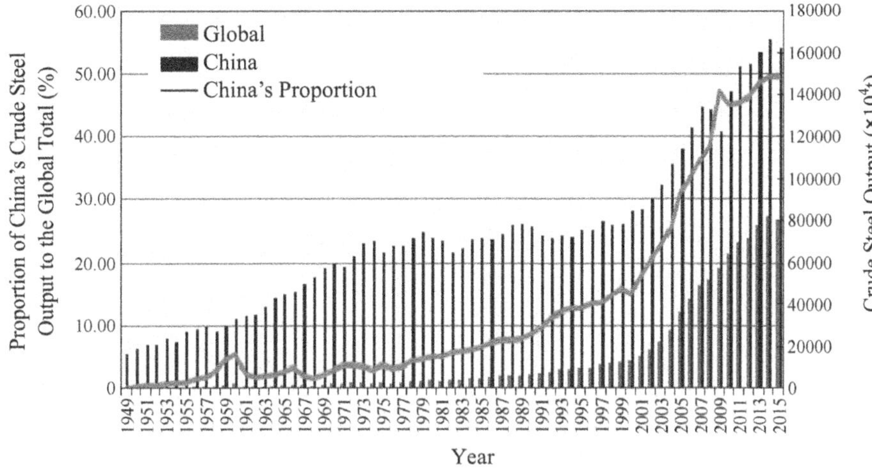


**Fig. 3.3** Comparison of China's crude steel output with the world

producer in the world, China has been ranking first for 21 years, its advantages are getting bigger and bigger, its status is getting more and more stable, and China will continue to maintain this for a long period of time in the future, which will last for more than 100 years. The comparison of China's crude steel output with the world [2] is shown in Fig. 3.3.

## 3.2 Stage Division of the Global Iron and Steel Development After the World War II

After the end of the World War II, the global iron and steel development has generally gone through three stages roughly since the mid-twentieth century (see Fig. 3.4) and is currently at the beginning of the fourth stage.

The first stage: From the end of the World War II to the early 1970s, the iron and steel industry experienced a long-term rapid growth dominated by the world powers. In 1950, global crude steel output was about 190 million tons. By 1974, the crude steel output had grown to a peak of 704 million tons at that time; that is, the output was increased by 3.7 times in 24 years, with an average annual growth rate of 5.6%.

The second stage: From the early 1970s to the end of the twentieth century, affected by two oil crises, the economic restructuring of the western countries, and the dissolution of the Soviet Union, the global iron and steel development was in a plateau period and generally maintained a fluctuating, steady, and low-speed upward trend. During that period, the global crude steel output was increased from 708 million tons from 1974 to 777 million tons in 1998, showing a trend of fluctuation and repeated

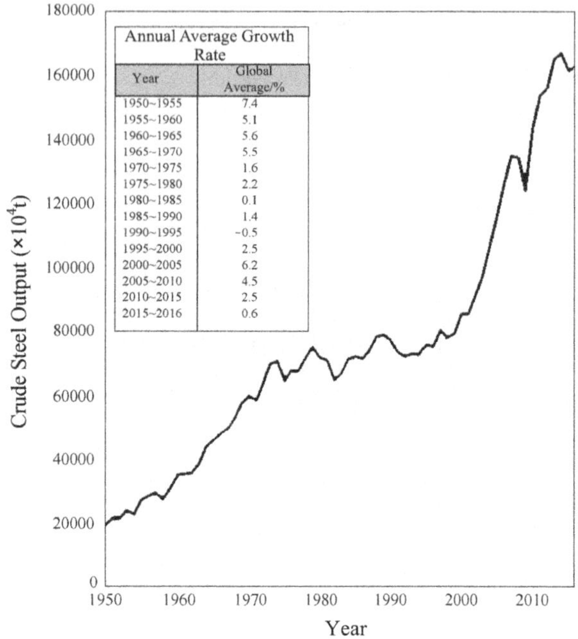


**Fig. 3.4** Development of the global iron and steel production after the World War II

increase and decrease. That is, in another 24 years, the global crude steel output had only increased by 10.4%, with an average annual growth rate of only 0.4%.

The third stage: From the turn of the century to 2014, the rapid development of the steel production and consumption in China led to a new round of development of the global iron and steel industry. Despite the impact by the international financial crisis, the global crude steel output was growing rapidly in more than 10 years, reaching a historical peak of 1.665 billion tons in 2014, and the steel output had more than doubled, with an average annual growth rate of 5.3%. Among them, the increase in China accounted for 85% of the total increase.

According to the World Steel Association, the global crude steel output in 2015 was 1.62 billion tons, a decrease of 2.8% compared to the previous year; in 2016, the global crude steel output was 1,628.5 million tons, an increase of 0.6% over the previous year; in 2017, the global crude steel output was 1691.12 million tons, an increase of 5.3% over the previous year. At present, the global economy is undergoing a tortuous recovery with deep adjustment, and the global steel demand shows a trend of fluctuating development. It is judged that the global iron and steel industry has entered the fourth stage since the World War II, is presenting a new round of plateau period, and will see a relatively long restructuring process: The steel consumption in developed countries and regions is relatively stable and that in China is slowing down with fluctuation. While the steel consumption in some countries in South Asia, Southeast Asia, and Africa will grow to a certain extent under the drive of industrialization and urbanization, they will offset the impact by the declining steel consumption in China.

## 3.3 Outstanding Contributions Made by China to Cutting Overcapacity of the Global Iron and Steel Industry

### *3.3.1 Overall Requirements and Arrangement for Cutting Overcapacity of China's Iron and Steel Industry*

The CPC Central Committee and the State Council attach great importance to "cutting overcapacity" and regard it as the first among the five major tasks of supply-side structural reform. In February 2016, the *Opinions on Cutting the Overcapacity of the Iron and Steel Industry to Realize A Turnaround* was issued and implemented, which requires to achieve the target of further reducing crude steel production capacity by 100–150 million tons in 5 years starting from 2016 on the basis of eliminating the backward steel production capacity in recent years. At the beginning of 2017, that target was clearly defined as 140 million tons (from 2016 to 2018). In order to ensure the orderly and solid progress of the overcapacity cutting work, China has established an inter-ministerial joint meeting system for easing overcapacity and researched and introduced supporting regulations and measures in eight aspects of awards, staff resettlement, finance and taxation, financing, land, environmental protection, quality, and safety. The provincial people's governments and the inter-ministerial joint meeting have signed target responsibility documents to give the tasks to specific enterprises and facilities on an annual basis. The provincial people's governments are generally responsible for cutting overcapacity of the iron and steel industry in the region. Iron and steel enterprises are the subject of responsibility, forming a work and institutional system for cutting overcapacity from the state level to the local level and then to enterprises.

### *3.3.2 Promotion Measures for Cutting Overcapacity of China's Iron and Steel Industry*

The year 2016 marked the beginning of cutting overcapacity of the iron and steel industry, and the inter-ministerial joint meeting dispatched special inspection and supervision teams and acceptance inspection teams to direct, supervise, and urge the implementation progress. The effective practices and advanced experiences of Hangzhou Steel, Panzhihua Steel, and other enterprises in terms of capacity reduction, staff resettlement, asset disposal, and transformation of products and production were summarized and promoted. Legislation-based methods were applied to reduce excess capacity according to the laws and regulations on environmental protection, energy consumption, quality, safety, and technology and on the basis of industrial policies. Three joint actions including eliminating backward capacity, clearing up illegal construction projects, and joint law enforcement had been organized and carried out. The violations of laws and regulations by Jiangsu Huada Company and Hebei Anfeng Company were seriously investigated and handled. Some regions had established a compensation mechanism for reducing excess iron and steel production capacity and established a market trading platform for iron and steel production capacity; some had established a comprehensive evaluation system for the production capacity of iron and steel enterprises based on environmental protection and energy consumption standards to reduce the excess capacity in turn according to the comprehensive evaluation.

The year 2017 marked the tough fighting against excess capacity reduction of the iron and steel industry, and the state proposed an annual target of reducing 50 million tons of excess capacity in the iron and steel industry. The members of the Inter-Ministerial Joint Meeting jointly issued the *Opinions on Well Cutting Overcapacity of the Iron and Steel and Coal Industries in 2017 to Realize A Turnaround*, which requires the forced and accelerated exit of the backward and non-compliant capacity through more intensified marketization and legalization and highlighted strict standards, guides the orderly exit of inefficient production capacity, and encourages the reduction of excess capacity through mergers and acquisitions. Five associations including the China Iron and Steel Association issued the *Opinions on Supporting the Strike against "Substandard Steel" and Defining the Range of Power Frequency and Medium Frequency Induction Furnaces*, which clearly specifies the definition criteria for "substandard steel". Since May of 2017, the inter-ministerial joint meeting had sent inspection and supervision teams to various places to conduct special inspections and supervisions on the banning of "substandard steel". Meanwhile, the satellite-based remote sensing technology is adopted to monitor the overcapacity cutting of iron and steel industry and ensure real cutting of overcapacity.

### 3.3.3 Achievement of China's Efforts to Ease Excess Production Capacity of the Iron and Steel Industry

1. The excess capacity has been really reduced

On the basis of eliminating the backward steelmaking capacity of 94.8 million tons during the 12th Five-Year Plan, in 2016, China completed the annual target of 45 million tons in cutting overcapacity ahead of schedule, and the reduced crude steel production capacity in the whole year was over 65 million tons, 202,000 workers were resettled, and the utilization rate of crude steel capacity had steadily increased. In 2017, the country has completed the goal of reducing the capacity of crude steel by 50 million tons. Since the release of the *Opinions on the Development of the Iron and Steel Industry to Solve the Overcapacity Problem* (No. 6 [2016] by the State Council), China has reduced the crude steel production capacity by 115 million tons.

2. The economic performance of the iron and steel industry has improved

Since 2016, with the implementation of the supply-side structural reform of China's iron and steel industry, the effect of "cutting overcapacity" has begun to appear, positive changes have appeared in the market, and the iron and steel industry has been running steadily. In 2016, the iron and steel enterprises included in the key statistics realized a profit of 30.4 billion yuan, a year-on-year increase of 108.3 billion yuan, and the comprehensive price index of steel products rose from 56.37 points at the beginning of the year to 99.51 points. From January to November of 2017, the iron and steel enterprises included in the key statistics realized a profit of 157.8 billion yuan; the operating conditions of the iron and steel industry continued to improve, and market confidence was significantly enhanced.

3. Iron and steel enterprises rely on innovation drive to enlarge effective supply

"Cutting overcapacity" has pushed iron and steel enterprises to establish new development concepts and rely on innovation drive to improve effective supply. The materials developed by Ma'anshan Steel and Taiyuan Steel for wheels and axles of 350 km/h high-speed trains have undergone 600,000 km operation test, laying a foundation for the localization of the wheels and axles of high-speed trains. Anshan Steel has broken through the width limit of the duplex stainless steel plates and realized the localization and independence of key equipment and materials for nuclear power in China. Xingcheng Special Steel was awarded the "Global Performance Excellence Award" and its 250 mm thickness EH36 steel plates have been successfully applied to the first mobile pilot oil drilling platform of "Offshore Oil Drilling Rig 162". HBIS Group has acquired the Smederevo Steel Plant, which achieved profitability by the end of the year through the implantation of advanced technology and management. The merger of Baosteel and Wuhan Iron and Steel into China Baowu Group will have a profound impact on the steel competition pattern in China and even the world.

### *3.3.4 China Makes Outstanding Contributions to Cutting Overcapacity of the Global Iron and Steel Industry*

China's promotion of supply-side structural reform and implementation of "cutting overcapacity" in the iron and steel industry demonstrate the role as the largest steel power and establish an international image of a responsible giant. As the largest steel producer and consumer in the world, China's iron and steel industry has a pivotal influence on the world. In the face of steel overcapacity, which is a global and international problem, China never slides over, dares to undertake, and acts firstly. With the superiorities of the socialist market economy system with Chinese characteristics, China has promoted the stable elimination of the excess iron and steel capacity of 100 million tons basis, which is unprecedented over the industrial history of the world; moreover, the Chinese government has put forward and participated in the establishment of G20 Global Iron and Steel Overcapacity Forum, thus playing a model role in addressing the global iron and steel overcapacity. Only taking the year of 2016 as an example, the crude steel output in the world was 1.6285 billion tons with a production capacity availability of 69.3%, and the reduced iron and steel overcapacity in China was over 65 million tons, contributing 1.9% in increasing the global iron and steel production capacity availability, while the contribution made by the other regions was −2.3%.

## 3.4 Joins Hands to Face Challenges

China's steel exports have shown a rapid growth in recent years. As the largest steel exporter in the world, China's steel exports have met various trade protection policies of some countries while meeting the needs of international customers and have become an important issue and hot topic in international affairs. How can China's steel exports be looked upon objectively and fairly? How can the global iron and steel industry work together to meet the challenges? They are the issues which need the practitioners in the global iron and steel industry to study together.

### *3.4.1 Looking at China's International Steel Trade in "Six Aspects"*

Firstly, positively looking at China's international steel trade China's steel exports play a remarkably positive role in not only effectively satisfying the real demand of the downstream international customers but also strengthening the competition vitality of the international iron and steel market. In regard to market competition, the improved international competitiveness of China's steel products is a result of the selection by a vast number of international customers.

Secondly, dialectically looking at China's international steel trade, the iron and steel industry shows a property of high globalization from raw materials, production to products. China's steel output accounts for half of the global total, and the steel consumption accounts for nearly half of the global total. The global steel production center and consumption center are highly coincided in China, which is the most outstanding feature of the global iron and steel industry. China's iron and steel industry plays a major role in the world, and China's steel product market represents the international steel product market.

Thirdly, historically looking at China's international steel trade, when looking back at the history, China had long been one of the major net steel importers in the world and firstly became a net steel exporter till 2006. From the perspective of cumulative volume, China had not achieved the basic balance of cumulative steel exports and imports until 2014 and had become a net steel exporter till 2015.

Fourthly, comprehensively looking at China's international steel trade China's steel product export volume is far higher than that of other countries, but the proportion of exported part is less than 15% of the total steel output in China, far lower than the world average level of 30% and also lower than that of major iron and steel producers like Japan, Germany, and South Korea. China's steel products are still mainly used for satisfying domestic demand rather than expanding exports, nor does China hope to solve the overcapacity problem by relying on exports.

Fifthly, objectively looking at China's international steel trade China's steel export growth is only a superficial phenomenon, and the underlying cause is the improved competitiveness of China's steel products, for which the primary impetus comes from the arduous efforts that Chinese iron and steel enterprises have made in aspects such as reducing cost, improving quality, strengthening services, and improving technology. In the fields of most products, China's steel products have surpassed or caught up with international counterparts or have greatly shortened the gap with them.

Sixth is to look at the international trade of Chinese steel products in development. For a certain period to come, China's steel products will be more competitive in the aspects of quality, service, and brand, and China's steel exports will remain at a high level for a long period of time, which is decided by the internal development law of the global iron and steel industry and could not be transferred by the narrow interests or unilateral will of any party.

### 3.4.2 Positively Facing the International Steel Trade Protectionism with the Confidence of a Powerful Country and the Responsibility of a Major Power

Through the hard work by several generations of the people working in the iron and steel industry, China's iron and steel industry has developed into a superior industry with strong global competitiveness. China should abandon the ideological concept of pinch pennies and anger, and positively face the international steel trade protectionism with the confidence of a powerful country in concept and the responsibility of a major power in action.

1. In concept, China should build up the confidence of a powerful country when facing international steel trade protectionism

As an important basic industry in China, iron and steel are the grain of industry and the backbone of the country. Since the founding of the People's Republic of China, especially since the reform and opening-up, China's iron and steel industry has grown from small to large and from weak to strong. Under the drive of both scale growth and quality improvement, the qualification of the iron and steel industry has improved significantly, basically satisfying the development needs of various industries in the national economy including construction, machinery, energy, and automobile industries, and playing an irreplaceable role in building a manufacturing power and carrying out international capacity cooperation. Competitiveness with certain comparative advantages has been formed, and its international status has been greatly improved, which has profoundly changed the structure of the international iron and steel industry, enjoying pivotal influences. In dealing with the international steel trade protectionism, Chinese counterparts in the iron and steel industry and related parties should have the confidence of a powerful country (specifically including product confidence, talent confidence, system confidence, technology confidence, management confidence, etc.). China's iron and steel industry is neither afraid of the bias or ill-intentioned trade protection of certain countries, nor care about the babies-like care given by them to their weak iron and steel enterprises, because they are meaningless compared to the contributions and effect made by China's iron and steel industry to the global iron and steel industry. Therefore, China will put up with them and not allow them to distract China's iron and steel industry in achieving higher-level development.

2. In action, China should shoulder the responsibility of a major power when facing the international steel trade protectionism

In 2015, China's steel export growth rate was 19.9%, and the total export volume was 112 million tons [3], accounting for about a quarter of the global steel trade volume, 2.7 times that of Japan which is the second largest steel exporter, and equivalent to the sum of the steel export volume of Japan (ranking second), EU (ranking third), and South Korea (ranking fourth); thus, China should shoulder the responsibility of a major power in the international steel trade. It is mainly reflected in three aspects: First, China should deal with unreasonable international steel trade protection policies from the perspective of maintaining the normal order of the international steel trade in the world, instead of responding to suits from the perspective of simply safeguarding China's steel export interests; second, China should turn from passiveness to initiative to intervene in advance in the analysis of the current status and situation of international steel trade, study the advantages and disadvantages and competitiveness characteristics of the exported steel products of various countries, investigate some potential trade disputes, and take proactive measures to make adjustment in order to adapt to relevant requirements; third, China should proactively organize relevant parties to study the establishment of new international trade orders and new rules for international steel trade, for instance, creating diversified bilateral and multilateral dialogue and cooperation systems for steel trade, and discussing with key countries to explore a long-term stable, mutually beneficial, and win-win model steel trade system of model significance.

### 3.4.3 Strategic Recommendations on China's Steel Import and Export

Considering China's strategy for manufacturing power and the development of iron and steel industry at home and abroad, it is suggested that the overall strategy for

China's steel import and export in the coming period is to give full play to the comparative advantages of the industry, actively participate in international competition, and steadily promote the strategic shift of China's international steel trade from "quantity" to "quality".

(1) Strengthening regulated competition and promoting "quality-orientated" export. For the "quality-orientated export" strategy in the iron and steel industry, China shall know that there are things must be done and things must not be done to gradually expand the export volume and export proportion of "quality products, superior products, and excellent products", which requires regulating the competition of enterprises in terms of product quality, service level, market reputation, and brand building, so as to enhance the overall competitiveness of the iron and steel industry.
(2) Increasing independent innovation and promoting "optimization-orientated" import. For the "optimization-orientated import" strategy in the iron and steel industry, China shall force imported products and related enterprises to further "optimize services, optimize quality, and optimize prices" through strengthening independent innovation and expanding the market.
(3) Establishing the self-confidence of a powerful country and assuming the responsibilities of a major power to actively respond to trade protection. China shall face up to international trade frictions, actively respond to them in accordance with laws and regulations, and reasonably cooperate with international customers and competitors to maintain and even build a good new competition order in the international steel market.

### 3.4.4 "Three Joints" Initiatives for the Global Iron and Steel Industry

1. Jointly building a fair and ordered international steel trade environment

A fair and ordered international steel-trading environment is conducive to the healthy development of the global iron and steel industry, and also conforms to the interests of all steel-producing countries and regions. Governments, industrial organizations, and enterprises in relevant countries and regions should not rely solely on trade protection as a means of safeguarding their backward iron and steel producers. They should give full play to their respective advantages and work together with other countries to build a fair and ordered international steel-trading environment, so as to improve their own competitiveness while promoting the healthy and sustainable development of global iron and steel industry.

2. Jointly undertaking the historical task of resolving the overcapacity of the global iron and steel industry

At present, both China and other countries (regions) are facing the severe challenge of overcapacity in the iron and steel industry. The iron and steel industry in China has done a lot of fruitful work in cutting overcapacity. During the 12th Five-Year Plan, the cumulative eliminated steelmaking capacity was about 94.8 million tons. During the 13th Five-Year Plan, China still attaches great importance to and actively cuts the overcapacity of the iron and steel industry. Based on the elimination of backward production capacity in recent years, China will reduce the crude steel production capacity by 100 million to 150 million tons in five years. China has been made, is making, and will continue to make significant contributions to resolving the overcapacity of the global iron and steel industry. The iron and steel industry in the USA and other relevant counterparts should abandon narrow thinking and join hands with the iron and steel industry in other countries, including China's iron and steel industry, to jointly undertake the historical task of resolving the overcapacity of the global iron and steel industry.

3. Jointly promoting the development and application of iron and steel materials worldwide

As the mostly used structural material and important functional material, iron and steel materials have made unparalleled contributions to the development of the global economy and the progress of human civilization. In the foreseeable future, iron and steel materials will continue to play an important role in the development of the global economy due to their excellent performances, abundant resources, greenness and low carbon, low cost, and mature system, but they will inevitably face a more powerful challenge of other advanced alternative materials. Therefore, the practitioner in the global iron and steel industry should seek common ground while reserving differences, put attention, people, and property in how to better meet the changing needs of customers, and effectively promote the development, application, and market development of the iron and steel materials worldwide.

## References


1. Zhang, X. P., & Zhang, Q. Y. (2005). Development trend of the global iron and steel industry and its impact on China's iron and steel industry. World Regional Studies.
2. China Iron and Steel Industry Association. (2001–2016). China steel statistics. Beijing: China Iron and Steel Industry Association.
3. Statistics of the General Administration of Customs, P.R. China.

# Part II
# Reduction, Innovation and Transformation

# Chapter 4
# Reduction

## 4.1 Connotation and Origin of Reduction

The connotation of the reduction of the iron and steel industry means: China's economic development has entered into a new normal, with the economic growth showing medium and high speed, the industrial structure moving toward mid-to-high end and the shifting to new economic growth drives the consumption intensity of steel products per unit of GDP tends to decline, and the total steel consumption has entered a declining zone after the peak arc zone. Despite the fluctuations and repetitions, the general trend of reduction-based development has become quite clear, which will lead to decline in total energy consumption, total pollutant emission, and total resource consumption, have a far-reaching effect on the process restructuring, technical equipment structure, product structure, energy structure, resource structure, organizational structure, and industrial layout, and put forward new requirements for industrial development mode and development drives. And the "cutting overcapacity" through supply-side structural reform is a master key to recognizing, adapting, and leading the new normal of the reduction in the iron and steel industry. Therefore, from the perspective of "cutting overcapacity", the reduction of the iron and steel industry is elaborated mainly in two aspects: one is to eliminate backward production capacity and promote industrial upgrading; the other is to resolve excess capacity and help the industry to get out of trouble, which both refer to steel equipment and production capacity. The former focuses on equipment, while the latter focuses on production capacity, but the main subject is ironmaking and steelmaking equipment.

The elimination of backward production capacity of China's iron and steel industry has been implemented for a long time. As early as the mid-1950s, there was a discussion in Anshan Steel: "Is it better to have an open hearth furnace or a converter?" That debate lasted for decades. In 1963, China's first 30 t converter was put into operation in Shougang. After that, Benxi Steel and Panzhihua Steel also adopted converter steelmaking process. Compared with open hearth furnaces, oxygen converters showed powerful advantages in terms of tap-to-tap cycle, fuel consumption,



X. C. Li, *The Road Map of China's Steel Industry*,
https://doi.org/10.1007/978-981-15-2074-7_4

investment and operating costs, and product varieties. Therefore, the converter steelmaking process was subject to a rapid development, which accelerated the elimination of open hearth furnaces.

In order to stop low-level redundant construction, speed up the restructuring pace, and promote the upgrading of production processes, equipment and products, the former State Economic and Trade Commission issued the *Catalogue of Eliminating Outdated Production Capacity, Processes, and Products* (*First Batch*) [1] in January 1999, which officially opened the prelude to the elimination of backward production capacity.

## 4.2 Reduction History of China's Iron and Steel Industry

The elimination of backward iron and steel production capacity in China can be roughly divided into the following four stages.

(1) Initial stage (1999–2004): At this stage, the backward production processes of open hearth steelmaking were mainly eliminated in order to improve the technical level and production efficiency of steelmaking, and significantly reduce resource and energy consumption.
(2) Promotion stage (2005–2009): At this stage, the backward production capacity of electric furnaces and converters of 20 t and below was mainly eliminated in order to carry out restructuring, upgrading and total quantity control of the iron and steel industry, improve the level of process technology and equipment, and implement clean production and sustainable development of the iron and steel industry.
(3) Normalization stage (2010–2014): At this stage, the backward production capacity of blast furnaces below 400 $m^2$ and converters and electric furnaces of 30 t and below was mainly eliminated, while the certification criteria for backward equipment had not been enhanced, and the total quantity control and energy conservation and emission reduction had been carried out in a parallel way, with restructuring, transformation, and upgrading as the main directions and independent innovation and technological transformation as supports, to promote the transformation of the iron and steel industry from large to strong.

    The focus of the first three stages of the reduction work was to eliminate the backward equipment and production capacity. The main means were taking volume and capacity as the certification criteria for backward equipment and adopting the "single solution" work mode for mandatory promotion. Although that work mode was controversial, the effect was more significant due to its higher operability.
(4) Resolving stage (2015–2020): At this stage, the backward equipment elimination task with volume and capacity as the main criteria had been basically completed, and the reduction work of the iron and steel industry entered into the

new stage of supply-side structural reform to resolve excess capacity. The resolving stage is characterized by implementing more reasonable policy measures, emphasizing the compliance with laws and regulations, focusing on the combination of reduction and improvement as well as passiveness and activeness, which is expected to achieve a positive progress.

## 4.3 Analysis of the Measures and Their Effects on Reduction at Each Stage

### *4.3.1 Initial Stage (1999–2004)*

1. Policy Measures

In January 1999, the former State Economic and Trade Commission issued the *Catalogue for Eliminating Outdated Production Capacity, Processes, and Products* (*First Batch*) clearly classifying the open hearth furnaces as backward production process equipment to be completely eliminated by the end of 2000.

In December 1999, the former State Economic and Trade Commission issued the *Catalogue of Eliminating Backward Production Capacity, Processes, and Products* (*Second Batch*) [2] stipulating the elimination of converters of 10 t and below, side-blown converters, electric furnaces of 5 t and below as well as normal frequency furnaces for producing low-quality steel bar or open ingots before the end of 2000, and the elimination of the converters of 10–15 t (included), electric furnaces of 5–10 t (included) as well as cupola furnace steelmaking process before the end of 2002. That was the first time to classify the backward steelmaking process equipment according to their nominal capacity, which had promoted the improvement of the steelmaking process equipment level in China and played an important role in eliminating backward steelmaking capacity.

2. Achievements

In December 28, 2001, the steelmaking plant of Baotou Iron and Steel Company eliminated the last 500 t open hearth furnace for production, officially announcing that the open hearth steelmaking process had exited the historical stage, and also marking the end of the first step of eliminating backward production capacity. Its characteristic is that the innovation and upgrading of process technology have played a decisive role. The high efficiency and energy saving of the converter have promoted the voluntary elimination of outdated process equipment by enterprises and achieved the successful completion of the goal of eliminating backward production capacity.

### 4.3.2 Active Promotion Stage (2005–2009)

Since 1996, China's crude steel output has been ranked first in the world. By 2005, the crude steel output and production capacity exceeded 300 million tons [3] and 400 million tons, respectively, continuously marking a record high. The gross domestic product (GDP), fixed asset investment, and crude steel output growth rate of China during the 10th Five-Year Plan are shown in Fig. 4.1. The rapid growth of crude steel output had not only met the needs of national economic and social development, but also brought massive consumption of resources and energy. In particular, China's iron and steel industry was subject to a low level, repeated construction, low industrial concentration as well as a large gap with international advanced levels in terms of technical level and material consumption. Moreover, because of low added value of products and low comprehensive competitiveness of enterprises, the situation of being big but not strong had lasted for a long time. Therefore, it must change the development model of the iron and steel industry and achieve its sustainable development through the elimination of backwardness, technological upgrading, and restructuring.

1. Policy Measures

Since 2005, the state has successively issued a series of policies and documents on eliminating backward production capacity to define backward production capacity and vigorously promote their elimination.

(1) *The Development Policy for Iron and Steel Industry* promulgated by the National Development and Reform Commission [4] in July 2005 pointed out that China would speed up the elimination and prohibit the construction of backward technological equipment such as blast furnaces and electric furnaces

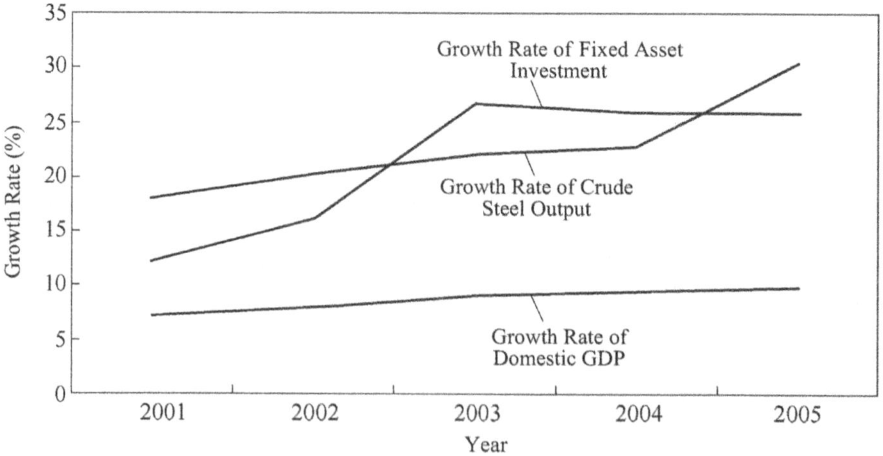


**Fig. 4.1** Growth rate of GDP, fixed asset investment, and crude steel output in China during the 10th Five-Year Plan (unit: %)

of 300 $m^3$ and below, iron melting-based steelmaking facilities, converters, and electric furnaces of 20 t and below, and medium frequency induction furnaces.

(2) *The Catalogue of Industrial Restructuring Guidance* (2005 Edition) (No. 40 Order of NDRC in 2005) promulgated by the National Development and Reform Commission in December 2005 [5] again listed the normal frequency and medium frequency induction furnaces for producing low-quality steel bar, ingots or casting billets, the iron melting-based steelmaking facilities, converters of 15 t and below (excluding ferroalloy converters) as well as electric furnaces of 10–20 t (excluding electric furnace for mechanical casting) into the catalogue to be eliminated immediately; the converters of 10 t and blow (excluding ferroalloy converters) and electric furnaces of 10–20 t and below (excluding electric furnace for mechanical casting) into the catalogue to be eliminated by the end of 2005; the converters of 20 t and blow (excluding ferroalloy converters), electric furnaces of 20 t and blew (excluding electric furnaces for high-alloy steel and mechanical casting) into the catalogue to be eliminated by the end of 2006; and the blast furnaces of 200–300 $m^3$ (included) into the catalogue to be eliminated by the end of 2007.

(3) *The Decision of the State Council on Issuing and Implementing the Interim Provisions on Promoting Industrial Restructuring* promulgated in December 2005 (No. 40 of the State Council in 2005) [6] pointed out to limit and eliminate backward production capacity and prevent blind investment and low level repeated construction, and effectively promote the optimization and upgrading of industrial structure. For the eliminated projects, the investment shall be prohibited, and all regions, departments, and related enterprises must take effective measures and eliminate them within the prescribed time limit. For those who fail to complete elimination as scheduled, the local governments at various levels shall order them to suspend production or shut down them according to relevant state laws and regulations.

(4) *The Notice on Accelerating the Restructuring of Overcapacity Industries* promulgated in March 2006 (No. 11 of the State Council in 2006) [7] first put forward that there was a significant excess of iron and steel production capacity, and the contradiction between resource and environmental constraints was conspicuous; thus, restructuring must be carried out. It shall give play to the basic role of resources allocation by the market, make full use of the power of the market to drive competition and promote the survival of the fittest, formulate more stringent standards for environment, safety, energy consumption, water consumption, comprehensive utilization of resources, quality, technology, scale, etc., to advance access thresholds. Through restructuring, transformation, and elimination, the pace of restructuring would be accelerated, and blast furnaces of less than 300 $m^3$ and converters and electric furnaces of less than 20 t would be eliminated.

(5) *The Notice on Controlling Total Quantity, Eliminating Backwardness and Accelerating the Restructuring of the Iron and Steel Industry* issued in June 2006 (No. 1084 of NDRC in 2006) [8] pointed out that the capacity of small converters and electric furnaces of 20 t and below shall be 55 million tons

among the steel production capacity of 420 million tons formed at the end of 2004, accounting for 13.1% of the total capacity; and it also clearly defined that backward capacity of converters and electric furnaces of 20 t and below would be mainly eliminated before 2007.

(6) *The Notice of the State Council on Printing and Distributing the Comprehensive Work Plan for Energy Conservation and Emission Reduction* promulgated in June 2007 (No. 15 of the State Council in 2007) [9] pointed out that energy conservation and emission reduction should be the focus of the macroeconomic regulation and control at that time and the breakthrough point and main fulcrum for economic restructuring and growth mode transformation to resolutely inhibit the excessive growth of high energy consumption and high-pollution industries, strictly control the construction of new high energy consumption and high-pollution projects, and devote more efforts to eliminate backward production capacity in power and iron and steel industries. The Notice was also attached with a list of the eliminated backward production capacity during the 11th Five-Year Plan, which quantified the elimination of outdated equipment and clearly pointed out the elimination of backward ironmaking and steelmaking capacity of 100 million tons and 55 million tons, respectively, during the 11th Five-Year Plan, including elimination of 30 million tons and 35 million tons, respectively, by the end of 2007.

(7) *The Notice of the National Development and Reform Commission on the Work of Shutting Down and Eliminating Backward Production Capacity* in the Iron and Steel Industry issued in October 2007 (No. 2761 of NDRC in 2007) required that, in response to some problems, the elimination standards of backward iron and steel production capacity shall be strictly implemented; backward equipment must not avoid elimination by means of transformation from elimination category into limited one; the certification for Dismantling Backward Iron and Steel Production Capacity and Equipment shall be issued; and the law enforcement and verification should be enhanced, so as to ensure the actual implementation of the shutdown and elimination of backwardness.

(8) *The Notice of the National Development and Reform Commission on the Strict Prohibition of the Transfer of Backward Production Capacity* issued in October 2007 (No. 2792 of NDRC in 2007) pointed out that the production process technology, equipment, and products that were explicitly eliminated by the state should not be imported, transferred, produced, sold, used or applied, and backward production capacity should not be transferred. Moreover, backward equipment and facilities should be completely abolished to resolutely prevent the outflow of backward production capacity. Backward process equipment that could not be transformed or upgraded must be eliminated. The list of enterprises that had eliminated backward process equipment should be announced on the local major media and submitted to the public for supervision.

(9) *The Plan for Restructuring and Revitalization of the Iron and Steel Industry* promulgated in March 2009 (No. 6 of the State Council in 2009) [10] clearly required that new production capacity should be strictly controlled; the steel projects simply for new production capacity or expansion be not approved or

supported; and all projects must be based on the elimination of backwardness. By the end of 2010, the eliminated production capacity of blast furnaces of 300 $m^3$ and below was 53.4 million tons, and that of converters and electric furnaces of 20 t and below was 3.2 million tons; by the end of 2011, the blast furnaces of 400 $m^3$ and below, and the converters and electric furnaces of 30 t and below were eliminated, corresponding to the elimination of backward ironmaking capacity of 72 million tons and backward steelmaking capacity of 25 million tons. In regions where large steel plants are built based on the elimination of backward production capacity and those where conditions permit, it was planned to raise the criteria to eliminate the backward production capacity of blast furnaces of 1,000 $m^3$ and below and corresponding steelmaking capacity.

(10) *The Notice of the State Council on the Approval of the Options about the Suppressing Overcapacity in Some Industries and Repeated Construction to Guide the Healthy Development of the Industries submitted by the Development and Reform Commission and Other Departments* (No. 38 of the State Council in 2009) [11] issued in September 2009 pointed out that, making full use of the pressure by market forces to accelerate the structural adjustment and technological progress by eliminating backwardness, joint restructuring and relocation of urban steel plants, under the premise of reducing or not increasing production capacity, so as to promote the transformation of the iron and steel industry from largeness to mightiness. The steel projects simply for new capacity and capacity expansion would no longer be approved and supported. It was strictly forbidden various regions to build steel projects on their own by avoiding the supervision and approval of the national environmental protection, land and investment authorities in the name of eliminating backward production capacity. By the end of 2011, the blast furnaces of 400 $m^3$ and below, and the converters and electric furnaces of 30 t and below were resolutely eliminated.

(11) In November 2009, the Ministry of Industry and Information Technology issued the *Notice on Splitting and Implementing the Task of Eliminating Backward Production Capacity in 2009*, which pointed out that, by the end of 2009, it plans to eliminate backward ironmaking and steelmaking capacity of 21.13 million tons and 16.91 million tons, respectively, over the country, of which the task of eliminating backward production capacity was arduous for Jiangsu Province and Hebei Province.

2. Achievements

That stage was the period in which China's iron and steel industry had developed at the highest speed, the energy conservation and emission reduction had achieved remarkable results, and the iron and steel industry had effectively met the needs of economic and social development. The state's policy measures on eliminating backwardness had been steadily advanced, and the elimination of backwardness had achieved certain results.

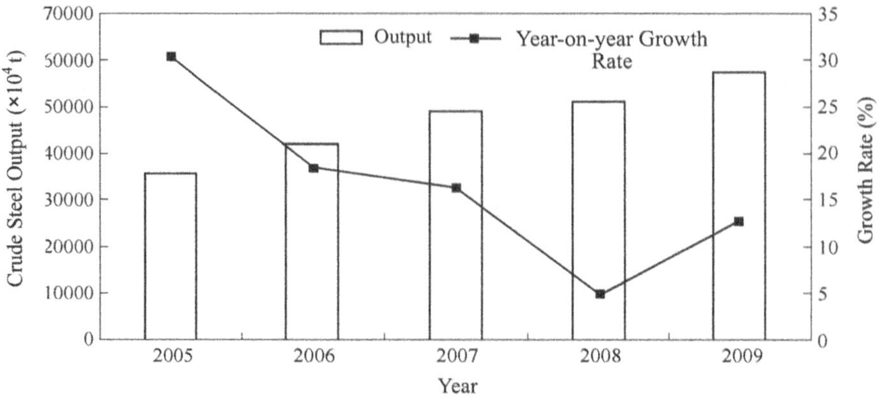


**Fig. 4.2** Status and growth rate (%) of China's crude steel output (unit: 10,000 t) from 2004 to 2009

(1) Crude steel output continued to grow, but the growth rate slowed down significantly. China's crude steel output growth rate dropped sharply from 30.4% in 2005 to 12.6% in 2009 with a decrease of 17.8%. During that period, the growth rate of crude steel production capacity decreased from 24.6% to 11.4% with a drop of 12.2%, and the crude steel output utilization rate was maintained within a reasonable range of 80% or more (up to 89.1% in 2006), as shown in Fig. 4.2.
(2) The technical level of steelmaking equipment had been significantly improved. The converters and electric furnaces of 20 t and below had been basically eliminated. During the 11th Five-Year Plan, the proportion of large-scale steelmaking equipment in key large- and medium-sized iron and steel enterprises in China had been significantly improved. The converters of 100 t and above and the ultra-high-power electric furnaces of 50 t and above had basically reached the advanced level of similar equipment abroad. The proportion of advanced production capacity had been continuously improved, and they had become the main equipment for steelmaking production in China, as shown in Fig. 4.3.
(3) The energy conservation and emission reduction had achieved remarkable results. In 2010, the energy conservation and emission reduction indexes of key iron and steel enterprises included in the statistics were improved comprehensively. The comprehensive energy consumption per ton of steel fell to 605 kg of standard coal; the consumption of freshwater was reduced to 4.1 $m^3$; and the sulfur dioxide emissions were reduced to 1.63 kg, a reduction of 12.8%, 52.3%, and 42.4%, respectively, compared to those in 2005. The comprehensive utilization rate of solid waste was increased from 90% to 94%.

During that period, due to the lack of effective assessment methods for the elimination of backward production capacity, together with the strong drive by market demand, enterprises mainly aimed at guaranteeing production and increasing production, and their enthusiasm for eliminating backward production capacity was not

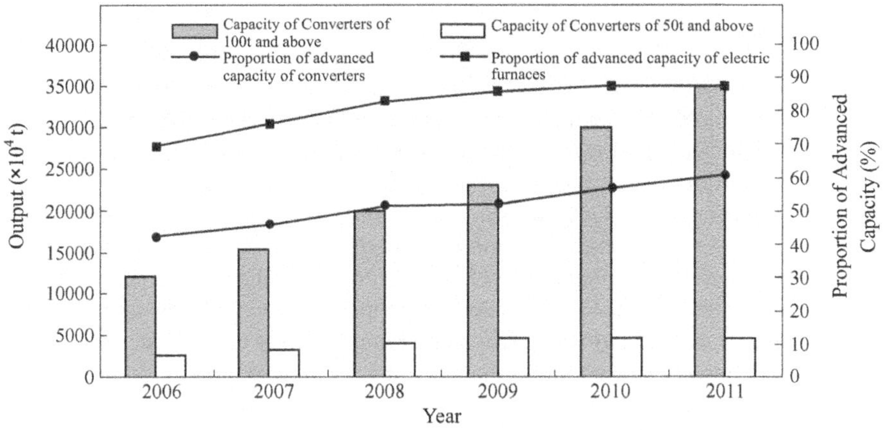


**Fig. 4.3** Changes in China's steelmaking capacity during the 11th Five-Year Plan (unit: 10,000 t, %)

strong; thus, the capacity replacement was basically stopped, and the overall progress of eliminating backward production capacity was slow.

### *4.3.3 Standardization Stage (2010–2014)*

1. Policy Measures

The "12th Five-Year Plan" period was a period for standardizing the elimination of backward production capacity. As the state further strengthened the efforts to eliminate backward production capacity, relevant supporting measures such as the assessment methods for elimination of backward production capacity and the incentive central financial funds had been successively introduced, and great attentions were paid to resolving excess production capacity, especially the issuance of the Notice of the State Council on Further Strengthening the Work of Eliminating Backward Production Capacity (No. 7 of the State Council in 2010) was of epoch-making significance in the elimination of backward production capacity, resulting in substantial achievements in eliminating backward production capacity.

(1) *The Notice of the State Council on Further Strengthening the Work of Eliminating Backward Production Capacity* (No. 7 of the State Council in 2010) [12] issued in February 2010 pointed out that, it must give full play to the role of the market, adopt more effective measures, and comprehensively apply law, economy, technology, and necessary administrative means to further establish and improve a long-term mechanism to eliminate backward production capacity, so as to ensure the targets of eliminating backward production capacity are achieved on schedule.

(2) *The Notice of the State Council on Further Efforts to Ensure the Realization of the Energy Conservation and Emission Reduction Targets During the 11th Five-Year Plan* (No. 12 of the State Council in 2010) promulgated in May 2010 required to enhance the elimination of backward production capacity to eliminate backward ironmaking and steelmaking capacity by 25 million tons and 6 million tons, respectively, in 2010.
(3) *The Opinions of the General Office of the State Council on Further Strengthening Energy Conservation and Emission Reduction to Accelerate the Structural Adjustment of the Iron and Steel Industry* (No. 34 of the State Council in 2010) promulgated in June 2010 pointed out: improving the withdrawal mechanism of backward production capacity and giving full play to the basic role of market in allocating resources, enhancing strict tax collection and management, cleaning up and correcting local tax incentives for iron and steel enterprises, and striving to create a market environment that promotes fair competition between enterprises and withdrawal of backward production capacity; improving and implementing policies on land use and differential electricity price, enhancing the implementation of differential electricity prices, substantially increasing the price increase standard for differential electricity prices, and further increasing the production costs of backward production capacity.
(4) In June 2010, the Ministry of Industry and Information Technology issued the *Regulations on the Production and Management of the Iron and Steel Industry*, which put forward clear requirements for the product quality, environmental protection, energy consumption, comprehensive utilization of resources and process equipment of iron and steel enterprises, and required that environmental protection, energy consumption and equipment level rather than only equipment level shall be taken into account in the elimination of backward production capacity.
(5) In October 2010, the *Guidance Catalogue for the Elimination of Outdated Production Process Equipment and Products in Some Industrial Sectors* (2010 Edition) issued by the Ministry of Industry and Information Technology further clarified that the blast furnaces of 300 $m^3$ and below, the steelmaking converters and electric furnaces of 20 tons and below, the normal frequency and medium frequency induction furnaces for producing low-quality steel bar and carbon steel (excluding steel ingots for mechanical casting) as well as the electric furnaces of 5,000 kVA and below for producing high-alloy steel must be eliminated before the end of 2010; the blast furnaces above 300 $m^3$ but not exceeding 400 $m^3$ and the steelmaking converters and electric furnaces above 20 tons but not exceeding 30 tons must be eliminated before the end of 2011.
(6) In January 2011, eighteen departments including the Ministry of Industry and Information Technology and the National Development and Reform Commission jointly issued the *Notice on Printing and Distributing the Implementation Plan for the Elimination of Backward Production Capacity* (No. 46 of Ministry of Industry and Information Technology in 2011), which clearly defined that the evaluation, assessment, reward, and punishment system for job objective

responsibility on elimination of backward production capacity shall be established and improved according to the general requirements of clear objectives, sound organization, responsibilities in place, measures in place, supervision in place and level-by-level evaluation, and the responsibilities of local governments and enterprises shall be implemented, so as to ensure the successful completion of the goal of eliminating backward production capacity.

(7) *The Outline of the Twelfth Five-Year Plan for National Economic and Social Development of the People's Republic of China*, promulgated in March 2011, stated that the metallurgical industry should base itself on domestic demand, strictly control aggregate expansion, optimize variety structure, and make new progress in product R&D, comprehensive utilization of resources, energy conservation and emission reduction, etc. The elimination of backward production capacity shall be enhancing, and excess capacity shall be reduced and diverted.

(8) *The Guidance Catalogue for Industrial Structure Adjustment* (2011 Edition) issued by the National Development and Reform Commission in March 2011 once again pointed out that the normal frequency and medium frequency induction furnaces for producing low-quality steel bar, carbon steel and stainless steel, the converters of 30 tons and below (excluding ferroalloy converters), the electric furnaces of 30 tons and below (excluding mechanical casting furnaces) and steelmaking process by means of iron melting must be eliminated before the end of 2011. Meanwhile, the *Guidance Catalogue for Industrial Structure Adjustment* (2005 Edition) was abolished.

(9) In October 2011, the *Twelfth Five-Year Development Plan for the Iron and Steel Industry* was officially released, which clearly pointed out that it should further promote energy conservation and emission reduction, reduce the energy consumption, carbon dioxide emission and water consumption per unit added value in iron and steel enterprises according to the general requirements of national energy conservation and emission reduction and local decomposed task indicators, and decrease the total sulfur dioxide emissions. At the same time, it pointed out that it should continue to strengthen the elimination of backward production capacity, and resolutely eliminate the normal frequency and medium frequency induction furnaces for low-quality steel bar and carbon steel (excluding steel ingots for mechanical casting), the steelmaking converters of 30 tons and below, the steelmaking electric furnaces of 15,000 kVA and below (30 tons and below), high-alloy electric furnaces of 5,000 kVA and below (nominal capacity of 10 tons and below) as well as other backward production capacity.

(10) In December 2011, the State Council issued the *Industrial Transformation and Upgrading Plan (2011–2015)* (No. 47 of the State Council in 2011), which clearly required to control newly added capacity and general expansion, as well as improve the overall quality of the iron and steel industry from aspects of technical transformation and modification, backwardness elimination, mergers and acquisitions and circular economy.

(11) In October 2013, the State Council issued the *Guidance on Resolving the Contradictions of Serious Overcapacity* (No. 41 of the State Council in 2013),

which proposed "digesting a batch, transferring a batch, integrating a batch, and eliminating a batch" of excess capacity. Through five years of hard work, the total production capacity of the iron and steel industry shall be compatible with environmental carrying capacity, market demand, and resource guarantee. The layout shall be coordinated with the regional economic development. The total capacity of reduced iron and steel production shall be over 80 million tons, and the capacity utilization rate shall reach a reasonable level.

(12) In December 2013, the State Council issued the *Government-approved Catalogue of Investment Projects* (2013 Edition) (No. 47 of the State Council in 2013), which required that, for the projects in the industries with severe overcapacity such as steel and electrolytic aluminum, relevant departments of the State Council and the local governments should strictly control the newly increased production capacity in accordance with the requirements of the guidance on resolving the contradiction of severe excess capacity.

(13) In November 2014, the State Council issued the *Government-approved Catalogue of Investment Projects (2014 Edition) (No. 53 of the State Council in 2013)*, which implemented the filing system for investment projects in the iron and steel industry, which were no longer included in the approved catalogue. It required to strictly implement the No. 41 document of the State Council in 2013 for the iron and steel projects, and work together to promote the resolution of the contradiction of serious excess capacity.

2. Achievements

Under the unified arrangement and leadership of the Party Central Committee and the State Council, the members of the Inter-Ministerial Coordination Group for the Elimination of Backward Production Capacity had effectively cooperated, gradually established a work system, continuously improved policy measures, and achieved substantial progress in eliminating backward production capacity, which optimized industrial structure, promoted energy conservation and emission reduction, effectively driven up the transformation and upgrading of the iron and steel industry.

(1) Alleviated contradiction of excess capacity. From 2010 to 2014, the eliminated ironmaking capacity was 118 million tons and eliminated steelmaking capacity was 89.66 million tons [13]. The timely elimination of those backward production capacity had alleviated the contradiction of excess capacity to some extent and promoted industrial restructuring and energy conservation and emission reduction, and reserve spaces regarding land, energy, resources, market space and environmental capacity for local development.

(2) Optimized industrial structure. In the coking industry, the most small coke ovens with a coking chamber height less than 4.3 m haven been eliminated. The production capacity ratio of large advanced coke ovens with a coking chamber height of 5.5 m and above has increased from 30% to 40%. In the iron and steel industry, the production capacity ratio of blast furnaces of 1000 $m^3$ and above has been increased from 33% to 50% only in Shandong Province.

(3) Enhanced industrial energy conservation and emission reduction. Since 2010, a large number of outdated equipment featuring high energy consumption and high emissions have been eliminated, which directly promoted industrial energy conservation and emission reduction. According to the calculation by Sichuan Provincial Energy Conservation Supervision Center, since 2010, the accumulated elimination of backward production capacity in Sichuan Province has exceeded the energy-saving capacity of more than 10 million tons of standard coal, which make about 30% contribution for achieving the province's industrial energy-saving target. Eliminating backward production capacity has become an important support and means for achieving the targets of energy conservation and emission reduction around China.
(4) A long-term mechanism for eliminating backward production capacity has been established and improved.

The first is to establish an organizational coordination and implementation system. An Inter-Ministerial Coordination Group for the Elimination of Backward Production Capacity led by seventeen departments including the Ministry of Industry and Information Technology, the National Development and Reform Commission, the Ministry of Supervision, and the Ministry of Finance was established to study and solve major problems and coordinate the overall work. All local governments have also set up coordination (leadership) teams with provincial government leaders or heads of leading departments as team leaders and participated by responsible personnel of relevant departments to decompose and implement the objectives and tasks, clarify the division of responsibilities, formulate work plans, and promote the elimination of backward production capacity under unified leadership.

The second is to improve award and incentives policies and measures. The Ministry of Industry and Information Technology, the Ministry of Finance, and the Energy Bureau have formulated the *Measures for the Management of Central Financial Incentive Funds for Eliminating Backward Production Capacity*. In accordance with the principle of "rewarding on those who have eliminated backward production capacity on time or in advance", an annual reward threshold has been set to guide enterprises to actively eliminate the backward production capacity, and thus a working mechanism that links the reward funds with the elimination target tasks, which has strongly supported the work carried out in various places.

The third is to comprehensively apply legal means, economic means, and necessary administrative means. The Ministry of Industry and Information Technology has formulated the *Guidance Catalogue for the Elimination of Outdated Production Process Equipment and Products in Some Industrial Sectors* (2010 Edition), which clearly defines the standards of judgment and clarifies the elimination period. The Ministry of Industry and Information Technology and the State Electricity Regulatory Commission have jointly issued the *Notice on Further Strengthening the Supervision of Power Energy Conservation and Emission Reduction and Doing a Good Job in Eliminating Backward Production Capacity* to urge power supply enterprises to do a good job in cutting off or limiting power supply in accordance with the laws to prevent the resurgence of backward production capacity. Relevant departments have increased enforcement regarding land use, environmental protection, production licenses, industrial and commercial registration, etc., in order to standardize market fair competition and promote backward production capacity to be eliminated through market competition.

The fourth is to establish inspection, assessment, and supervision mechanisms. With the approval by the State Council, the members of the Inter-Ministerial Coordination Group have formulated the *Implementation Plan for the Assessment of Elimination of Backward Production Capacity*, which incorporates the completion of the target tasks into the local government's performance assessment system, so as to improve the urgency and execution ability of local governments.

The fifth is to play the supervision role of social media. After the target tasks were given, the Ministry of Industry and Information Technology, the provincial (district, municipal) governments, and industrial and information administration departments, respectively, announced to the public on the official Web site and mainstream media the list of enterprises that had eliminated backward production capacity, outdated equipments and backward production capacity have been removed; after the equipment was dismantled, the situation was announced to the public again. The supervision role of the society and the media was fully played to further ensure the implementation of the target tasks and policy measures.

### *4.3.4 Resolving Stage (2015–2020)*

1. Basic Judgment

By the end of 2014, the production capacity ratio of blast furnaces of 400 $m^3$ and below in the iron and steel industry was about 1%, and that of converters of 30 tons and below was 0.6%. The industrial structure was optimized, and the equipment level had been significantly improved. The elimination task of outdated equipment identified by volume and capacity standards had been basically completed, and the reduction work of iron and steel industry had entered a new stage of removing excess capacity.

2. Background

As early as 2013, the *Guiding Opinions on Resolving the Contradictions of Severe Excess Capacity* (No. 41 of the State Council in 2013) proposed to reduce the total steel production capacity by more than 80 million tons through "four batches" (i.e., digesting a batch, transferring a batch, integrating a batch and eliminating a batch). And the volume and capacity of outdated equipment were not mentioned therein.

In June 2015, the National Development and Reform Commission and the Ministry of Industry and Information Technology issued the *Notice on Printing and Distributing the Opinions on Cleaning Up Illegal Projects in Iron and Steel, Electrolytic Aluminum and Shipbuilding Industries* (No. 1494 of NDRC and MIIT in

2015), which gave specific opinions on the illegal projects that were under construction or had been completed, and required that the main responsibility should be properly undertaken, and the excess capacity should be removed unswervingly.

In November 2015, the 11th meeting of the Central Financial and Economic Leading Group put forward the proposal of "supply-side structural reform" to promote the same. It is a major innovation that adapts to and leads the new normal of economic development, and the key for the transformation of economic development mode and the strategic adjustment of economic structure. To solve the deep-seated structural contradictions and problems in the current economic and social development, efforts must focus on the supply-side structural reform. The main purpose of supply-side structural reform is to release new demands and create new supplies. The Central Economic Work Conference made it clear that the primary task for 2016 was to actively and steadily resolve excess capacity. The purpose was to free precious resources from those industries with severe excess capacity and limited growth space and "zombie companies", so as to streamline supply, increase effective supply, and create new productivity.

Therefore in February 2016, the State Council issued the *Opinions on Resolving the Excessive Capacity of the Iron and Steel Industry to Realize the Development Out of Difficulties* (No. 6 of the State Council in 2016) [14], which required a focus on promoting the supply-side structural reform of the iron and steel industry based on the principles of acting under the pressure of marketing forces, enterprises as main bodies, organization by local governments, support from central government, highlighting key points, complying with laws and regulations, comprehensively applying market mechanisms, economic means and legal means, adapting to local conditions, applying policies by categories, and treating both the symptoms and the root causes, in order to actively and steadily resolve excess capacity, and it also proposed to reduce the crude steel production capacity by 100 million to 150 million tons in 5 years by means of withdrawal according to laws and regulations and initiative withdrawal under guidance.

Document No. 6 issued by the State Council in 2016 was an important regulation after the reduction of China's iron and steel industry entering into the resolving stage, and the whole industry has entered the substantive operational stage of resolving excess steel capacity.

3. Supporting Policies

The supporting policies of the document No. 6 issued by the State Council in 2016 were very comprehensive, and eight special supporting policy documents were formulated, respectively, by the Ministry of Finance, the Ministry of Human Resources and Social Security, the Ministry of Land and Resources, the Ministry of Environmental Protection, the People's Bank of China, the State Administration of Taxation, the General Administration of Quality Supervision, Inspection and Quarantine, the State Administration of Work Safety, the China Banking Regulatory Commission, and the Coal Supervision Bureau in respects of financial and taxation support, staff resettlement, land, environmental protection, quality and safety, namely

- Opinions of the Ministry of Land and Resources on Supporting the Iron and Steel Industry to Resolve Excess Capacity to Realize the Development Out of Difficulties (No. 3 of MLR in 2016)
- Opinions of the Ministry of Environmental Protection, the National Development and Reform Commission and the Ministry of Industry and Information Technology on Supporting the Iron and Steel Industry to Resolve Excess Capacity to Realize the Development Out of Difficulties (No. 47 of MEP in 2016)
- The Opinions of the Seven Departments including the Ministry of Human Resources and Social Security on Doing a Good Job in Staff Relocation in the Process of Resolving the Excess Capacity of the Iron and Steel Industry to Realize the Development Out of Difficulties (No. 32 of Ministry of Human Resources and Social Security in 2016)
- Opinions of the General Administration of Quality Supervision, Inspection and Quarantine on Resolving the Excess Capacity of the iron and steel industry to Realize the Development Out of Difficulties (No. 193 of General Administration of Quality Supervision, Inspection, and Quarantine in 2016)
- Opinions of the State Administration of Work Safety and the State Administration of Coal Mine Safety on Supporting the Iron and Steel Industry to Resolve the Excess Production Capacity to Realize the Development Out of Difficulties (No. 38 of SAWS and SACMS in 2016)
- Opinions of the Ministry of Finance and the State Administration of Taxation on Resolving the Excess Capacity of the Iron and Steel Industry to Realize the Development Out of Difficulties (No. 151 of the Ministry of Finance in 2016)
- Opinions of the People's Bank of China, the China Banking Regulatory Commission, the China Securities Regulatory Commission, and the China Insurance Regulatory Commission on Supporting the Iron and Steel Industry to Resolve Excess Capacity to Realize the Development of Deprivation (No. 47 of the People's Bank of China in 2016)
- Notice of the Ministry of Finance on Printing and Distributing the Measures for the Management of Supplementary Funds for Structural Adjustment of Industrial Enterprises (No. 253 of the Ministry of Finance in 2016).

Among them, document No. 253 issued by the Ministry of Finance in 2016 pointed out that the central government would actively support the iron and steel and coal industries to cut overcapacity. The main measures include the establishment of a special fund for structural adjustment of industrial enterprises by the central government, with a total scale of 100 billion yuan; the implementation of relevant tax preferential policies for the iron and steel and coal industries continued the implementation of the steel export tax refund policy, such as the abolition of imported steel bonded tax under processing trade, the implementation of tax incentives for the comprehensive utilization of resources through waste heat recovery power generation by iron and steel enterprises; the implementation of fiscal and taxation accounting support policies for steel and coal enterprises' restructuring and bankruptcy, including tax incentives, land transfer income policies, and financial accounting systems; the implementation of financial policies for steel and coal enterprises to resolve excess capacity, support financial enterprises to dispose of non-performing assets in a timely manner, support eligible projects through special construction funds, and support iron and steel, coal and other industries to "go abroad" through export credit insurance.

4. Achievements

In February 2016, the State Council issued the *Opinions on Resolving the Excess Capacity of the Iron and Steel Industry to Realize the Development Out of Difficulties* (No. 6 of the State Council in 2016) and then the Ministry of Finance, the Ministry of Human Resources and Social Security, the Ministry of Land and Resources, the Ministry of Environmental Protection, and other central ministries and commissions issued a series of supporting documents, which formed a complete "1 + 8" policy system for cutting overcapacity. At the same time, a number of special actions were launched, such as "eliminating backward production capacity", "eliminating construction projects", and "energy consumption inspection". Under the great attention and unified deployment of the Party Central Committee and the State Council, leading by the central ministries and commissions, cooperation by local governments at all levels, and participation and joint work by industry associations, enterprises, consulting agencies and other parties, the crude steel production capacity was reduced by 65 million tons in 2016, exceeding the established target of 45 million tons by 44.4%.

Not only 201,000 employees have been resettled stably and orderly, and the industry's operating conditions had been improved (in 2016, the total profit of the members of the China Iron and Steel Association was 30.3 billion yuan, turning losses into profit, marking an profit increase of about 108.3 billion yuan compared with that in 2015); meanwhile, the industry transformation and upgrading had been promoted. Not only Baosteel and Wuhan Steel were merged as Baowu Iron and Steel Group Co., Ltd. to build a steel aircraft carrier with strong international competitiveness through strong joint and optimized structure, but also the restructuring of other enterprises (Anshan Steel and Benxi Steel) is under research and in progress. In addition, the fair competition of the market order has been initially restored, effectively curbing the unfair market order of "bad money drives out good".

In February 2017, China Iron and Steel Association issued the *Opinions on Supporting the Strike Against "Substandard Steel" and Defining the Range of Power Frequency and Medium Frequency Induction Furnaces* (No. 23 of CISA in 2017), which clearly specified the definition criteria for "substandard steel" produced by medium frequency induction furnaces. In April, the inter-ministerial joint meeting for resolving excess capacity of iron and steel and coal industries to realize the development out of difficulties under the leadership of the National Development and Reform Commission issued the *Opinions on Resolving the Excess Capacity of the Iron and Steel and Coal Industries to Realize the Development Out of Difficulties in 2017* (No. 691 of NDRC in 2017), which clearly clarified the key tasks of cutting overcapacity in 2017, such as resolutely banning "substandard steel" and continuing to reduce the crude steel production capacity by 50 million tons.

From the release of document No. 6 by the National Development and Reform Commission in 2016 to the end of October 2017, the target of cutting overcapacity by 115 million tons was successfully completed, accounting for more than 80% of the target of 140 million tons during the 13th Five-Year Plan period. At the same time, more than 700 enterprises producing "substandard steel" had been completely eliminated, involving over 100 million tons of production capacity.

Meanwhile, in terms of production capacity structure, the utilization rate of crude steel production capacity had rebounded sharply and gradually returned to a reasonable range. In terms of organizational structure, the crude steel output CR10 in 2016 was 35.87%, an increase of 1.69% over 2015. From January to May 2017, the crude steel output CR10 was 35.76%, which was basically the same as that in 2016. In terms of layout structure, blast furnace No. 2 at the Zhanjiang Iron and Steel Project of Baowu Group was completed and put into operation in July 2016, coastal projects like Shougang Jingtang Project Phase-II and Tangshan Bohai Steel Group were under construction, and the relocation projects of Jinan Steel (which had achieved full production shutdown in July 2017) as well as Taihang Steel in Wu'an, Jinan Steel and Shijiazhuang Steel in Hebei Province were in progress; in terms of product structure, compared with 2015, the proportion of long products decreased by 2.5% from January to May 2017, the proportion of sheet and strip products increased by 3.3%, and the proportion of pipes decreased by 0.6%. In particular, the elimination of "substandard steel" has effectively improved the quality level of China's rebar products, which has significantly improved the variety structure of rebar. The production ratio of rebar of Grade III and above has been significantly improved. In 2016, the proportion of production of rebar Grade III and above by the members of China Iron and Steel Association reached 97.1%, an increase of 4.9% over 2015.

## 4.4 Situation Faced and Work Prospects

### *4.4.1 New Situation of Supply-Side Structural Reform in the Iron and Steel Industry*

The iron and steel industry in China has already possessed strong international competitiveness and will play an important role in building a strong manufacturing country and the international capacity cooperation and will lead the global iron and steel development for a long time. However, it should also be noted that presently the iron and steel industry is facing serious challenges of severe overcapacity and weak innovation capability. It is necessary to grasp the law of development, accelerate innovative development, resolutely implement the requirements of supply-side structural reform, and effectively promote transformation and upgrading in order to achieve sustainable development.

From an international perspective, the global economy has undergone a tortuous recovery from the deep adjustment. The global steel demand has entered a plateau period, showing a trend of stable but fluctuating development. The overcapacity of iron and steel has become a global common problem. In particular, the trade protectionism in the international steel market is spreading, and the competition will be more intense; meanwhile, the rules system for international investment and trade will be accelerated, and the opportunities and challenges for international iron and steel capacity cooperation and international trade will coexist. A new round of scientific and technological revolution and industrial transformation are in the ascendant. The industrial form, production management, and development mode of the global iron and steel industry are undergoing an unprecedented profound transformation. The steel materials and other materials move toward a general trend of mutual competition and collaborative integration. The financial attributes of staple commodities such as iron ore and coking coal have increased, their price fluctuations have intensified, the risk of investment in the global mining industry has increased, and uncertainties have increased significantly. Relying on the strategy of reindustrialization, the advanced steel powers strengthen the scientific and technological innovation and the strategic layout in the frontier domains to occupy the commanding heights in the mid- and high-end steel markets; some emerging economies take shares from the ordinary steel market by relying on their advantages of low-cost factors such as labor; thus, the risks and challenges faced by China's iron and steel industry faces are increasing.

From the domestic perspective, the economic growth under the new normal is at a medium-to-high speed, and periodical contradictions and structural contradictions coexist, of which the structural problems are the main contradiction, and the total steel consumption and consumption intensity tend to decline. In addition, China's development mode is extensive, the unbalanced, uncoordinated, and unsustainable problems are still outstanding, resource constraints are tightening, and the deterioration of the ecological environment has not been fundamentally reversed. The iron and steel industry urgently needs to improve its innovation capability and accelerate adjustment and upgrading. However, the fundamental that China's economy will sustain long-term growth remains unchanged. In the process of achieving a jump from low-level supply–demand balance to a high level supply–demand balance, improving quality and structure of the steel product supply will form a new engine for China's iron and steel industry to move toward the mid-to-high end.

### *4.4.2 New Problems in the Supply-Side Structural Reform in the Iron and Steel Industry*

In the next few years, the problem of excess capacity in the iron and steel industry will still exist in China. The reduction work will still face the grim situation of internal and external troubles. Resolving excess capacity will be no longer only the scope of the iron and steel industry, but will rise to the height of the national foreign policy and economic development policy.

1. Resolving excess production capacity under the pressure of intensified international trade disputes

According to data from the Ministry of Commerce of China, there were 37 cases of trade frictions encountered by China's iron and steel industry in 2015 (an increase of 37% over the previous year), and the amount involved was 4.7 billion US dollars. The number of global trade friction incidents against China's steel exports has soared, and trade protection measures have also diversified, including single or combined means such as anti-dumping and countervailing duty investigations, tariffs, non-tariffs, and safeguard measures.

China's steel products are not export-oriented instead of meeting domestic demand, but they are more recognized on the international market because of their good quality, prices, good scale, service, and brand, so China's steel exports still remain at a high level even in the face of increasingly harsh export environment. The increase in export volume can alleviate the contradiction of excess capacity to a certain extent. However, as the situation of foreign trade of steel products deteriorates further, the reduction from the source is the solution.

2. USA and European countries frequently applying big stick diplomacy

The global economic downturn and shrinking demand are the root causes of excess capacity in the steel industry currently. Some countries in Europe and the USA ignore this basic fact and the positive measures taken and the great sacrifices made by the Chinese government to resolve the excess capacity of the iron and steel industry. Instead, they frequently apply trade protection measures as the main means of responding to the crisis and make indiscreet remarks or criticisms against China.

At the end of 2015, nine iron and steel associations in the USA and Europe issued joint statements for several times, stating that the global iron and steel industry was suffering from a crisis of excess capacity and China's iron and steel industry was the main influence of that issue and opposed China automatically obtaining market economy country qualification by December 2016.

In March 2016, EU issued the policy document titled Maintaining Sustainable Employment and Growth in the European Iron and Steel Industry, proposing that the EU would further adopt trade remedy and strengthen proactive regulation supervision on the basis of 37 anti-dumping and countervailing measures on imported steel products. In the past few months, the European Commission has conducted three investigations into Chinese steel plants and imposed punitive tariffs on two types of steel products imported from China.

In April 2016, the US Department of Commerce accused China of taking no action to reduce excess capacity at the Brussels Steel Conference and pointed out that, if China would not take immediate action to reduce excess steel production capacity, the USA and other "victimized countries" would have no choice but adopt trade measures to protect their domestic iron and steel industry and workers from damages. That was a naked threat. In the same month, the US Steel Corporation filed an application with the US International Trade Commission to request 337 investigations on the carbon steel and alloy steel products exported to USA from about 40 Chinese steel companies, including Hebei Iron and Steel Group Corporation and Shanghai Baosteel Group Corporation, and issued a permanent exclusion order and prohibition.

Excess capacity of iron and steel industry is a global problem. It is the common challenge and common responsibility of the relevant economies. Only when more countries adopt positive measures that are in line with their own national conditions and development stages, and strengthen policy communication and coordination among them, can we gradually resolve the problem of excess capacity in the iron and steel industry and achieve mutual benefit and win-win. It is foreseeable that the USA and EU will continue to ignore the efforts made and losses incurred by China in resolving excess steel production capacity and continually make indiscreet remarks or criticisms on that issue to force China to increase its efforts in cutting overcapacity through political and diplomatic means.

3. Domestic difficulty in cutting overcapacity is increasing in China

First, the equipment to be reduced is becoming larger. After more than a decade of eliminating outdated equipment in the iron and steel industry, a large and significant decline in the proportion of small blast furnaces and small converters has been achieved. Further, capacity reduction will inevitably face more and more large-scale equipment, and thus, the promotion work will face greater challenges.

Second, the process flow tends to be continuous and integrated, and the degree of dependence between processes is significantly increased. Taking Hebei Province as an example, in addition to the outdated equipment identified by the state in the previous stage, there are more independent ironmaking and steelmaking enterprises, their overall equipment level is low, and the processes have less influence on each other. However, most of the equipment to be reduced in the next few years belongs to the steel joint enterprises in operation. The process capacity is basically matched, and their continuous and integrated production features are more obvious. After partial or complete dismantling of ironmaking or steelmaking equipment, the corresponding systems of sintering, pelletizing, rolling, and oxygen generation, lime, power generation, etc., (partially including coking) will be significantly affected until the entire production line is shut down.

Third, the relationship between corporate equity, creditor's rights and debts is complex, and the amount of debt is large. Therefore, the risk of social instability is increased due to cutting overcapacity. The nature of Chinese iron and steel enterprises is complex, including central government enterprises, provincial enterprises, and collective, private, foreign, joint venture, and many other types. Credits and debts are also difficult to be sorted out. In addition, the total amount of debts of iron and steel enterprises is tens of billions yuan, even exceeding one billion yuan, featuring large volume. The cutting overcapacity work may lead to the withdrawal of some enterprises. The eight policies as safeguard measures do not provide specific support for debt and fixed asset losses and only provide certain financial support for capacity and personnel resettlement.

It can be seen that after the "slimming" of China's iron and steel industry, the equipment is becoming more and more large, and the process capacity matching is more reasonable; thus, the losses brought by cutting overcapacity is greater. The whole body will be touched in case one aspect is changed, and the enterprises' resistance emotions may increase obviously. Therefore, the cutting overcapacity work will face enormous challenges.

### *4.4.3 New Requirements for Supply-Side Structural Reform of the Iron and Steel Industry*

To solve the problem of China's medium and long-term economic development, the fundamental solution is to promote supply-side structural reform. This is a path and barrier that is forced to go through. It is also the trend of the times and the situation. How can the supply-side structural reform be practically promoted in the iron and steel industry? In general, it shall resolutely implement the five development concepts of "innovation, coordination, greenness, openness, and sharing", and focus on the five key tasks of "cutting overcapacity, reducing excess inventory, deleveraging, reducing cost, and strengthening areas of weakness" to intensively analyze the problems existing in the industry development and then explore solutions based on that.

According to the requirements of those five development concepts, the iron and steel industry has a certain extent of disparity in several aspects. For example, in terms of innovation, China's iron and steel industry still has such problems as insufficient long-term investment in independent innovation, lack of original technology, weak collaborative innovation in production, education and research and application as well as repeated allocation of innovative resources. In addition, there are still other uncoordinated problems such as unreasonable industrial layout and poor connection between steel manufacturing, service and market demand. The above-mentioned "cutting overcapacity, reducing excess inventory, deleveraging, reducing cost, and strengthening areas of weakness" are the five key tasks of supply-side structural reform. Among them, cutting overcapacity is in the first place, and the iron and steel industry is a typical representative industry with severe overcapacity, and it is also the focus of cutting overcapacity. Therefore, cutting overcapacity is given the first priority for promoting supply-side structural reform in the iron and steel industry. However, while focusing on cutting overcapacity in the iron and steel industry, it could not neglect the requirements by other key tasks on the iron and steel industry. The objective analysis indicates that, as an important part of the real economy, a lot of achievements still can be achieved in the iron and steel industry in terms of deleveraging, reducing cost, and strengthening areas of weakness. The debt problem of some enterprises is outstanding, and thus the risk of inducing regional financial problem is increasing dramatically. In recent years, the iron and steel enterprises are subject to increasing and overwhelming rigid expenditure pressure in terms of financial expenses, which has become a critical factor influencing their market competitiveness. At the same time, the weaknesses of China's iron and steel industry in terms of technological innovation, quality manufacturing, intelligent manufacturing, service-oriented manufacturing, green manufacturing, and international development also need to be strengthened.

### *4.4.4 Late Outlook*

In July 2017, Premier Li Keqiang pointed out in the report of the fourth grand inspection: resolving excess capacity must be unswervingly promoted to prevent resurgence. Those who violating regulations must be resolutely punished according to laws, relevant supervision authorities must be seriously accountable for weak supervision, as well as orders and proclaim prohibition must be enforced strictly. It can be seen that the determination of the Party Central Committee and the State Council is clear and the attitude is clear. As an important part of the supply-side structural reform of the iron and steel industry, the cutting overcapacity work will continue and run through the entire 13th Five-Year Plan. Next, in addition to continuing to implement the three special actions including joint law enforcement, a series of comprehensive measures such as dynamic management of steel standard conditions, capacity reduction and replacement as well as disposal of zombie enterprises will be taken in order to achieve the core targets of the iron and steel industry in cutting overcapacity, development out of difficulties, transformation and upgrading as well as enhanced competitiveness.

Among them, reduction is an inevitable trend in the development of China's iron and steel industry. Reduction is not only to resolve excess capacity, but also to reduce steel production. It is an inevitable choice for controlling incremental and revitalizing stocks and is also important for adjusting and optimizing the structure of China's iron and steel industry. For the iron and steel industry, the development of reduction will be a long-term process for process flow adjustment, survival of the fittest, diversified development and innovation development; for the iron and steel enterprises, they shall reduce and gradually withdraw the products of inefficient production lines, optimize the structure of product production lines, enhance the quality of large-scale products, focus on developing high-end and featured products, actively combine national and regional major strategies, intervene in relevant major projects in advance, and strive to build their own strategic product systems, so as to achieve the development of total amount reduction and high-quality and large-scale products, the continuous improvement of operating benefits, and ensure that enterprises have sustainable development with quality and benefits.

## 4.5 Industrial Practices of Reduction

The development of reduction is an important development stage for China's iron and steel industry. The clear definition and adaption to the development of reduction will play an important role in the future development of the industry, government management and enterprises' transformation and upgrading. As a staff department of government agencies, a leader of industry development and a think tank for enterprise planning, China Metallurgical Industry Planning and Research Institute (hereinafter referred to as MPI) has done a lot of detailed work on the basic research of reduction development and transformation and upgrading, different stages of policy drafting and revision to the implementation as well as different levels of government, industry and enterprises, which have made positive and important contributions to the practical promotion of supply-side structural reform in the iron and steel industry.

The practices of MPI in promoting industry reduction are detailed in Table 4.1.

**Table 4.1** Practices of MPI in leading industry reduction

| No. | Type | Main content | Typical cases |
|---|---|---|---|
| 1 | Basic research of industry | Focus on current situations, problem analysis, and industry situation judgment | Analysis of the Current Situations of China's Steel Production Capacity, Analysis of Resolving Capacity of China's Iron and Steel Industry, Research on the Policy Development History and Trend of China's Iron and Steel Industry, Review and Outlook of the Development of China's Iron and Steel Industry, Development Plan for Pakistan International Capacity Cooperation Park of Hengsheng Casting Industry, Special Supervision and Research on Resolving Excess Capacity in the Iron and Steel industry, Research on Bilateral Strategic Planning and Docking Cooperation Between China and Saudi Arabia, Research Bilateral Cooperation Planning Between China and Iran, Research on Capacity and Investment Cooperation Planning Between China and Jamaica, and Research on Capacity Cooperation Planning of Major Countries in Africa |
| 2 | Participation in policy drafting and revision | Analysis of and research on policy objectives, key tasks, supporting measures, organizational structure, etc. | Drafting and revision of 5th Five-Year to 12th Five-Year Development Plans for the Iron and Steel Industry, Adjustment and Upgrading Plan of Iron and Steel Industry (2016–2020), Notice on Printing and Distributing Implementation Measures for the Replacement of Production Capacity of Some Industries with Serious Excess Capacity, Guidance Catalogue of National Development and Reform Commission on Amending Industrial Structure Adjustment (2011 Edition), Opinions on Supporting the Strike of "substandard Steel" and Defining the Range of Power Frequency and Medium Frequency Induction Furnaces, Development Policy for China's Iron and Steel Industry, Standard Conditions for the Iron and Steel Industry, etc. |

(continued)

**Table 4.1** (continued)

| No. | Type | Main content | Typical cases |
|---|---|---|---|
| 3 | Policy implementation | Guide enterprises to carry out reduction development, transformation and upgrading according to the requirements of national industrial policies, and participate in the inspection and implementation of policies | Implementation Plan for Optimized Structure of Reduction System of Anshan Steel Group, proposal for Sino Egypt 5MPTA Steel Project (Phase-I), Implementation Plan for Reduction, Development Out of Difficulties and Transformation of Chuanwei Group, Overall Plan for the Optimization of the Merger and Reorganization of the Iron and Steel Industry in Hebei Province, Development Plan for the Iron and Steel Industry in Anyang City, Application Report for Integration into the Park, Reduction and Upgrading Project of Hebei Baoxin Iron and Steel Group, Implementation Plan for Limited Production of Iron and Steel Industry During Heating Season in Tangshan City, Development Plan for Transformation and Upgrading of Huainan Hongtai Steel Company, Consulting Services for China–Pakistan (Teng'ao) Two-Park Development Project, Capacity Replacement Plan of Ningbo Iron and Steel Co., Ltd., and Transformation and Upgrade Pilot Program of the Iron and Steel Industry in Tangshan |

## References


1. Website of the Ministry of Environmental Protection. Catalogue (First Batch) for Eliminating Backward Production Capacity, Processes and Products [1999-01-22]. http://kjs.mep.gov.cn/qjsc/zcbz/199901/t19990122_89240.shtml.
2. Chinese Government Network. Catalogue (Second Batch) for Eliminating Backward Production Capacity, Processes and Products [1999-12-30]. http://www.gov.cn/gongbao/content/2000/content_60014.htm.
3. Website of the Bureau of Statistics. National Data. http://data.stats.gov.cn/easyquery.htm?cn=C01.
4. Website of Chinese Government. Steel Industry Development Policy [2005-07-08]. http://www.gov.cn/flfg/2006-01/17/content_161597.htm.
5. Website of the National Development and Reform Commission. Guidance Catalogue for Industrial Structure Adjustment (2005 ed.) [2005-12-02]. http://www.ndrc.gov.cn/fzgggz/fzgh/Zcfg/200512/t20051222_65963.html.
6. Website of Chinese Government. Decision of the State Council on Issuing and Implementing the Interim Provisions on Promoting Industrial Structure Adjustment [2005-12-21]. http://www.gov.cn/zwgk/2005-12/21/content_133214.htm.
7. Website of Chinese Government. Notice of the State Council on Accelerating the Structural Adjustment of the Industries with Excess Capacity [2006-03-20]. http://www.gov.cn/zwgk/2006-03/20/Content_231376.htm.
8. Website of Chinese Government. Notice on Total Volume Control, Eliminating Backward Capacity and Accelerating Structural Adjustment of the Iron and Steel Industry [2006-07-01]. http://www.gov.cn/zwgk/2006-07/17/content_337825.htm.

9. Website of Chinese Government. Notice of the State Council on Printing and Distributing the Comprehensive Work Plan for Energy Conservation and Emission Reduction in the "13th Five-Year Plan" Period [2017-01-05]. http://www.gov.cn/zhengce/content/2017-01/05/content_5156789.htm.
10. Website of Chinese Government. Plan for Adjustment and Revitalization of the Iron and Steel Industry [2009-03-20]. http://www.gov.cn/zwgk/2009-03/20/content_1264318.htm.
11. Website of Chinese Government. Notice about the Approval by the State Council of Several Opinions of the National Development and Reform Commission on Suppression of Excess Capacity and Repeated Construction in Some Industries and Guiding Healthy Development of the Industries [2009-09-29]. http://www.gov.cn/zwgk/2009-09/29/content_1430087.htm.
12. Website of Chinese Government. Notice of the State Council on Further Strengthening the Elimination of Backward Production Capacity [2010-04-06]. http://www.gov.cn/zwgk/2010-04/06/content_1573880.htm.
13. Website of Ministry of Industry and Information Technology. Announcement No. 69 of 2015 by the Ministry of Industry and Information Technology [2015-11-23]. http://www.miit.gov.cn/newweb/n1146285/n1146352/n3054355/n3057292/n3057307/c4439007/content.html.
14. Website of Chinese Government. Opinions of the State Council on Resolve Excess Capacity to Achieve Development Out of Difficulties in the Iron and Steel Industry [2016-02-04]. http://www.gov.cn/zhengce/content/2016-02/04/content_5039353.htm.

# Chapter 5
# Greenness

## 5.1 History Review and Status Analysis

### *5.1.1 History Review*

The iron and steel industry is the fundamental industry for China's economic development. Becoming a powerful country in iron and steel industry is the dream of many generations of iron and steel practitioners in China. China's iron and steel industry has experienced a development out of nothing and grew from small to large, from weak to strong, making a major contribution to China's economic development. At the same time, the iron and steel industry is also a major source of energy, resource consumption, and pollutant emissions. The environmental protection in the iron and steel industry has been highly valued and concerned by the government for a long period. The environmental protection in the iron and steel industry has developed along with the development of the iron and steel industry. It has experienced the development process from nothing to something, from individual links to the whole industry, and from extensive development to green one. It is an important part of the transformation and upgrading of the iron and steel industry and sustainable development.

From the 1970s to 1990s, the development of China's iron and steel industry gradually entered the normal track and began to introduce, digest, and absorb foreign advanced technologies. The technical level of the iron and steel industry was rapidly improved. The typical examples were the introduction of 1700 mm rolling mill in Wuhan Steel, the comprehensive introduction in Shanghai Baosteel and Tianjin Seamless Steel Pipe Factory, as well as the technical transformation and upgrading in other old iron and steel enterprises. In 1996, China's crude steel output broke through the 100 million tons mark for the first time and became the world's largest steel producer.

During this period, in response to the international energy crisis, China put forward the policy of "Laying equal stress on energy development and conservation and giving priority to conservation", integrating energy conservation into national economic and



X. C. Li, *The Road Map of China's Steel Industry*,
https://doi.org/10.1007/978-981-15-2074-7_5

social development plans, and promoting the energy conservation in the iron and steel industry. During the 6th Five-Year Plan period, the main energy-saving task of the iron and steel industry was to carry out energy-saving publicity and education, set up institutions and teams throughout the industry, strengthen management and establish systems, and reduce energy loss and waste. The energy management of the iron and steel enterprises has gradually embarked on a scientific and institutionalized track. After entering the 7th Five-Year Plan period, the main work content was embodied in "Three Turns", that is, the emphasis of energy-saving work should be shifted from the energy conservation of individual equipment and process to the energy conservation of overall enterprises; the management mode of energy conservation should be shifted from experience management to modern management to improve the management level and efficiency; the energy-saving management system should be shifted from exclusive energy-saving department to the comprehensive management on the labor division of entire enterprise management system.

During this period, China began to participate in important international environmental conferences, pay attention to environmental pollution caused by industries, and strengthen environmental legislation. In June, 1972, the United Nations Conference on Human Environment in Stockholm, Sweden, approved the "*Declaration on the Human Environment*". In June, 1992, the United Nations Conference on Environment and Development in Rio de Janeiro, Brazil, presented a strategy for sustainable development. In 1973, the government promulgated *Tentative Standard on Discharge of Three Industrial Wastes* (GBJ4-1973), which is China's first environmental protection standard. In December, 1989, China promulgated the *Environmental Protection Law*. China's environmental protection laws and standards system had been gradually improved. During this period, the awareness of environmental protection in China's iron and steel industry was gradually strengthened and the environmental management system was improved. While introducing advanced production equipments and technologies, advanced environmental protection technologies and equipments had also been introduced for iron and steel industry, and the level of environmental protection management and pollution prevention technologies had been comprehensively improved. What was worth mentioning was the import and assimilation of the integrated wastewater and waste gas pollution prevention processes and technologies, and the engineering design of environmental protection facilities, equipment manufacturing, technology research and development in iron and steel industry had made great progress. The environmental protection achievements for the iron and steel industry were gradually taking lead in China's industry and were widely used by other industries.

In the twenty-first century, China's iron and steel industry entered a stage of sustained and rapid development, and our crude steel production has always ranked first in the world. After years of import and assimilation, structural adjustment, and closing down outdated production facilities, the level of steel production equipment has increased rapidly. During this period, the understanding regarding the international environmental crisis had been deepened and the awareness of environmental protection had been raised to a higher level. In August, 2002, the World Summit on Sustainable Development in Johannesburg, South Africa, proposed three pillars of economic growth, social progress, and sustainable development of environmental protection. In June, 2012, the United Nations Conference on Sustainable Development in Rio de Janeiro, Brazil, proposed a green economic concept for sustainable development. During this period, the environmental protection work for China's iron and steel industry gradually led to a formation of a complete environmental management system, comprehensive implementation of the internationally accepted environmental protection quality management system, improved environmental management institutions of iron and steel enterprises, and standardized environmental behavior of iron and steel enterprises. At the same time, the iron and steel industry vigorously implemented the transformation of production equipment and technologies and constructed large-scale energy-saving devices, such as the development of continuous casting, increasing coal injection ratio, construction of TRT and sintering machines, and waste heat recovery equipment for blast furnace hot blast stoves. In the process of in-depth study of Handan steel's experiences, energy-saving administrators introduced economic value and began the exploration of "Energy Economy" energy conservation, focusing on analyzing the impact of changes in energy consumption indicators on enterprise profits, analyzing key parts of energy conservation and profit increasing potential, and analyzing and predicting the impact of energy price changes on enterprise profits, so as to directly show the link between energy conservation and production costs, and promote the in-depth development of energy conservation.

Since the beginning of the 11th Five-Year Plan period, China's iron and steel industry has experienced both scant demand and overproduction, and gradually shifted from extensive operation to green, information-based, and diversified management and from long-term rapid incremental development to green transformation and upgrading development. The concept of energy conservation and environmental protection in the iron and steel industry has been gradually updated and enriched. Promoting clean production, developing recycling economy, practicing low-carbon economy, and implementing advanced environmental concepts such as resource conservation and environmental friendliness have become the technological approaches for green transformation and upgrading of the iron and steel industry. And the targets of decreasing total discharge of major pollutants and energy consumption per unit of GDP have been taken as the binding indicators for economic and social development. The iron and steel industry has adopted a series of strong policy measures in strengthening target responsibility, adjusting industrial structure, implementing key projects, promoting technological progress, strengthening policy incentives, and strengthening supervision and management. Propelled by the government, the energy saving and emission reduction of the iron and steel industry have entered a new stage of development, especially a large number of advanced and mature energy-saving technologies have been widely promoted and applied. During this period, the sudden haze weather continuously covering China's major economically developed areas triggered public concern about PM2.5 governance, which caused the government and the public to attach great importance to environmental protection and forced environmental protection to become a hot spot again. The government attached great importance to and comprehensively promoted environmental protection work. Environmental protection as a national policy had become a prerequisite for China's social and economic development. In January, 2015, the government promulgated a new environmental protection law. For the first time in more than 20 years, the environmental protection law was comprehensively revised. At the same time, the government had successively issued "*Ten Measures on Air Pollution Prevention*", "*Ten Measures on Water Pollution Prevention*", and "*Ten Measures on Soil Pollution Prevention*" to strengthen the environment supervision. In September, 2013, the government issued the "*Air Pollution Prevention and Control Action Plan*", which is "*Ten Measures on Air Pollution Prevention*". In April, 2015, the government issued the "*Water Pollution Prevention Action Plan*", which is "*Ten Measures on Water Pollution Prevention*". On May 28, 2016, the government issued the "*Soil Contaminant Prevention and Control Action Plan*", which is "Ten Measures on Soil Pollution Prevention". In October, 2012, the government implemented eight pollutant discharge standards for steel and coking industries, which promoted the overall upgrading of environmental protection for iron and steel enterprises, and the environmental development in green transformation and upgrading of the iron and steel industry entered the fast lane.

2016 is the first year of the 13th Five-Year Plan Period. It is also a crucial year for the national supply-side reform to promote the accelerated process of cutting over capacity in the iron and steel industry, the continuous strengthening of ecological civilization construction, and the leap-forward development of the environmental protection and the green development concept in the iron and steel industry. With the acceleration of the strategic deployment on environmental protection and ecological civilization construction as one of the important goals, the 13th Five-Year Plan continues to regard the iron and steel industry as a major concern, which is a traditional industrial pollution source with high energy consumption process, intensive fixed-source pollution discharge. On July 16, 2016, the *White Paper on Environmental Protection of China's Iron and Steel Industry (2005–2015)*, compiled by China Metallurgical Industry Planning and Research Institute (hereinafter referred to as MPI), was released to systematically analyze the external environment and green development path of energy conservation and environmental protection work.

In order to fulfill the goal of "West Lake Blue" pointed out during the G20 Summit, the prelude to the production suspension and restrictions of the iron and steel industry in the surrounding radiation area was unveiled. In 2016, the Central Environmental Inspection Group carried out two times special inspections on the key iron and steel enterprises in several provinces and autonomous regions in China, especially on the discharge of major pollutants and the installation and operation of automatic pollutant monitoring facilities of iron and steel enterprises. The inspection accelerated the process of environmental protection upgrading and reconstruction in Hebei, Jiangsu, Shandong, and other areas with intensive existence of iron and steel enterprises. At the same time, it became an important starting point for the implementation of the *Opinions of State Council on Cutting Overcapacity in the Iron and Steel Industry to Realize the Development out of Poverty*. The intensity of "*Environmental Protection Storm*" continued to increase after consecutive days of heavy pollution in Beijing, Tianjin, and Hebei. The Ministry of Environmental Protection dispatched 13 inspection groups to conduct special assessments on early warning plans prepared by municipal governments of the red and orange alert cities and pay special attention to the degree of implementation of the plan by the iron and steel enterprises. Hebei provincial government issued air pollution prevention and control dispatching order for the first time. Frequent large-scale production suspension and restrictions due to heavy pollution weather were therefore enabled in areas with intensive iron and steel production capacity like Hebei and Shandong.

It is safe to say that China's iron and steel industry has made outstanding achievements in energy conservation, environmental protection, and green development. Since the 1970s, the energy conservation and environmental protection in the iron and steel industry have experienced tremendous changes in the past 50 years. The iron and steel industry has gradually developed from the stage of import and assimilation to the current endogenous development, which is oriented with independent innovation and supplemented by the introduction. The energy conservation and environmental protection in the iron and steel industry have made remarkable achievement. The Beijing-Tianjin-Hebei and surrounding "2 + 26" cities, the Yangtze River Delta, and the Pearl River Delta regions were also required to complete the application for sewage discharge permissions by the end of 2017. The iron and steel companies in other regions were required to complete the application for the same by the end of 2018. As an important starting point for rationalizing environmental supervision procedures and reshaping the environmental supervision system, the sewage discharge permission system has become a trump card for promoting environmental protection in the iron and steel industry. Echoing with the implementation of the sewage discharge permission system, the *Environmental Protection Tax Law* was implemented on January 1, 2018. All pollutant discharge enterprises are required to fulfill their environmental tax liability without exclusions, and the pollution discharge fees that had been collected for nearly 40 years eventually stepped down from the stage of history. From "Fee" to "Tax", environmental protection supervision will embark on a more standardized and equal path.

At the same time as the policy continued to deepen, the iron and steel industry also ushered in the stricter version of "The Most Stringent Environmental Standards in History" in 2017. On the basis of a series of new emission standards for the iron and steel industry implemented on January 1, 2015, the Draft Amendment for Comments on Fugitive Emission Standards for the Iron and Steel Industry and the Special Emission Limits for Sinter and Pellet Plants were published on June 13, 2017, further tightening emission thresholds. An appropriate transition period is given to existing enterprises, and fugitive emission control measures will be implemented on January 1, 2019. In order to ensure the fugitive emission control measures for the "2 + 26" cities situated on the Beijing-Tianjin-Hebei, air pollution transmission channel comes to effect on the pollution control in the autumn and winter of 2017, and the 2017 air quality improvement target determined by the *Air Pollution Prevention Action Plan was achieved* was achieved; the "2 + 26" cities were required to implement it from October 1, 2017. The goal is to strive to reach the overall target of reducing the total amount of pollutant discharge by more than 15%, which was determined in the *Adjustment and Upgrade Plan of Iron and Steel Industry* (*2016–2020*) issued by the Ministry of Industry and Information Technology, at the end of the 13th Five-Year

Plan period, and improve all the actual problems caused by insufficient environmental carrying capacity in the iron and steel production capacity densely distributed areas such as Beijing, Tianjin, Hebei, and the Yangtze River Delta Region. The major green transformation and upgrading projects for constructing sheds and bins for stockyard, installing desulfurization and denitrification facilities for coke oven flue gas, and providing multi-pollutant synergy treatment for sintering (pelleting) flue gas have been gradually put in place to achieve green sustainable development of the industry [1].

### *5.1.2 Current Status of Green Development of China's Iron and Steel Industry*

1. Energy Conservation and Consumption Reduction

   (1) Continuous optimizing of energy consumption indexes.
   During the 35 years from 1980 to 2015, the comprehensive energy consumption indexes of the iron and steel industry were continuously optimized. The comprehensive energy consumption per ton of steel (standard coal) dropped from 2,040 to 572 kg, down 72%, of which, in the 20 years from 1980 to 1999, the comprehensive energy consumption per ton of steel in the whole industry dropped from 2,040 to 1,240 kg, down 39.22%. Since 2000, according to the statistics conducted on large and medium-sized iron and steel enterprises, the comprehensive energy consumption (standard coal) dropped from 906 kg (in 2000) to 572 kg (in 2015), with a reduction rate of 36.87% (Fig. 5.1).

   Compared with the new revision of "*Energy Consumption Limits per ton of Crude Steel*" and "*Energy Consumption Limits per ton of Coking Product*" in 2013, the compliance rate of energy consumption enterprises in coking process was 91%, in sintering process was 93%, in ironmaking process was 91%, and in converter process was 42%.

   (2) Production process flow becoming more efficient.
   During the 30 years since 1980, especially in the 1980s and 1990s, the process of the iron and steel industry was gradually becoming more continuous, compact, and less quantitative due to a series of production structural adjustment and process flow optimizations such as replacing open hearth furnace with oxygen converter, replaced die casting with continuous casting, replaced multi-heating forming with one-heating forming.

   In 1988, the production technology scheme of "Continuous Casting-Oriented" was established in China's iron and steel industry, and the development of continuous casting system was vigorously promoted. By 1999, the continuous casting ratio in iron and steel industry had reached 78.62%. By 2014, the continuous casting ratio in national key iron and steel enterprise had reached 99.71%.

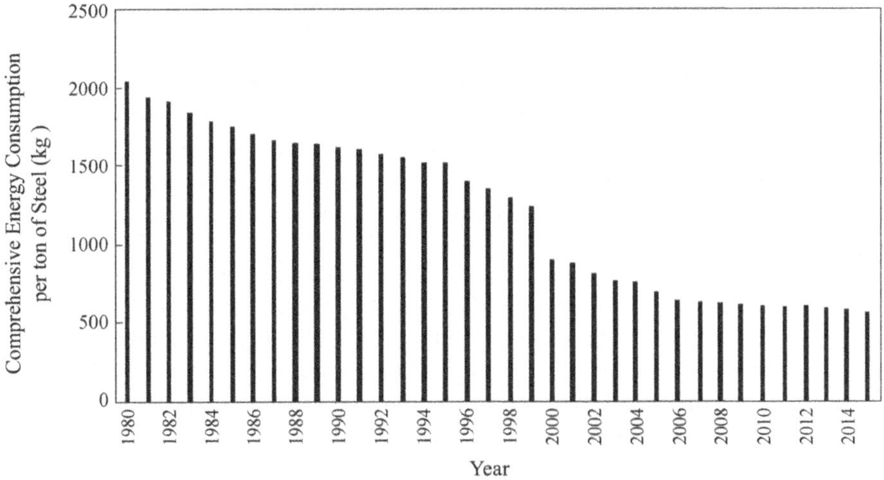


**Fig. 5.1** Changes in comprehensive energy consumption per ton of steel in the iron and steel industry (1980–2015)

In the middle and late 1990s when large-scale oxygen top-bottom combined blowing BOFs were put into operation in Baosteel, Wuhan Steel, Anshan Steel, and Shougang, the steelmaking process with converter entered a period of rapid development. By 1999, the proportion of steelmaking with converter in China had reached 82.7%. By the end of 2002, steelmaking process with open hearth furnace had been eliminated nationwide.

At present, in the China's key statistical iron and steel enterprises, the capacity of coke ovens higher than 5 m accounts for 48% of the total coking capacity, capacity of blast furnaces more than 1,000 $m^3$ accounts for 65% of the total ironmaking capacity, and capacity of 100 t and above converter accounts for 59% of the total converter steelmaking capacity.

(3) A large number of advanced and mature technologies have been widely applied.

With the development of China's iron and steel industry, the iron and steel industry's energy-saving technology has also achieved rapid development. Coke dry quenching, dry-type dedusting, waste heat recovery for sinter plant, dry-type differential pressure power generation (TRT), high-efficiency PCI, regenerative combustion, fully gas-fired power generation, hot charging and hot transporting, and other key common technologies have been widely promoted and applied. At present, the waste heat recovery power generator for sintering machine has gained more popularity and been used on more than 20% sintering machines, exceeding the development target proposed by the Ministry of Industry and Information Technology in 12th Five-Year Plan. The popularity of TRT in the iron and steel industry has reached nearly 100%, and that of coke dry quenching technology has reached 85%. At the same time, the world's largest single-machine low-calorie combined gas and steam generator set, high-pressure and ultra-high-pressure fully gas-fired power generator set, waste heat recovery and utilization technology for sintering machine, saturated steam power generation technology, etc., are available in China, and their technologies are at the leading level of the world. The number of waste heat recovery power generator for sintering machine, CDQ, and TRT devices in iron and steel industry ranks first in the world.

(4) The efficiency of secondary energy recovery and utilization is continuously improving. Increasing the secondary energy recovery and utilization of enterprises is the main way to reduce the amount of purchased energy and achieve energy conservation and emission reduction. In 2016, the utilization rate of BF, coke oven, and converter gas in iron and steel enterprises was still increasing, and the loss rate was generally decreasing. Compared with 2015, the utilization rate of coke oven gas in 2016 decreased slightly by 0.75 percentage points; the utilization rate of BF gas was basically flat; the utilization rate of converter gas increased by 0.7 percentage points; the recovery of converter gas per ton of steel increased by 2.68%; the captive power ratio of enterprises increased slightly. A certain breakthrough has been made in the recovery and utilization of medium- and low-temperature waste heat resources generated by fume from coke oven uptakes, large flue of sintering machine, BF slag granulating water, and reheating furnace in steel rolling plant.

(5) The level of energy management is constantly improving.
The energy-saving management of the iron and steel industry began with energy consumption measurement, statistics, and the establishment of energy consumption index system. In 1982, the *Interim Provisions on Calculation Methods for Energy Balance and Energy Consumption Indexes of Iron and Steel Enterprises* was issued, which enabled iron and steel companies to take the lead in normalization and standardization of energy statistics and energy consumption index; in terms of process energy-saving management, 17 "Process Energy-Saving Regulations" have been successively formulated and process energy-saving upgrades have been carried out since 1979 to effectively promote energy conservation of process and enterprise. In recent years, with the promotion and application of energy management centers and energy management systems in the iron and steel industry, the iron and steel industry has gradually moved toward into systematic and delicacy management.

In general, the energy-saving management mode of the iron and steel industry has undergone a transition from experience management to modern management, and the energy-saving management system has undergone a transition from a single energy-saving department to an integrated management system of the entire enterprise. Hence, the overall energy-saving management level has been continuously improved.

2. Waste Gas Treatment

With the rapid development of China's iron and steel industry, air pollutant emissions have increased with the increase in steel output. As one of the major industrial discharge sources, the iron and steel industry has actively adopted advanced emission reduction technologies in recent years to narrow the gap with foreign advanced iron and steel industry. At the same time, it has also developed internationally advanced waste gas treatment technologies through independent research and development, so as to reduce the pollutant discharge amount per ton of steel continuously. According to the *Statistics on Environmental Protection in China's Iron and Steel Industry*, the total amount of sulfur dioxide emissions by key statistical enterprises in 2015 was 471,721.69 t, and the sulfur dioxide emissions per ton of steel decreased from 2.75 kg (in 2005) to 0.85 kg; the total amount of smoke dust emissions per ton of steel was 450,750.11 t, and the smoke dust emissions per ton of steel also dropped from 2.11 kg (in 2005) to about 0.81 kg [2] (Figs. 5.2 and 5.3).

In particular, the latest emission standards for the iron and steel industry, which was implemented on January 1, 2015, have significantly tighten the original emission limits, promoted the construction of the flue gas desulfurization facilities with sintering machine charge end, facilitated the upgrading and reconstruction of dedusting equipment, and further reduced discharge amount of waste gas pollutants. The current control technologies and application status of main waste gas pollutants are as follows:

(1) Flue Gas Desulfurization Technology for Sintering Process and Its Application Status.

The iron and steel industry is one of the major industrial $SO_2$ emission sources, in which the $SO_2$ emission amount from sintering process accounts for more than 70%

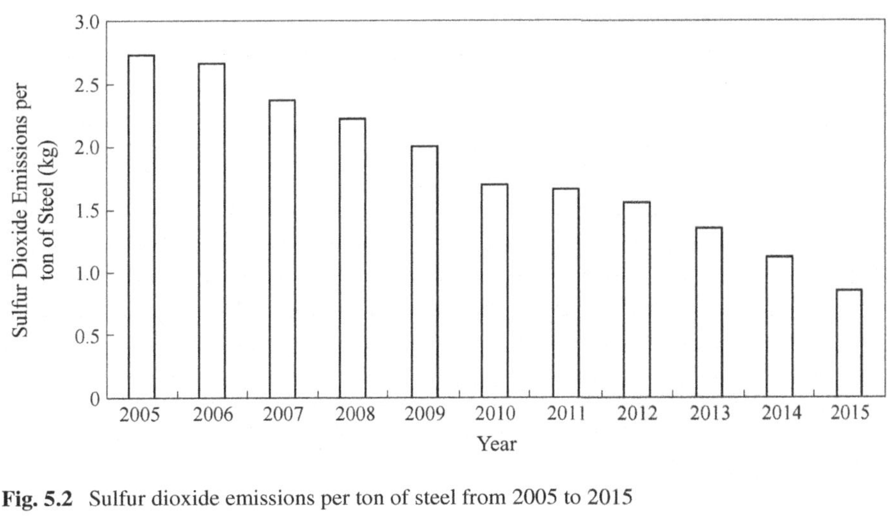


**Fig. 5.2** Sulfur dioxide emissions per ton of steel from 2005 to 2015

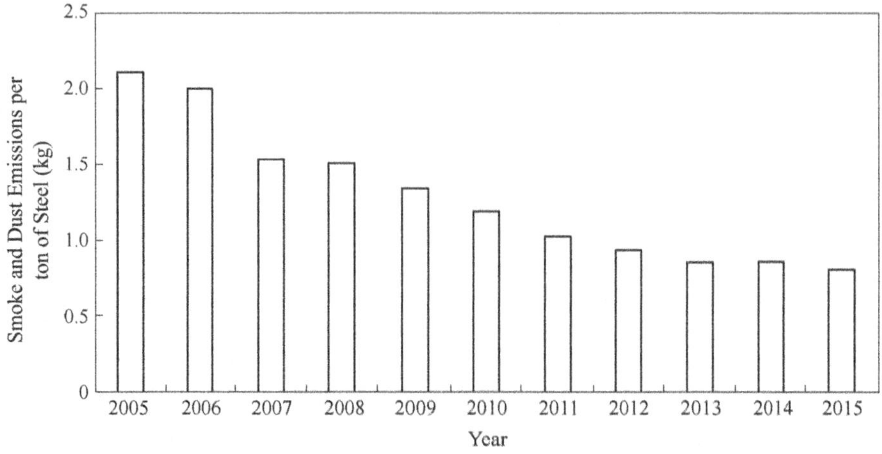


**Fig. 5.3** Smoke and dust emissions per ton of steel from 2005 to 2015

of the total emission amount of iron and steel enterprises. Since government listed the reduction of total amount of $SO_2$ contaminant by 10% as a binding indicator in the national 11th Five-Year Plan, iron and steel enterprises have carried out large-scale flue gas desulfurization work for sintering process. During the 12th Five-Year Plan period, the pace of constructing flue gas desulfurization facilities for sintering process was accelerating. Since 2007, the flue gas desulfurization work for sintering process in the iron and steel industry has entered a substantive implementation stage. A number of flue gas desulfurization equipment for sintering machine were successively completed and put into operation in Jinan Steel, Liuzhou Steel, Sanming Steel, Shijiazhuang Steel, Baosteel, and Ma'anshan Steel. By the end of 2015, the area of sintering machines equipped with desulfurization facilities reached 138,000 $m^2$, and the installation rate was as high as 88%.

The flue gas desulfurization technology for sintering process is mainly classified as wet-type desulfurization and (semi) dry-type desulfurization, as shown in Fig. 5.4. Of which, the wet-type desulfurization technology includes limestone/lime-gypsum method, magnesia method, ammonia-sulfur ammonium method, double alkali method, and the like. (Semi) dry-type desulfurization technology includes circulating fluidized bed method (CFB), rotary spray method (SDA), dense phase dry tower method, new desulfurization and dedusting integration technology (NID), MEROS method, and activated carbon method (AC). All above-mentioned methods can be found in the sintering process of China's iron and steel industry. At the same time, after years of exploration and long-term practice, limestone/lime-gypsum method and circulating fluidized bed method have certain advantages in terms of operational stability and economy, so they are recognized and selected by iron and steel enterprises, gradually forming into the mainstream of flue gas desulfurization process for sintering machine.

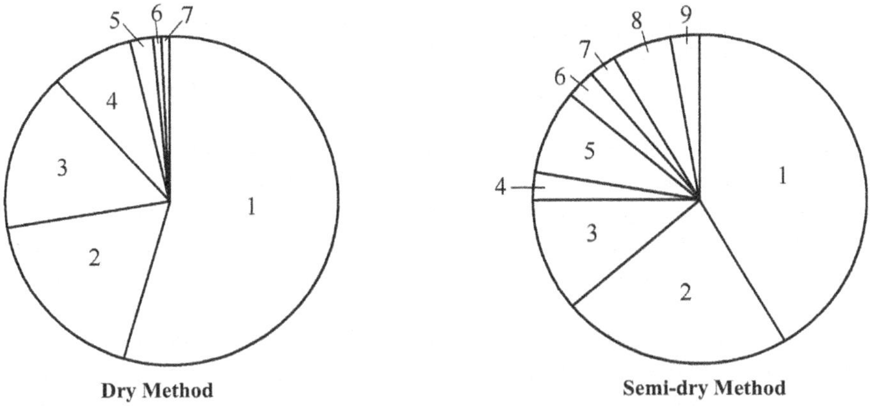


1- Limestone/lime-gypsum method, 54.5%
2-Ammonia-ammonium sulfate method, 17.9%
3-Double alkali method, 15.7%
4-Magnesium oxide method, 8.2%
5-Organic amine method, 2.2%
6-Steel slag method, 0.7%

1-Circulating fluidized bed method, 41.7%
2-Dense phase dry tower method, 22.2%
3-Rotary spray method, 11.1%
4-Activated carbon method, 2.8%
5-ENS method, 8.3%
6-MEROS method, 2.8%
7-NID method, 2.8%
8-LEC method, 5.6%
9-Simple semi-dry method, 2.8%

**Fig. 5.4** Proportion of different desulfurization processes for sintering machine

Although the first phase of constructing the flue gas desulfurization facilities for sintering machine charge end in iron and steel industry was basically completed, due to the new environmental emission standards implemented on January 1, 2015 (the $SO_2$ emission concentration shall not exceed 200 mg/m$^3$, special emission limits shall be less than 180 mg/m$^3$), most of the desulfurization facilities built before 2012 would fail to meet the current emission standards due to insufficient margin (they were designed according to the standard of $SO_2$ emission concentration not exceeding 600 mg/m$^3$). So, in the next stage, some of them need to be upgraded and reconstructed.

(2) Current Application Status of Smoke and Dust Control Technology.

At present, the smoke and dust pollution control technologies applied in the iron and steel industry mainly include bag filter, electrostatic precipitator, wet-type electrostatic precipitator, electrostatic fabric filter, wet-type deduster, and mechanical deduster.

Bag filters are currently and most widely used in the iron and steel industry. The application rate of bag filter in the iron and steel industry has reached more than 95%, and bag filter is widely used for the dust sources in the stockyard, coke oven, lime kiln, BF stock house and cast house, hot metal pre-treatment, pig casting machine, secondary flue gas system of converter, electric furnaces for steelmaking and rolling plant.

The dedusting efficiency of the bag filter is as high as 99.9–99.99%, and its operation is stable, and it is not affected by the specific resistance of dust. There is no limitation on the dust type, concentration (dust that can be treated at a concentration up to 1,000 g/m$^3$ under the standard state), and particle sizes. Bag filters have good adaptability to a wide range of load fluctuation conditions caused by different process conditions. The air capacity ranges from 1000 m$^3$/h to more than 2,000,000 m$^3$/h. Maintenance and repair of filter bags and bag cages can be operated online. When replacing damaged filter bags, different filtering chamber can be changed over at any time without shutdown. Bag filter is highly efficient and stable. By means of setting different process parameters and adopting various filter bag materials, the emission concentration (standard state) can be stably controlled below 20 mg/m$^3$ or even the ultra-low limit of 5 mg/m$^3$, and a large amount of raw materials and other materials can be recycled and raw material utilization is improved. Therefore, bag filters have been widely used in the iron and steel industry.

Since the electrostatic precipitator is known for its low collection efficiency for fine dust with a particle size less than 10 $\mu$m, the proportion of electrostatic precipitator used in the iron and steel industry is gradually declining with the further tightening of particulate emission limit in the latest environmental protection standards. Now, it is mainly used for high-temperature and high-humidity occasions where bag filter is not suitable, that is, flue gas from sintering (pelleting) machine charge end and gas from converter.

The electrostatic precipitator enjoys high dedusting efficiency up to 99.0%, high temperature resistance up to 350 °C, low running resistance of only 200–300 Pa, and high air capacity up to 1,000,000 m$^3$/h. However, because its processable specific resistance of dust ranges from 10$^4$ to 10$^{10}$ $\Omega \cdot$ cm and is greatly affected by the change of physical and chemical properties of dust, fine particles are easy to bond onto the counter electrode. After working for a long period, the internal components are deformed, the electric field changes, and the dedusting effectiveness will decrease. However, there are many kinds of smoke and dust in the iron and steel industry, and the specific resistance changes greatly, which sometimes is beyond the processable range of precipitator. Moreover, the great variation of some process conditions will also made its dedusting efficiency unstable and difficult to maintain the discharge concentration (standard state) below 50 mg/m$^3$ for a long time under complicated working conditions. Therefore, the proportion of electrostatic precipitators applied in the iron and steel industry is reduced.

The wet-type electrostatic precipitator is mainly used in the dusty flue gas treatment after the wet flue gas desulfurization for sintering machines in the iron and steel industry. But its equipment investment cost is high, and its investment economic benefit and operating cost need to be continuously improved in the process optimization and through market competition. The wet electrostatic precipitator uses liquid to flush the surface of the dust collecting poles for cleaning, which can effectively collect fine particles (PM2.5 dust, $SO_3$ acid mist, aerosol), heavy metals (Hg, As, Se, Pb, Cr), organic pollutants (polycyclic aromatic hydrocarbons, dioxins), and the like. As far as the operating wet ESPs are concerned, the effect of the wet-type electrostatic precipitator put into operation can meet the national special limit of 40 $mg/m^3$ on dust emission concentration (standard state) at the sintering machine charge end and even remains stable in the range of 20 $mg/m^3$.

Electrostatic fabric filter is a new type of dedusting technology that has emerged in recent years. It integrates the advantages of high temperature resistance of electrostatic precipitator and high dedusting efficiency and wide application range of bag filter. It is applied in the sintering flue gas dedusting process and the transformation of existing electrostatic precipitator. At present, the composite dedusting process with the pre-electrostatic precipitator and post-bag filter is popularly used in the iron and steel industry. By the action of pre-dusting and high-voltage electric field, the temperature will be reduced, sparkle will be extinguished, and large-size dust will be captured; consequently, 80–90% of dust can be collected. Thereafter, the flue gas passes through the guiding device and enters the inlet chamber of bag filter, so the particulate matter in the flue gas can be effectively removed by means of external filtering. With the advantage of relatively low operating resistance and low startup and operation costs, this technology has been successfully applied in the sintering process of Shaanxi Steel Group and Laiwu Iron and Steel Co., Ltd.

3. Water-Saving and Wastewater Treatment

   (1) Changes in Water Consumption.

   Under the current background that freshwater resources are in shortage and the low-carbon economy gains more attention worldwide, water-saving work has attracted more and more attention. The iron and steel industry is a major water consumer. In order to save water resources and respond to the requirements on sustainable development of the iron and steel industry, the iron and steel industry has made effective efforts on water-saving work over more than ten years, which has improved the water efficiency and gradually reduced the water consumption in the iron and steel industry. China's crude steel output increased rapidly from 353 million tons in 2005 to 804 million tons in 2015, with an increase of 2.28 times. But the freshwater consumption in the iron and steel industry decreased from 3.74 billion cubic meters in 2005 to 3.31 billion cubic meters in 2015, with a decrease of 11.5%. The proportion of water consumption in the industrial water consumption of the country remained at around 2.5%. Accordingly, the freshwater consumption in the iron and steel industry was effectively controlled, as shown in Fig. 5.5.

   China's key statistical iron and steel enterprises adopted comprehensive water-saving technologies and measures such as optimizing water systems, innovating water-saving production processes, circulating water efficiently, and recycling wastewater, collecting and utilizing rainwater and urban reclaimed water, so the water efficiency was improved in an all-round

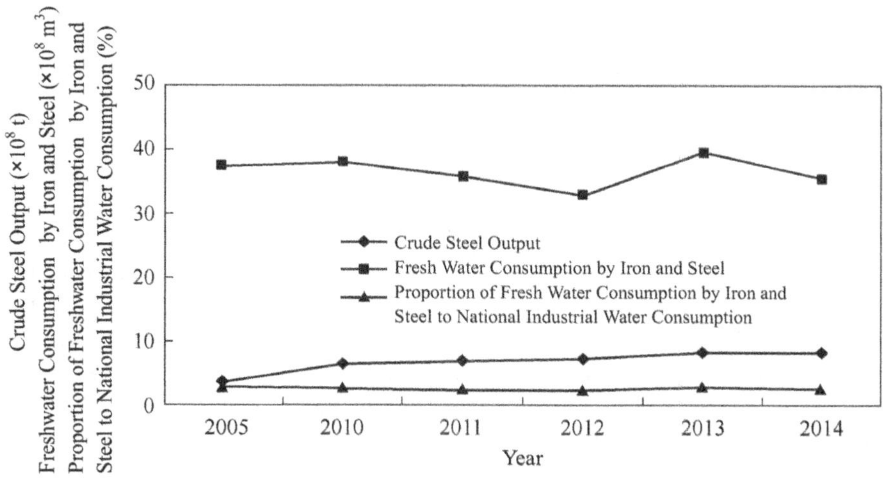


**Fig. 5.5** Changes of freshwater consumption in the iron and steel industry from 2005 to 2015

way. The water consumption was 3.30 cubic meters per ton of steel, which was 19.7% lower than that in 2010. The water-saving level in China has taken the lead in the world.

According to the *Statistics on Environmental Protection in China's Iron and Steel industry* [2], the changes on freshwater consumption per ton of steel in the key statistical iron and steel enterprises (2005–2016) are shown in Fig. 5.6.

However, there are certain differences in the freshwater consumption per ton of steel by different enterprises. Taking the large-scale iron and steel complex as an example, in 2014, the advanced level of freshwater consumption in large-scale iron and steel complex reached 2.15 cubic meters per ton of steel (in TISCO), and the backward level of the same

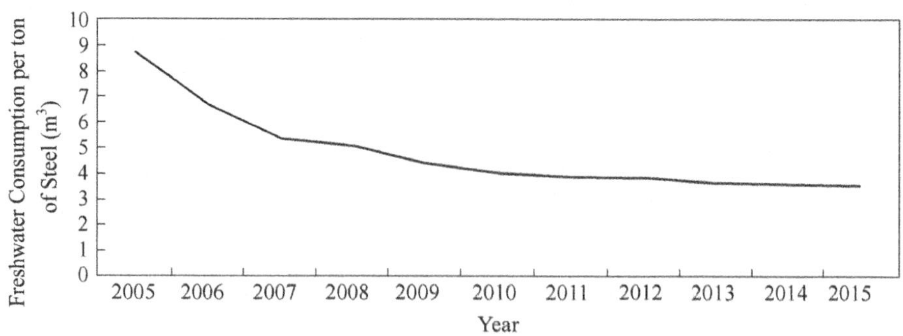


**Fig. 5.6** Changes on freshwater consumption per ton of steel in the key statistical iron and steel enterprises from 2005 to 2015

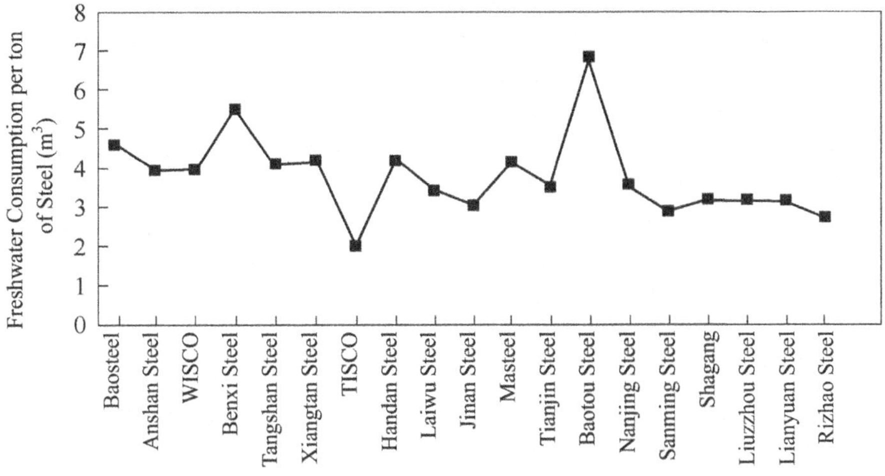


**Fig. 5.7** Freshwater consumption per ton of steel in some iron and steel complex in 2015

was 5.69 cubic meters per ton of steel (Baotou Steel), with the difference of 2.5 times. In 2015, the freshwater consumption per ton of steel in some large-scale iron and steel complex is shown in Fig. 5.7.

(2) Changes in Water Discharge.

In 2015, the industrial wastewater discharge all over the country was 19.95 billion tons, with a decrease of 17.9% compared with that in 2005 (24.31 billion tons). China's key statistical iron and steel enterprises in 2016 discharged 441 million tons of wastewater, which was 63.6% lower than that in 2005 (1.21 billion tons), with a decrease of 81% per ton of steel. At the same time, the proportion of wastewater discharged from the iron and steel industry to the national industrial wastewater discharge has gradually declined (Figs. 5.8 and 5.9).

(3) Changes in the Reuse Rate of Water.

The circulating water supply system is widely applied for the production water in the iron and steel industry to improve the recycling rate of water. The circulation rate of the open indirect cooling water circulating system is above 97%, and that of the turbid direct cooling water circulating system is above 96%; on the basis of water recycling, the cascade water supply system is set up for reasonable cascade application, according to the quality of sewage from the circulating water system. At the same time, more attention is attached to wastewater reuse; the sewage of the whole plant is reused after being treated to improve the reuse water rate of the whole plant.

The water recycling rate of key statistical iron and steel enterprises increased from 94.27% in 2005 to 97.7% in 2016, with an increase of 3.43 percentage points, and the water recycling rate increased year by year, as shown in Fig. 5.10.

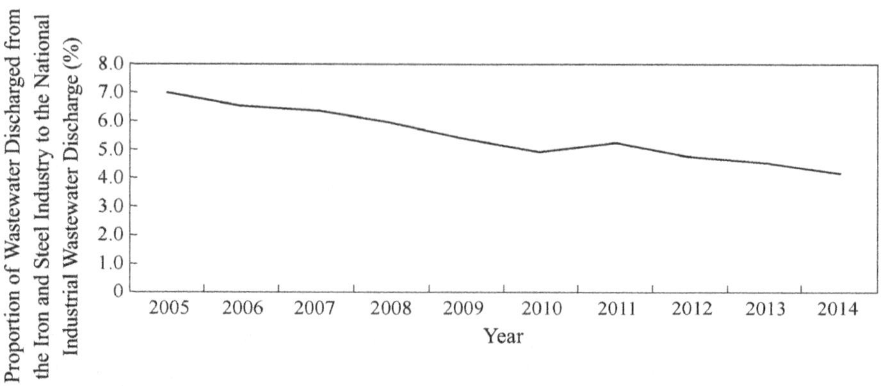


**Fig. 5.8** Proportion of wastewater discharged from the iron and steel industry to the national industrial wastewater discharge from 2005 to 2014

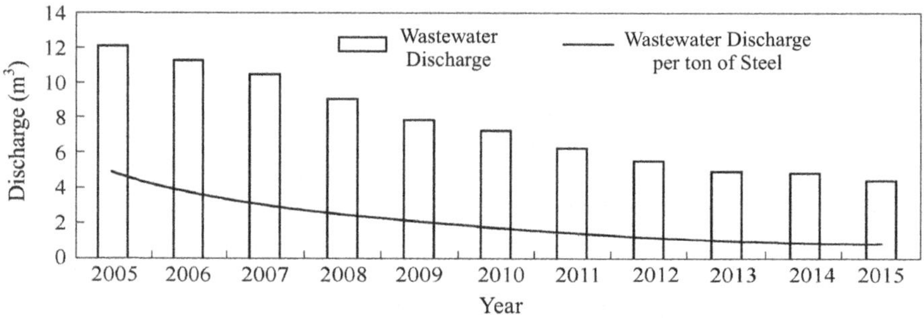


**Fig. 5.9** Proportion of wastewater discharged from the key statistical iron and steel enterprises to the national industrial wastewater discharge from 2005 to 2014

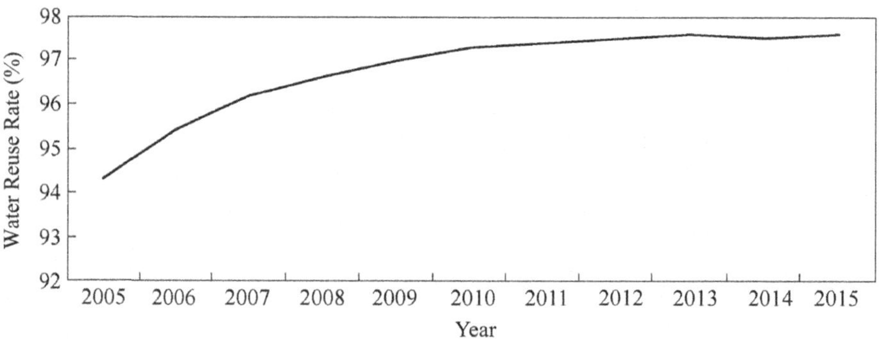


**Fig. 5.10** Changes on water reuse rate of key statistical iron and steel enterprises from 2005 to 2015

(4) Changes on the Discharge of Major Pollutants in Wastewater.
The main pollutants per ton of wastewater discharged from key statistical iron and steel enterprises in China, such as COD and ammonia nitrogen, are reduced year by year. The COD and ammonia nitrogen emissions per ton of steel were reduced from 0.247 and 0.024 kg in 2005 to 0.024 and 0.0015 kg in 2015, respectively (Figs. 5.11 and 5.12).

(5) Wastewater Treatment Technologies and Measures.
The gradual reduction of wastewater discharge and the pollutants in it by China's iron and steel industry is attributable to the improvement of wastewater treatment technology. Some of our technologies have taken the lead internationally. The main current wastewater treatment technologies and measures adopted in the iron and steel industry are:

1) Technologies for Reducing Wastewater Generation.
Mainly include: dry-type BF gas dedusting technology, CDQ technology, primary dry-type dedusting technology for converter, vaporization

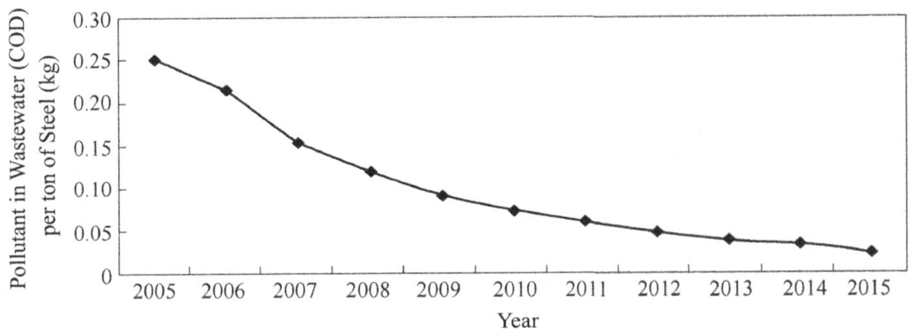


**Fig. 5.11** Changes on pollutant in wastewater (COD) per ton of steel discharged from key statistical iron and steel enterprises

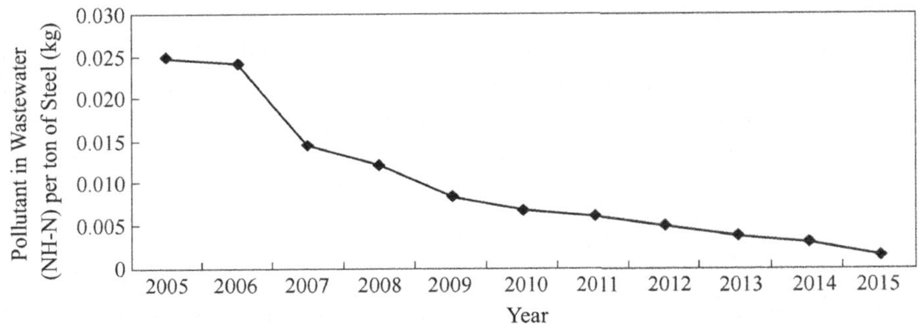


**Fig. 5.12** Changes on pollutant in wastewater (NH-N) per ton of steel discharged from key statistical iron and steel enterprises

cooling technology for reheating furnace, and spray cooling technology for CCM.

At present, the proportion of operating dry-type BF gas dedusting technology in China's iron and steel industry has exceeded 90%; 20% converters have employed primary dry-type dedusting technology; application rate of CDQ technology has reached more than 95% (in the key statistical iron and steel enterprises); that of vaporization cooling technology for reheating furnace and spray cooling technology for CCM has exceeded 90%.

The application of these technologies not only greatly reduces the water consumption, but also greatly decreases the amount of wastewater generated, thereby relieving the stress on wastewater treatment in iron and steel enterprises, improving the reuse rate of wastewater, and reducing the wastewater discharge.

2) Recycling Technology of Process Water.
The water used for various processes of the iron and steel industry will be recycled after treatment. The water recycling technologies, including that used in open indirect cooling water circulating systems and that used in turbid direct cooling water circulating systems for equipment, are adopted to reduce the wastewater discharge. Mainly include:

High-efficiency cooling technology is used in open indirect cooling water circulating systems for equipment, mainly including cooling tower with water collector, evaporative air cooling or air cooling technology, etc., and the current water recycling rate exceeds 97.5%.

High-efficiency precipitation and filtration technology is used in turbid direct cooling water circulating system, mainly referring to the turbid direct cooling water circulating system in the steel rolling process and steelmaking and continuous casting process of the iron and steel industry. This part of water will be recycled after being precipitated and filtered. In the past, the treatment process, such as cyclone precipitation, secondary horizontal flow precipitation, pressure filtration, and cooling, was mainly used, while now the technologies such as cyclone precipitation, chemical de-oiler or rare earth disk, high-efficiency filter, and recycling use after cooling are mainly used to improve treatment effect and cycle utilization rate, and the current cycle utilization rate is as high as 98% or more.

Another technology is used in the water circulation system for wet-type primary dedusting for converter. The wastewater treatment technology is gradually evolved from the original amplitude flow sedimentation technology to the high-efficiency inclined tube (plate) sedimentation technology, which improves the water treatment efficiency and increases the dedusting water recycling rate from 94% to about 97%.

High-efficient water quality stabilization technology—in the iron and steel industry, by adopting high-efficiency water quality stabilization technology and improving the automation level of the dosing system, the concentration multiple of the circulating water system is increased from 1.5 times to about 2.5 times, even up to 4.5 times, thus greatly reducing the amount of sewage discharged from the circulating water system.

3) Treatment Technologies for Wastewater from Coke Oven.
The common activated sludge treatment technology was originally used for coke oven in China's iron and steel industry and was the traditional way for biological wastewater treatment used for coke oven. Today, it is still mainly adopted aboard. However, due to the stricter requirements on environmental protection in China, the COD and ammonia nitrogen in wastewater treated by this technology cannot meet the new emission standards. Therefore, through continuous technological innovation, many technologies are developed and they include A/O, $A_2/O$, $A/O_2$, O-A/O methods, namely the integration technology of "Intensified Denitrification Technology + High-Efficiency Coagulation + Ozone Catalytic Oxidation", Fenton oxidation wastewater treatment technology. The wastewater treated by them can basically meet the requirements of new direct emission standards, with the COD not more than 80 mg/L and ammonia nitrogen concentration not more than 10 mg/L. The treated water is mainly sent to the BF system and used as the makeup water for slag granulating water system. In some enterprises, it will be reused after further treatment in the integrated sewage treatment plant.
In order to save water and reduce the content of COD and ammonia nitrogen in the wastewater discharged from enterprises, some enterprises intent to treat this part of the water using advanced treatment technology, that is, the treated phenol cyanide wastewater will be reused for production water system after being further treated, and a small amount of concentrated discharge will be used for BF slag granulation. The treatment process includes pre-treatment of filtered water + ultrafiltration + nanofiltration + reverse osmosis. However, due to the high operating cost, it is still difficult to promote the advanced treatment and reuse technology alone used for coke oven wastewater in the iron and steel industry.

4) Treatment Technology for Wastewater from Cold Rolling Plant.
The wastewater from cold rolling plants can be classified as acid–alkali wastewater, oil- and emulsion-contained wastewater, and chrome-contained wastewater depending on the source and composition.
Acid–alkali wastewater is generally treated by means of neutralization method; oil- and emulsion-contained wastewater is difficult to be treated and often requires a combination of various methods, such as gravity separation, air flotation, chemical method, biological method, membrane method, and adsorption method; the chrome-contained wastewater is normally treated by means of reduction precipitation method.

The chemical method and ultrafiltration method are commonly used for the oil-contained wastewater and emulsified oil from cold rolling plant. In order to thicken waste emulsion to the maximum extent, two-stage ultrafiltration is generally applied.

The chromium-contained wastewater is generally subjected to two-stage reduction, followed by neutralization aeration and flocculation precipitation.

The treated water in some enterprises will be reused for the water system of cold rolling plant, but most of enterprises will sent it to integrated wastewater treatment plant for further treatment before being reused.

5) The Technology on Comprehensive Wastewater Treatment and Reuse [3].

The comprehensive wastewater treatment and reuse in the iron and steel industry is an important way to save water and reduce the wastewater discharge. By the end of 2016, the proportion of supporting construction of integrated wastewater treatment plants in key statistical iron and steel enterprises has reached more than 75%, and the water reuse rate of the same has been greatly improved. Some enterprises have basically achieved "zero" emissions of production wastewater [1].

The comprehensive wastewater treatment processes used in the iron and steel industry during the 10th Five-Year Plan and 11th Five-Year Plan period mainly included flocculation, sedimentation, and filtration for eliminating SS and COD, but no salt was removed. With tighter requirements on water preservation, the concentration ratio of circulating water system in many enterprises has been continuously increased, and the salt content of recycled water has increased. If the recycled water is mixed with the original industrial freshwater as the makeup water for clean circulating water system, the salt content of the entire water supply system will be increased, which will result in serious scaling and corrosion of equipment, affecting the service life of the equipment, and also having a certain impact on the water safety for the production process. In order to improve the comprehensive wastewater reuse rate of iron and steel enterprises and ensure the recycled water quality, the technology adopted has been developed into a high-efficiency coagulation, sedimentation, and filtration, followed by further advanced treatment. At present, the method of "high-density clarifier + V-shaped filter + membrane" is proved to be the best process technology for comprehensive wastewater recycling in the iron and steel industry. The popularity rate of this technology is above 50%.

The common membrane separation technologies are mainly classified as micro-filtration (MF), ultrafiltration (UF), nanofiltration (NF), reverse osmosis (RO), and electro dialysis (EDI). Ultrafiltration (OF) and reverse osmosis (RO) are the main approaches of the membrane method in the iron and steel industry.

4. Comprehensive Utilization of Solid Waste

In recent years, as the country accelerates the construction of ecological civilization, and the iron and steel industry pays more and more attention to the comprehensive utilization of resources and the development of circular economy, the solid waste reduction, comprehensive utilization rate, and utilization level in China's iron and steel industry have increased steadily. The changes of solid waste generation and comprehensive utilization rate in key statistical iron and steel enterprises of CISA members from 2005 to 2015 are shown in Fig. 5.13.

As shown in the Fig. 5.13, the solid waste generation per ton of steel in the China's iron and steel industry decreased from 628 kg/ton in 2005 to 585 kg/ton in 2015, and the comprehensive utilization rate of solid waste resources increased from 94.8% in 2005 to 97.5% in 2015. In 2015, the total solid waste generation by China's iron and steel industry was about 470 million tons, of which BF slag was 240 million tons, steelmaking slag was 100 million tons, and Fe-contained dust slurry was 50 million tons [2]. Affected by factors such as the market of products made of solid waste and the different conditions of smelting raw materials, the comprehensive utilization rate of solid waste resources is generally edging up, while the total solid waste generation per ton of steel is generally declining.

In addition, the output scale and enterprises strength on comprehensive resources utilization in China's iron and steel industry are continuously enhanced, and the industrial structure of comprehensive resources utilization continues to improve.

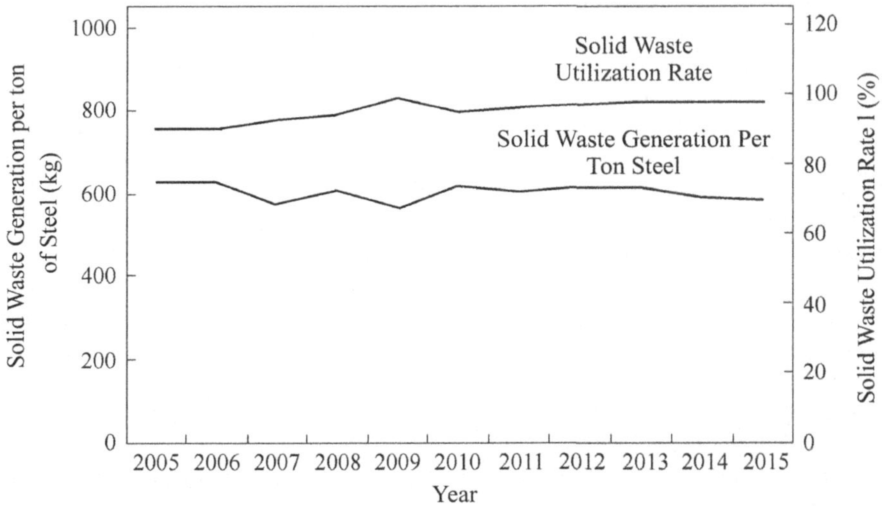


**Fig. 5.13** Changes of solid waste generation and comprehensive utilization rate in key statistical iron and steel enterprises of CISA members from 2005 to 2015

5. Deficiency in Energy Conservation and Emission Reduction

China's energy-saving technology and equipment level have been in the leading position in the world, but the gap on energy management level between China and advanced countries is still large. The leadership of the enterprise still does not pay enough attention to energy conservation work, the level of energy conservation management institutions is low, the number of full-time energy conservation management personnel is few, the basic capacity of energy conservation management is weak, and the usage rate of energy measuring meters is generally insufficient. Most enterprises do not have a complete energy management system, and the level of automation and information management is still low.

Conventional energy-saving measures such as CDQ, BF top pressure power generation, waste heat power generation of sinter plants, and waste heat and energy generation are gradually popularized, but independent research and development and innovation of new energy-saving technologies are difficult and slow, and energy-saving technologies in China, such as coal moisture control for coke oven, waste heat recovery of uptakes, waste heat utilization of steelmaking slag, and low-carbon metallurgy, still fall behind those in other advanced countries.

The waste gas and wastewater treatment as well as comprehensive utilization of solid waste in China's iron and steel industry has made significant achievements. The compliance rate of waste gas and wastewater discharge, the reuse rate of water, the pollutant discharge per ton of steel, and the recycling rate of solid waste have been greatly improved [1], and some of them rank among the international level, but there are still some problems in the industrial waste gas and waste water treatment and solid waste utilization of China's iron and steel industry:

In the treatment of waste gas, firstly the enterprise's fugitive dusty gas treatment needs to be strengthened, the management level of desulfurization facilities needs to be improved, and the targeted control technology of nitrogen oxides needs to be further researched and developed; secondly, the waste gas pollutants, such as heavy metals and dioxins, are not put under control.

In the treatment of wastewater, firstly, the treatment technology needs to be further developed and researched, such as water quality stabilization technology with high efficiency and without secondary pollution, special low-cost wastewater treatment technology—treatment technology of wastewater from coke oven and cold rolling plant; secondly, wastewater treatment efficiency needs to be further improved, the wastewater treatment efficiency of some iron and steel enterprises is still low, and they are still operating with extensive treatment methods. There are still room for improvement in the application of membrane treatment technology, reverse osmosis desalination technology, and urban wastewater reuse technology in the iron and steel industry.

Solid waste resources in China's iron and steel industry are mostly roughly processed into low value-added building materials. The level and added value of comprehensive solid waste resources utilization are not high. Many bulk solid wastes such as BF slag and tailings from steelmaking slag are sold out mostly for producing sub-grade materials, blocks, cement, and other building materials. The most Fe-contained dust slurry will be returned to production for simply reutilization, and the classification and high-efficiency utilization of Fe-contained dust slurry with different components are still not realized. The industrial concentration and technical level of the comprehensive solid waste resources utilization in China's iron and steel industry need to be improved. The number of leading, backbone, and pillar enterprises in solid waste utilization enterprises is few; moreover, their technical level and scientific and technological strength in comprehensive resources utilization are poor. Their awareness of comprehensive solid waste resources is weak, and their management mechanism and related policies and standard systems are not perfect.

## 5.2 Development Environment and Policy Orientation

### *5.2.1 Historical Background of "Greenness"*

Since the 1950s, the environmental pollution and ecological damage have become increasingly serious. Extreme weather such as hurricanes, droughts, and blizzards has occurred frequently. Environmental problems such as acid rain, desertification, sharp decline of forests, and extinction of species have shown a trend of globalization and internationalization. In response to the current severe environmental situation, the first UN Environment Conference held in Nairobi in June, 2014, passed 16 decisions and resolutions to promote actions by the international community to deal with air pollution, illegal wildlife trade, marine plastic waste and chemicals, hazardous wastes, and other major environmental issues. The international community becomes increasingly aware that green development and sustainable development are the trend of the times in today's world and the inevitable choices for coordinating economic growth, social development, and environmental protection.

With the sustained and rapid growth of China's economy in recent years, environmental problems in the industrialization process of developed countries for decades or even hundreds of years have broken out intensively in China, reflecting in the frequent outbreaks of smog, red tide, unqualified drinking water, cadmium rice, and other pollution incidents, which are the warnings that the current environmental situation in China is very serious and has seriously affected the sustainable development of the Chinese nation. The deterioration of environmental and ecological issues requires the transformation in development model, seeking a balance between environment and economic benefits, achieving green development, and ensuring the ecological environment to provide the environmental services needed for human well-being while promoting economic growth.

Since the official announcement of ecological civilization construction proposed in the 17th National Congress of the Communist Party of China, the understanding on green development has gradually deepened and the theoretical system of greenness has become more and more perfect. In March, 2015, in the *Opinions on Accelerating the Construction of Ecological Civilization* reviewed and approved by the Political Bureau of the CPC Central Committee, the goals of "Coordinated Promotion of New

Industrialization, Urbanization, Informationization, Agricultural Modernization and Greenness" are clearly stated. The proposal of "New Five Modernizations" takes the "Greenness" to the level equivalent to political tasks that affect people's well-being and the long-term future of the nation. The 5th Plenary Session of the 18th CPC Central Committee regarded "Greenness" as one of the five development concepts proposed, which means that green development has been part of the will of the party and the state and officially became the party's and the country's ruling philosophy, indicating the determination of the Chinese government to firmly follow the path of green development.

The green development will be incorporated into the 13th Five-Year Plan as a national strategy, which will guide the coordinated development of China's future economy, society, environment, and resources. To promote sustainable development, we must rely on green development. Only by persisting in green development we can promote social innovation and achieve structural adjustment of the economy and transformation of growth models. Green development has become the essence and core content of the 13th Five-Year Plan, marking the advent of the new normal of China's economic and social development into a comprehensive greenness.

### *5.2.2 Constraints by Energy and Environmental Policy*

"Greenness" is not only a slogan or a concept, but also needs various laws, systems, and policies as guarantees. In order to implement "Greenness" during recent years, the CPC Central Committee and the State Council issued the *Overall Plan for the Reform of Ecological Civilization System* in September, 2015, which formulated a systematic and comprehensive plan for "Greenness". The relevant departments of the Chinese government also issued a series of laws and regulations on energy and environmental protection, which have played a key guiding role in accelerating green development.

1. Energy Policy

*Several Opinions of the General Office of the State Council on Further Strengthening Energy Saving and Emission Reduction to Accelerate the Structural Adjustment of the Iron and Steel Industry* (No. 34 [2010] of the General Office of the State Council), *Catalogue of Backward Production Equipment and Products to be Eliminated in Some Industrial Sectors*, and *Codes and Conditions for Production and Operation in Iron and Steel industry*. According to the national plan on eliminating backward production capacity, the backward production capacity shall be eliminated on time to promote industrial upgrading and technological upgrading and improve energy efficiency. A series of regulations have been made on energy consumption and comprehensive utilization of resources, environmental protection, and production scale for the iron and steel enterprises, which have great practical significance to change the current development model of China's iron and steel industry, guide the healthy development of the iron and steel industry, save energy and reduce consumption, control pollution, and reduce emission as well as eliminating backward production capacity.

In the *Comprehensive Work Plan on Energy Conservation and Emission Reduction during 13th Five-Year Plan* period (No. 74 [2016] of the State Council)", the overall goal of energy conservation and emission reduction during the 13th Five-Year Plan period is clearly stated, and the goal is decomposed and assigned to each province (autonomous regions, municipalities directly under the Central Government), major industries and sectors, and key areas for air pollution control, which is the programmatic document for guiding the in-depth development of national energy conservation and emission reduction work in the new era. First of all, it is necessary to further highlight the binding effect of the "Double Control" target in terms of total volume and intensity. The "Double Control" target of total energy consumption and intensity, presented in the Outline of the 13th Five-Year Plan on the basis of energy intensity control objectives and as a prescriptive measure for the construction of ecological civilization, shall be implemented by striving to reduce pollutant emissions from the source, so as to force the transformation of economic development mode. In addition, the structural optimization and upgrading should be taken as the key task of energy conservation and emission reduction in the new era, and shall be continuously expanded by the three starting points of energy conservation and emission reduction, such as structural adjustment, technological progress, and management improvement. During the 13th Five-Year Plan period, as China's major high energy-consuming industries entered the stage of saturation and pursued "Cutting Overcapacity" development, the potential to significantly improve energy efficiency through technical modifications continues to decrease, and structural optimization and upgrading becomes the key task of energy conservation and emission reduction in the new era. The structural optimization and upgrading not only includes the three-industry structural adjustment, but more importantly, the internal structure optimization of the industry, the upgrading of product structure, the improvement of quality and added value, which not only contribute to energy saving and emission reduction, but also are consistent with the fundamental requirements of supply-side structural reforms. The Work Plan focuses on the transformation and upgrading of traditional industries, the accelerated development of emerging industries, and the adjustment and optimization of energy structure. It stated the tasks to optimize the industry and energy structure, reflected the maximum potential of energy conservation and emission reduction, and also contained a huge potential space for cultivating new kinetic energy for economic growth.

*China's National Climate Change Programme, China's Policies and Actions for Addressing Climate Change, Work Plan for Controlling Greenhouse Gases Emissions during the 12th Five-Year Plan period, Action plan of Industrial Sector for Addressing Climate Change* and other policies on addressing climate change. The urgency of coping with climate change in the future is becoming more and more intensive, and the binding force is becoming stronger. The dependence on purely political and policy approaches will be shifted to economic means and market means, and the integration with economy, energy, environmental protection, and resource development will be strengthened. The unified national carbon emission trading market is being planned. Each enterprise will allocate a certain quota. If the emission quota is insufficient, it must be purchased from the market. Carbon emissions will effectively affect the economic benefits of each enterprise. From the perspective of the assessment system, the basic control path from controlling intensity to total quantity to controlling absolute quantity is very clear. In both the international and domestic levels, long-term policy signals to enterprises have been conveyed, and iron and steel companies must be prepared accordingly!

China signed the "*Paris Agreement*" and promised to actively carry out domestic greenhouse gas emission reduction work, strengthen international cooperation on climate change, and demonstrate the great determination and responsibility of the global climate governance power. As an active participant in the global response to climate change, in order to achieve the above goals, China has declared its action targets for addressing climate change through addressing the National Independent Contribution Document on Climate Change (*Strengthening Action to Address Climate Change—China's National Independence Contribution*), including as follows: Carbon dioxide emissions will reach its peak in around 2030, and $CO_2$ emissions per unit of GDP will fall by 60–65% compared with that in 2005, the non-petrochemical energy consumption will account for about 20% of primary energy consumption, and forest reserves will increase by 4.5 billion cubic meters compared with that in 2005.

The iron and steel industry is one of the most important basic industries of the national economy and plays an important supporting role in social and economic development. China is the world's largest iron and steel producer and consumer, and has a significant influence in the world iron and steel community. At the same time, its production structure is still dominated by the long process technology of BF converter based on ores and coking coal, and the industry is still the main industry of carbon emissions. Therefore, China's iron and steel industry will play an important role in fulfilling the goal of addressing climate change.

The 13th Five-Year Plan period is a crucial period for China to build a well-off society in an all-round way. The iron and steel industry has also entered a downward period after the production and consumption peak, in which deep adjustment of the market structure is undergoing, transformation and upgrading reaches decisive point, and historical opportunity for innovation and development presents. Facing the increasingly severe energy-saving and carbon-reduction situation at home and abroad, the iron and steel industry and enterprises in China should address climate change with a more active attitude.

First is to accelerate the establishment and improvement of the unified national carbon emission trading market, and promote and protect the industry's ability to save energy and reduce emissions.

According to the road map announced by the National Development and Reform Commission, the establishment of a unified national carbon emission trading market is divided into three phases. The first phase is the preparation phase from 2014 to 2015, during which the establishment of relevant laws and regulations, the development of technical standards, and the formulation of quota allocation methods will be carried out. The second phase is the operational improvement phase from 2016 to 2020, which is also the initial phase of the unified national carbon emission trading market. During the period, the National Development and Reform Commission will fully implement and improve the unified national carbon emission trading market. The third phase is the expansion phase after 2020, during which the participation scope of enterprises and the trading categories in the carbon emission trading market will be expanded, and the possibility of interfacing with other international pilots will be explored.

At current stage, the three sectors of power, cement, and electrolytic aluminum have announced carbon allocation schemes, which will take the lead in entering the carbon emission trading market. The iron and steel industry is one of the eight key emission industries in the first batch to be included in the national carbon emissions trading market. The iron and steel industry is also at a critical stage in the promotion of the carbon emission trading market. Accelerating the establishment of a unified national carbon emission trading market for the iron and steel industry is the key task in next step. The carbon emissions will be allocated to enterprises in the form of quotas, which will effectively affect the economic benefits of each enterprise.

The second is to accelerate the realization of goal of reaching the carbon emission peak in the iron and steel industry by improving energy efficiency and resolving overcapacity.

The government actively promoted supply-side structural reforms and proposed five major tasks of "Decapacity, Destocking, Deleveraging, Cost Reduction, and Short-Boarding", which played an important role in the structural adjustment and transformation of the iron and steel industry.

It was proposed in the No. 6 [2016] Document issued by the State Council in February, 2016, to reduce the crude steel production capacity by 100 million to 150 million tons in five years from 2016 and further accelerate the pace of iron and steel industry in cutting overcapacity. At the same time, iron and steel enterprises should actively promote systemic energy conservation and cost reduction, and continuously accelerate the research and development, innovation, and industrialization of advanced energy-efficient utilization technologies.

Cutting overcapacity and reducing the systemic cost are not only the important starting points for iron and steel companies to turn losses into profits, but also accelerates the realization of carbon emissions peak target of China's iron and steel industry. In the next stage, improving energy efficiency and resolving overcapacity should remain an important direction for low-carbon sustainable development of iron and steel companies.

The third is to optimize the energy consumption structure and increase the proportion of scrap utilization.

Energy transformation and low-carbon development have become a worldwide trend. In the low-carbon development of iron and steel enterprises in the future, adjusting and optimizing energy structure and reducing the consumption of petrochemical energy are the directions that enterprises should constantly pursue.

As we all know, the short process with EAF mainly feeds on the renewable resources—scrap steel—which not only reduces the consumption of reductant carbon, but also reduces the energy consumption for chemical reaction, so it has obvious advantages in reducing energy consumption and carbon dioxide emissions. In general, the energy consumption of the short-flow steelmaking process with EAF is half of the long process of BF converter, and the carbon dioxide emission is about one-third of the same.

The proportion of EAF steel in China's iron and steel industry is less than 10%, and it has been at a relatively low stage of development. As the accumulation of scrap resources in China increases, the short-flow steelmaking process with EAF feeding on scrap will face greater development opportunities. In addition, the iron and steel industry can also work more in the development of new energy and renewable energy according to the actual conditions, including more beneficial attempts on solar energy and wind energy.

The fourth is to actively research and develop and prepare advanced low-carbon smelting technology to achieve scientific and technological support.

Looking to the future development situation, the *Paris Agreement* proposes to achieve net and zero greenhouse gas emissions in the second half of the twenty-first century, which means that the low carbonization transformation of the iron and steel industry has become an inevitable trend. In order to achieve a significant reduction in carbon dioxide emissions, it is necessary to develop new low-carbon smelting production technologies, and the iron and steel industry will also undergo revolutionary changes.

Iron and steel enterprises should keep a watchful eye on and strengthen the research and introduction of advanced low-carbon smelting technology at home and abroad, and at the same time accelerate the research and development, prepare and application of advanced low-carbon smelting technology that meets the needs of China's iron and steel industry transformation and development.

2. Latest Environmental Protection Policy

It was proposed in the 18th National Congress of the Communist Party of China to vigorously promote the construction of ecological civilization, strive to build a beautiful China, and realize the sustainable development of the Chinese nation. In order to promote the construction of ecological civilization, the related environmental protection laws and regulations have been introduced. In recent years, the government has successively issued a series of policies and measures such as the *Air Pollution Prevention Action Plan, Interpretation for Several Issues Concerning Applicable Law on Handling Criminal Cases of Environmental Pollution,* and *Notice on Strengthening Environmental Supervision and Enforcement by the General Office of the State Council.* In particular, since January 1, 2015, *the new Environmental Protection Law*, which alleges more severe punishment, has officially been implemented, and *the Interim Measures for Continuous Penalty by the Environmental Protection Administration* and *Interim Measures for the Administration of Environmental Protection to*

*Restrict Production and Stop Production* in conjunction with *the new Environmental Protection Law* were implemented at the same time. Therefore, environmental violations such as excessive discharges will be severely punished, such as "Penalty on a daily basis and without Ceiling, Security Detention and Criminal Responsibility". At the same time, the *Air Pollution Prevention and Control Law* was officially implemented on January 1, 2016. *The Water Pollution Prevention Action Plan* has been released and implemented, and *the Soil Contaminant Prevention Action Plan* has also been completed. In the face of increasingly strict and perfect environmental laws and regulations, "Greenness" has become an inevitable choice for the iron and steel industry to meet environmental challenges.

2017 is the pivotal year of implementing the *Air Pollution Prevention and Control Action Plan* in the first phase assessment. There are many challenges such as meteorological conditions and the increase of pollutants, mounting pressure on smog control. For the iron and steel industry, especially urban iron and steel enterprises, the pressure on enterprises to restrict and stop production as well as upgrade and reconstruct environmental protection facilities is unprecedented. On March 23, 2017, the Ministry of Environmental Protection, the National Development and Reform Commission, the Ministry of Finance, the National Energy Administration, and the People's Governments of Beijing, Tianjin, Hebei, Shanxi, Shandong, and Henan jointly issued the *Work Plan for Air Pollution Prevention and Control in Beijing, Tianjin and Hebei and Surrounding Areas in 2017*, involving "2 + 26" cities including Beijing, Tianjin, Shijiazhuang, Tangshan, Langfang, Baoding, Cangzhou, Hengshui, Xingtai, and Handan in Hebei Province, Taiyuan, Yangquan, Changzhi, and Jincheng in Shanxi Province, Jinan Zibo, Jining, Dezhou, Liaocheng, Binzhou, and Heze in Shandong Province, as well as Kaifeng, Anyang, Hebi, Xinxiang, Jiaozuo, and Puyang in Henan Zhengzhou. In the plan, it is clearly stated that for key cities such as Shijiazhuang, Tangshan, Handan, and Anyang, the steel production capacity during the heating season is limited to 50%, which is calculated based on the production capacity of BF and evidenced by the actual electricity consumption of the enterprise.

At the end of 2016, the General Office of the State Council officially promulgated the *Implementation Plan on Permit System for Controlling Pollutant Discharge* (No. 81 [2016] of the General Office of the State Council) [4], established the discharge permit system as the core system for China's fixed-source environmental management, and announced that the new discharge permit system for the iron and steel industry is on the stage of history. The new pollutant discharge permission system replaced a number of environmental protection functions and systems including "Environmental Statistics", "Pollution Charges", "Total Quantity Control", "Environmental Monitoring", and "Environmental Standards" to realize the "Management with one Permit". The environmental protection department will check the enterprises against all items in the pollutant discharge permit. Any enterprises failing to meet the requirement and thus causing violation to the environmental laws will be punished continuously on a daily basis and subjected to production restriction or suspension, business termination, close-down, etc.

*The Environmental Protection Tax Law*, which is in line with the new discharge permit system, was also officially implemented on January 1, 2018, and stresses the principle of "To Decrease Tax by Reducing Discharge", which means the enterprises with less pollutant discharge are levied with lower tax. This new law also boosts the progress of the iron and steel enterprises in strengthening environmental protection layout since the iron and steel industry discharges large quantity of pollutants. Thus, the more environmentally friendly enterprises survive while the inferior ones are eliminated, and the phenomenon of "Bad Money Drives out Good Money" will not happen. The implementation of the new discharge permission system will be combined with the environmental tax law to complete the issuance of pollutant discharge permits for all fixed pollution sources, including the iron and steel industry, and basically establish a control pollutant discharge permission system with complete legal system, scientific technical system, and efficient management system, so as to implement the whole process and coordinated control over multiple pollutants for fixed pollution sources, and realize the "Management with one Permit" featuring systematization, scientification, legalization, refinement, and informationization. It can be foreseen that the introduction of environmental protection tax will bring the iron and steel industry into a "Green Tax" system, further promoting the green transformation and development of China's iron and steel industry.

## 5.3 Case Analysis

### *5.3.1 HBIS Group Tangsteel Company*

1. Concept of Greenness

HBIS Group Tangsteel Company upholds a good environmental protection concept and strictly implement the principle of "Production Giving Way to Environmental Protection" and regard the environment as its lifeline. The production and benefits are regarded less important than environment; the idea of "Survival First and Environment First" is established. The environment of the factory should be completely clean, and it is better to sacrifice the output rather than sacrifice the environment and better to sacrifice the short-term benefits rather than sacrifice the environment.

In addition, Tangsteel Company focus on the use of secondary energy as much as primary energy, taking indicator management as the core and the refined management as the main approach, relying on the energy environment control center, to realize the visualization, digitalization, controllability, responsibility, and accountability of energy and resource management, thus further reducing energy consumption and improving energy and resource recovery, conversion, and utilization efficiency.

2. Measures of Greenness

Since 2008, Tangsteel Company have carried out all-round technical transformation aiming at structural adjustment and elimination of backward capacity, eliminating five lime kilns with backward technology, three 60 $m^2$ sintering machines, and one 8 $m^2$ shaft furnace, shutting down and dismantling three ordinary long product production lines and EAF steelmaking production lines, such as rolling plant No. 3 and rolling plant No. 4. At the same time, three 450 $m^3$ BFs in the southern district were dismantled and replaced by a technologically advanced large-scale 3,200 $m^3$ BF. All the above-mentioned emptied areas after dismantlement have been greened and beautified. The elimination of these backward production capacity and equipment has solved the problem of high energy consumption and pollution of the enterprise from the source and has improved the energy saving and emission reduction of Tangsteel Company to a new level.

Tangsteel Company vigorously implemented the projects for energy-saving and emission reduction transformation. Under the circumstances that production and operation were facing great difficulties and adhering to the principle of no reduction in projects, funds and management effort, the company was determined to promote pollution and emission reduction, committed to clean production and environmental management, and invested 3.18 billion yuan to implement 32 energy-saving and emission reduction transformation projects, which already achieved outstanding economic, environmental, and social benefits.

Through the implementation of clean production, Tangsteel Company have strengthened the energy-saving and emission reduction work, realized the green transformation of the enterprise, and took the lead in embarking on a development path of integrating development, benign interaction, and harmonious coexistence with the city. Therefore, it was reputed by the China Iron and Steel Association as "The Cleanest Steel Plant in the World". In recent years, many industry leaders, well-known enterprises, social leaders, and international manufacturers came to Tangsteel Company for visiting and learning from them, and Tangsteel Company have become a benchmark for the companies in the iron and steel industry to learn from.

3. Effect of Greenness

Through the implementation of the two comprehensive measures of "Actively Eliminating Backward Production Capacity to Solve High Energy Consumption Pollution from the Source" and "Strengthening the Construction of Energy-Saving and Emission Reduction Transformation", Tangsteel Company have achieved remarkable results which improved main environmental protection indicators up to the international first-class level.

(1) Discharge of Wastewater per ton of Steel.
The discharge of wastewater per ton of steel is reduced from 0.93 t in 2006 to 0.61 t in 2015, down 33%, as shown in Fig. 5.14.
(2) $SO_2$ Emissions per ton of Steel.
$SO_2$ emissions per ton of steel is reduced from 2.3 kg in 2006 to 0.7 kg in 2015, down 69.9%, as shown in Fig. 5.15.
(3) Emission of Smoke and Dust per ton of Steel.
The emission of smoke and dust per ton of steel is reduced from 1.8 kg in 2006 to 0.36 kg in 2015, down 80.2%, as shown in Fig. 5.16.

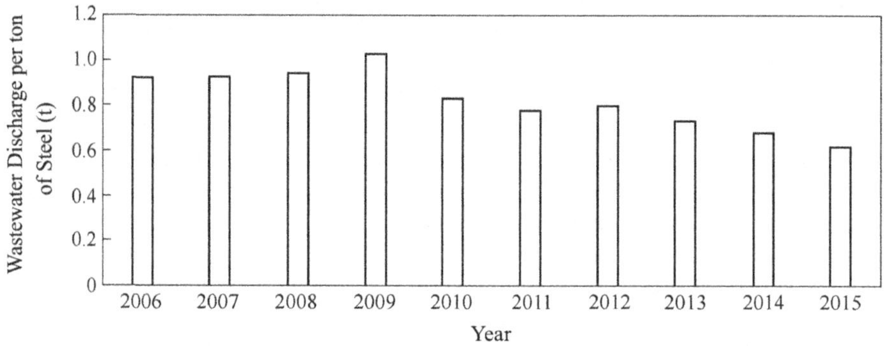


**Fig. 5.14** Wastewater discharge per ton of steel from 2006 to 2015 of Tangsteel Company

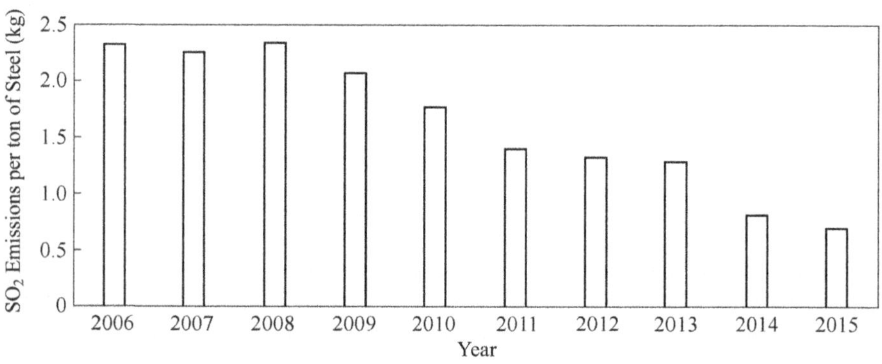


**Fig. 5.15** $SO_2$ emissions per ton of steel from 2006 to 2015 of Tangsteel Company

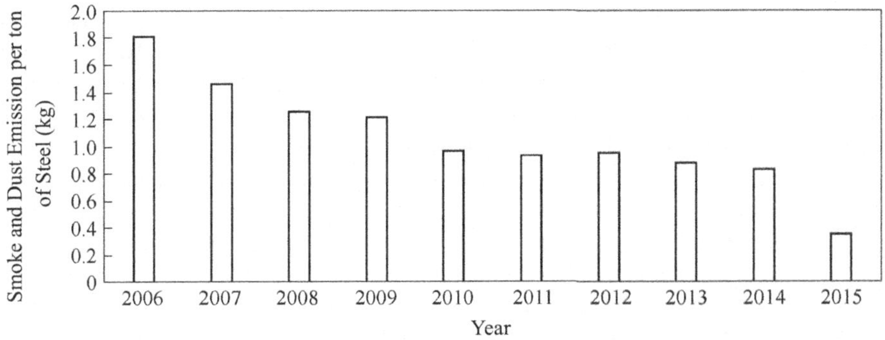


**Fig. 5.16** Smoke and dust emission per ton of steel from 2006 to 2015 of Tangsteel Company

### *5.3.2 Taiyuan Iron & Steel (Group) Co. Ltd. (TISCO)*

1. Concept of Greenness

Based on the concept innovation and driven by self-dependent innovation, Taiyuan Iron & Steel (Group) Co. Ltd. (hereinafter referred to as TISCO) adheres to the path of energy conservation, emission reduction, circular economy, and green development, continuously improves the allocation of energy conservation and emission reduction and enhances performance, and fulfills the harmonious integration of steel plant and city.

Focusing on the most competitive development strategy, TISCO has established the "1124" green development model, namely establishing a concept—steel plants and cities live in harmony and join efforts in development; establishing a goal—to build a demonstration factory of energy saving and recycling economy in the metallurgical industry; relying on two innovations—technological innovation and management innovation; expanding four functions—product manufacturing, energy conversion, waste disposal, greening and landscaping, which creates a new path of harmonious development between inland steel plant and cities [5].

2. Measures of Greenness

TISCO resolved to eliminate all old coke ovens, small BFs, small EAFs, and backward smelting and rolling equipment and adhere to self-dependent innovation and system integration. Accordingly, TISCO has built the world's most advanced 7.36 m coke oven, 450 $m^2$ sintering machine, 4,350 $m^3$ BF, 160 t ultra-high-power EAF, 180 t AOD furnace, 180 t converter, 180 t LF and slab caster, 2,250 mm hot rolling mill, and wide steel cold rolling line, so that it has upgraded the whole process technology and main equipment, featuring large-scale, high-efficiency, energy-saving, and environmental protection equipment, and has become the single largest stainless steel enterprise with the highest level of process equipment and the most complete varieties and specifications.

In terms of green innovation, TISCO took the lead in building China's first set of activated carbon desulfurization, denitration, dioxins, heavy metals, and dust removal and acid production system for flue gas from sintering machine, China's first steel slag fertilizer production line, China's first set of coke oven gas desulfurization and acid production system, a municipal wastewater treatment system, a large-scale membrane industrial wastewater treatment system, and an environmental monitoring center (including pollution sources, atmospheric environment, plant boundary noise, video surveillance, etc.) in the steel plant, as well as became the first iron and steel enterprise in launching the campaign to control PM2.5. The construction of these energy-saving and environmental protection and circular economy projects has created environmental and economic benefits for TISCO and laid the foundation for TISCO to build into a demonstration plant for energy conservation, emission reduction, and recycling economy in the metallurgical industry.

3. Effect of Greenness

TISCO has achieved remarkable results through the elimination of backward production capacity and using innovative energy conservation and environmental protection technologies, which has made it a successful urban iron and steel enterprise. Its main environmental indicators have been improved significantly.

(1) $SO_2$ Emissions per ton of Steel.
The $SO_2$ emissions per ton of steel decreased was reduced from 3.75 t in 2006 to 0.44 t in 2015, down 88.2%, as shown in Fig. 5.17.
(2) Emission of Smoke and Dust per ton of Steel.
The smoke and dust emission per ton of steel decreased from 1.15 kg in 2006 to 0.12 kg in 2015, down 88.8%, as shown in Fig. 5.18.

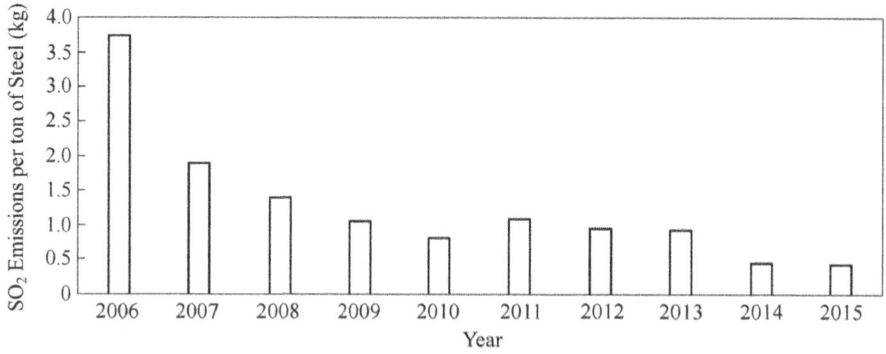


**Fig. 5.17** $SO_2$ emissions per ton of steel from 2006 to 2015

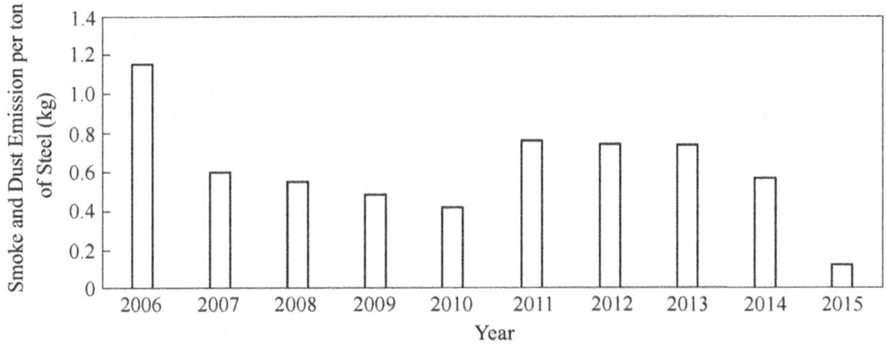


**Fig. 5.18** Smoke and dust emission of TISCO from 2006 to 2015

### *5.3.3 Shougang Jingtang United Iron & Steel Co. Ltd.*

1. Concept of Greenness

The concept of greenness of Shougang Jingtang United Iron & Steel Co. Ltd. (hereinafter referred to as Shougang Jingtang) can be summarized as "The Industry based on Environmental Protection, Waste Utilization or Treatment". In terms of concept, Shougang Jingtang raised the environmental protection work from the initial pollution control to the level of developing sunrise industry. The idea on environmental protection industrialization of Shougang Jingtang is to realize the comprehensive utilization of internal environmental resources.

As a very large state-owned enterprise, Shougang Jingtang has made positive contributions to environmental protection and fulfilled its corporate social responsibility. In this context, the Shougang Green Action Plan (2015–2016) came into being. Guided by the spirit of the 18th National Congress of the Communist Party of China and the spirit of the 3rd and 4th Plenary Sessions of the 18th Central Committee, this plan serves to thoroughly implement the new ideas and requirements of the government on ecological civilization construction, strengthen the process control and the whole process management of environmental protection projects to establish and improve foundation work, strictly implement work feedback mechanisms and other measures, strengthen environmental protection, comprehensively improve the environmental management level of Shougang, and vigorously promote the green development, circular development, and low-carbon development of Shougang.

2. Measures of Greenness

Shougang Jingtang has realized the high-efficiency energy conversion of waste heat and residual pressure in accordance with the whole process energy conversion system based on circular economy, which completes the industrial chain of circular economy and is exerting significant economic and social benefits.

First, the industrial chain of comprehensive utilization of seawater, which was promoted by seawater desalination, was initially formed. For the first time in China, the thermal low-temperature multi-effect desalination technology is applied, leading to the construction of a desalination plant with a daily output of 50,000 tons of freshwater and saving of a lot of freshwater resources. Following the principle of "Energy Cascade Utilization", the model of steam-electric-water cycle has been realized with the power generation capacity of 340 million kW · h, which is the best embodiment of recycling operation. The seawater desalination forms an industrial chain together with downstream salt industry, and concentrated brine produced by seawater desalination is supplied to nearby Tangshan Sanyou Chemical Co., Ltd. The second is to provide energy products to the society. Recycling waste heat resources generated during the production will not only satisfy the demand of enterprise itself, but also the surrounding enterprises. At present, it has realized the supply of heating water to some enterprises in the industrial zone. The production capacity of the equipment will be fully developed to realize the out-selling of oxygen, nitrogen, argon, and hydrogen. At present, 20 out of more than 50 varieties of energy products are sold, which have fulfilled the function of serving the society. The third is the utilization of solid waste resources. Through the integration of high-efficient recycling, reclamation, and productization technologies of solid waste such as waste slag and dust slurry, the enterprise realizes zero emission and enhances the value of solid waste recycling products through deep processing. Various solid wastes, such as BF slag, steelmaking slag, coal ash, collected ash, and scale, can be 100% recycled.

The environmental protection work shall be vigorously strengthened to promote the construction of "Two-Oriented" enterprises. There are 128 sets of waste gas treatment facilities and 8 sets of wastewater treatment facilities in Shougang Jingtang. In order to strengthen the monitoring on pollution sources, an environmental online monitoring system was constructed. At present, 14 online flue gas monitoring systems for captive power plants, sintering desulfurization system, dedusting system at BF cast house, and secondary dedusting system for steelmaking have been networked with the system of Municipal Environmental Protection Bureau.

3. Effect of Greenness

Shougang Jingtang was among the first resource-saving and environmentally friendly enterprises in the iron and steel industry listed in pioneer enterprises through implementation of a series of energy-saving and emission reduction measures. In October 2012, it passed the ISO14001 environmental management system certification audit of TÜV Rheinland. In June 2013, it was awarded as one of 100 Demonstration Enterprises for Satisfying Environmental Protection Standard in Tangshan. In 2014, it successfully passed the environmental protection acceptance of the first phase by the Ministry of Environmental Protection, was awarded the "Clean Production and Environmental Friendly Enterprise" by China Iron and Steel Association, and was identified by Hebei Province as the first benchmarking enterprise for clean production in the province.

Shougang Jingtang led the way in environmental protection when it was constructed. Therefore, its main environmental indicators have not changed much. However, on the basis of its excellent performance, its smoke and dust indicators per ton

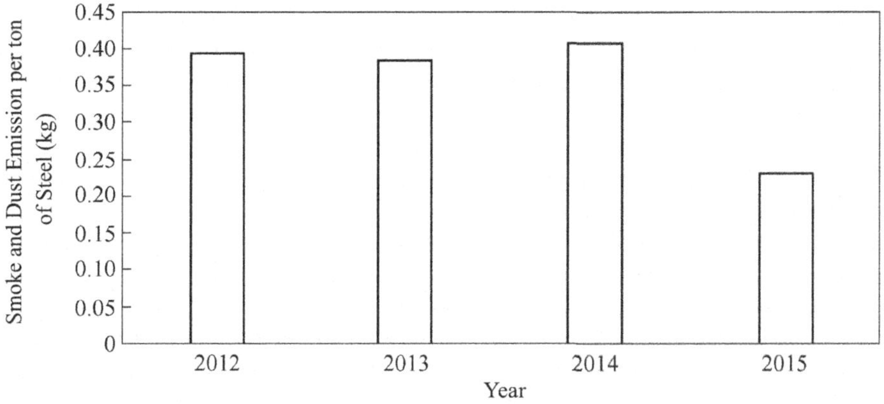


**Fig. 5.19** Diagram of smoke and dust emission per ton of steel by Shougang Jingtang from 2012 to 2015

of steel are greatly improved, decreasing from 0.39 kg in 2012 to 0.23 kg in 2015, and the smoke and dust emission per ton of steel was reduced by 41.5%, as shown in Fig. 5.19.

### *5.3.4 Baoshan Iron & Steel Co. Ltd. (Baosteel)*

1. Concept of Greenness

In 2009, Baoshan Iron & Steel Co. Ltd. (hereinafter referred to as Baosteel) proposed an environmental management strategy, rethinking the original business philosophy from an environmental perspective, exploring specific practical methods combining environmental benefits and competitiveness, implementing a "full life cycle" management of products, and operating a new environment concept through all aspects of business operations, from raw material purchase to product design, production, marketing, consumption, and waste recycling. Specifically, Baosteel's environmental management is composed of three parts: green manufacturing, green products, and green industry [6].

At the same time, Baosteel adheres to the national ecological civilization construction strategy and Baosteel's environmental management strategy and practices the concepts of green manufacturing, circular economy, and sustainable development. In accordance with Baosteel's green manufacturing development plan, Baosteel continues to maintain a high proportion of R&D investment and technological transformation investment, continuously improves resource utilization efficiency and manufacturing technology level, develops and promotes more and higher-level environmentally friendly steel products, and creates a more ecological and pleasant garden factory area to achieve common development with cities and communities. It has achieved the goal of living in harmony with society and developing in environment friendly manner, thus being the driver of the green industry chain and the best practicer of environmental friendliness.

2. Measures of Greenness

Baosteel has been committed to building itself into a clean factory with international advanced level by adopting the world's advanced energy-saving and emission reduction technologies, controlling pollutant emissions from sources, improving energy and resource utilization, and continuously improving.

In order to achieve this goal, Baosteel adopted a series of important energy-saving and emission reduction measures, such as CDQ technology, coal moisture control technology for coke oven, sensible heat recovery technology of sinter, waste heat recovery technology for flue gas from sintering machine, BF top pressure power generation technology, BF oxygen enrichment and high PCI rate technology, gas purification and recovery technology for converter, and gas-steam combined cycle power generation technology.

In the aspect of air pollution control, the high-efficiency bag filter and electrostatic precipitator, as well as effective fugitive dust control measures, were adopted to reduce smoke and dust emissions; by controlling the sulfur content of the raw fuel and installing flue gas desulfurization facilities in the sintering machine and power plant, the $SO_2$ emissions were controlled.

By implementing the dry-type dedusting technology for BF converter, CDQ technology, and water-saving technologies such as water supply by different quality, reuse of reclaimed water, rain and sewage diversion, recycling of rivers around plant, and satisfied wastewater discharge after treatment, freshwater consumption and wastewater discharge were reduced. Through technological innovation and management optimization, the reclamation of solid by-product resources back to the production process was continuously strengthened, the channels for socialized comprehensive utilization were expanded, and the added value was enhanced to continuously improve the comprehensive utilization rate of solid by-product resources.

Ironmaking process technology other than BF was actively explored. In 2007 and 2010, two of the world's largest COREX3000 were built. Baosteel also developed a green and low-carbon process technology with Baosteel's independent intellectual property rights in thin-strip continuous casting, sintering flue gas reutilization and emission reduction, and steelmaking slag drum treatment.

For a long time, Baosteel has mainly reduced carbon emissions and pollutant emissions in the steel production process through implementing energy-saving and emission reduction actions within the company to drive the overall progress of the industry. And through the development of circular economy and increasing the circulating utilization of by-product resources, the consumption of natural resources was reduced.

3. Effect of Greenness

Baosteel has adopted a series of important energy-saving and emission reduction measures, which has achieved remarkable results and improved main environmental protection indicators significantly, so it has become a clean factory with international advanced level.

(1) Discharge of Wastewater per ton of Steel.
    The discharge of wastewater per ton of steel is reduced from 2.78 t in 2006 to 0.86 t in 2015, down 69.1%, as shown in Fig. 5.20.
(2) $SO_2$ Emissions per ton of Steel.
    $SO_2$ emissions per ton of steel were reduced from 1.98 kg in 2006 to 0.44 kg in 2015, down by 77.6%, as shown in Fig. 5.21.
(3) The smoke and dust emission per ton of steel was reduced from 0.89 kg in 2006 to 0.14 kg in 2015, a decrease of 84.4%, as shown in Fig. 5.22.

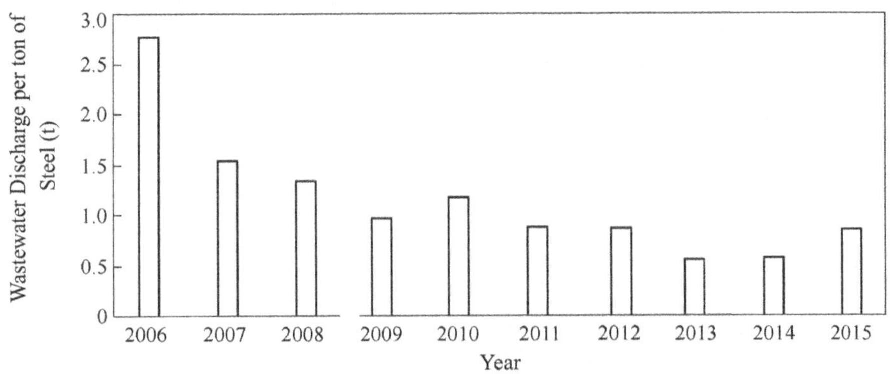


**Fig. 5.20** Wastewater discharge per ton of steel from 2006 to 2015 of Baosteel

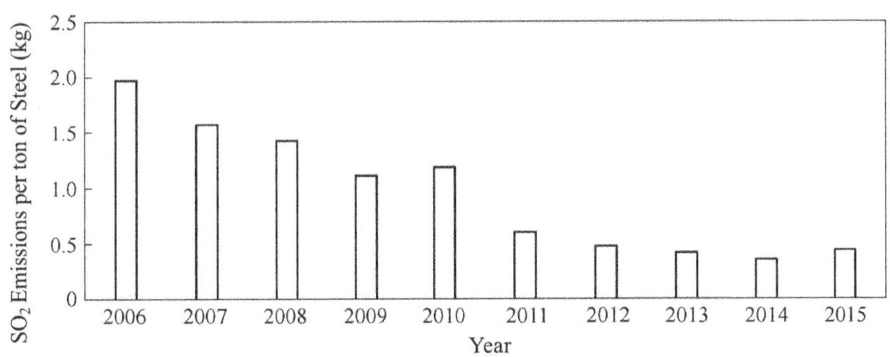


**Fig. 5.21** $SO_2$ emissions per ton of steel from 2006 to 2015 of Baosteel

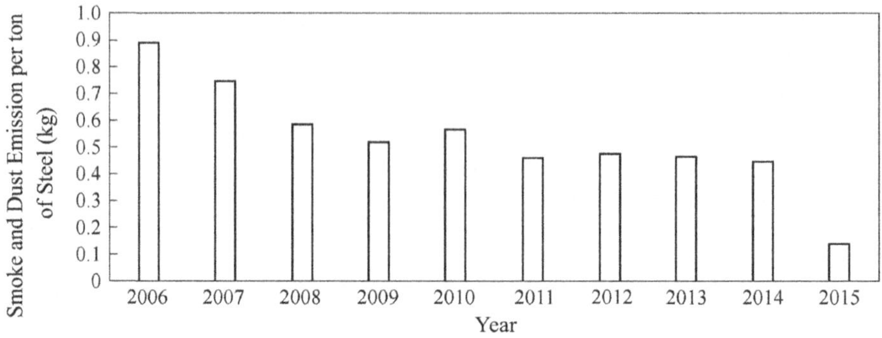


**Fig. 5.22** Smoke and dust emission of Baosteel from 2006 to 2015

### *5.3.5 Delong Iron & Steel Co. Ltd.*

1. Concept of Greenness

Responding to the national decision-making and deployment of ecological civilization construction, Delong Iron & Steel Co. Ltd. (hereinafter referred to as Delong Steel) closely follows the development concept of innovation, coordination, green, openness, and sharing, implements the requirements of supply-side structural reform, and relies on the ideas of "quality, high-end, green, and ecological" to build a permanent benchmarking enterprise of the industry. Delong Steel insists on taking the air pollution prevention and control action as the opportunity. In accordance with the provincial and municipal governance requirements for five major industries, it invested 800 million yuan in total and completed 14 outstanding environmental protection projects with high standards. With the courage of the strong man and the determination of being first-class enterprise, Delong Steel has interpreted the legend turn from traditional enterprises to a green iron and steel enterprise and has become a model of environmental protection in the industry. The North China Inspection Center of the Ministry of Environmental Protection invited more than 30 batches of researchers from Anyang, Tangshan, Baoding, Handan, and other cities to visited Delong Steel; Vice Chairman of the Standing Committee of the National People's Congress-Shen Yueyue and Vice Chairman of the National Committee of the Chinese People's Political Consultative Conference-Ma Peihua led a group of environmental protection experts to inspect Delong Steel and fully affirmed the achievement; On May 15, 2016, Xingtai City Environmental Protection Work Conference was held in Delong Steel, and 100 key enterprises of the city came to visit and learn from Delong Steel, who is a model of "Green Production".

2. Measures of Greenness

    (1) Great Investment and High-Standard Treatment.
    Delong Steel insisted on taking the smoke and dust emission reduction as the key to pollution control, and effectively controlled the emission of

smoke and dust by means of dismantling, sealing, improving, and replacing facilities causing pollution to achieve clean production. "Dismantling" is to remove backward production capacity. On the basis of the previous shutdown and prohibition, three more 70 $m^3$ white lime kilns, two 35 t converters, and two 410 $m^3$ BF were newly dismantled, avoiding 280 t $SO_2$ emission, 1,320 t smoke and dust emission, and 148 t nitrogen oxides emission. "Sealing" means that the materials storing and transportation are completely enclosed. By introduction of international advanced European standards and the investment of more than 100 million yuan, three fully enclosed stockyards were newly built and connected to the production workshop through underground corridors, so that the raw materials are sealed in the enclosed circumstance all the time from storage to transportation, which will avoid fugitive dust emission up to 2,000 t, thus greatly improving the environment of the plant. For example, the fully enclosed stockyards have a construction area of 26,000 $m^2$, which is equivalent to the summation of four football fields, a 12-story building, and a medium-sized community. The whole stockyard is of bolt-connected spherical grid structure and without column, and it is the largest stockyard in the province with the largest volume and highest standard. "Improving" means improving the production process. By improving the dedusting systems for sintering, ironmaking, steelmaking, and other process with the domestic leading technologies, the emission of particulate matter is greatly reduced. Taking the ironmaking process as an example, the investment was nearly 20 million yuan. The cast houses of No. 2 and 3 BF were fully enclosed and flattened, the emission concentrate of particulate matter emission was reduced to less than 20 mg/$m^3$, which is far below the national emission standard of 30 mg/$m^3$ and subverted people's impression on traditional ironmaking, such as hot metal splashes, heat waves, intensive fume and smoke. "Replacing" means replacing the old energy with clean one. In order to reduce emissions from transportation vehicles in the plant, Delong replaced all internal transportation vehicles with new energy-fuel gas vehicles, which not only saved 25% of energy, but also reduced harmful emissions by more than 85%. At the same time, Delong actively accepted social supervision, established a real-time monitoring and display system for pollutant discharge by enterprises, and publicly disclosed the discharge of particulate matter, sulfur dioxide, and nitrogen oxides on the displays in public places and main road displays, which was welcomed by the people.

(2) Reduce Resource Use and Reduce Environmental Impact. Delong consumes 14,000 t production water and produces 7,000 t wastewater per day. Relying only on the groundwater, besides wastewater is discharged, will cause serious damage to the local water ecology. Delong intended to solve this problem from the two aspects of water consumption and sewage treatment and strove to solve this problem internally and externally. Externally, surface water and pressure groundwater will be introduced. Delong invested 23 million yuan to build the Zhuzhuang-to-Delong Water Diversion Branch

Project based on the Zhuzhuang-to-Xingtai Water Diversion Project, laying 13-km-long water pipelines to transport water from Zhuzhuang Reservoir to the plant area, with the annual water transporting capacity of 5 million tons, which satisfied the production demand and ended the history of consuming groundwater. Internally, reclaimed water and recycled rainwater are utilized. The company invested 20 million yuan to implement the expansion and reconstruction of circulating pools and the soft water system upgrading project for steelmaking system. After the upgrading, the capacity of the pool will be expanded from 6,500 to 11,500 $m^3$ and all industrial wastewater and part of rainwater will be reclaimed and recycled, so the sewage will be zero emission. In particular, the wastewater from the plant, which is treated by nearly ten processes such as reverse osmosis technology, can reach the national first-class water quality standard and can be directly drunk. At the same time, Delong also has vigorously developed water-saving technology, making the water consumption per ton of steel dropped to 1.57 tons, which reaches the domestic advanced level.

3. Effect of Greenness

Delong Steel is committed to creating the "Garden-Style Factory". In developing the enterprise, it is committed to the development path of coexistence with the eco-city. It has made unremitting efforts to prove to the society: "Iron and steel can be warm and steel plant can be also green". After years of in-depth management, under the principle of maintaining near-zero emissions of wastewater, the main environmental indicators of atmospheric pollutant emissions have dropped significantly.

(1) $SO_2$ Emissions per ton of Steel from 1.59 kg in 2006 to 1.08 kg in 2016, $SO_2$ emissions per ton of steel decreased by 32.1%, and this is especially significant in the context of equipment upgrades, as shown in Fig. 5.23.

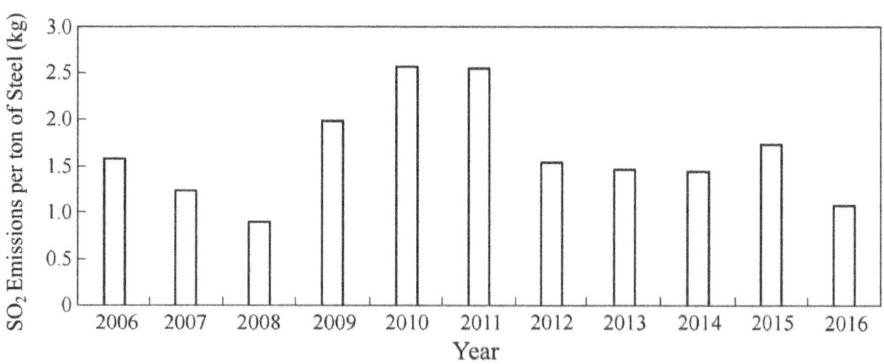


**Fig. 5.23** $SO_2$ Emissions per ton of steel in Delong Steel from 2006 to 2016

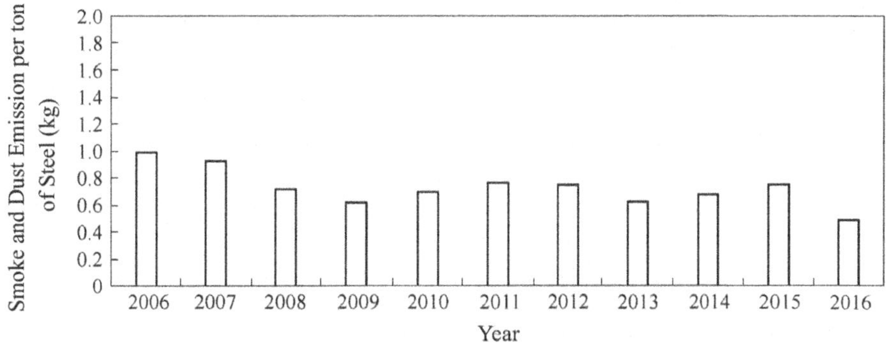


**Fig. 5.24** Fume and smoke emission per ton of steel in Delong Steel from 2006 to 2016

(2) Emission of Smoke and Dust per ton of Steel from 0.99 kg in 2006 to 0.49 kg in 2016, the fume and dust emissions per ton of steel decreased by 50.5%, as shown in Fig. 5.24.

### *5.3.6 Pohang Iron & Steel Co. Ltd. (POSCO)*

1. Concept of Greenness

Since the establishment of the plant, Pohang Iron & Steel Co. Ltd. (hereinafter referred to as POSCO) has attached great importance to environmental protection, advocated the development strategy of green industry, and put forward the slogan of "Clean and Green". In recent years, POSCO has focused its business strategy on environmental protection and actively taken active actions, such as establishing an environmental management system at the group level, minimizing emissions, improving eco-efficiency, piloting in low-carbon green growth, and publicizing environmental management results. The company plans to establish a "POSCO Certified Partner (PCP)" system to upgrade the environmental management to a new level. After these years of unremitting efforts, POSCO has become a veritable green iron and steel enterprise. Its greenness concept is mainly reflected in the following three aspects:

(1) Paying attention to environmental protection investment.
Since the establishment of the company, while investing in the construction of steel plants and the procurement of equipment, the company has been spending 9% of the total investment every year to purchase environmental protection equipment correspondingly, thus ensuring the simultaneous development of environmental protection and enterprise production.

(2) Paying attention to the development of environmentally friendly technologies
POSCO has always attached great importance to the development of environmental technologies. The new ironmaking technology FINEX is an environmentally friendly smelting technology developed by POSCO. With this technology, the equipment investment cost can be reduced to 92% of that for BF ironmaking, the steel production cost is reduced to 83%, and the pollutants such as dust and $SO_2$ generated during raw materials processing are reduced to one-tenth of that in BF ironmaking process.

(3) Developing a high-standard environmental indicator system.
In order to coordinate development with the surrounding environment and live in harmony with nearby residents, POSCO has set its own environmental standards, which is far higher than the standards promulgated by the Korean government. Within the area of more than 20 $km^2$ around the Pohang and Gwangyang plants, there are totally 106 atmospheric monitoring points, which provide the real-time monitoring results to the surrounding residents, so as to gain their support and recognition.

2. Measures of Greenness

   (1) Implementing Green Product Design.
   The greenness implemented by POSCO advocates that in the process of product design, the environmental protection must be considered to reduce resource consumption and achieve sustainable development strategies, meanwhile the commercial interests of POSCO should be also considered to reduce costs and potential responsibility risk, thus increasing competitiveness. The specific approach is to pay attention to the future modifiability, upgrade accessibility and variable product possibility in product design, providing a substantial opportunity to reduce solid waste pollution. In product design, one shall achieve the goal of renewing the product by simply redesigning some parts, so as to reduce solid waste. In the product design, it should also be considered to make sure that less materials or more energy-saving ingredients are consumed during the production. Priority should be given to replace the more toxic raw and auxiliary materials with the non-toxic, low-toxic, and less polluting raw and auxiliary materials to prevent the harm of raw materials and products to humans and the environment.

   (2) Implementing the Overall Green Production Process.
   To control the greenness during the whole process, it requires POSCO to adopt the production technology generating less waste or even no waste at all, as well as efficient production equipment, and minimize the use of toxic and harmful raw materials. Risk factors and toxic and harmful intermediate products in the production process should be minimized, and easy and reliable operation and control methods are adopted. Good hygiene practices, hygienic standard operating procedures, and hazard analysis and critical control points are established. Material recycles are achieved, and a comprehensive quality management system is established. Production organization

is optimized, and necessary pollution control measures are implemented to achieve clean and efficient utilization and production.

(3) Implementation of Material Optimization Management.
Material optimization management is an important part of POSCO's implementation of greenness. Choosing materials carefully and measuring the life cycle are the important aspects for improving material management. POSCO focuses on the reuse and recyclability of materials to be selected and implements rational closed circulation of materials, mainly including material flow during recycling of raw materials and products, that during product use, and that during product manufacturing processes flow.

(4) Paying Attention to the Development of Green Technology.
The development and application of environmentally friendly technologies can reduce pollution and save costs, which is a two-pronged thing for POSCO. The commercial promotion and application of these technologies have brought huge benefits to POSCO and iron and steel enterprises around the world.

3. Effect of Greenness

In order to reduce the total amount of air pollution emissions, POSCO first sets a target emission concentration for each equipment and then managed the pollutant emissions in real time. It also put cleaning equipment into operation to reduce pollution and effectively managed the total amount of air pollution discharged through those equipment. At the same time, POSCO also launched the "Plan on Dust Emission Reduction" to reduce the dust generated in the stacking, transferring and transportation, and replaced the aging sprinkler pipes and expanded the windbreaks, and took many dust reduction measures, such as enclosing, for transferring facilities. Through several years of efforts, remarkable results were achieved, and the main environmental indicators have improved significantly.

(1) $SO_2$ Emissions per ton of Steel.
The $SO_2$ Emissions per ton of Steel is reduced from 0.7 kg in 2007 to 0.56 kg in 2014, down 20%, as shown in Fig. 5.25.

(2) Emission of Smoke and Dust per ton of Steel from 0.22 kg in 2007 to 0.10 kg in 2014, the smoke and dust emission per ton of steel decreased by 60%, as shown in Fig. 5.26.

(3) $NO_x$ Emissions per ton of Steel from 1.22 kg in 2007 to 0.89 kg in 2014, $NO_x$ emissions per ton of steel decreased by 27%.

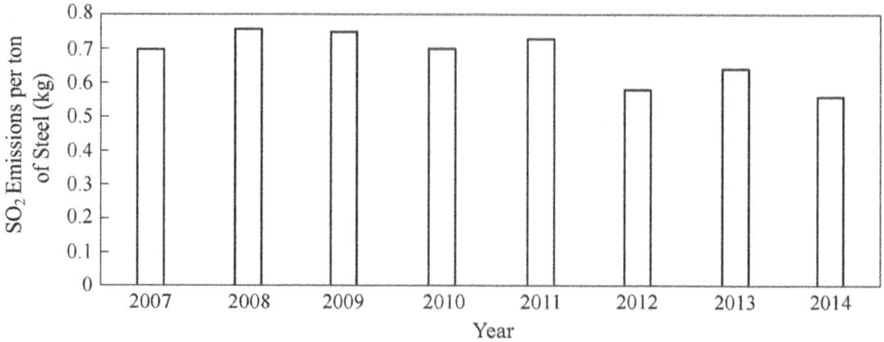


**Fig. 5.25** Diagram on $SO_2$ Emissions from 2007 to 2014 of POSCO

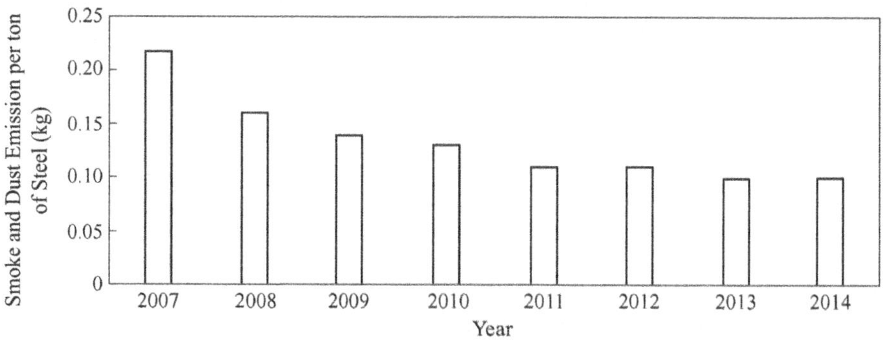


**Fig. 5.26** Smoke and dust emission of POSCO from 2007 to 2014

## 5.4 Prospect and Path Analysis of Greenness Trend

### *5.4.1 Prospects for Greenness*

The iron and steel industry has experienced the initial stages featuring the end-of-pipe treatment, energy conservation and emission reduction, and clean production. The industry will seize the important strategic opportunity of the 13th Five-Year Plan for the transformation of the economic development mode and the promotion of the green development model to achieve sustainable development with reduced cost and improved efficiency, environmental protection, and high quality, in accordance with the requirements of promoting ecological civilization construction.

Green development is a brand-new business philosophy and model in response to the requirements for promoting ecological civilization construction. As the largest and most widely used raw material producing industry, the iron and steel industry is the most promising and most qualified practitioner of green business. Building green iron and steel enterprises with high efficiency, low consumption, low emissions, and beautiful environment will be the fundamental goal of the iron and steel industry to achieve green development. Greenness transformation will become an important magic weapon to resolve the three challenges: "Resource-energy constraint, environment-ecological constraint and market-brand constraint". Thanks to the leverage effect of green development and ecological civilization construction, under guidance of far-sighting strategic goals, inertial thinking shall be overcome, strategic reflection shall be promoted, and the right direction and ways shall be found to gradually help the iron and steel industry reborn and out of the winter through innovation and reformation.

## 5.4.2 Path Analysis

To achieve a high level of green development, China's iron and steel industry has no other options but adhering to the Six-in-One green development concept of "Green mine, green procurement, green logistics, green manufacturing, green products, and green industry" (Fig. 5.27) so as to gradually achieve the overall upgrade in green manufacturing, product greening, and corporate greening. China's steel industry

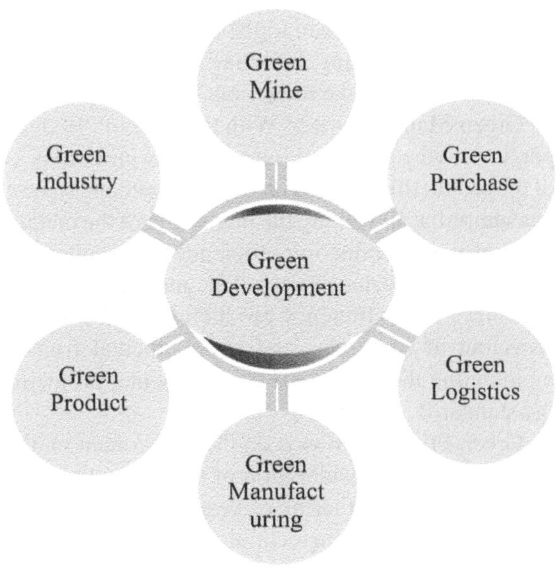


**Fig. 5.27** Six-in-One green development

should strive to achieve a higher level of green transformation in the innovation breakthrough of industrial ecological chain with the vision of development and strategic thinking.

Green Mining: In accordance with the government's requirements on the green mines, the concept of ecological civilization should be implemented throughout the development and utilization of iron mines, and the scientific and orderly mining should be strictly implemented to enhance the utilization efficiency of iron ore resources. At the same time, soil and water conservation and land reclamation should be carried out in the whole process of mine development, the disturbance to the mining area and surrounding environment should be minimized to ensure that the mine is always in harmony with the surrounding environment and integrated into the sustainable development track of society.

Green Procurement: The green procurement concept shall be established and integrated into the business strategy, and the green procurement guideline shall be prepared and continuously improved, and the procurement standards, procurement system, and procurement process shall be optimized and revised, and the items imposing less environmental impact shall be preferred in procurement, and the upstream suppliers for the iron and steel industry shall be encouraged to integrate environmental improvement performance into the whole process of business management, and a green procurement performance evaluation system shall be established, so as to consciously take up social responsibility and realize the transition from green procurement management to green supply chain management.

Green Logistics: Based on all logistics links throughout the whole iron and steel industry chain, with the aim of reducing pollution to the environment and reducing resource consumption, using advanced logistics facilities, management, service, and equipment technologies, on the premise of considering overall and long-term interests, a full-process green logistics operation system for transportation, warehousing, loading and unloading, handling, packaging, processing, and distribution should be established to achieve reduction, efficiency, and cleanliness of logistics.

Green Manufacturing: With the goal of "Reducing consumption, increasing output, improving efficiency, and reducing emissions", taking the reducing of the impact of the production process on the environment as the core, focusing on reducing energy consumption, and taking the increasing of the metal yield rate as the guideline, relying on improving production efficiency and supporting by strengthening recycling, the cost should be strictly controlled, and the quality and efficiency should be improved to achieve high-efficiency production, consumption reduction, minimum emissions, recycling of resources, and environmental friendliness on the basis of resolutely maintaining the red line of environmental protection and energy consumption up to the standard.

Green Products: Based on the assessment on the whole product life cycle and in comprehensive consideration of the energy consumption and environmental impact of steel products in terms of design, production, processing, application, and recycling, the investment in technology, capital, and manpower shall be strengthened, and the steel products with less environmental impact and less energy consumption in the whole life cycle of products shall be developed and produced, and the output and proportion of energy-saving, material-saving, and environmentally friendly green products shall be increased to provide high-quality green products to the society and provide green solutions to the customers.

Green Industry: By virtue of the collaborative innovation capability of manufacturer-college-institute-user, the foreign advanced technology in the fields of comprehensive utilization of resources and energy conservation and environmental protection shall be introduced. The transformation of self-dependent intellectual property rights shall be strengthened, and the high-end research and development strength of universities and research institutes shall be integrated, despite of the difficulties, so as to build a first-class domestic green technology industry export service platform and a professional operation management platform for project investment and financing, thus achieving the Three-in-One green industry development of technology output, design and construction, operation and management.

## 5.5 Industrial Practices of Greenness

China Metallurgical Industry Planning and Research Institute (hereinafter referred to as MPI) possesses a team of various specialties and experienced experts and has established a Green Development and Research Center for Iron and Steel jointly with the Environmental Engineering Evaluation Center of the Ministry of Environmental Protection. In recent years, MPI has fully played its role as the government's advisor on formulating the green development policy for the iron and steel industry, as the leader on green development of iron and steel industry, and as the think tank on green development of iron and steel enterprises. MPI has done a lot of work in promoting green development of iron and steel industry and has made gratifying achievements in promoting clean production, strengthening pollution control, realizing green transformation and sustainable development, creating ecological factories, and promoting harmonious integration with the city, as shown in the Table 5.1.

**Table 5.1** Practices of MPI in promoting greenness of iron and steel industry

| No. | Type | Main contents | Typical cases |
|---|---|---|---|
| 1 | The government's advisor on formulating the green development policy for the iron and steel industry | MPI has contributed to the formulation of standards, technical policies, and industrial policies regarding energy conservation, environmental protection, low-carbon and green development for the National Development and Reform Commission, the Ministry of Industry and Information Technology, the Ministry of Environmental Protection, and the China Iron and Steel Association | To mention a few, *National Implementation Scheme and Plan on Pollution Reduction of Iron and Steel Industry*, *Special Plan for Sintering Flue Gas Desulfurization in the Iron and Steel Industry*, *Recent Action Plan and Long-Term Strategic Planning on Dioxin Reduction of China's Iron and Steel Industry*, *Study on Feasible Technology of Pollutant Control and Verification Method of Pollutant Discharge Permit of China's Iron and Steel Industry*, *Manual of Determination of Fixed Pollution Sources Intensity (Iron and Steel)*, *Special Report on the Status of Carbon Dioxide Emissions in China's Iron and Steel Enterprises*, *Research on the Countermeasures for Promoting the Relocation of Urban Steel Plants for Environmental Protection and the Layout Optimization of the Iron and Steel Industry*, *Management System and Policy Research on Industrial Water conservation*, *Research on the Policy of Closing Down Outdated Production Facilities and Environmental Protection in Iron and Steel Industry*, etc. |

(continued)

**Table 5.1** (continued)

| No. | Type | Main contents | Typical cases |
|---|---|---|---|
| 2 | The think tank on green development of iron and steel enterprises | MPI has provided consulting services in energy conservation planning, environmental planning, clean production planning, green development action plans, and eco-factory establishment planning for iron and steel enterprises in their efforts of performing energy-saving diagnosis and environmental diagnosis | *Clean Production Promotion Plan of Hua Ling Xiangtan Steel*, *Plan on "Two Model" Demonstration Enterprise of Tangshan Steel*, *Energy Conservation Planning for Wuhan Steel, Benxi Steel, Laiyang Steel and Other Iron and Steel Enterprises*, *Energy Audit of Nanjing Steel, Wuhan Steel, and Hengshui Pipe and Other Iron and Steel Enterprises*, *Sustainable Development Report and Planning for Shougang Relocation*, *Environmental Protection Planning for Fangda Special Steel, Nanjing Steel, Shaanxi Steel and Xiangtan Steel* and *Circular Economy Planning for Anshan Steel, Wuhan Steel and Shagang* |
| 3 | The leader on green development of iron and steel industry | MPI has led the green development of iron and steel industry by proposing and promoting advanced green development concepts, establishing a green development benchmark, and promoting advanced pollution control technologies in the industry | MPI has taken the lead on promoting full implementation of sintering flue gas desulfurization and the concept of environmental cost in the iron and steel industry, on issuing Green Rating of China's Iron and Steel Enterprise, on establishing a green enterprise benchmark, on issuing *White Paper on Environmental Protection of China's Iron and Steel Industry*, so as to guide the society to accurately understand the environmental protection work of China's iron and steel industry |

## References

1. Ministry of Industry and Information Technology. Notice of the Ministry of Industry and Information Technology on Printing and Distributing the Adjustment and Upgrade Plan of Iron and Steel Industry (2016–2020) [2016-11-14]. http://www.miit.gov.cn/n1146285/n1146352/n3054355/n3057569/n3057573/c5353862/content.html.
2. China Iron and Steel Association. (2005–2015). Environmental Protection Statistics of China's Iron and Steel Industry (Internal Information).
3. Cheng, J.J., & Qin, F.X. (2016). Water-saving technology for iron and steel industry (Vol. 55). Nanjing Hohai University Press.
4. Baosteel Group. (2010–2015). Social Responsibility Report of Baosteel Group (Internal Information).
5. TISCO. (2010–2015). Social Responsibility Report of TISCO (Internal Information).
6. Office of the State Council. Implementation Plan for Permit System of Controlling Pollutant Discharge [2016-11-21]. http://www.gov.cn/zhengce/content/2016-11/21/content_5135510.htm.

# Chapter 6
# Coordination

The connotation of "coordinated development" of the iron and steel industry means consistent industrial policies, uniformed policy standards, operable policy implementation, a fair and competitive market environment and guides members in the iron and steel industry to a common platform to fully compete and healthily develop. Therefore, the book focuses on the coordination of the iron and steel industry from the perspective of industrial policy guidance and the establishment of a fair market environment. It mainly includes three aspects: first, the improvement of the industrial policy system and the establishment of a long-term mechanism for cutting overcapacity; second, to promote competitiveness as the core and promote the reorganization of iron and steel enterprises; focus on supporting the combination of dominant iron and steel enterprises and the reorganization of the enterprises with specific market-featured products to create world-class steel enterprise groups with international competitiveness; the third is to resolutely crackdown on the production and sales of substandard steel and counterfeit and illegal authorized products, so as to force the production capacity that does not meet the requirements of laws, regulations, industrial policies, and relevant standards to exit from the market. In this way, can we create a fair and competitive market environment, restore the market in good order, and promote healthy development of the iron and steel industry.

## 6.1 Symptom of Disordered Iron and Steel Industry

The year 2000 has witnessed the rapid development of China's economy and accelerating urbanization, thus drastically expanding market demand for iron and steel products and momentous development of iron and steel industry. For example, during the 10th Five-Year Plan period, China's economy was growing at an average annual rate of nearly 9.8%, and the urbanization rate was increased by 1.36% per year. The crude steel output was boosted by 2.75 times during the same period, with an average annual growth rate of 22%. During the 11th Five-Year Plan period, the average annual growth of China's economy reached almost 11.4%, and the urbanization rate was



X. C. Li, *The Road Map of China's Steel Industry*,
https://doi.org/10.1007/978-981-15-2074-7_6

enhanced by 1.39% on an annual basis. During the same period, the crude steel output was increased by 1.81 times with an average annual growth rate of 12.6%. During those years, about 50 million tons of iron and steel production capacity was added every year. What is more notable is the fact that in 2012 the increased volume was 89.56 million tons and in 2013 the number even reached 120.97 million. The long-term extensive growth has spawned a large number of illegal production capacities that violates environmental protection, quality, safety, and land regulations and has seriously disrupted the market order. Moreover, the mechanisms for supervision and penalties as well as exit of backward production capacity were not well-established, the inefficient capacity and zombie enterprises were hard to be expelled based on the market selection, the industry lacked sufficient self-discipline, and the market competition was in chaos. All these have aggravated the vicious competition in the market.

A disordered market was manifested in three aspects specifically: First, the number of enterprises is huge; however, the degree of industrial concentration is low; and the homogeneous competition is becoming increasingly fierce. The decisive role of market in resources' allocation cannot be effectively exerted, which makes the reduction of production, salary, and staff, and even the suspension of production becomes the new features of the development of the iron and steel industry. For example, the suspended production capacity of China reached 93 million tons in 2015, with major large- and medium-sized iron and steel enterprises suffering a total loss of 64.534 billion yuan throughout the whole year, and particularly the main business loss amounted to 112.663 billion yuan, with a scale of losses of over 50%; the second is the common fact that enterprises in the iron and steel industry extensively used excessively low prices to seize market shares, as many say: "The enterprises are hard to live and even hard to die, because they are in a vicious price war and hold the last breath on low price. That is so common in the entire iron and steel industry and the biggest obstacle in the way of capacity reduction". For example, the cost of stainless steel scraps, the main raw material for ordinary stainless steel, is 16,000 yuan/ton, plus 1,500–2,000 yuan/ton as the processing cost, requiring at least 17,500–18,000 yuan/ton to produce finished products. But during a period of time, it has become a common phenomenon that the steel plants only ask for 17,000 yuan/ton. Third, in order to get huge profit, some enterprises produce, sell, and use the substandard steel in large quantity, or illegally use labels of other brands and counterfeit them in production and sales. The existence of the huge numbers of substandard steel in market aggravates overcapacity, and what is more serious is that the fake and poor quality steel products directly incur the effect of "bad money drives out good". For example, Henan Province has the annual re-bar sales of about 8 million tons from Anyang Steel; however, actually Anyang Steel only produces 1.5 million tons of re-bar each year.

## 6.2 Ways to a Coordinated Iron and Steel Industry

### 6.2.1 *Guarantee of Policy System for Iron and Steel Industry*

In the process of rapid development of China's iron and steel industry, fueled by the continuous technological progress and the gradually upgraded energy conservation and emission reduction standards, China's iron and steel industrial policy system has been unceasingly improved and developed, covering all aspects including eliminating backward and excessive production capacity, enterprises' merger and reorganization, industrial planning, standardized market access, environmental protection, energy-saving and resources guarantee, thus guiding the industry to develop in an ordered way and promoting the industrial transformation and upgrading. Among them, the elimination of backward and excessive production capacity is significant throughout the whole course of the industrial development and also becomes the epitome of the development of policies for the entire iron and steel industry. Especially, consistent improvement of comprehensive criteria and requirements on environmental protection, energy consumption, safety, quality, etc., has become an essential need for the development of enterprises in the industry and legitimate elimination and reduction has become the most effective and fair means to resolve the excessive capacity.

Under the new normal, China's economic growth has transformed from high speed to medium–high speed. With the transformation of economic structure, development impetus, and modes, the consumption intensity of steel per unit of GDP has decreased significantly; the consumption of steel has been over the peak value and started to reduce. The "double reduction" of production and consumption indicates that China's steel industry has entered a new era of development with reducing output. The serious overcapacity has increasingly become a prominent contradiction and the root of many problems in China's economic development. In December 2015, the Central Economic Work Conference proposed "cutting overcapacity, destocking, deleveraging, reducing corporate costs, and shoring up weak spots", ranking "cutting overcapacity" as the top of the five major tasks and especially emphasizing the decapacity of the steel industry as an important area and pilot to promote supply-side structural reform. *Opinions on Resolving Excessive Production Capacity of the Iron and Steel Industry to Realize Profitable Operation and Development* (No. 6 [2016] of the State Council) and eight supporting documents were released, together with special actions to curb illegal construction projects, eliminate backward production capacity, and carry out joint law enforcement, to form a relatively complete "1 + 8 + 3" industrial policy system, opening a new chapter in the healthy and orderly development of the iron and steel industry.

The supply-side structural reform is not based solely on equipment size as the basis for cutting overcapacity, but relies on strict enforcement on environmental protection, energy, quality, safety, and technology, defining the five bottom lines which are mandatory to force the capacity entities that do not comply with laws and industrial standards to exit from the market. That guides the iron and steel enterprises to gain development space from environmental protection, to get benefits from energy saving, to seize market with superior quality and renowned brand, to guarantee their development with safety production, and to lay development foundation with advanced process and technology and leads the iron and steel enterprises to a healthy and sustainable path of development. The schematic diagram of policy system for cutting overcapacity in China's iron and steel industry [1] is shown in Fig. 6.1.

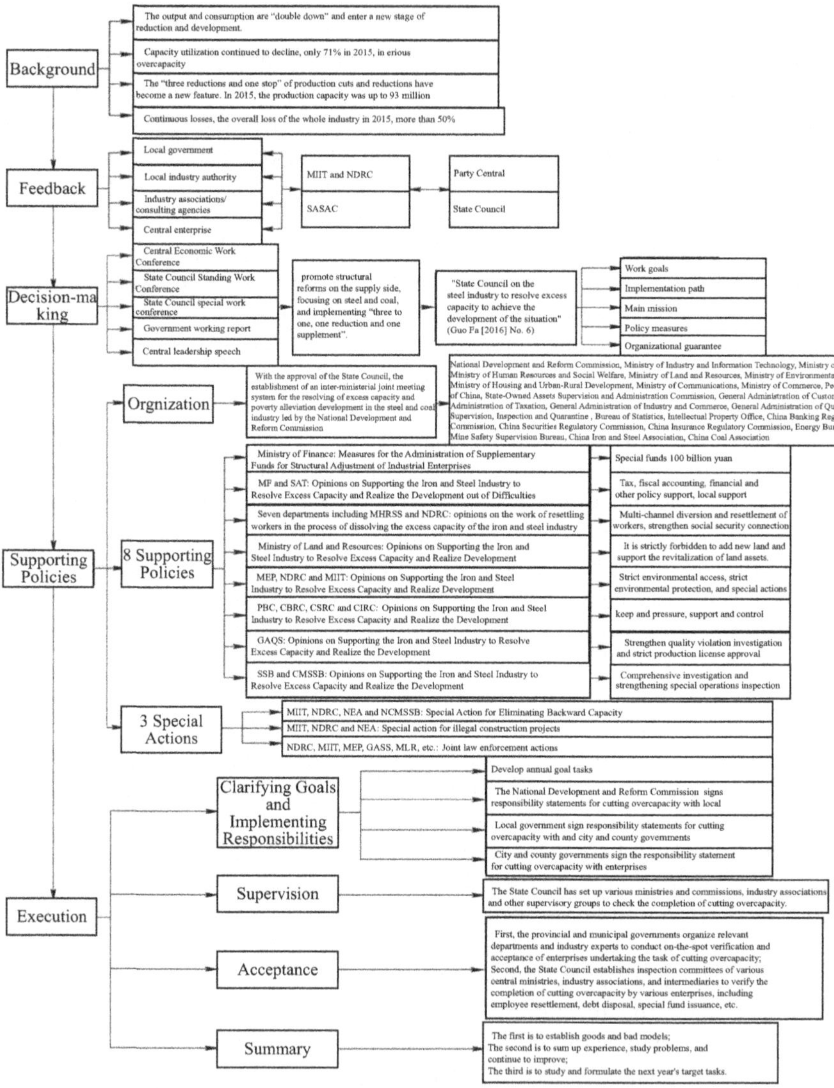


**Fig. 6.1** Schematic diagram of policy system for cutting overcapacity in China's iron and steel industry

1. Problems in Policy System of Iron and Steel Industry

In the current policy system of iron and steel industry, there are four problems that are urged to be resolved.

First, the standards are not uniform. For example, the *Guidance Catalogue for Industrial Restructuring (2011)* (Revised in 2013) and the *Standard Conditions for Iron and Steel Industry* (Revised in 2015) are inconsistent with the nominal capacity of new converters and electric furnaces; inconformity also emerges in calculation of production capacity on the same specification between the *Notice on Issuing of Methods to Implement Capacity Replacement in Industries with Seriously Excessive Production Capacity by Ministry of Industry and Information Technology* and the *Notice on Issuing of Eliminating Illegal Projects in Iron and Steel, Electrolytic Aluminum and Shipbuilding Industries by National Development and Reform Commission and Ministry of Industry and Information Technology* as well as other inconsistencies.

The second is the relationship between the long-term mechanism and the special action plan. At present, there is no sufficient or complete long-term mechanism guiding healthy development of the iron and steel industry, but is led by a series of specific actions sush as backward capacity elimination, illegal projects cleanup, joint law enforcement, which resulted in huge cost on the one hand, and only can be deemed as short-term behavior instead of promoting the development of iron and steel industry in the long run and overall.

The third is the establishment of a fair and competitive market environment and the issue of giving play of energy efficiency. State-owned iron and steel enterprises have taken on more social responsibilities which especially are the historical burdens. That means they have played a more active role in taxation, environmental protection, staff resettlement, etc. While private steel enterprises have already supported the other half of China's iron and steel industry, although many have been blamed for poor environmental protection, low product quality and tax payment, etc., the private enterprises have paid more in financing and getting loan compared with the state-owned enterprises.

The fourth is the problem of multiple management authorities and overlapped function. In September 1956, the Ministry of Metallurgical Industry was established. It is the functional department under the State Council in charge of the national metallurgical industry. The functions of governmental authorities and enterprises are separated. With the principles of grasping macro issues while letting free the micro ones, the Ministry has simplified its structure and transformed its functions to manage the metallurgical industry in industrial planning, coordination, supervision, service, etc. In March 1998, the State Council's institutional reform program reorganized the Ministry of Metallurgical Industry into the State Bureau of Metallurgical Industry and became the administrative body under the administration of the State Economic and Trade Commission, responsible for the management of metallurgical industry. In 2001, nine state bureaus, including the State Bureau of Metallurgical Industry under the administration of the State Economic and Trade Commission, have been officially canceled in the reform of national authorities and organizations. The Bureau's administrative functions were incorporated into the State Economic and Trade Commission, and those of the other state bureaus were incorporated into the relevant departments and bureaus under the State Economic and Trade Commission. In March 2003, the institutional reform plan of the State Council was approved in the first session of the Tenth National People's Congress; the Ministry of Foreign Trade and Economic Cooperation and the State Economic and Trade Commission were thus revoked; instead, the Ministry of Commerce was established. The State Bureau of Metallurgical Industry was revoked in this reform. The functions of the former Ministry of Metallurgy were dispersed to the State-Owned Assets Supervision and Administration Commission, the China Iron and Steel Industry Association, the Ministry of Industry and Information Technology, and the National Development and Reform Commission. At this point, the era of only one department managing the industry has become the history, and the industry has started to be managed by multiple authorities. Since then, under the background of rapid national economic growth, local governments and enterprises have been in a fever of the iron and steel industry due to its features such as being capital-intensive, labor-intensive, and linkage-intensive, initiating an irreversible chaos in the rapid industrial development. During this period of time, the major objectives including the iron and steel industrial restructuring, capacity regulation, merger and reorganization failed to be well carried out or completed. Until the release of No. 7 Document of the State Council in 2010, an inter-ministerial coordination group for the elimination of backward production capacity, which was jointly participated by 18 departments, was established. Provincial coordination groups were established in various places. The unified management, unified office, and unified administrative work mode once again achieved great efficacy as various tasks were completed smoothly. On the basis of the Inter-Ministerial Coordination Group for the Elimination of Backward Capacity in 2015, an Inter-Ministerial Coordination Group for Resolving Overcapacity has been established, which continues to play an important role in cutting overcapacity and promoting enterprises to be profitable.

2. Improvement of Policy System of Iron and Steel Industry

The essential of perfecting the iron and steel industry system and giving full play to the industrial policy and guidance is to establish a fair market environment under the premise of unifying the standards, improving the construction of industrial mechanism and system and continuing to insist unified management. In other words, we should establish a fair and competitive market environment by means of rigorous law enforcement in aspects such as environmental protection, energy consumption, quality, safety and taxation, taking advantages of differentiated financial policies (that means giving fair treatment to the state-owned enterprises and the private enterprises, while imposing differentiated policies on the "zombie" and illegal enterprises). This shall be done under the joint supervision of government and industrial administrative authorities so as to better utilize the decisive function of the market in resource allocation.

(1) Strictly Control New Production Capacity. Controlling new production capacity is the key to solving the disorderly development of China's iron and steel industry, and it is also an important measure to optimize the stock based on mergers and reorganizations, resource integration, adjustment, and upgrading. Two tasks have to be completed for controlling of new production capacities: One is reduction and replacement of production capacity, and the other is rigorously cracking down on illegal construction projects.

1) Revision of Capacity Replacement Scheme. *The Notice on Issuing of Methods to Implement Capacity Replacement in Industries with Seriously Excessive Production Capacity by Ministry of Industry and Information Technology* (No. 127 [2015] of Ministry of Industry and Information Technology) [2] clearly stipulates that equivalent or reduction replacement shall be carried out with replacement scheme for construction of industrial projects that suffering serious overcapacity; in the areas with sensitive environment such as Beijing-Tianjin-Hebei, Yangtze River Delta, and Pearl River Delta, reduction replacement shall be implemented. The document is valid until December 31, 2017, and revised timely in accordance with the industrial development. *The Adjustment and Upgrading Plan for Iron and Steel Industry (2016–2020)* requires that it is forbidden to have net increase in iron and steel making capacity nationwide; reduction replacement of production capacity has to be carried out for structural adjustment and modification projects; reduction replacement has to be done for the projects scheduled or under construction that have been approved by the national government and filed in local authorities. However, the proportion of the reduction was not clearly defined.

The following issues have to be noted in the revision of capacity replacement scheme:

The first is to unify the standards for capacity calculation. In June 2015, the National Development and Reform Commission and the Ministry of Industry and Information Technology issued the *Notice on Issuing of Eliminating Illegal Projects in Iron and Steel, Electrolytic Aluminum and Shipbuilding Industries by National Development and Reform Commission and Ministry of Industry and Information Technology* (No. 1494 [2015] of the National Development and Reform Commission and Ministry of Industry and Information Technology), put forward clear suggestions to dispose the illegal projects that were under construction or constructed as well as projects out of the scope of disposal, and presented assessed production capacity. There are several problems here. First, there is no specific equipment corresponding to the assessed production capacity. Even if the equipment data can be found in the documents filed in the provincial, municipal, and county authorities, it is still impossible to clearly define the capacity of each blast furnace or converter; second, part of iron or steel making production capacity of enterprises is missed in the assessment; the third issue is that the production capacity assessment is way different from that stipulated in the "Sheet for Calculation of Production Capacity" in the *Notice on Issuing of Methods to Implement Capacity Replacement in Industries with Seriously Excessive Production Capacity by Ministry of Industry and Information Technology* (No. 127 [2015] of Ministry of Industry and Information Technology), puzzling the local government, enterprises, assessment organizations, and specialists and generating many different viewpoints; another situation is that the Ministry of Industry and Information Technology holds a different opinion with the National Development and Reform Commission, which incurs confusion in the scheme for production capacity replacement.

Therefore, the criteria to be implemented for the production capacity replacement scheme should be clarified. At the same time, the Sheet for Calculation of Production Capacity should be revised and improved. Another solution is to make holistic evaluation based on the Sheet for Calculation of Production Capacity, with comprehensive consideration of the upstream and downstream process flow and the selected process and technical route instead of solely consideration of the production capacity of a single blast furnace or converter.

The second is to further clarify the criteria defining the legitimate production capacity. *The Adjustment and Upgrading Plan for Iron and Steel Industry* (*2016–2020*) clearly stipulates that replacement of production capacity must not be applied to the production capacity abandoned before (including) 2015, the backward production capacity, the production capacity that is listed in the reduction task and the production capacity subsidized and supported by policy. But the problem is it is not clarified what kind of capacity can be used for replacement. First is the production capacity presented in Document No. 1494 legal and can it be replaced as long as it does not belong to the "four forbidden situations" stipulated in the *Adjustment and Upgrading Plan for Iron and Steel Industry* (*2016–2020*)? Secondly, Document No. 1494 also clarified how to treat the projects "not covered in the scope of disposal"; that is, for the projects completed and put into operation before 2005, the relevant regions shall study and handle the registration procedures by themselves. At that time, many local governments and enterprises failed to report this part of production capacity in time due to various reasons. The country began to conduct a general survey of the national iron and steel production capacity in the second half of 2016 in order to comprehensively understanding the iron and steel production capacity of China. If this part of production capacity is not in the "four forbidden situations", can it be used for replacement? Third are the three batches of more than 300 iron and steel enterprises that comply with the specified requirements announced by the Ministry of Industry and Information Technology as well as the enterprises whose access announcements have been revoked at the beginning of 2017 allowed for replacement if they are not within the "four forbidden situations"?

Therefore, the above situation should be further clarified so as to unify the standards and requirements.

The third is to break the regional restrictions on trading of production capacity. The core target for the replacement of production capacity is to improve the technology and equipment level of the entire iron and steel industry by means of modification, upgrading and reduction replacement, practically integrating cutting overcapacity with transformation and upgrading. To optimize the distribution of the iron and steel production capacity is a significant direction for the development of the industry, and the trading of the production capacity is an important means. However, the capacity replacement scheme has always been controversial, especially in inter-regional trading. The problem is a lack of holistic planning covering the whole nation. In other words, all local governments and authorities are assigned the tasks to cut overcapacity with different intensity based on regional capacity proportion, and the total production capacity of iron and steel in each region is not permitted to increase but only to decrease, thus restricting the capacity within areas where it is supposed to be transferred out and banning the desired capacity from being brought in.

Therefore, except for some areas where the environmental capacity is extremely sensitive, trading of production capacity should focus on whether the national capacity is reduced by capacity replacement, rather than setting a hard threshold only to stem the increasing of production capacity in the region.

The fourth is to establish a capacity replacement database. The lack of indicators for production capacity and the asymmetry of capacity replacement data result in uncertainties in practical outcomes of overcapacity eliminating in the replacement scheme and possibility of repeated elimination. Besides, there is no means to carry out precise and strict supervision, especially for plans of inter-regional capacity replacement; it is even harder to provide effective supervision. In this case, some enterprises may commit illegal operation.

Therefore, the national capacity trading platform should be given full play to and the capacity replacement database file should be completed to supervise the industry in a more accurate and effective way.

The fifth is to improve the evaluation part in the capacity replacement scheme. Choose a professional third-party organization to evaluate the capacity replacement scheme to find out whether the production capacity is legal, the replacement scheme is scientific and reasonable, and the replacement process is well-established. The government has selected a number of professional organizations through setting thresholds for the local government authorities or enterprises to choose them as the third-party evaluation organizations. If such organizations commit illegal behaviors in evaluation, they will be deprived of the qualification for evaluation and punished rigorously.

2) Rigorously Crackdown on Illegal Construction. In December 2016, the Standing Committee of the Political Bureau of the CPC Central Committee debriefed a report from the State Council on the investigation on Jiangsu Huada Iron and Steel Co., Ltd. and Hebei Anfeng Iron and Steel Co., Ltd. for their illegal operation as well as the handling work and stressed the cutting overcapacity in the iron and steel and the coal industries as significant contents in both deepening of supply-side structural reform and implementation of "cutting overcapacity, destocking, deleveraging, reducing corporate costs, and shoring up weak spots". The investigation revealed that Jiangsu Huada Iron and Steel Co., Ltd. was producing and selling the substandard steel, and Hebei Anfeng Iron and Steel Co., Ltd. was building iron and steel projects without prior approval. The Committee urged to thoroughly investigate and punish the two companies as well as the persons in charge and report the results to the public.

In order to enforce the disciplines of the party and the country and ensure effective implementation of policies from the Central Government, the Party Central Committee and the State Council decided to rigorously investigate and dispose these two cases and the persons in charge. First, the Provincial Governments of Jiangsu and Hebei provinces were instructed to make in-depth self-inspections and report to the State Council. Second, the 111 responsible persons in Jiangsu and 27 persons in Hebei, including the vice provincial governors, were investigated for dereliction of duties. Third, Jiangsu Province was instructed to thoroughly dispose the illegal operations in the whole province including sales of the substandard steel and the new production capacity; Hebei Province was instructed to dismantle, within a time limit, all outdated blast furnaces with capacity less than 1,000 $m^3$ and converters with capacity less than 100 tons belonging to Hebei Anfeng Iron and Steel Co., Ltd. Fourth, the investigation and punishment on Jiangsu Huada Iron and Steel Co., Ltd. and Hebei Anfeng Iron and Steel Co., Ltd. were reported to the public nationwide. Fifth, the State Council would organize special supervision and rectification on backward production capacity in the coal, iron and steel, cement, glass, and other industries. These two issues should be regarded as negative examples for warning and education, helping to ensure smooth progressing of the tasks of cutting overcapacity and elimination of backward production capacity.

In December 2016, the Office of Inter-Ministerial Meeting on Resolving Overcapacity and Profitable Development for National Iron and Steel and Coal Industry reported issues of Jiangsu Delong Nickel Industry Co., Ltd. illegally constructing iron and steel projects [3]. The investigation showed that in 2013, Jiangsu Delong Nickel Industry Co., Ltd. implemented an expansion project with annual capacity of 300,000 tons of nickel alloy and illegally built an 80-ton electric furnace, four 60-ton AOD furnaces, and related supporting facilities to produce stainless steel. Jiangsu Delong boldly broke the laws and regulations for its illegal construction under the background of the Party Central Committee and the State Council vigorously promoting the supply-side structural reform, repeatedly giving orders to curb illegal production capacity, and all relevant parties focusing on resolving the overcapacity in the iron and steel industry. While completely dismantling the illegally constructed stainless steel smelting equipment, the relevant responsible persons were seriously held accountable.

The zero-tolerance attitude, the sustained high-pressure policies, and the strict accountability for illegal construction have completely eradicated some opportunistic local governments and enterprises and further unified people's thinking and built consensus. That has laid a solid foundation for improving the supply-side structural reform as well as healthy and orderly development of the iron and steel industry.

(2) Environmental Protection. The iron and steel industry features heavy emission which always draws common and intensified attention concerning environmental protection from the whole society. After years of unremitting efforts, the iron and steel enterprises have invested a lot of money in environmental protection and the entire industry has seen an obvious improvement in energy-saving and emission reduction. The total discharge of major pollutants has been effectively controlled, and the industry has stood at the starting line of green development. However, the total quantity of energy consumption and pollutant emission accompanied by the large production capacity accumulated because of the rapid development of the iron and steel industry remains too high. According to statistics, in terms of air pollutant emissions, sulfur dioxide emission from ferrous metal smelting and rolling industries accounts for 12% of total industrial sulfur dioxide emission, and nitrogen oxide emission accounts for 10% of total industrial emission, and fume (dust) emission accounts for 32% of the total. That means the iron and steel industry causes the most severe dust particle pollution in the air. In 2015, the emissions of sulfur dioxide, fume, and industrial dust from China's major iron and steel enterprises were 472,000 tons, 153,000 tons, and 297,000 tons, respectively. The revision and improvement of environmental protection policies and the development of innovative model of enforcement of environmental protection laws are not only important means to improve regional environmental quality and promote the city-industry integration involving iron and steel enterprises, but also significant measures to create a fair and competitive market environment and promote the transformation and upgrading of the iron and steel industry.

1) Strictly implement the new environmental impact assessment law (hereinafter termed as "the EIA Law") and the new pollutant discharge permit system to comprehensively improve the supervision on environmental protection. The first is the new environmental impact assessment law. The latest revised *Environmental Impact Assessment Law* was reviewed and approved at the 21st Meeting of the Standing Committee of the Twelfth National People's Congress on July 2, 2016, and officially implemented since September 1 of the same year. *The EIA Law* has been revised in nine aspects, of

which the emphasis is laid on more severe punishment for illegal construction without prior official approval; namely, a penalty of 1–5% of the total investment was imposed. At the same time, it has simplified the administrative review and approval of EIA for some projects and strengthened the EIA on planning. The new version of *Environmental Impact Assessment Law* stipulates that "entities commencing construction projects at their own without officially approved statements or report tables on environmental impacts of construction projects or fail to re-submit documents for approval or review on the statements or report tables as per Article 24 under this Law before commencement shall be ordered to stop construction by the competent department of environmental protection above county level. The involved entities shall be fined an amount of 1–5% of the construction project's total investment and ordered to restore the site. The management personnel holding direct responsibilities and other directly responsible persons shall be imposed with administrative sanctions". The revision changes the convention that all illegal construction projects, no matter large or small, are punished similarly within an identical range; instead, the specific construction status will be considered as the basis to intensify the punishment on large-scaled construction projects that commence the work without prior approval. Moreover, the involved parties may be ordered to restore the site other than to terminate construction and pay a fine. That will cause huge economic losses to those illegal construction projects that have already been commenced.

At the same time, the subjects of law enforcement that are responsible for illegal construction project without permission have been changed from the competent environmental protection departments which are entitled to review and approve the EIA Statements to those above the county level. That has stressed the territorial principle in market supervision and law enforcement and changed the situation that the local environmental protection departments have no enforcement power over the projects approved by the higher environmental protection departments. In addition, the revision imposes more severe punishment on the planning departments for their illegal behaviors as it specifies that "any persons in direct charge of the planning authorities who fail to organize or practice fraud in EIA or misconduct duty and thus make the EIA way inconsistent with the facts shall be imposed with disciplinary sanction according to the law by the higher authorities or the supervisory departments, so shall the others with direct liability". The new version of *Environmental Impact Assessment Law* also stipulates that the EIA administrative examination and approval are no longer a precondition for the approval of the feasibility study report or the project approval. The EIA approval and feasibility study report approval or project approval shall be carried out simultaneously, but still, they have to be completed before commencement of project. At the same time, it stipulates that "the EIA Statement and the Statement Form of the construction project shall be submitted by the construction unit in accordance with the provisions of the

State Council for approval by the environmental protection administrative department with the power of examination and approval". "China conducts administration of recordation of environmental impact registration form". This deepens the streamline administration and institute decentralization and optimizes the approval process as the examination and approval of water and soil conservation plan by the competent administrative department are no longer a precondition for EIA. The revision of *Environmental Impact Assessment Law* further embodies the reform philosophy in EIA approval of simplifying the preliminaries and strengthening the supervision during and after the approval. That helps to boost administrative efficiency and gives full play to the macro-control. The introduction of the new EIA regulations has built up a more creative and vital market environment as it has shortened the procedure and time for approval and saved time and financial costs for enterprises.

The second is the new pollutant discharge permit system. As the *Plan for Implementation of Permit System for Controlled Pollutant Discharge* (No. 81 [2016] of the State Council) required, the management of pollutant discharge permits has been official launched since 2017 in thermal power, papermaking, and iron and steel industries. Meanwhile, in order to promote the air pollutant control in Beijing-Tianjin-Hebei Region, the management of elevated pollutant discharge permits is piloted in some cities within the Region.

On December 28, 2016, specific work targets were set in the *Notice on Launching Management of Pollutant Discharge Permits in Thermal Power and Papermaking Industries and Elevated Pollutant Discharge Permits in Beijing-Tianjin-Hebei Region* (No. 189 [2016] of MEP, NDRC, and MWR) officially released by the Ministry of Environmental Protection. The notice requires to complete the application of pollutant discharge permits by enterprises in thermal power and papermaking industries as well as the relevant examination and approval work before June 30, 2017; hence, the permits shall be used as the basis for environmental supervision and enforcement. The "1 + 2" key cities (Beijing, Baoding, and Langfang) which are located in the major air pollutant spreading channel in the Beijing-Tianjin-Hebei Region shall be the pilot cities where application, approval, and issuing of the elevated pollutant discharge permits for enterprises in the iron and steel and the papermaking industries shall be completed before the same date. From July 1, 2017, the relevant enterprises in operation are entitled to discharge pollutant only with the permits and shall establish their own systems for self-supervision, information disclosure, recording, and periodical reports. The pollutant discharge permits shall be issued in the pilot cities to the iron and steel complexes that have at least two or more processes including coking, sintering, pelleting, ironmaking, steelmaking, and steel rolling. Hence, a new pollutant discharge permit system for the iron and steel industry has been presented officially. The system improves the supervision on the pollutant discharge nodes and the way of calculating the discharge amount in all production procedures of the enterprises, instead of only to supervise the total amount of pollutant discharge as before, which is ambiguous, making the supervision on pollutant more professional and specific.

The new pollutant discharge permit system replaces a number of environmental protection functions and systems including "environmental statistics", "pollution charges", "total quantity control", "environmental monitoring", and "environmental standards", to realize "management with one permit". The environmental protection department will check the enterprises against all contents of the pollutant discharge permit. Any enterprises fail to meet the requirement and thus causing violation to the environmental laws will be punished continuously on a daily basis with production restriction or suspension, business termination, close-down, etc. The new pollutant discharge permit system is the core means to supervise and regulate the point source discharge control. It is important for improving the procedures of environmental supervision and regulation and the environmental supervision system, becoming an indispensable impetus to promote environmental protection.

2) Fully enforce the *Environmental Protection Tax Law* to raise the standard for environmental protection in the iron and steel industry. On December 25, 2016, the *Environmental Protection Tax Law* was voted to be promulgated at the 25th session of the Standing Committee of the 12th National People's Congress. *The Environmental Protection Tax Law*, as China's first specific law promulgated to promote ecological civilization construction, has come into effect since January 1, 2018. The nationwide enterprises discharging pollutants are required to fulfill their environmental tax liability without any difference, and the pollution discharge fees that had been implemented for nearly 40 years eventually became the history. From "fee" to "tax", environmental protection supervision is becoming more standard and fair. *The Environmental Protection Tax Law* stresses the principle of "to decrease tax by reducing discharge". That means the enterprises with less pollutant discharge are levied with lower tax. This new law also boosts the progress of the iron and steel enterprises' upgrading and reconstruction for better environmental profit since the iron and steel industry discharges large quantity of pollutants. Thus, the more environmentally friendly enterprises survive while the inferior ones are eliminated, and the phenomenon of "bad money drives out good money" will not happen. According to the *Environmental Protection Tax Law*, the enterprises, public institutions, and other production operators that directly discharge taxable pollutants into the environment shall be the taxpayers. The quantity of taxable pollutant discharge is the basis for taxation, in which the quantity of air and water pollutants is determined as per the pollution equivalent weight converted by the discharge amount, that of the solid pollutants are determined according to the discharge amount of solid wastes, and the noise shall be determined according to decibels exceeding the national standards. After the environmental

protection fee is transformed to tax, the collection department has been changed from the environmental protection departments to the tax authorities. Both of them cooperate with each other to establish a taxation mode featured with "declared by enterprises, levied by tax authorities, supervised by environmental protection departments and shared information". Besides, the environmental protection tax shall be part of the local tax and included to the general public budget to effectively curb local protectionism. As the Version 2.0 of the pollutant discharge fee, the environmental protection tax is not targeting at collecting fiscal revenue. Instead, it aims at building a better production and living environment for the people by reducing pollutant discharge from enterprises with the tax lever. As a major financial and taxation measure to promote ecological civilization construction, the environmental protection tax has improved China's "Green Tax" system. That means to provide taxation incentives to the iron and steel enterprises that actively practice green development while increasing the costs for environmental protection for those discharge pollutants illegally, thus to propel the iron and steel industry to upgrade and transform.

3) Vertical management and regularized supervision of environment protection shall be strictly carried out, and high-pressure situation of law enforcement for environmental protection shall be maintained continuously. 2016 was the year suffering from the most severe smog and witnessed the most rigorous iron and steel production restriction for environmental protection in history ever. The enterprises, especially for the urban steel plants, are experiencing an increasing impact from environmental protection. The environmental protection departments are conducting more regularized special supervisions. At present, the environmental protection cost of the key iron and steel enterprises in the statistics occupies almost 10% of the iron and steel production cost. The growing cost for pollutant discharge lays an increasing burden on the iron and steel industry that is in its "ice age" with low profit. That urges the iron and steel industry to upgrade and transform from a primitive way of development that sacrifices the environment for the benefit to an upgraded green industry mode. By doing this, the backward production capacity can be eliminated faster and the iron and steel industry will embrace an accelerating economic restructuring and transformation in the development mode. In such trend and with the breakthroughs of high-end technologies in the iron and steel industry over the recent years, revisions are supposed to be done on 8 prevailing standards which are officially enforced on October 1, 2012, on pollutant discharge including the *Emission Standard of Air Pollutants for Sintering and Pelletizing of Iron and the Steel Industry* and *Emission Standard of Air Pollutants for Iron Smelt Industry* (GB 28663–2012) in later the 13th Five-Year Plan period. That will bring the industry more rigorous standards for the air pollutant discharge such as particulate matters, sulfur dioxide, and nitrogen oxides as well as water pollutant discharge including COD and ammonia nitrogen from the key pollution nodes. As a general trend, green development requires less pollutant

discharge from fix sources in the iron and steel industry, a reduced total quantity of new pollutant discharge, synergy of the *Environmental Protection Tax Law* and the new pollutant discharge permit system in supervision and a profound policy guarantee with impeccable environmental protection laws and regulations. Fulfillment of all these requirements will boost the entire iron and steel industry to eliminate the backward production capacity and realize upgrade and transformation to green development.

Since 2017, the Ministry of Environmental Protection has been carrying on conducting activities for the Year of Environmental Protection Law Enforcement on the basis in the previous two years. The methodology shall be insisted on with the connotation of problem-oriented and multi-party coordination, strict supervision on both government and enterprises, and severe punishment on violators and standardized law enforcement. In 2017, the Ministry of Environmental Protection continued to complete inspection in the 15 provinces where the Central Environmental Inspection has not been done. Hence, the environmental protection inspection has covered all over China. Another important direction of work that needs to be followed in the future by the environmental protection departments is to be free from the outdated system and to strengthen the capability of law enforcement at the grassroot level. That means, in general, to gradually dissolve the entanglement of interests between the local governments and the environmental protection authorities that leads to unfair law enforcement, assisted by the reform in vertical management of environmental protection monitoring and supervision authorities below the provincial level. In the future, the Ministry of Environmental Protection will also improve the relevant laws and regulations to make them more practical and specific in their implementation. The Ministry of Environmental Protection will continue to put more efforts in law enforcement. That entails "zero tolerance" to violations to environmental protection laws and to maintain extremely strict to environmental supervision and law enforcement. By doing this, the environmental protection laws will be carried out pragmatically. *The Instruction on Elimination of Backward Production Capacity Legally with Comprehensive Standards* released recently gives full play to the new environmental protection law and supporting documents as very effective means to help eliminate the backward production capacity and further promote the supply-side structural reform.

(3) Taxation and Financial Credit. The taxation and financial credit policies not only reflect how an enterprise takes social responsibility, but also concern the survival and development of an enterprise. Moreover, they are significant means to build up a fair market where the good ones are promoted while the outdated are wiped out.

First, the irrational taxes, fees and fund policies of iron and steel industry have to be modified.

On the one hand, preferential policies for comprehensive utilization of resources shall have to be modified to practically encourage the iron and steel enterprises to develop a self-recycled way of comprehensive resource utilization. The current tax policy is based on the sales of products from comprehensive utilization of resources. Especially, the favorable corporation income tax can be reduced by 10% of the sales of products from comprehensive utilization of resources as specified. However, most of the comprehensive utilization products of the iron and steel enterprises, such as waste heat generation and recycling of industrial wastewater, are recycled internally instead of being sold in the market. Therefore, although the enterprises have invested a lot, still they cannot get benefits from the preferential policies. The recommendation is to carry out the taxation for comprehensive resource utilization as per the real situation and to optimize it with differentiated calculation mode. That means to convert the actual output of the comprehensive utilization products of the enterprises that are internally recycled and can be calculated with the current market price into sales revenue; for those recycled internally and unquantifiable, the actual production cost of the comprehensive resource utilization products can be taken as the sales revenue to calculate the amount of deduction in the current payable income tax.

Another work is to eradicate the irrational funds in the iron and steel industry. The iron and steel enterprises use the gas, waste heat, and residual pressure internally to generate power, which belongs to comprehensive resource utilization that can be exempted from major hydraulic construction funds, renewable energy development funds, etc., as per the previous policies. However, in 2010, the Ministry of Finance issued the *Interim Procedures for Collection, Utilization and Management of National Major Hydraulic Construction Funds*, starting to levy major hydraulic construction funds, renewable energy development funds, etc., which de-energized the iron and steel enterprises in the development of comprehensive resource utilization. It is suggested that the above-mentioned comprehensive utilization projects of resources for the iron and steel enterprises should continue to be exempted from major hydraulic construction funds and renewable energy development funds. Besides, the replacement of sales tax to value-added tax has to be accelerated, and the financing and insurance expenses of iron and steel enterprise shall be included within the range of input tax deduction. The iron and steel enterprises shoulder heavy financial burden due to their huge capital demands, high financing amount, and insurance expenses. However, contradictorily, the input taxes of their current interest and insurance expenses are not deducted; instead, all such costs have to be paid by the enterprises, which imposes heavy pressure of taxes and fees on them. Therefore, the recommendation is that the iron and steel enterprises shall be entitled to the deduction of input tax for financing and insurance expenses.

Second, the export rebate rate of the steel products with high added value shall be adjusted.

- Pickled Sheet
  The pickled hot-rolled sheets used as the finished products for special purposes or the raw material to be re-processed are featured with different properties and purposes to the normal hot-rolled plates, which makes them high-value-added products. Therefore, it is improper to apply the prevailing export tax rebate policy which equals the pickled sheets to the normal hot-rolled plates. It is recommended to lift the export rebate rate from 0 to 9% for pickled sheets with the following five tax identification numbers: 72082500, 72082610, 72082690, 72082710, and 72082790.
- Coated Steel Strip
  The coated steel strips narrower than 600 mm are widely used for specific purposes and essentially different from the ordinary hot-rolled narrow strip. Therefore, they are high-value-added products. It is recommended to lift the export rebate rate from 0 to 9% for coated steel strips with the 6 tax identification numbers including 7212.
- Steel Wire
  The iron and steel wire products come from the normal wire rod products after pulling, drawing, coating, and other processes and have much higher product added value than the regular wire rods. Therefore, it is recommended to encourage developing such products by increasing the export rebate rate from 0%–9% to 13% for iron and steel wire products with the 5 tax identification numbers including 7217.
- Stainless Steel
  Although China has long been supporting the development of stainless steel products, the low export rebate rate for the stainless steel bars and rods makes them weak in price competition in the international market. Therefore, it is recommended to lift the export rebate rate from 9 to 13% for stainless steel strips with the 6 tax identification numbers including 7220 and that from 5 to 13% for stainless steel bars, profiles, and wires with the 6 tax identification numbers including 7222 and 7223.
- Alloy Steel
  Alloy steel is the major special steel products for special purposes. They are high added value products which are manufactured with high technology especially for tool steel, oriented silicon steel, etc., and should be encouraged with preferential financing and tax policies. Hence, it is recommended to lift the export rebate rate from 0%–9% to 13% for alloy steel products with the 21 tax identification numbers including alloy steel sheets (72253000, 72259100, 72259200, 72259910, 72259990), alloy steel strips (72262000, 72269100, 72269199, 72269200, 72269910, 72269920, 72269990), alloy steel bars and sections (72271000, 72272000, 72279090, 72282000, 72286000, 72287010, 72287090), and alloy steel wires (7229200, 72299090).
- Tool Steel and Oriented Silicon Steel
  The export rebate rate for tool steel and oriented silicon steel with the 4 tax identification numbers should be raised from 13% to 15% including the

tool steel (72264010, 72269110) and the oriented silicon steel (72251100, 72261100).

Third, protective or suppressive measures shall be taken according to different conditions, and the credit policies have to be more differentiated. The Instruction on Supporting Industrial Restructuring and Resolving Overcapacity promulgated by China Banking Regulatory Commission in 2014 raised measures such as to clarify the differentiated credit standards, carry out green credit policy, support industrial restructuring and structural optimization. However, some authorities enforce the policies only to differentiate the overcapacity industries and the non-overcapacity industries while failing to differentiate the enterprises in the overcapacity industries. That, to some extent, weakens the policy effect. Therefore, the banks are recommended to strengthen the enforcement of the differentiated credit policies, upgrading the industrial structure while resolving the overcapacity:

On the one hand, it supports the investment for the transformation and upgrading projects of iron and steel enterprises with preferential loan rates. The iron and steel enterprises will be supported in the following projects: projects to improve the product quality and upgrade the products structure; projects to extent the industrial chain and for specialized production; environmental protection and energy conservation projects; intelligent projects; the major industrial technology innovation projects; overseas projects under the Belt and Road Initiative; merging and reorganization projects to optimize the industrial organizational structure.

On the other hand, it supports the development of the legal and superior enterprises. To clean up and rectify projects is an important work to resolve overcapacity and the basis to cooperate with other supporting policies, which have been well promoted. The relevant departments and institutes are suggested to take this as an important reference when enforcing the policies. Loan should be released with general market-based interest rates to the enterprises and projects getting legal approval and recording as well as those with strong innovative ability, comprehensive competitiveness, and good benefit. The projects and enterprises that fail to follow laws and regulation to get approval and recording as well as those with substandard pollutant discharge or commits fatal environmental pollution accidents shall be deprived of financial supports or released with loan with penalty interest rate.

Fourth, the iron and steel enterprises shall have the privilege of direct power purchase as larger consumers or be supported for integrative development of coal, power, and steel. The legal and advantage iron and steel enterprises, as large consumers, shall be better supported with more customized power supply in a larger proportion. The areas with abundant coal power shall give full play to the resource and power advantages for the integrative development of coal, power, and steel.

Fifth, transportation in the iron and steel industry shall be charged with preferential freight. Preferential freight or that for the shortest path shall be applied for the transportation of iron ores, steel products, cokes, etc.

Sixth, studies on adjusting and reducing taxes for large iron mines belonging to iron and steel enterprises. Tax reduction measures such as adjusting the collection conditions and standards of resource assessment tax and resource compensation tax have to be studied on and proposed for the large iron mines belonging to iron and steel enterprises to reduce repetitive taxing. Besides, the range of added value tax deduction needs to be widened to relieve the enterprises' burdens.

Seventh, the debt restructuring of the iron and steel enterprises shall be supported to optimize the financing mode. The iron and steel enterprises with heavy debts shall be encouraged to restructure the debt through debt-to-equity swap as a means to relieve the enterprises' burdens and by lessening the enterprises' burdens of repaying capital with interest to liberate the capital for equipment upgrading, technical modification, new product development, etc. Moreover, the enterprises shall be encouraged to optimize their financing solutions by means of transfer of trust capital, bills and stock rights, issuing enterprise bonds and funds and multiple other ways to relieve the financial burden from the increased loan as a result of over-dependence on banks.

Eighth, national construction bonds shall be applied to support the development of the iron and steel industry. The national construction bonds of 50 billion to 100 billion yuan are issued to better support the comprehensive resource utilization, energy conservation, environmental protection and research and development of new technology in iron and steel industry.

(4) Energy Saving. The growth of iron and steel production increases energy consumption on an annual basis. According to statistics, in 2015, the total energy consumption for China's ferrous metal smelting and rolling was about 700 million tons of standard coal, accounting for 27% of the total of the national industrial energy consumption; the total consumption of coals (including coking industry) was 700 million tons, which weighs 18% of the total national industrial coal consumption; totally 5.796 million kilo-Watt hours of electric power was consumed, taking up 12% of the total national industrial power consumption and 9% of the total national power consumption. Since energy is constraining the industry, we have to promote the construction of energy conservation system, especially the supervision and regulation of energy consumption, and use differentiated power and water prices as per the law. These have become the indispensable measures to save energy in the entire industry and create a fair market environment.

1) Establish Regulation and Law Systems for Green and Low-Carbon Development.

The first is the establishment of the *Climate Change Law* and the supporting laws. Climate change is a significant challenge to the whole world as well as the biggest issue for the iron and steel industry in the twenty-first century as it is the industry that discharges large quantity of carbon. At present, China is promoting the enacting of the *Climate Change Law* and a series of supporting laws and regulations. As a major carbon discharger, the iron and steel industry also makes regulations specifically for itself, such as the *Iron and Steel Industry's Action Plan to Cope with Climate Change* and the *Regulations for Carbon Trading in Iron and Steel Industry*. For example, after the promulgation of the *Energy Conservation Law* in 1998, a series of supporting regulations have been formulated, including the *Regulations for Energy Conservation of Major Energy-consuming Units*, the *Regulations for Power Conservation*, the *Regulations for Energy Conservation in Civil Buildings*, and the *Regulations for Certification of China's Energy Conservation Products*.

The second is to establish supporting regulations and laws for the green manufacturing system. To promote green development and economy is an important development strategy for industry. In 2016, the Ministry of Industry and Information Technology issued documents to build a green manufacturing system. The Ministry defined the main content of the green manufacturing system including green factories, green products, green park areas, and green supply chains. Besides, the Ministry has proposed to forge the green manufacturing system to be a benchmark of the manufacturing industry's green transformation and upgrade and a power to take the lead in the international competition. In the next few years, a series of well-targeted regulations are promulgated under such general guidelines to steer the enterprises to better development such as to provide implementation plans for the construction of green factories, green products, green park areas, and green supply chains.

2) Improve Construction of Standard Systems for Energy Conservation and Low-Carbon Development.
   The energy conservation and low-carbon standards are the basis for the national energy conservation system and effectively support to resolve the overcapacity and promote energy conservation and discharge reduction. At present, China has promulgated more than 10 codes concerning the iron and steel industry including the energy-saving quota standard, the energy efficiency standard for energy-consuming equipment, and the energy-saving design standard, which are indispensable for resolving the overcapacity in the iron and steel industry. In the future, more codes will be established or revised to supplement the standard system covering energy conservation of production equipment, supervision, and management of energy conservation, energy measurement, energy administration and audit, etc. The improvement of energy conservation and low-carbon standard system will be more and more important in the future sustainable development of the iron and steel industry for the industrial restructuring and upgrade as well as legal resolving the overcapacity.
3) Accelerate Green Transformation of Traditional Industries and Its Demonstration and Promotion.

The intensifying energy-saving and low-carbon constraint urges a faster development of industrial restructuring, green transformation, and its demonstration and promotion for the traditional industries represented by the iron and steel industry. The specific implementations include industrial energy efficiency improvement action, transformation to a clean production process, transformation to efficient energy utilization and low-carbon development, and transformation of traditional industries with advanced and suitable technology. All these have been included in a series of documents promulgated during the 13th Five-Year Plan period. The above-mentioned contents of green transformation are put into practice with detailed plans to formulate the regulations including the Guide to Actions for Green Transformation and Upgrading of Iron and Steel Industry and List of Recommended Advanced technology Suitable for Energy Conservation and Low-Carbon technology.

4) Promote Construction of Supervision and Evaluation System and Mechanism.
For the time being, the Chinese government delegates its approval authority to local governments, streamlines its administration, and stresses assessment management. All these are also reflected in the trend of the development of energy policies. The recent energy assessment regulation is a good example. *The Measures to Appraise Energy Saving Efficiency for Investment Projects of Fixed Assets* has revised many contents of the previous Decree No. 6. The revision cancels the energy-saving assessment and appraisal by the National Development and Reform Commission which was one of the preconditions to approve a project and proposes to strengthen the supervision during and after the law enforcement including the dynamic supervision and regulation on national energy-saving appraisal, the management of information and credit of law and regulation violators nationwide, and the acceptance for implementation of appraisal comments on energy-saving work.
Since energy conservation is an important means to resolve the overcapacity issue in the iron and steel industry as per the laws, the assessment and management mechanism and the assessment management have to be improved continuously besides making more strict energy consumption standards considering the development tendency in the future.

5) Use Economic Means to Assist Resolving Overcapacity in Iron and Steel Industry.
In 2017, the National Development and Reform Commission and the Ministry of Industry and Information Technology jointly issued the *Notice on Operation of Price Means to Promote Structural Reform of Supply Side of Iron and Steel Industry*, emphasizing to use price measures such as differentiated power prices to help resolving the overcapacity issues in the iron and steel industry. The development trend tells us that to use diverse economic means to help resolving the iron and steel industry's overcapacity issue is an important means to establish and improve the market mechanism for energy-saving and discharge reduction. That means to build up mechanisms

for paid usage, budget management, investment and financing, etc., through improving the mechanism for trading of power utilization right, pollutant discharge right, and carbon discharge right.

### *6.2.2 Accelerate Merging and Reorganization of Iron and Steel Enterprises*

Merger and reorganization are an importance means to effectively integrate the enterprises' resources, boost their development, and enhance their competitiveness as well as an important path to resolve the conflicts brought by the production overcapacity, optimize the industrial structure, and improve the development quality and profits.

The Party Central Committee and the State Council have attached great importance to the merger and reorganization in the iron and steel industry. They have, in response to the problems of inadequate service systems and mechanisms, difficult cross-regional and cross-ownership restructuring, hard financing and heavy burden, proposed the strategy of "more mergers and organization instead of bankruptcy and liquidation" as the guideline, under which to use market-based means to well cope with the enterprise debt and non-performing asset of banks, put into practice the financial and tax policy that the financing institutes to cancel the bad debts after verification and improve the financial and tax policies that support the financial institutes to better deal with the debt-repaid assets. Especially, they have raised the philosophy of "to innovate and develop some, to restructure and reorganize some, and to swipe out some" to promote the restructuring of the state-owned enterprises. These have eradicated mechanism obstacles for the iron and steel industry's debts restructuring and bankruptcy liquidation and boosted the restructuring among iron and steel enterprises and among the upstream and downstream in the industrial chain.

1. Current Situation of Iron and Steel Industry Concentration

From CR4 and CR10, we can find out that the concentration of China's iron and steel industry has been through a typical and complete "wave" (see Figs. 6.2 and 6.3). The increasing mergers and reorganizations since 2001 have witnessed the birth of some iron and steel group companies including Pancheng Steel, Panchang Steel, and Dongbei Special Steel until 2006. However, since China's crude steel output has been soaring up during this period of time, for example, the year-on-year growth in 2005 was 29.5%. The growth is always higher than that of CR4 and CR10 average scale during the same period, which results in a continuous decreasing of industrial concentration.

In 2007, the merger and reorganization of the iron and steel enterprises embraced a boom. Especially, breakthroughs have been made in the reorganization of cross-regional enterprises: Baosteel merged Bayi Steel and Shao Steel, Wuhan Steel merged Kunming Steel and Liuzhou Steel, Shougang Steel merged Shuicheng Steel, Changzhi Steel and Tonghua Steel, Panzhihua Steel merged Xichang New Steel,

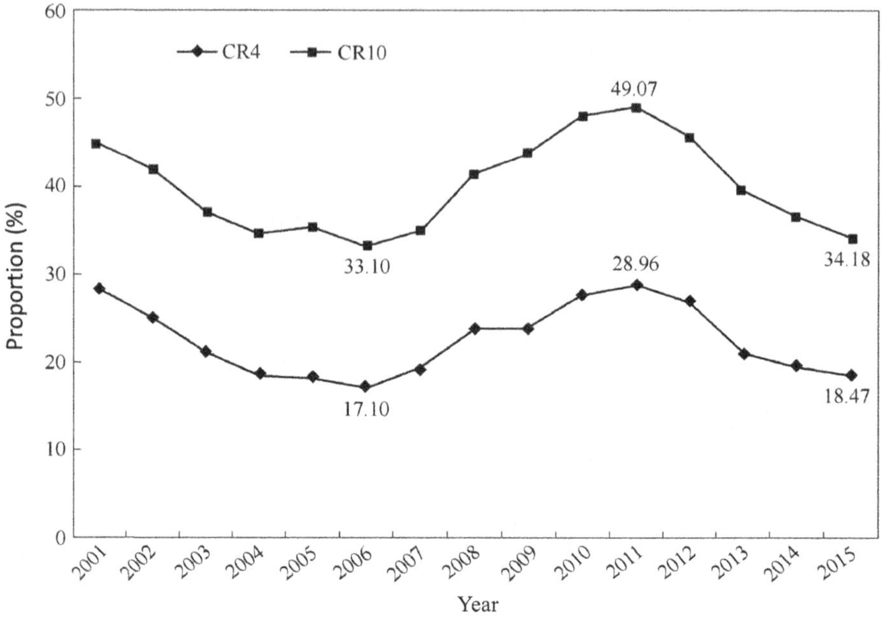


**Fig. 6.2** Change of CR4 and CR10 percentage in China's iron and steel industry during 2001–2015

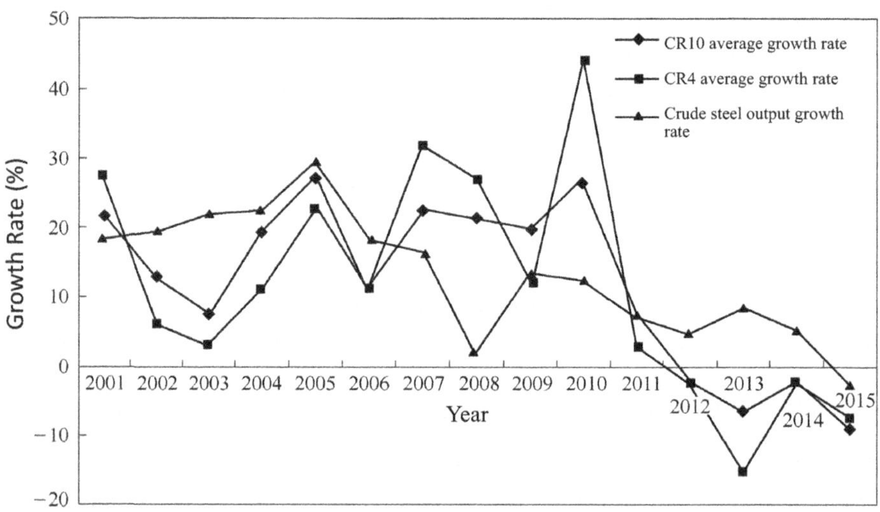


**Fig. 6.3** Change of growth rate for output of China's CR4/CR10 and crude steel during 2001–2015

Anshan Steel merged Panzhihua Steel, Shagang Steel merged Yongxing Steel, Xixing Steel, etc. Besides, intra-regional iron and steel groups were massively emerging such as Shandong Steel Group, Hebei Steel Group and Bohai Steel group and until then China's iron and steel industry has been led by multiple iron and steel giants such as Baosteel, HBIS, Wuhan Steel, Anshan Steel, Shagang Steel and Shandong Steel. At the same time, the iron and steel industrial pattern is more clear: Baosteel serves for the east, south, and northwest China market, Wuhan Steel sets its roots in the middle and southwest China market, HBIS Steel bases on the north China market, Shandong Steel serves for the east China market, and Anshan Steel is for the northeast and southwest China market. Due to the steep expansion of the scales of the leading iron and steel enterprises, CR4 and CR10's average scales had been growing rapidly from 2007 to 2011, which in 2011 reached 49.60 million tons and 33.615 million tons, respectively, with a year-on-year growth also surpassing the growth of national crude steel output. In 2010, the growth of the average scale of CR4 reached 44.3% and that of CR10 reached 26.9%, which stably promoted the concentration of the iron and steel industry. In 2011, China's CR10 concentration of the iron and steel industry reached the peak—49.07%. Refer to Fig. 6.4.

After 2012, Baosteel, Wuhan Steel, Anshan Steel, and other large central enterprises ceased their steps toward reorganizations that increase production capacity, and some of them have even been divided. HBIS and Anyang Steel failed in reorganization in the mode of "progressive merger of stock rights" with the private iron and steel enterprises within the region and ended up with the private enterprises' quitting. These have drastically slowed down the merger and reorganization in China's iron and steel industry and opened the door to the "deep water zone" and "exploration stage" for China's reorganization practically. The good point is during this

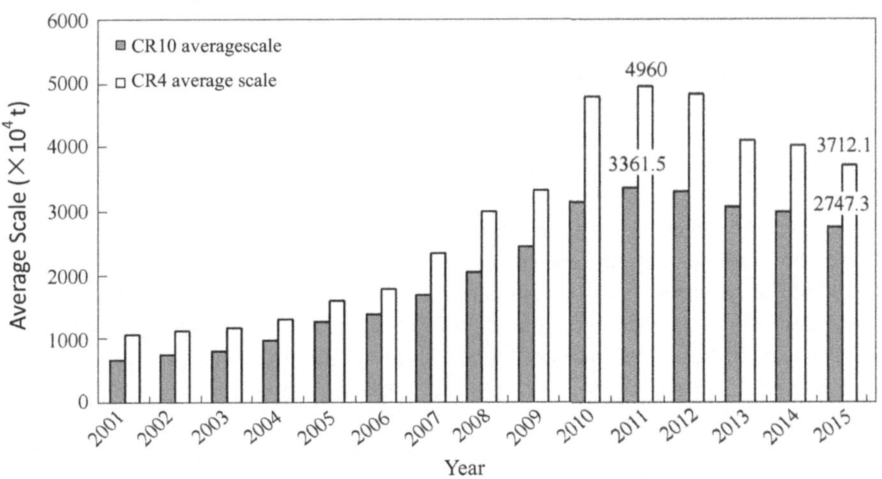


**Fig. 6.4** Change of CR4 and CR10 average scales of China's iron and steel industry from 2001 to 2015

period of time, policies including *Instructions to Promote Merger and Reorganization of Major Industries and Enterprises* and *The State Council's Opinions on Further Improving a Market Environment for Merger and Reorganization of Enterprises,* and the plans for local iron and steel industrial restructuring have guided and promoted a favorable progress in the merger and reorganization in the regional iron and steel industry. For instance, Shandong Province has promulgated the *Plan for Implementation of Merger and Reorganization of Iron and Steel Enterprises in Shandong Province*, which proposed to establish 5 regional iron and steel groups including Zibo, Weifang, Laiwu, Linyi, and Binzhou. Until now, four regional iron and steel groups have been reorganized including Linyi Dexin Iron and Steel Group, Shandong Zouping Iron and Steel Group, Shandong Jingte Iron and Steel Co., Ltd., and Zibo Qixin Iron and Steel Group Co., Ltd. In 2014, Hubei Province promulgated *Plan for Iron and Steel Industrial Restructuring in Hubei Province* that proposed to combine and reorganize the 23 iron and steel enterprises in the province and set up another 5 iron and steel groups including Jinshenglan Steel, Lijin Steel, Danfu Steel, Xinxin Steel, and Jinzhou Qunli Steel. At present, Jinshenglan Steel and Lijin Steel Group have been merged and prepared for reorganization.

Besides, many large iron and steel enterprises accelerate their overseas allocation of resources and capacity and the establishment or improvement of their production and trade service systems in order to build themselves the most competitive group enterprises in the world that are capable to participate in international competition and cooperation in a wider range, more extensive area, and higher level. For example, in June 2014, Ma'anshan Steel acquired the high-speed rail component manufacturer SAS Valdunes, France, in the form of asset acquisition; in October 2014, Anshan Steel purchased, by means of increasing the capital and share, Lianzhong (Guangzhou) Stainless Steel Co., Ltd. under Taiwan E United Group, China, and held 60% of its stock rights; in November 2014, Hebei Steel bought out 51% of the stock rights of the world's largest iron and steel trader, Duferco International Trading Holding.

During this period of time, basically 31 iron and steel enterprises with an annual capacity of 10 million tons of products have been set up. The concentration of these enterprises tends to decrease continuously. In 2015, the 31 enterprises produced 487 million tons of crude steel which occupies 60.6% of the national total output. The average production scale is 15.73 million tons. The growth of CR4 and CR10 average scale slowed down continuously and tended to touch the bottom. Both of them were slower than the growth of the national crude steel output at the same period of the previous year. The concentration of the national iron and steel industry was experiencing a constant decrease. In 2015, the CR10 only accounted for 34.2%. See Fig. 6.5. But on the other hand, it also shows that large iron and steel enterprises have played an important role in resolving overcapacity and actively reducing the crude steel output.

The General Secretary has urged "to honor a very sincere and earnest promise". According to this, Hebei Steel Group officially acquired Smederevo Steel Plant, Serbia, on June 22, 2016, and made it a sample project of Sino-Central East European capacity cooperation; on August 27 of the same year, Wen'an Steel under Hebei Xin wu'an Iron and Steel Group signed a cooperation memo with China Metallurgical

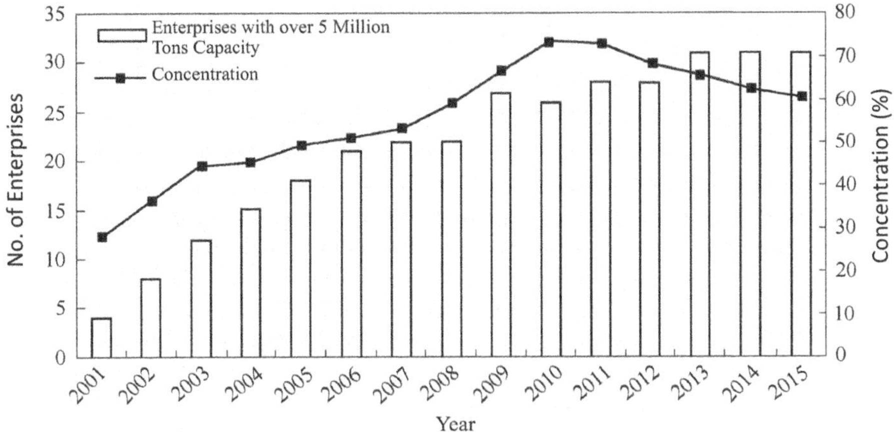


**Fig. 6.5** Changes of number and concentration of China's iron and steel enterprises with a capacity of over 5 million tons from 2001 to 2015

Group Corporation (MCC) to invest in building an iron and steel project in Malaysia which became the first Chinese private iron and steel enterprise built in a country along the Belt and Road and praised by the Minister of Industry of Sarawak, Malaysia, as "another significant milestone in Sarawak's development history"; at the beginning of December, China Baowu Steel Group Co. Ltd. was officially founded in Shanghai, symbolizing the setting out of a new flagship of China's iron and steel industry.

In 2016, China had 21 enterprises whose crude steel output was over 10 million tons and a total output of 418 million tons of crude steel. Baowu Group contributed 63.8056 million tons of crude steel in that year, taking up 7.89% of China's total crude steel output, and became a super enterprise group with the largest iron and steel capacity in China and the second largest worldwide. It is precisely because of the establishment of Baowu Group that China's iron and steel industry concentration CR10 (Baowu Group, HBIS Group, Shagang Group, Ansteel, Shougang Group, Shandong Steel, Masteel Group, Jianlong Group, Valin Group, Bengang Group) and CR4 (Baowu Group, HBIS Group, Shagang Group, Ansteel) terminated their downward trend and were increased by 1.69% and 3.20% to reach 35.87% and 21.67%.

2. Policy Orientation for Merger and Reorganization of Iron and Steel Industry

    (1) First Comprehensive Policy Document for Iron and Steel Industry.
    In July 2005, the National Development and Reform Commission has promulgated the *Development Policy for Iron and Steel Industry* (No. 35 by NDRC) approved by the State Council through discussion in the executive meeting of the State Council. The policy clearly proposed to restructure the iron and steel industry through merger and reorganization, expand the backbone enterprise groups that have comparative advantages and improve the

concentration of the industry. By 2010, the number of iron and steel enterprises has decreased by a large margin. The top 10 iron and steel groups produce more than 50% of the national total output, and the percentage is supposed to be over 70% by 2020. The iron and steel enterprises are encouraged to be developed into groups and reorganized by means of combination, merger and reorganization, mutual stock holding, etc., to reduce the number of iron and steel production enterprises and thus realize the restructuring, optimization, and upgrading of the iron and steel industry. In particular, the documents made a point that to support and encourage the qualified large enterprise groups to conduct cross-regional merger and reorganization; by 2010, the industry shall form two giant enterprise groups with a capacity of 30 million tons and several other giants with a capacity of 10 million tons, which have international competitiveness.

China's policies have guided and promoted the large iron and steel enterprise groups to achieve substantial development and even breakthroughs in their cross-regional reorganization and merger. By 2010, all the five giant iron and steel groups had produced more than 30 million tons of products, respectively, including HBIS, Baosteel, Wuhan Steel, Anshan Steel, and Shagang Steel, in which HBIS Group had produced 52.86 million tons of products, Baosteel had produced 44.4951 tons of products, and Anshan Steel has produced 40.2816 tons of products. Wuhan Steel, HBIS, Shougang Steel, and Shagang Steel had all succeeded to double their crude steel output, in which HBIS Group achieved a growth rate of 177.4% and that of Wuhan Steel Group was 165.6%. Besides, 8 iron and steel enterprise groups had respectively produced more than 10 million tons of crude steel including Shougang Steel, Bohai Steel, New Wu'an Steel, Baosteel, Ma'anshan Steel, Shandong Steel, Anyang Steel, and Valin Steel. The production scale is beyond the expected target. See Fig. 6.6.

(2) First Special Policy for Merger and Reorganization.

In August 2010, the State Council issued the *Opinions on Promoting Merger and Reorganization of Enterprises* (No. 27 [2010] by the State Council), clearly stating to further implement the adjustment the revitalization plans for key industries, and strengthen and expand the superior enterprises. Automobile, iron and steel, cement, machinery manufacturing, electrolytic aluminum, rare earth, etc., are the key industries in which the superior enterprises have to be promoted in their combination, cross-regional merger and reorganization, overseas acquisition and investment and cooperation to improve the industrial concentration for large-scale and intensive operation. The backbone enterprises with independent intellectual property rights and world-renowned brands have to be fueled for their development, and a batch of large-scale enterprise groups with international competitiveness should be set up. By doing this, the industrial structure can be optimized and upgraded.

On the one hand, the objectives have to be clarified; on the other hand, the system obstacles in the way to merger and reorganization of enterprises have

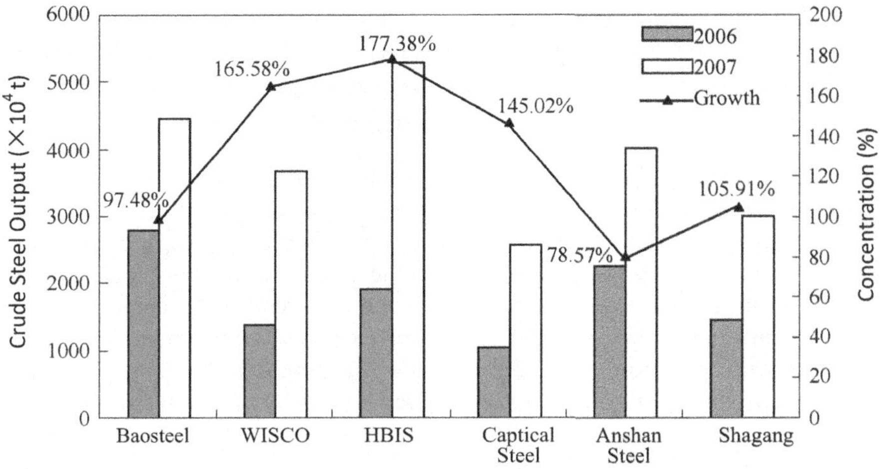


**Fig. 6.6** Comparison of crude steel output of some large iron and steel enterprise groups in China from 2006 to 2010

to be wiped out. The eight supportive policies have to be put into practice including preferential taxation policy, more financial fund investment, more financial support, to support the enterprises in independent innovation and technological progress, to allow full play to the capital market's function of promotion, to improve the relevant land management policies, to well handle the credit and debt as well as staff resettlement issues, and to deepen the enterprise system reform and management innovation. Besides, to improve the management and service for merger and reorganization has to be stressed.

More importantly, led by the Ministry of Industry and Information Technology, a Enterprise Merger and Reorganization Work Coordination Team is built up, with participation of the National Development and Reform Commission, the Ministry of Finance, the Ministry of Human Resources and Social Security, the Ministry of Commerce, the People's Bank of China, the State-Owned Assets Supervision and Administration Commission, the State Administration of Taxation, the China Banking Regulatory Commission, China Securities Regulatory Commission, etc., to make overall plans and coordinate the merger and reorganization of enterprises, study and resolve major issues occurred during the enterprises' merger and reorganization, elaborate the relevant policies and the supporting measures, coordinate the relevant local governments and enterprises for good organization and implementation, and clarify *Table of Work Division for Merger and Reorganization of Enterprises*.

(3) First Special Policy for Overcapacity Cutting in Iron and Steel Industry.
In February 2016, the State Council released *Opinions on Resolving Overcapacity for Profitable Development of Enterprises* (No. 6 [2016] by the State Council) which clearly indicated to encourage the qualified enterprises

to conduct cross-industrial, cross-regional, and cross-ownership reductive merger and reorganization, and the enterprises in the major steel-producing provinces were in the priority list of merger and reorganization.

(4) First Special Policy for Merger and Reorganization for Iron and Steel Industry.
*The Instruction on Promoting Iron and Steel Industry's Merger and Reorganization and Disposal of Zombie Enterprises* released by the State Council in September 2016 was the top-level design for cutting overcapacity and structural optimization and adjustment of the iron and steel industry and another document that urged specific targets and requirements to merger and reorganization following the *State Council's Opinions on Promoting Merger and Reorganization of Enterprises*, the *Instructions to Promote Merger and Reorganization of Major Industries and Enterprises* (No. 16 [2013] by China Federation of Industrial Economics under the Ministry of Industry and Information Technology). Moreover, it is the first Special Policy for Merger and Reorganization of the Iron and Steel Industry. The policy was more targeted to highlighting the key points, implementation step by step, flexible adjustments as per local realities, differentiated solutions for different enterprises. It emphasized more on resolving the current problems and eliminating the root causes and establishing a long-term mechanism for selecting the superior and eliminating the inferior enterprises to actively yet prudently revolve the overcapacity and promote the industrial reorganization.

3. Analysis of Characteristics and Obstacles of Merger and Reorganization in Iron and Steel Industry

(1) Analysis of characteristics.
1) The work is mainly based on the government's promotion and with the help of the enterprises themselves.
For a long term since the reform and opening-up, all mergers and reorganizations were conducted by the government. Since the 16th National Congress of the Communist Party of China, the State-Owned Assets Supervision and Administration Commissions of all levels have manifested increasing importance in the mergers and reorganizations. The government's promotion presents the investors' will, which becomes the key for a successful combination and reorganization. Many cross-regional reorganizations and local combinations are promoted by the local State-Owned Assets Supervision and Administration Commissions. The reorganization between the state-owned enterprises is driven by the government in most cases. The cross-regional reorganizations are such as the alliance between Baosteel and Shanghai Metallurgical Holding Group Corporation to reorganize Xinjiang Bayi Steel and Zhejiang Ningbo Steel Plant, the reorganization of Wuhan Steel to Kunming Steel and the Shougang Steel holding shares of Shuigang Steel; the local mergers include the reorganization of Wuhan Steel to Echeng Steel, the reorganization of Panzhihua Steel to Chengdu Seamless Pipe

Plant, Chengdu Steel Plant and Changcheng Steel Plant, the reorganization of Ma'anshan Steel to Hefei Steel, the Hebei Steel set up from combination of Tangshan Steel and Handan Steel and the Shandong Steel established from the combination of Jinan Steel, Laiwu Steel, and Shandong Metallurgical Corporation.

At the same time, the combination and reorganization mainly based on enterprises are accelerating. The acquisitions conducted by enterprises themselves such as the acquisition and reorganization among private enterprises are usually through the capital market.

2) Combination of asset transfer and market operation.
The acquisition and reorganization among the state-owned enterprises over the recent years have presented two different ways to those between the state-owned giants and the local state-owned enterprises: the asset transfer and the market transactions, while those between the state-owned enterprises and the private enterprises are completed through a market-based mode.

The ever-deepening reform and improving market economic system have diversified financing means and expanded the stock market, which brought more flexible and diverse means to acquisition and reorganization of the state-owned enterprises. Administrative transfer is evolving into various other means for combination and reorganization. In addition, mergers and acquisitions through various capital transactions such as increase in capital and share, equity replacement, overall listing, and consolidation by merger are becoming more common.

The acquisition and reorganization between the state-owned giants and the local state-owned enterprises are completed through asset transfer such as the Shougang Steel holding shares of Shuicheng Steel and the reorganization between Jianlong Steel and the new Fushun Steel. Some acquisition and reorganization are carried out through directional increase in capital and share such as the acquisition of Kunming Steel by Wuhan Steel, the capital increase of Anshan Steel to Tiantie Cold-Rolled Sheet Plant, the reorganization of Baosteel to Xinjiang Bayi Steel, and the acquisition of Baosteel to Ningbo Steel Plant, as well as acquisition of equity replacement, such as Baosteel's acquisition of Ningbo Steel Plant. Cases of overall listing include the listing of Valin from the alliance of Xiangtan Steel, Lianyuan Steel, and Hengyang Steel Tube and the overall listing of Nanjing Steel's capital with its reorganization by Fosun International. Some are through mergers such as the merger between Tangshan Steel Group and Handan Steel and the mergers among Jinan Steel, Laiwu Steel, and Shandong Metallurgical Corporation.

Another form of reorganization is that the large enterprises found a joint venture company (new assets), such as Shougang Jingtang Ltd. Iron & Steel Co., Ltd., Guangdong Steel, Guangxi Steel and Chaoyang Anlin Iron and Steel Co., Ltd. under Anshan Steel Group from merger of

Anshan Steel and Linyuan Steel. Acquisitions in the form of increment of cash include Ma'anshan Steel's acquisition of Hefei Steel, Valin's acquisition of Wuxi Steel, Shagang Steel's acquisition of Huai'an Steel, Henan Yonggang Steel, Xinrui Special Steel and Jiangsu Yonggang Steel, Jinxi Steel's acquisition of Xinyi Steel, etc.

3) Attach equal importance to intra-regional reorganization and cross-regional reorganization.
In recent years, the large-scale superior iron and steel enterprises all over China are taking an active part in and dominate the acquisition and reorganization of local state-owned enterprises. In some cases, the large enterprises initiate the acquisition and reorganization of local state-owned enterprises; in other cases, some local state-owned enterprises also actively transfer the controlling interest to the large enterprises such as the reorganizations of Baosteel to Xinjiang Bayi Steel and Ningbo Steel Plant, Wuhan Steel to Echeng Steel, and that of Ma'anshan Steel to Hefei Steel. While the large enterprises are reorganizing the local state-owned enterprises by merger and acquisition, the intra-regional merger and reorganization of the state-owned enterprises, especially those in the same province, are striding forward. Cases of regional enterprise alliances such as the merger between Tangshan Steel and Handan Steel and that among Jinan Steel, Laiwu Steel, and Shandong Metallurgical Corporation are aiming at forging large and superior enterprises to take the lead in China's iron and steel industry. While for the state-owned enterprises which cannot be strengthened within the province, many provinces and cities conduct nationwide merger and reorganization and some of them take the first steps actively to transfer the controlling interest to the large state-owned enterprises for better development, for which a good example is Shougang Steel holding shares of Shuigang Steel.

4) Extend horizontal reorganization to vertical reorganization.
As the main form of the acquisition and reorganization of the state-owned enterprises in recent years, the enterprises' horizontal reorganization includes not only the merger and reorganization among the enterprises in the same industry, but also the reorganization and integration of the same or similar business segments within the enterprise or among enterprises.

The enterprises' vertical reorganization means to improve the industrial chain. After the horizontal reorganization of iron and steel enterprises to form a large enterprise group, the next step is to strengthen the competitiveness by tightening the grasp to resources and the market. That means to merge with the resource enterprises to set foot in the upstream and to extend to the downstream for closer cooperation with the customers. The reorganization and strategic cooperation alliance with the raw material supplier upstream and the customers' downstream complete an industrial chain which goes across industries, regions and areas, being an effective way for vertical reorganization.

Iron ores and cokes are the lifeblood for iron and steel enterprises to ensure production and maintain their development. Therefore, they are the key contents for developing the raw material supply chain system. The recent years have witnessed active implementation of the Go Out policy by China's large enterprises by means of direct investment in the overseas market and acquiring the foreign iron ore companies or holding their shares. For example, Anshan Steel holds shares of Gindalbie Metals, Australia, and Valin holds shares of Fortescue Metals Group Ltd. Domestically, they exchange shares with the large coal (coke) suppliers or set up joint ventures to develop resources together. For example, Wuhan Steel joins its venture with Pingdingshan Coal Group, Wuhan Steel holds shares of Pingdingshan Coal Resource Development Co., Ltd., and Pingdingshan Coal holds shares of the coal chemical industry of Wuhan Steel. The sales industrial chain has been remolded to a long-term stable sales of iron and steel products to the customers. To establish a fixed long-term cooperative relationship with the downstream terminal customers is the best choice for iron and steel enterprises to maintain a favorable product demand and a stable marketing channel. For example, Baosteel holds shares in two of CSIC's major shipbuilding base construction projects, and Anshan Steel and Dalian Huarui Heavy Industry Group Co., Ltd. have joined their venture to build a ship plate distribution center. That has enhanced the integral and comprehensive competitiveness of the enterprises.

5) Combine the merger and reorganization with the enterprise's reform and restructuring.

In recent years, some state-owned enterprises have combined acquisitions and reorganizations with enterprise restructuring by actively attracting state-owned enterprises, foreign-funded enterprises, or private enterprises as strategic investors to accelerate the shareholding reform in groups' level.

Features of acquisition and reorganization of the state-owned enterprises show that on the one hand, the acquisition and reorganization of China's enterprises are much more market-based, which is reflected in the improvement of the market's function and the enterprises' dominance in the acquisition and reorganization; on the other hand, the acquisition and reorganization of Chinese enterprises are characterized with economic transformation period, which means that the function of market and enterprises has not been given full play to acquisition and reorganization and market-oriented acquisition and reogranization is required to be promoted further.

(2) Analysis of Obstacles.

1) Institutional Obstacles.

In China, the mode of "centralized ownership and level-to-level management" is applied in the state-owned property management system.

To be specific, all nominal state-owned enterprises are actually categorized into the central government enterprises, the provincial state-owned enterprises, and the municipal state-owned enterprises. The essence of such system is that the government manages the enterprises and the government manages the properties, human resources, and operations as an organic whole. The iron and steel enterprises are of diversified ownership including the state-owned enterprises, the private enterprises, the shareholding enterprises, the foreign-funded enterprises, etc. Under this system and mechanism, merger and reorganization of enterprises reallocate the benefit and power among governments and among the different reorganized bodies. Due to the differences among various interest demands and management systems, there are contradictions between the central and local governments and among different local governments in the distribution of fiscal and taxation and income from the state-owned assets. The complex benefit-based relationships have, to some extent, impeded asset restructuring. Therefore, it is very difficult to reorganize the assets. Moreover, such administrative division system is more likely to obstruct cross-regional merger and reorganization.

2) Obstacles of Investors' Intention.

   Most of China's large iron and steel enterprises are state-owned enterprises. Therefore, the government's willingness to invest determines the success or failure of reorganization. The state-owned enterprises are under the management of governments in asset, human resources, and operation. Local governments are very powerful in deciding the senior executives of the reorganized enterprises, whereas the enterprises themselves are not allowed to make such decision. The government's performance appraisal indicators often restrict the reorganization of enterprises.

   Today, the reorganization of enterprises can thoroughly embody the investors' intention. However, the merger and reorganization of the iron and steel enterprises in China are not completely capital market-oriented, so they cannot fully reflect the enterprises' intention of reorganization. The investors' right of personnel administration cannot be guaranteed, making the enterprise reorganization lack of authoritativeness. The enterprises cannot well cooperate even though they are combined together, let alone integration after the reorganization and its synergy.

3) Obstacles from Regional Protectionism.

   At present, local governments have become a major obstacle to the cross-regional reorganization of iron and steel enterprises. The local governments are more likely to be indulged in departmental selfishness, which imposes negative impacts on the reorganization. The iron and steel industry is divided into regions. Most of the local governments grant new projects for better political achievements and GDP, regrettably considering only short-term benefits. Iron and steel enterprises are big taxpayers in almost every province. The regional protectionism stems the merger and reorganization.

The regional protectionism, however, stimulates the reorganization of enterprises within the region. There are less conflicts in reorganization of enterprises within the same province or of the same ownership system where the reorganization optimizes allocation of resources in the region; however, in order to solve the historical pending issues, the industrial reorganization is more likely to be restricted within a province instead of becoming alliances between giants. The regional protection forms barrier to foreign enterprises, which objectively protects the regional backward production capacity, but is not conductive to exert the synergistic effect of the reorganization.

4) Obstacles from Benefit Distribution.

The tax distribution system and the transfer payment system are the prevailing financial and taxation systems in China today. The tax distribution system means to categorize taxes into the central government revenue, the local government revenue, and the central and local government-shared revenue. In such a system, the benefit distribution among governments acts as a huge barrier for cross-provincial and cross-regional merger and reorganization of iron and steel enterprises.

When studying the issues related to merger and reorganization, local governments and enterprises are worried about the unified development plan after the merger and reorganization. If the enterprise business after merger and reorganization is not in the local area, it will affect the income and fiscal revenue of the region. Therefore, they are only willing to engage in loose corporation groups and are unwilling to engage in merger and reorganization with assets as the bond so that the substantive effects of reorganization cannot be fully achieved. Coordinating the local interests and overall interests and establishing the reasonable interest compensation and transfer payment mechanism are the key to promoting the reorganization between giants across cities, provinces, and regions.

5) Barriers to Supporting Policy Measures.

Due to historical reasons, many iron and steel enterprises, especially the state-owned large-scale iron and steel enterprises, have a large number of historical problems, such as the large number of personnel, the enterprises burdened with social, the plant-run large-scale collectives, and the heavy burdens of auxiliary companies. In response to these problems, the state has successively introduced some policies. At present, schools, hospitals, and public security bureaus run by state-owned enterprises are gradually being transferred to local governments. In 2002, the State Economic and Trade Commission and other eight ministries and commissions issued the *Implementation Measures on the Separation of the State-Owned Large and Medium-Sized Enterprises' Main and Subsidiary Separation, Subsidiary Industry Restructuring and Distributing Surplus Employees*, and the auxiliary industry restructuring work is being smoothly implemented. On November 6, 2005, the State Council also approved the guidance opinions on the pilot reforms of the plant-run collectives in the northeast region and requested that it shall be actively and properly resolved on the premise of ensuring stability. Only by well handling, the separation of the main and the auxiliary businesses can the efficiency of the subject after reorganization be fully leveraged. However, the merger and reorganization of iron and steel enterprises are an important and pressing task so that the implementation of supporting policies constitutes a major issue.

To the major problems of reorganization, there have not been any comprehensive policies, including a series of coherent systems in finance, social security, laws, environmental protection, etc. For example, fiscal levy of government at all levels should be balanced to solve the issue of benefits distribution between the central and local governments and among local governments in this regard. The first thing that needs to be done is about the central and local finance and taxation policies, which will shift the value-added tax from production type to the consumption type, and the benefits between the central and local governments and among the local governments in the reorganization should be redistributed.

6) Lack of Internal Impetus for Restructuring.
At present, China's iron and steel industry is in a period of development, and most enterprises have relatively good returns. Self-development and expansion are the main means for enterprises to grow bigger and stronger so that the internal impetus of merger and reorganization is insufficient. The enterprises show much impetus only under the circumustances of fierce market competition, high operating pressure and difficulty for survival and development. Most enterprises would seek reorganization with large enterprises for financial support only when the market is stagnant or when in the crisis of lacking funds and broken capital chain. This kind of reorganization is a short-term expediency without long-term development goals, which will affect the effects of reorganization.

4. Development Concepts of Mergers and Reorganization in the Iron and Steel Industry

Through years of development, China's iron and steel industry has made tremendous progress and remarkable achievements. It has already achieved the "quantitative catching up" in terms of the scale, but has long been criticized due to the problems of "being big but not strong" like the weak independent innovation ability, uneven development, decentralized operation, disordered competition, and low benefit level. This indicates that it is even harder to achieve a "qualitative leap" from catching up to leading in competitiveness. In essence, the so-called big and strong are not contradictory but complementary to each other. As scale expansion is the basic law of the development of the iron and steel industry, being "big" is an advantage that can be comprehensively promoted from the overall industrial perspective. This has been proved by practice; however, to break through the bottleneck of "being strong", targeted goals and measures must be made. Stronger enterprise will bring about stronger industry. To this end, a more efficient way is to make breakthroughs with a focus on the key enterprises. The core is to build a world-class cluster of iron and steel enterprises with strong international competitiveness and to form a coordinated development pattern featuring the super iron and steel enterprises groups as the leader, giant iron and steel enterprises groups as the key, and the regional iron and steel enterprises groups as the support. Meanwhile, a layout structure of "a network with one belt and several points" will be established to make concentrations in the coastal areas and industrial parks and extension to inland markets. Such efforts will promote the transformation and upgrading of the iron and steel industry and achieve sound and sustainable development.

Iron and steel enterprises should continue to be problem-oriented and solve the outstanding issues. In response to the problems like unsound service systems, institutions and mechanisms, difficult cross-region and cross-ownership reorganization, hard financing and heavy burdens, some enterprises, with the market-based means, have properly addressed the debts and non-performing assets, implemented the fiscal policies of bad debts canceling and verification, and improved financial institutions' supporting policies for increasing the disposal of debt-repaid assets by taking the strategic ideas of "more M&As and less bankruptcy liquidation" as the guidance, the strategic arrangement of "innovatively developing a batch, reorganizing a batch and pulling a batch out of the market" as the direction, the reorganization of regional iron and steel enterprises, combination of the strong ones and international cooperation as the emphasis as well as the supporting policy and risk prevention as the pillar. These have cleared up the barriers hindering the debt restructuring and bankruptcy liquidation in the iron and steel industry, boosted the structural optimization and industrial clustering, and promoted the efficient development with less quantity and more specialty. Iron and steel enterprises should also achieve the "double optimization" in organizational structure and industrial layout; "double reduction" in the number of enterprises and total emissions; "double improvement" in the quality and efficiency of enterprise development.

To promote the merging and reorganization in the iron and steel industry and achieve orderly development, we should adhere to the "six basic principles": First is to adhere to market-based reforms and provide good coordination and guidance. The decisive role of market in resource allocation should be leveraged, and administrative intervention should be cut down. Coordination and guidance of the government should be strengthened to continuously promote administration streamlining and power delegation and create a good external environment. Second is to adhere to the main body of the enterprise and implement their responsibilities. The responsibility of the main body should be strengthened. With a scientific and rational incentive and restraint mechanism, the initiative and creativity of the leaders, innovative talents and management personnel at all levels in the enterprise will be fully mobilized. Third is to strengthen policy bottoming and ensure social stability. Strengthening bottoming ability of policies will help to achieve a reliable disposal of enterprises exit. The role of the party organizations, trade unions, and workers' congress of the enterprise should be fully leveraged to do a good job in ideológical work in an in-depth manner so that employee relocation will be properly addressed and social stability will be maintained. Fourth is to conduct overall planning, key implementation, and hierarchical advancement. By fully considering the industrial characteristics and realities of the enterprise, top-level design should be strengthened with overall planning. Enterprises should establish an evaluation system for major projects, in which a batch of projects will be cultivated after being planned and a project will only be pushed forward under mature conditions. The fifth is to adapt to local conditions, make tailored measures for enterprises, and provide special support. Enterprises should strengthen the experience summary, project demonstration, and promotion and set up a "closed-loop" special working group for key projects, namely to take one-to-one responsible and support for each project, to press ahead one project by putting it into real operation, and to drive forward a batch of projects to take the lead. Sixth is to intensify market supervision and build a fair environment. The working methods will be transformed from mainly depending on the administrative measures to a comprehensive utilization of laws and regulations, economic means and necessary administrative measures. The defined standards should be shifted from mainly depending on the equipment scale and process technical standards to the comprehensive standards in environmental protection, energy consumption, water consumption, quality, safety, technology, and others. A market-based, legalized, and normalized work promotion mechanism will be established by consolidating supervision and law enforcement so as to build a level-playing market environment and promote the orderly competition, mutual promotion, and common improvement of both the state-owned and private iron and steel enterprises.

5. Suggestions for Merger and Reorganization in Iron and Steel Industry

    (1) Maintaining existing strategies and deepening merger and reorganization. Taking building a strong world-class iron and steel enterprise with Chinese characteristics as the core, the mergers and acquisitions of the enterprises shall be supported, guided, and encouraged in an orientation of forming a highly concentrated and efficient industrial organization with fine division of labor. On the one hand, we should promote strong alliances centering on the advantageous products. With the goal of forming the world-class competitiveness and innovation capability in automotive sheets, electrical steel, home appliance plates, shipbuilding plates, pipeline steel, oil well pipes, and other products, iron and steel enterprises should be given support in divesting ineffective and inefficient production capacity and participating into mergers and reorganization with high-quality assets so as to create a world-class iron and steel enterprise with Chinese characteristics. On the other hand, we should focus on supporting the reorganization of iron and steel enterprises in the region. In the process of cutting overcapacity, due to the tight time schedule and heavy tasks, some local governments adopted the method of apportioning to let enterprises cut down their capacity according to specific proportion. Such practice may cause the mismatching

between the process and capacity of all enterprises in the region, which would greatly increase the production cost and substantially undermine the guidance for enterprises' competitiveness. Accelerating the regional reorganization should have government guidance to become an internal impetus of the enterprises. Therefore, an industrial pattern dominated by superior and strong enterprises along with the coordinated development of small- and medium-sized enterprises (being professional, specialized, and innovative) can be formed to optimize the market-based environment and accelerate the transformation and upgrading. At the same time, focus will be put on supporting the strong alliance of large iron and steel enterprise groups and the "international capacity cooperation". We should give full play to the leading role of large iron and steel enterprise groups in independent innovation, layout optimization, standardized management, resource integration, and the leading role of the iron and steel industry in the "Belt and Road" Initiative and international capacity cooperation to create a batch of world-class iron and steel enterprises with strong international competitiveness and Chinese characteristics as well as several backbone iron and steel enterprises with specialized division of labor. The ultimate goal is to comprehensively and continuously meet the needs of national economic and social development.

(2) Making top-level design and clarifying development goals. In order to achieve the development goals of mergers and reorganizations, follow the rules of market-based operation, and give full play to the government's policy guidance and management services, merger and reorganization of China's iron and steel industry needs to be designed from the top level from three perspectives, namely building international competitiveness, regional advantages, specialized integration. Meanwhile, development goals targeting at the scale and quantity of various types of enterprises should be proposed. In response to the clear favorable policies in such fields as the coordination mechanism for state-owned property transactions, standardized operations, direct power supply of large customers, resource security, logistics system optimization, technological structure adjustment, merger and acquisition tax, inefficient asset withdrawal mechanism, land use and employee resettlement, we need to guide and encourage local authorities and enterprises to actively push forward under the unified deployment of the state.

(3) Preparing merger and reorganization plan in categories. According to the national arrangement on merger and reorganization, the plans of which should be prepared respectively according to the categories, including the plans prepared by the large-scale enterprise group of the central government enterprise or state-owned system, the competent local department with regional advantages, and the agencies entrusted by the Ministry of Industry and Information Technology. We should actively guide the regions and companies to determine the target enterprises of mergers and reorganization according to their own development strategy plans and the requirements

of national industrial policies, and the plans of mergers and reorganizations should be formulated in a scientific manner with meticulous design, investigation, and discussion combining the conditions of the macroeconomy, industry, and enterprises. In choosing the target enterprise, it should be conducive to realizing the complementary advantages of resources and maximizing the synergy effect of resources.

(4) Strengthening integration after reorganization. Enterprises should attach great importance to the integration of various factors and resources after reorganization and strengthen the integration of personnel, culture, and management. Taking the merger and reorganization as an opportunity, enterprises shall deepen the reform of the management system and actively carry out the transformation of the corporate system and the shareholding system to further improve the corporate governance structure. In combination with realities, business process reengineering can be conducted in personnel, finance, procurement, sales, production, R&D, and other links to optimize the allocation of production factors such as land, capital, technology, and talents as well as innovate management models and achieve complementary advantages and deep integration.

(5) Implementing policy measures. All local authorities and enterprises shall conscientiously implement the *Opinions of the State Council on Promoting Mergers and Acquisitions of Enterprises*, the plans in categories, and the measures promoting mergers and reorganization in the areas of finance, taxation, financial services, credits and debts, staff placement, land and mineral resources allocation to support enterprises in this regard. All local departments in industrialization and informatization, development and reform, finance, human resources, social security, land and resources, commerce, state-owned assets, industry and commerce, and others should study and introduce specific measures to promote mergers and reorganization of enterprises by combining the actual conditions of the region. Priority should be given to supporting enterprises conducting mergers and reorganization to make technological transformation, encourage enterprises to strengthen and innovate management, and enhance their comprehensive competitiveness.

(6) Creating a good environment. Adhere to market-based operations, fully respect the willingness of enterprises, fully mobilize the enthusiasm of enterprises, and guide and encourage enterprises to voluntarily carry out mergers and reorganization. Clean up, revise, and abolish all policies, regulations, and practices that are not conducive to enterprises' mergers and reorganization. In particular, the local regulations restricting mergers and reorganization by non-local enterprises should be resolutely eliminated. Actively explore inter-regional interest sharing mechanism for cross-regional merges and reorganization. Under the premise of not violating the relevant national policies and regulations, a fiscal and taxation benefit-sharing agreement can be sign among regions after the merger and reorganization according to the scale of assets and profitability of the enterprise so as to properly solve the

problem of the attribution of statistical data such as the added value after mergers and reorganization and realize results sharing.

(7) Doing a good job in management services. It is recommended that local governments and relevant departments should urge enterprises to strictly implement the relevant laws and regulations as well as the national industrial policies concerning mergers and reorganization, standardize operational procedures, strengthen information disclosure, prevent and control insider trading, and prevent moral hazard. Encourage foreign-funded enterprise to participate in the restructuring, mergers and reorganization of domestic enterprises via equity participation, M&A, and other means. Meanwhile, security review on domestic enterprises made by foreign-funded enterprise should be strengthened to safeguard national security. For enterprises' mergers and reorganization that meet the statutory reporting standards for concentration of undertakings, it is necessary to conduct anti-monopoly review on such concentration. All regions should strengthen the guidance for mergers and reorganization of enterprises, study and formulate opinions to promote the enterprises' mergers and reorganization in the region, and combine this issue with enterprise restructuring, management innovation, technological transformation, and elimination of backwardness. Channels for exchanging information should be broadened to establish a public service platform which can promote enterprises' mergers and reorganization. We should make the intermediary services for mergers and reorganization more professional and standardized and focus on introducing and cultivating specialized talents who are familiar with M&A business, especially the cross-border M&A business so as to actively provide intermediary services in market information, strategic consulting, legal adviser, financial consultant, land valuation, assets evaluation, property rights transactions, financing intermediaries, independent audits, enterprise management, etc. Experience of typical examples should be summarized and advocated with strengthened publicity and guidance.

Enterprises should implement mergers and reorganizations in a planned manner, thoroughly study the possible contradictions and problems that may occur in mergers and reorganization, and attach great importance to the risks of market, finance, staff resettlement, cross-border M&A, and others in this process. We should also strengthen risk management, identify risk factors, assess risk intensity, properly formulate corresponding response plans and measures, and build a sound risk management system.

### 6.2.3 *Resolutely and Continuously Crack Down on "Substandard Steel"*

Enterprises producing MF furnace "substandard steel" generally do not have the secondary refining and inspection process. Such billets (ingots) and steel products have poor quality performance and obvious safety hazards. Although a few enterprises are equipped with refining equipment, the equipment is rarely used for the sake of cost, but for examination. Generally, these enterprises apply for and obtain licenses of hot-rolled steel bar production in the name of independent steel rolling enterprises, but actually they engage in productions of melted scraps, continuous casting billets, and steel products. At the same time, there are no invoices for procurement and sales in their operation and such enterprises produce substandard steel by means of OEM and counterfeiting. These violations of laws and regulations involving tax evasion and illegal profit-making not only seriously disturb the market order but also exacerbate vicious competition, resulting in the phenomenon of "bad money driving out good money" so that they must be thoroughly eradicated and banned.

As early as in December 1999, the former National Economic and Trade Commission issued the *Catalogue of Eliminating Backward Production Capacity, Processes and Products* (*Second Batch*), which clearly stipulated that the industrial frequency furnace for the production of substandard steel or ingots should be eliminated before the end of 2000. After that, substandard steel produced by MF furnace is listed as one to be eliminated in the *Steel Industry Development Policy*, *Guidance Catalogue for Elimination of Backward Production Processes, Equipment and Products in Some Industrial Industries*, and *Industrial Structure Adjustment Guidance Catalogue*, etc.

Due to the hidden production of substandard steel, the driving of market interests, and the "tolerance" of the local government for economic development, it has not been completely eliminated in a long time or even went out of control with sprawling production. In addition, the incompleteness and ambiguity of the relevant policies also give these enterprises the opportunity to exploit the loopholes. For example, the policy only stipulates that the MF furnace cannot produce substandard steel, plain carbon steel, stainless steel and construction steel, but not covers other steel grades (such as low alloy steel and high-quality carbon structural steel). At the same time, due to the huge smelting capacity of MF furnace, issues related to employee resettlement, credits, and debts are also very prominent. It is difficult for local governments, especially counties and towns, to resolutely dismantle them and put it into practice.

On September 12, 2016, the Office of the Inter-Ministerial Meeting of the National Iron and Steel Industry to Resolve the Excessive Production Capacity and the Pursue Development issued the *Notice on Illegal Production and Sales of Small Steel Plants in Xinyi, Jiangsu Province,* requiring that the provincial government should to attach great importance to its negative impact incident, investigate the causes and seriously hold those involved accountable. At the same time, all relevant regions are required to conduct a comprehensive general survey on the illegal construction of steel projects and the illegal production and sales of substandard steel products in their respective jurisdictions. This has opened the prelude of severely cracking down on the "substandard steel" produced by MF furnace.

In February 2017, China Iron and Steel Association, Chinese Society for Metals, China Foundry Association, China Special Steel Enterprises Association, Stainless Steel Council of China Special Steel Enterprise Association, Central Iron and Steel Research Institute, University of Science and Technology Beijing, China Metallurgical Industry Planning and Research Institute, China Metallurgical Information Standardization Institute, and other units as well as some iron and steel enterprises organized experts in the industry to put forward "the opinions on supporting the fight against 'substandard steel' and defining the scope of use of power frequency and medium frequency induction furnaces". This means that induction furnaces used as smelting equipment to produce all kinds of casting products in the foundry industry, the medium (power) frequency furnace that meets the following requirements in the production of special alloy materials, and the medium (power) frequency furnace only for melting ferrochrome, nickel–iron, and other alloys in production of stainless steel and high alloy steel are not in the list of shutting down. All medium (power) frequency furnaces' production lines that do not fall within the scope of the above three categories, regardless of their scale and the age of the production equipment and whether they are provided with liquid steel refining methods, continuous casting and rolling equipment, dust removal equipment, and other environmental protection facilities, are in the scope of "substandard steel" to be banned according to the laws. Equipment and facilities such as the main smelting equipment, transformers, dust hoods, operating platforms, and tracks shall be completely dismantled. Among them, enterprises that have been listed in the announcement list in accordance with *Iron and Steel Industry Standard Conditions* and *Foundry Pig Iron Enterprise Certification Standards* need to be immediately disqualified; the enterprises that have obtained the license to produce and sell billets (ingots) and steel products need to be revoked immediately with the license.

In April 2017, *Opinions on Well Resolving the Excessive Production Capacity of the Iron and Steel and Coal Industry in 2017 to Realize the Development of the Difficulties* of the Inter-Ministerial Meeting on Resolving Overcapacity and Profitable Development for National Iron and Steel and Coal Industry (NDRC [2017] No. 691) [4] clearly pointed out the illegal production capacity of "substandard steel" shall be resolutely and completely banned and eliminated. The standards defining "substandard steel" shall be implemented with reference to the *Opinions on Supporting the Crackdown on "Substandard Steel" and Defining the Application Range of Power Frequency and Medium Frequency Induction Furnace*s (China Iron and Steel Association [2017] No. 23). All local authorities and relevant central government enterprises shall completely dismantle the main equipment of medium (power) frequency induction furnace, transformers, dust hoods, operating platforms, and tracks used for the production of "substandard steel" by the end of June 2017 according to the law; enterprises that have dismantled the equipment shall be publicized on the Web sites of local governments to accept social supervision. The cut-off "substandard steel" capacity cannot be listed in the task of cutting overcapacity in each region in 2017 and cannot be supported by the central special fund policy. The employees involved should be resettled well by the local governments. The Inter-Ministerial Joint Meeting should organize inspection and acceptance and strengthen supervision to ensure complete removal as required and strictly prevent "shutdown instead of close", off-site transfer, and resurgence.

As of May 2017, the State Council has organized four rounds of special inspections for "substandard steel" and began the centralized spot check and acceptance in early August. The National Development and Reform Commission clearly stated at the mobilization meeting before the acceptance that if problems with bad social impacts like existing "substandard steel" enterprises which are not on the reported list, there are enterprises still producing "substandard steel" against the law, the related equipment of the enterprises used for producing "substandard steel" have not been dismantled, new capacity violating laws are increased are found, and joint investigation team will be set up to check the project compliance, capacity replacement, product quality, land use permit, environmental protection, construction permit, safe production, registration, bank credit, taxation, export, and others. If it is true, it will be reported to the State Council and the local governments at all levels, their departments and related enterprises will be strictly held accountable in accordance with the laws and regulations. In addition to the incident of Jiangsu Huada Company, defaults of relevant personnel in Beichen District Government of Tianjin, the District Industry and Information Technology Commission, Dazhangzhuang Town and Xiditou Town, and security incident of Yanshan Hongxin Metal Regeneration Co., Ltd. in Wenshan, Hunan Province, are typical negative examples. Eight responsible organizations and 20 responsible persons are held accountable in a serious manner. This has exerted great deterrent effect and brought about tangible results on the banning of "substandard steel". As of August 2017, more than 600 "substandard steel" enterprises with MF furnaces were banned according to laws, involving a production capacity of over 100 million tons.

Along with the effectiveness of cutting overcapacity in the iron and steel industry and the phased victory of eliminating "substandard steel" according to laws, China's iron and steel industry has achieved significant improvements in capacity utilization, production and operation efficiency, organizational layout optimization, and other areas. However, the substantial efficiency growth of the iron and steel industry may make those enterprises producing "substandard steel" take risks to resume the production. Therefore, continuous efforts shall be made in the establishment of a sound long-term mechanism, sustained high-pressure strikes, etc.

Firstly, we should further deepen the awareness. We will effectively strengthen the "awareness in four aspects" to ensure that the central government's decision-making arrangements are truly implemented. In the task of banning the "substandard steel", there should be no retreats, no conditions, and no discounts regardless of any costs.

Secondly, we should strengthen investigations to prevent resurgence. Always maintain a high degree of sensitivity and fully understand the complexity and arduousness on cracking down on the "substandard steel" to ensure the full coverage of its banning in accordance with the law. Investigations and inspection should be strengthened to prevent the resurgence.

Thirdly, we should enhance the supervision of related equipment and strictly prevent the off-site transfer of backward production capacity. For those which have not been fully dismantled, they must be completely dismantled as soon as possible in accordance with the "four thoroughness" requirements and be abandoned. Equipment such as MF furnaces and transformers which has been dismantled should have registration and be tracked in the future. Off-site transfer of backward production capacity is strictly prohibited.

Fourthly, we should establish a long-term mechanism and strengthen follow-up supervision. It is recommended to establish a long-term mechanism for the banning and elimination of the "substandard steel" and earnestly conduct the follow-up supervision of those enterprises to strictly prevent the resurgent and new increased production capacity. In particular, public opinion supervision shall be made good use with smooth channels for reporting by the masses. By intensifying the verification and feedback of the report information, enterprises producing "substandard steel" will have nowhere to hide. At the same time, supervision and investigation of steel products' circulation and application need to be reinforced so that no dead ends and exploited loopholes will exist on the entire industrial chain. In addition, relevant government departments should further strengthen the study and grasp of industrial policies and strictly restrict the introduction and landing of backward production capacity projects in attracting investment.

## 6.3 Industrial Practices of Coordination

Building a fair and orderly market environment and reshaping and consolidating market guidance are the orientation of national industrial policy and the common pursuit of the enterprises. By taking systematic learning, in-depth study, and thorough analysis of the industrial policies, China Metallurgical Industry Planning and Research Institute assumes the task of providing better service to the government, industry, and enterprises. With experience and active explorations accumulated in promoting the supply-side structural reforms of the iron and steel industry as well as its sound development, the institute helps to give clear development strategies, road maps, timetables and task books for industry authorities, local governments and enterprises in mergers, reorganizations, and orderly development. Such efforts help enterprises, and the industry adapts to grasp and guide the trend of sustainable development.

The practices of the institute in promoting the coordinated development of the industry are detailed in Table 6.1.

**Table 6.1** Practices of China Metallurgical Industry Planning and Research Institute in promoting coordinated development

| No. | Type | Main content | Typical cases |
|---|---|---|---|
| 1 | Tasks of ministries and commissions | Merger and reorganization status, problem analysis, solutions, supporting policies, development proposals, etc. | *Thoughts and Suggestions on Implementing Mergers and Reorganizations and Building World-Class Iron and Steel Enterprises with Chinese Characteristics*, *A Study on the Overall Thinking of Mergers and Reorganizations in China's Iron and Steel Industry*, *Comprehensive Research and Result Analysis of the Restructuring Path of China's Iron and Steel Enterprises* |
| 2 | Projects of local governments | Restructuring plan, supporting policies, key tasks, organizational framework, etc. | *Overall Plan for Merger and Reorganization and Layout Optimization of Iron and Steel Industry in Hebei Province*, *Optimization Plan for Merger and Reorganization in Liaoning Province*, *Iron and Steel Industry Adjustment Plan of Hubei Province*, *Iron and Steel Industrial Structure Adjustment Plan of Zibo City, Shandong*, *Development Plan for the Joint Restructuring, Transformation and Upgrading of Sichuan Local Metallurgical Groups*, *Development Plan for Iron and Steel Industry Adjustment and Upgrade of Yancheng*, *High-quality Steel Industry Development Plan of Handan* |
| 3 | Projects of enterprise | Propose concepts of capacity restructuring, debt restructuring, making trials, key breakthroughs, etc. | *Report on Speeding up the Cutting Overcapacity of Steel in Northeast China with Yingkou Steel as a Breakthrough*, *Working Plan for Providing Consulting Services for Rizhao Steel* |

(continued)

**Table 6.1** (continued)

| No. | Type | Main content | Typical cases |
|---|---|---|---|
| 4 | Special studies | System building, extension of meaning, key tasks, development initiatives, etc., and strive to build a new order for orderly development worldwide | *Suggestions on Building a Fair Competition Environment in the Iron and Steel Industry and Cultivating Strong Iron and Steel Enterprises, Strengthening Synergy for Reorganization of Baosteel and Wuhan Steel and Building the World's Iron and Steel Flagship, Building a World-Class Steel Enterprise Cluster to Promote the Second Catching-up of China's Iron and Steel Industry, China's Contribution and Initiatives in Resolving Global Steel Overcapacity, Discussion on the Implementation of Debt-to-Equity in Some Iron and Steel Enterprises* |

## References

1. Li, C., Jiang, X.D., Gao, S., et al. (2017). Research on the development history and trend of policies in China's iron and steel industry. Steel Planning and Research, 04, P104–P111. (internal material).
2. Department of Industrial Policy, Ministry of Industry and Information Technology. (2015). Compilation for the Exchange Meeting on Elimination of Backward Production Capacity (Vol. I, pp. 132–152) (internal material).
3. Website of the National Development and Reform Commission. Report of Small Steel Plants' Illegal Production and Sales in Xinyi, Jiangsu Province by the Office of Inter-ministerial Meeting on Resolving Overcapacity and Profitable Development for National Iron and Steel and Coal Industry [2016-09-12]. http://www.ndrc.gov.cn/xwzx/xwfb/201609/t20160912_818278.html.
4. Website of the Central People's Government. Opinions on Well Resolving the Excessive Capacity of the Iron and Steel and Coal Industry in 2017 to Realize the Development [2017-05-12]. http://www.gov.cn/xinwen/2017-05/12/content_5193214.html.

# Chapter 7
# Quality Focus

Quality focused development, which means attaching importance to brand building and product quality improvement, is an important guarantee for an enterprise to achieve sustainable development. Quality is the cornerstone of brand building, service is the guarantee of brand promotion, innovation is the source of brand continuation, and culture is the essence of brand promotion. Through technological innovation and cultural construction, a "four-in-one" quality system of standardization, inspection, traceability, and informatization will be established so as to comprehensively improve product quality and service quality and fully build a brand system integrating quality, service, innovation, and culture, thus creating a brand enterprise with strong comprehensive competitiveness. From the development experience of the domestic and international steel industry, quality focused development is an inevitable choice for China's iron and steel enterprises to overcome the current obstacles and achieve sustainable development in the future.

## 7.1 History Review and Status Analysis

### *7.1.1 History Review*

1. Development History of Brand Building of Domestic Iron and Steel Enterprises

A brand is a comprehensive logo of an enterprise and the goods or services it provides. Its purpose is to identify the product or service of a seller or a group of sellers and distinguish it from products and services of their competitors. As the core embodiment of the enterprise's excellent quality, a brand contains many factors such as trademark, attribute, name, packaging, price, history, reputation, and advertising method, which imply the quality and reputation of the enterprise and its goods or services. A brand is not only the quality commitment of an enterprise to consumers, but also an enterprise's trust level obtained from consumers. Although it is an intangible asset, the social awareness created by a brand can enable an enterprise



X. C. Li, *The Road Map of China's Steel Industry*,
https://doi.org/10.1007/978-981-15-2074-7_7

to own strong cohesiveness and diffusive power, which will finally transform into the strong support and momentum for an enterprise's development. For consumers, a strong brand can enhance consumers' desire to buy; for producers, the brand is a sign of success. Therefore, brand building together with product quality, technology, management, service, etc., constitutes the core competitiveness of an enterprise. As for the understanding of brand building, iron and steel enterprises, like most manufacturing enterprises, have experienced the process from a unitary perspective to diversified ones, from tangible recognition to intangible one, from a company's own action to the requirements of the national strategy. As the iron and steel enterprises have increased their awareness in this regard, brand building has entered a stage of growth after the stage of formation.

Since steel products are mostly circulated in the form of intermediate products and unlike automobiles and home appliances, which are consumer-oriented, many iron and steel enterprises believe that the relevancy between a brand and an enterprise is relatively low. Therefore, for a long time, most iron and steel enterprises in China have neglected brand building, neglected the law of brand generation, and did not pay attention to science and technology investment, talent cultivation, and management innovation. The brand exists only as a product name and logo. At this time, the trademark of the product is equivalent to the brand, and the narrow brand has been formed. We define it as the period of brand building.

Since China's accession to the WTO, iron and steel enterprises have found that brands are so important to enterprises after getting more involved in the international market. Especially, in recent years, against the backdrop of the overcapacity, rising production cost, fierce market competition, weak domestic demand in the iron and steel industry, iron and steel enterprises have increasingly recognized that, in order to change the passive situation and achieve steady management and sustainable development, they must focus on enterprise brand building and meet the needs of customers to the greatest extent and win market recognition. In fact, from a certain perspective, brand value has become the lifeline for the survival of iron and steel enterprises. A dozen years after this, iron and steel enterprises have been continually exploring the road of brand building that adapts to their own development. Enterprise brands have been endowed with more value, cultural content, and individuality. Enterprises are experiencing their changes from simply establishing product names and symbols to creating high-end featured products and brand services, achieving integrity management and standardized management, fulfilling social responsibilities, and other comprehensive brand building endeavors. In the past two years, General Secretary Xi Jinping's important guidance of "promoting the transformation from made in China to created by China, from China's speed to China's quality, and from Chinese products to Chinese brands" has pointed out the specific development orientation of Chinese brands in the new context. Along this path, Chinese iron and steel enterprises have been constantly discovering new highlights for brand building and actively exploring a distinguished road of brand building so as to realize the rapid growth of their enterprise brands.

2. Development History of the Products and Service Quality of Domestic Iron and Steel Enterprises

From the nineteenth to the twentieth century, the global iron and steel industry developed rapidly and once became one of the industries with the greatest growing capability in world's industrialization process. Unprecedented improvement has been achieved in terms of the output value, product structure, as well as industrial technologies. Entering the twenty-first century, iron and steel is still an irreplaceable raw material for human beings and an important indicator for measuring a country's overall national strength and industrial level. Growing from nothing, China's iron and steel industry has also experienced a gradual development process. Product types and quality of domestic iron and steel enterprises are constantly improving, and the scope and depth of services provided to the downstream industries are also increasing.

(1) Product quality development process. The quality of China's steel products has experienced a development process from the inability to meet domestic demand to the basic satisfaction, and then to the continuous improvement of product quality. It has obtained strong competitiveness even in the international market.

During the 10th Five-Year Plan period, China's iron and steel industry embarked on a rapid development track, and in the 11th Five-Year Plan period, China's steel product structure was further optimized with complete product varieties and improved product quality. Self-sufficiency rate of most steel varieties reached 100%. Significant progress has been made in key steel varieties, high-strength construction steel plates, high-performance pipeline steel, steel for large hydropower stations, high-magnetic-oriented silicon steel, rails for high-speed railway, alloy materials for spacecraft, high-strength re-bars for earthquake-proof construction, and other high-performance steel materials have supported the development of the national economy-related fields and guaranteed the smooth implementation of major national projects and key construction projects such as the Beijing Olympic Games, the Shanghai World Expo, the West-East Natural Gas Transmission, the Three Gorges Project, the Beijing-Shanghai High-speed Railway, Manned Spaceflight, and Post-disaster Reconstruction.

After entering the 12th Five-Year Plan period, the contradiction in the quantity of China's iron and steel industry has been greatly weakened. Product structure adjustment was no longer simply a quantitative increase or decrease, but had begun to focus on improving the quality of steel products, facilitating the transformation, and development of the downstream industries, as well as promoting resource saving, energy saving, and emission reduction. During this period, improving the quality, grade, and stability of iron and steel products with large quantity, and market coverage was taken as the top priority of product structure adjustment.

Since 2010, along with the accelerating transformation and upgrading of the downstream industries, the structure of China's steel products has been further optimized, and the localization rate of high-grade special steel has been significantly improved. For example, the localization rate of high-magnetic-oriented silicon steel for large transformers, ultra-high-strength steel plates for sedans, high-grade gear steel, and bearing steel for machinery, high-temperature and high-pressure resistant boiler tubes for power generation equipment, and other high-grade special steel products has been continuously improved. Reduction of steel usage has become the development trend of this industry. High-strength and material-saving steel products, represented by the Grade III deformed steel bar, high-strength automobile sheets, and high-strength shipbuilding plates, have seen an increase in the output and proportion. The steel product structure was, therefore, further optimized and the quality of steel products was further improved. Taking the high-strength re-bars in China as an example, we have witnessed the steady increase of its output and proportion. In 2004, the output of the high-strength re-bars of Grade III (HRB 400) and above was only 3.4 million tons, accounting for 5.8% of the national total re-bars output; but in 2015, this output has increased to 140 million tons and accounted for 92.2% of the total consumption. Output of the high-end products like automobile sheets, home appliance panels, and silicon steel in China has also grown substantially. In 2004, the output of the cold-rolled and galvanized automotive sheets in China was only 3.2 million tons, and by 2015, the output had reached 16.39 million tons. In 2005, China's oriented silicon steel output was only 132,000 tons, and the self-sufficiency rate was only 33.8%. By 2015, the output had reached 904,000 tons, achieving domestic self-sufficiency basically. The product quality and performance also saw significant improvement.

(2) The development history of service quality. For a long time, China's iron and steel industry has attached importance to production, construction, and operation while ignoring the coordination and cooperation with customers in the downstream industries, and attached importance to products manufacturing while ignoring providing services. Integration with the downstream industries is in low degree and the contacts among iron and steel products' R&D, production, and application are not close. Especially in the period before 2005, steel iron and enterprises often entitled themselves the "king in the industrial chain", with a weak sense of service and low level of after-sales service. Downstream customers generally had a lot of complaints about them.

The low service quality and service awareness in the iron and steel industry will affect the upgrading of steel products and the promotion of new products. For example, high-strength re-bars and other high-performance steel materials with mature production technologies have experienced a long promotion period; domestic industrial application of some high-end products like nuclear power evaporator tubes with production conditions and stainless steel furnace tubes for ultra-supercritical power station boilers has also experienced a relatively long process.

In recent years, the overall service level of China's iron and steel enterprises has been improved, but most enterprises only focus on improving production process equipment and increasing product quality without paying enough attention to the demands of the downstream customers. Some measures to bring relations between the two parties closer are not in place, so that there is still a certain gap in adapting to the development of manufacturing industry. Here, is an example. During the joint product R&D between the iron and steel enterprises and the downstream customers, Chinese enterprises lag far behind the foreign enterprises in terms of the initial involvement. In recent years, only a few domestic iron and steel enterprises like Baosteel have made real-time early involvement with downstream customers in the joint product R&D.

Iron and steel enterprises in China lack the initiative to cooperate with downstream customers while just passively accept their requirements for material performance. With the development of steel products lagging behind, iron and steel enterprises still cannot get involved in the early stage to obtain the first-hand demand information so as to better guide the R&D of steel products. Currently, ThyssenKrupp in Germany and Nippon Steel and JFE in Japan have achieved an all-around early involvement in the R&D of automobile sheets. Nippon Steel and JFE have already conducted cooperation in depth with automobile manufacturers in Japan, the USA, and Germany. The two enterprises have been involved in the design, selection, and stamping of steel products in the early stage of new models' R&D and design to guide customers in selecting steel.

It is precisely because of the importance of service quality that mindsets have been changed at both the national and enterprise-level during the 12th Five-Year Plan period, emphasizing that iron and steel enterprises should transform their service concepts, enhance service awareness, and establish strategic cooperation mechanisms with downstream customers so as to promote the transformation of Chinese iron and steel enterprises from steel producers to service providers.

3. Development Process of Quality Focused Enterprises

Quality focused development, which means attaching importance to brand building and product quality improvement, is an important guarantee for an enterprise to achieve sustainable development. Therefore, the development process of quality focused enterprises is also the process of brand building and product quality improvement.

(1) Brand building. From the perspective of brand, brand effectively differentiates the products and services, and reduces the risk and complexity of purchases, and spreads the benefits that products or services can provide so that recognition will be fostered and trust will be established. In the past, enterprises engaged in the consumer goods sector have placed more emphasis on brands than manufacturers in the industrial sector. The growing contradiction between homogenized products and individualized demand in the industrial product market indicates that, in the future, marketing competition will be a brand battle and a competition for brand ownership.

Enterprise brand strengths include brand stability, brand leadership, and international awareness. In the long run, to achieve success in all the three dimensions, the brand building generally employs the following five steps: brand planning, brand analysis, brand strategy, brand establishment, and brand auditing.

Brand planning. The key issue of brand planning is to strike a balance between sustainability and involvement. In order to achieve such balance, the following processes, steps, and procedures must be integrated within a enterprise: first, create an atmosphere for continuous transformation and the management should focus on the strategic orientation of the brand; second, determine the process of timely delivery of information, complete the competitiveness analysis report of brand positioning and brand recognition; third, develop relevant action plans based on the competitiveness analysis; fourth, ensure strong implementation; fifth, let everyone get involved into the plan.

Brand Analysis. On the basis of sufficient market research, the consumers, competitors, and the enterprise itself are analyzed in detail to determine and form the brand mission, characteristics, and brand value in line with its vision.

Brand strategy. Brand strategy formulation includes the brand's target market, brand positioning as well as brand architecture. Among them, the brand architecture includes three modes, which are company-led, product-led, and mixed-led.

Brand establishment. This includes establishing appropriate brand recognition, creating proper brand meaning, guiding right brand response, and fostering sound brand relationship with consumers.

Brand auditing. Brand auditing is designed to evaluate the strengths and weaknesses of a given brand or brand portfolio. The auditing usually consists of internal and external surveys.

(2) Quality construction. Quality concept is an important part of enterprise values and the core of brand awareness. Without a high-quality product as the foundation, it is impossible for an enterprise to build a famous brand. To create a famous-brand product, apart from promoting quality standards and ensuring the quality level, the key is whether the quality level of the product can meet the quality level demanded by consumers to the greatest extent, and whether an excellent quality culture has been formed. A sound quality culture will contribute to the long-term implementation of the brand strategy and is an integral part in the success of a quality focused enterprise. Creating a sound quality culture includes the following aspects:

The first is to establish a quality concept and formulate strict quality management standards; the second is to spread quality awareness through practical actions establish quality awareness through management tool innovation and implement quality awareness by relying on organizations; the third is to strengthen quality awareness through internationally accepted standard certification so as to improve the internal quality management level; the fourth is to gradually form its own unique quality management philosophy and quality culture, for example, Haier has formed its own featured High Quality Theory, OEC Management Mode, 6S Site Management Method, 6 Sigma Quality Management Method and so on; the fifth is the wide application of such quality culture.

Another example of Haier is that it has successfully merged many enterprises by using its enterprise culture, which is formed from the application of the intangible assets—quality culture. Quality culture has provided powerful support for enterprises to grow bigger and stronger.

### 7.1.2 *Status Analysis*

1. Brand Building Status of Domestic Iron and Steel Enterprises

For a long time, some iron and steel enterprises in China, such as Baosteel, Anshan Steel, and Taiyuan Steel, have been continuously implementing the brand image strategy based on enterprise culture, the top-quality strategy based on technological innovation, customer satisfaction strategy based on service innovation, and integrated spreading strategy based on core values. They have formed a brand cultivation system of their own and can guarantee to fulfill its commitment to the market and customers, thus establishing an enterprise brand. *China's 500 Most Valuable Brands* ranking list is evaluated by the World Brand Laboratory, chaired by the Nobel laureate in economic sciences and father of the euro, Robert Mundell, according to the financial analysis, consumer behavior analysis and brand strength analysis. The list has become an important basis for the intangible assets evaluation in many M&A processes with international authority. In 2016, (13th) *China's 500 Most Valuable Brands* list, five brands from the iron and steel industry rank among the Chinese national top brands based on financial, customer behavior, and brand strength analysis. See Table 7.1.

In 2013, on the basis of *Guiding Opinions on Accelerating the Brand Building of China's Industrial Enterprises* jointly issued by the Ministry of Industry and Information Technology and the National Development and Reform Commission, China Iron and Steel Association released the *Notice on Launching the Pilot of Iron and Steel Enterprises Brand Cultivation*, aiming to guide the enterprises which were implementing brand strategies to establish a comprehensive brand cultivation and management system, improve the ability and performance of brand cultivation, lead more enterprises to conduct brand cultivation in a more scientific way, and

**Table 7.1** Ranking of the *China's 500 Most Valuable Brands* in 2016

| Ranking | Brand name | Brand ownership | Brand value/$\times 10^8$ yuan |
|---|---|---|---|
| 41 | Baosteel | Baosteel Group Co., Ltd. | 887.73 |
| 56 | Anshan Steel | Anshan Iron and Steel Group Corporation | 447.18 |
| 104 | Shougang | China Shougang Group | 280.57 |
| 141 | Shagang | Jiangsu Shagang Group Co., Ltd. | 237.59 |
| 258 | Taiyuan Steel | Taiyuan Iron and Steel (Group) Co., Ltd. | 129.64 |

lay a foundation for accelerating the cultivation of a number of self-owned brands with international influence. Iron and steel enterprises such as Shagang, Baotou Steel, Liuzhou Steel, and Shenglong Metallurgy have been included into the list of the brand cultivation pilot enterprises of the Ministry of Industry and Information Technology. More steel enterprises in China have entered the stage of rapid growth in brand building with their brand cultivation system being increasingly mature and brand cultivation ability and performance improving significantly. But looking at the overall status of the brand building in China's iron and steel industry, although we have long been one of the world's largest iron and steel producer, well-recognized Chinese steel brands in the global market are rarely seen no matter in terms of plates, long products or steel pipes. Lack of brand influence is still one of the important issues facing Chinese iron and steel enterprises. In fact, most Chinese iron and steel enterprises are not failing to pay attention to brand building. The difference lies in how high the brand building is placed in each enterprise, and how to systematically manage the cultivation activities relating to brand value appreciation so as to make it more efficient with better effect. The problems are mainly manifested in the following three aspects: weak brand management awareness, insufficient planning for brand building system, and insufficient brand cultivation and promotion of key products.

China's iron and steel industry are in a critical period of deepening reform comprehensively. It is of more important strategic significance to promote steel brands building and improve the brand cultivation ability of iron and steel enterprises. Iron and steel enterprises should take honesty as the foundation, improving product and service quality as the core, promoting the value of products and services as the guide, and scientific brand cultivation methods as the orientation, pay attention to technical innovation and product quality assurance as well as scientifically and rationally determine the brand cultivation strategies, goals, and management methods, so as to continuously promote their brand building.

2. Current Product and Service Quality of Domestic Iron and Steel Enterprises

(1) Status of product quality. After nearly ten years of rapid development in China's iron and steel industry, the quality of steel products can basically meet the domestic demand.

During the 12th Five-Year Plan period, steel products quality has been greatly improved, effectively serving the transformation, and upgrading of the downstream industries. The number of steel products that have certified by the foreign advanced physical quality level and won the *Golden Cup Award* has reached more than 500, with their production volume accounting for about 40% of the total steel output. Breakthroughs have been made in the development of such key varieties as high-speed rails, steel pipes for 600 °C ultra-supercritical thermal power generating units, 690 U-pipes for nuclear power, nickel-based low-temperature steel for LNG ships, and steel for offshore platforms under level EH36. The third-generation automobile high-strength steel, high-grade-oriented silicon steel, T4003 stainless steel, and other products are in the world's leading level.

Re-bars products: Production ratio of high-strength re-bars and anti-seismic re-bars has increased significantly. Finished rolled re-bars have been put into systematic production and the highest strength can reach 1,080 MPa. Finished rolled re-bars PSB500, PSB785, PSB830, PSB930, and PSB1080 have been in volume production and widely used in the large-scale water conservancy projects, industrial and civil buildings, and the large and medium-span bridges for roads and railways. High-strength finished rolled re-bars with level of 1,230 MPa have entered the research and development stage.

Automobile sheets: All varieties of steel products for the automobile and home appliance can be provided by domestic large iron and steel enterprises represented by Baosteel, Wuhan Steel, and Anshan Steel in large quantities with stable and reliable quality. High-end cold-rolled automobile sheets for domestic and Korean, Japanese, German, and other joint-venture cars can be provided by Chinese enterprises. It is understood that most of the steel products used by China's leading home appliance enterprises such as Meidi, Gree, and Haier are domestically made.

High-quality special steel bar: Remarkable progress has been made in its development and production as the quality of bearing steel and gear steel has been closed to the international advanced level. Taking bearing steel as an example, the average oxygen content of CITIC Pacific Special Steel Group's bearing steel is $5.5 \times 10^{-4}\%$ with high purity of liquid steel and low composition segregation. Bearing made from such steel has a long fatigue life and the quality reaches the international advanced level. The bearing steel has been certified by many well-known bearing manufacturing enterprises (such as SKF) in the world. The oxygen content of the bearing steel products made by Northeast Special Steel can be controlled at $5 \times 10^{-4}\%$, the titanium content can be stabilized at less than $20 \times 10^{-4}\%$, the nitrogen content can be less than $60 \times 10^{-4}\%$, and Ds is below Grade 1.0.

Certainly, many problems still exist in the quality of domestic steel products, such as implementation standards of some domestic steel products are not strict enough so that it is difficult to guide products' optimization and upgrading; for some varieties with high technology content, their quality and performance stability have always had a certain gap with foreign products and the demand of domestic downstream customers; backward production capacity still exists in large quantity due to unsatisfying backward capacity elimination. In addition, basic research of China's iron and steel industry lacks a solid foundation. The technological progress of the iron and steel industry mainly relies on the introduction and absorption from foreign countries with insufficient supply of original innovation, lagging collaborative innovation between enterprises and customers, and inadequate R&D investment and application of cutting-edge technologies and key common technologies, which restrict the further brand quality improvement of iron and steel products.

In the future, the iron and steel industry should firmly establish quality awareness and build a brand system with quality as its core. Enterprises should actively adopt such quality improvement technologies as clean steel production, precise rolling, and consistent product quality management, and use tools and equipment with information and intelligent technologies to reduce human impact on quality control and improve the stability and consistency of actual product quality. Focusing on the quality benchmark, domestic enterprises should enhance their brand cultivation efforts and strive to produce more quality products in line with the international standards.

(2) Service quality. At present, some advanced iron and steel enterprises in China can be customer-centered with good service and quality while carrying out service standardized operations that can meet customers' requirements. Many iron and steel enterprises have extended their business to deep processing services, improved distribution, and logistics systems, provided standardized inventory management and timely delivery services, expanded sales channels, and strengthened technical support and after-sales services. They also provide material testing, steel saving, low-cost application technologies, and other services during processing. Some iron and steel enterprises even have opened up new service areas such as e-commerce and iron and steel Internet finance to promote service upgrading and create and share industrial chain value together with their customers.

Although the service awareness of iron and steel enterprises has been strengthened in recent years, the service quality still has a gap with the development of downstream customers. Initiative of China's iron and steel enterprises to cooperate with downstream customers is improving, but only some of them can make early involvement to obtain the first-hand information on demand to guide the R&D of steel products. Most iron and steel enterprises still lack deep cooperation with the downstream customers and few can participate into the design, model selection, and process test to guide the customers in selecting steel products in the early stage of R&D and design.

As a result, iron and steel enterprises should continue to strengthen the construction of the industrial chain. They should attach importance to establishing close cooperation with the downstream customers and timely understand or even guide their demands. With social progress, downstream customers do not only have new demands on the quality of steel products, but also put forward higher requirements for the relevant services. Iron and steel enterprises must keep in tight contact with customers so as to keep abreast of the changes in customers' demand. Facing increasingly fierce market competition, iron and steel enterprises should also push forward to the transform their service concepts and improve their customer management and service level. Compared with the products sales, providing services are of higher requirement and broader meaning, which covers the whole life cycle from product R&D, production, and application to recycling. Only by taking customers as the center and establishing an open service system that provides complete solutions for the downstream customers, iron and steel enterprises can continuously improve their service and take the lead in market competition.

3. Status of Quality Focused Development of Competitive Iron and Steel Enterprises

Today, the supply-demand gap of domestic steel products has been bridged, but problems in quality and varieties still exist. In particular, the development of key steel products in China lags behind, hindering further transformation, and upgrading of the iron and steel industry. Despite that China has become a net steel exporter, key steel products still heavily rely on imports. Steel products for some key areas such as offshore projects, aviation, and energy are not self-sufficient, which has brought bottlenecks to the development of the downstream industries. Domestic iron and steel enterprises with strong competitiveness like Baosteel, HBIS, and TISCO have strengthened their brand building and comprehensively improved product quality and service quality through technological innovation and enterprise culture construction. They have now embarked on the path of quality focused development.

Baosteel, as a new type of enterprise in China's reform and opening-up, has been shouldering the task of narrowing down the gap between the domestic iron and steel industry and the world's leading ones, realizing the leap-forward development and modernization of China's iron and steel industry, and meeting the demands for high-quality steel products in the national economic development. Through development of more than 30 years, the overall equipment and technical level of Baosteel has reached the international advanced standard, becoming the leader in the domestic iron and steel industry with its comprehensive competitiveness ranks among the top ones. Automobile sheets, home appliance panels, pipeline steel, electrical steel, ship plate, boiler and pressure vessel plate, steel for engineering and machinery, nuclear power steel, and other products produced by Baosteel have greatly met the needs of national economic development and provided strong support for the development of automobiles, home appliances, electricity, shipbuilding, energy, machinery, and other industries. It should be mentioned particularly that Baosteel has defined its automobile sheets as the strategic products. As it has successfully developed the 05 sheets, cold-rolled high-strength steel, and other automobile steels, Baosteel has established its leading position in the domestic automobile sheet field. With the world-class quality of the automobile sheets, the enterprise is now ranking among the first tier of the world's automobile sheets supplier.

(1) Market share. Currently, the market share of Baosteel's automobile sheets accounts for about half of the domestic market. The sales volume and market share of Baosteel's auto sheets have ranked the first in China for many years. Figure 7.1 shows the sales volume and market share of Baosteel's auto sheets in China in recent years [1, 2].
(2) Technology R&D. On the path of promoting quality focused development, Baosteel attaches great importance to the development of talent team and investment in technology research and development, and intensifies technological innovation, especially to enhance the reserves of future innovative products.

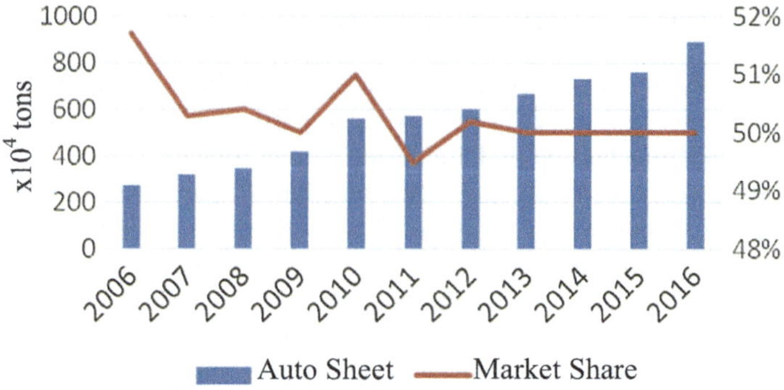


**Fig. 7.1** Sales volume and market share of Baosteel's auto sheets

R&D personnel. Baosteel boasts the most powerful R&D team for the development and application of automobile sheets in China. In 2016, the total number of R&D personnel in Baosteel was 1,060, accounting for 2.8% of the company's employees.

R&D investment. In 2016, Baosteel's R&D expenses amounted to 3.662 billion yuan with an input rate of 2.0%. In 2015, 623 items with intellectual property rights were newly approved in China, including 326 invention patents, 253 new trial patents, and 44 software copyrights.

Development of new products and new technology. In recent years, new product development has made outstanding progress in Baosteel. Global issuance was achieved in the third-generation automobile steel QP1180GA and medium-manganese steel; BW300TP new wear-resistant steel has been successfully applied in the production of CIMC' s mixer trucks, extending the machine service life by more than two times. S-class ultra-high toughness drill stem successfully realized drilling of 8038 meters, breaking the record of the deepest well in Tarim Oilfield. NSGO B27R080, B30R090, and coiled tubing CT110 achieved world debut. All these have strongly supported the development of equipment manufacturing industry and major engineering in China. Products from the first domestic industrialized demonstration production line of thin strip continuous casting and rolling independently integrated by Baosteel has been put into the market in batches, marking a major breakthrough in the technological innovation of China's thin strip continuous casting process.

(3) Scientific and technological achievements. Baosteel adheres to take the path of technological innovation. Development and industrialization of low-temperature NSGO manufacturing technology have been realized. Innovation and application of 600 °C ultra-supercritical thermal power unit steel tube have won the first prize in the National Prize for Progress in Science and Technology. Independent-integrated UOE welded pipe production process has won the first prize in Metallurgical Scientific and Technological Progress Award.

Achievements, such as the integrated technology R&D for advanced high-strength (super-high-strength) thin strip steel product and equipment, development, and industrialization of high-grade non-oriented silicon steel manufacturing technology, have won the first prize in Shanghai Scientific and Technological Progress Award.

(4) Standard system. In order to meet the needs on production, research and sales of automobile sheets, Baosteel has sped up the standardization of new technologies and new products by adopting the technical standard strategy and playing the leading role of standards. Through these efforts, it is striving to demonstrate and reflect Baosteel's technology in domestic and international standards, expand its influence and grasp the initiative in market competition. A series of enterprise standards have been developed in succession, and enterprise standard system for Baosteel's automobile sheets has been initially shaped. Under the support of the authorities in charge of metallurgical industrial standards, Baosteel has planned and prepared the national standard system for automobile steel, filling the gap of China in this regard. At the same time, through the implementation of the technical standard strategy, on the one hand, the industrialization of innovation results is achieved, which can reflect the value of innovation results, on the other hand, market competitiveness of Baosteel's automobile sheets is further improved.

(5) Quality culture. Taking Volkswagen as an opportunity, Baosteel proposed "two transformations", i.e., to transform the product quality from satisfying the standard to meeting customer's requirements, and put forward the concept of "standard + $\alpha$". $\alpha$ is the additional technical conditions of the customers apart from the supply standard. The second is to transform quality improvement from the closed type (solving the enterprise's own problems) to an open type, which means to put researches on the customer's needs and opinions on products in the first place, so that a "customer-centered" business philosophy has been gradually formed. With the enrichment of the "customer-centered" business philosophy, customer services have penetrated into the pre-sale, in-sale, and after-sale phases of products, including various aspects like product consulting and trial use, agent ordering, settlement, contract execution, supply channels, warehousing and transportation, cutting and distribution, KANBAN supply, disputes handling, technical support, customers' training, and early market intervention, so that customer service characteristics of Baosteel have been formed.

Baosteel advocates the "customer-centered" business philosophy and sticks to the strategy of "continuously meeting the needs of customers". With the idea of "focusing on customers, improvement, efficiency, value, and providing world-class products and services to society" which highlights the "consumer-centered" business philosophy, Baosteel takes making continuous improvement, quick response and efficiency increase as the starting point of all works and takes providing the society with the world's first-class products and services to maximize the value of the company and its social value as the foothold of all work, and these have become the guidelines for the company's sustainable development and continuous pursuit of excellence.

Based on the strategic goal of "becoming the world's most competitive iron and steel enterprise and the most valuable listed company worth investment", Baosteel takes improving competitiveness and achieving operational excellence as the striving direction of enterprise quality culture construction. The market "quick response" mechanism has been established. Through the implementation of ESI, it has established a continuous improvement mechanism for Six Sigma operations, implemented customization in large scale, created a learning-oriented organization, built a credit system, simplified departments, reorganized the business to achieve reengineering in a purpose to further improve work efficiency, management efficiency, and enhance Baosteel's core competitiveness.

(6) Brand core. Over the years, Baosteel has been committed to becoming "the best performance partner of leading automakers", which requires synchronization with the world-class quality assurance, execution capability, R&D level, and response speed. Therefore, "synchronization" has become a solid brand core of Baosteel's auto sheets. The brand meaning of "synchronization" fully reflects that Baosteel has been continuously made its auto sheets geared to the development of China's auto manufacturing industry, to the increasing demands of customers, to the continuous updating of the world's auto sheets technology, and to the comprehensive coordination and sustainable development of the society, which is the best interpretation of Baosteel's quality-oriented development.

## 7.2 Development Environment and Policy Orientation

### *7.2.1 Development Environment*

General Secretary Xi Jinping pointed out in the *Proposal of the Central Committee of the Communist Party of China on Formulating the 13th Five-Year Plan for National Economic and Social Development* that during the period of the 13th Five-Year Plan, China is still in a period of important strategic opportunities to achieve great development, but the connotation of the opportunity period will undergo profound changes. Specifically, the period of important strategic opportunities for China's development is shifting from "accelerating the development pace" to "changing the development mode", and from "rapid expansion of scale" to "improvement of developing quality and efficiency". This is not only related to the changes in China's own development, but also closely related to changes in the international environment. Recognizing the new characteristics of the current international and domestic economy and market environment will help the iron and steel enterprises to have a more comprehensive and profound understanding of the importance of the quality building in the iron and steel industry.

1. Quality building is the inevitable choice for the iron and steel industry to adapt to economic globalization

   (1) Medium and low-speed growth in the world economy accelerates economic globalization. After a "golden development period" of the world economy in the past few decades, the USA subprime mortgage crisis in 2008 triggered the global financial and economic crisis, bringing the world economy into a long period of deep adjustment. Both the developed and developing countries are facing profound pressure of adjustment in that the medium and low-speed growth of the economy has become a new normal state.

   According to the update report of *World Economic Outlook* released by the International Monetary Fund (IMF) in April 2017, under the support of the long-awaited periodical recovery in the active financial market, the manufacturing industry, and trade sector, the global economy in 2017 and 2018 registered a growth rate of 3.5% and 3.6%, respectively. In the medium to long term, the global economy will gradually emerge from the impact of the financial crisis. As it is estimated by the IMF, by 2020, global economic growth will reach 4.0%. The IMF believes that the downside risks facing the world economic growth include: first, rise of trade protectionism pose increased restrictions on global trade and immigration, which will damage the productivity and people's income and exert adverse influence on the market sentiment; second, long-term demand insufficiency of private demand and stagnant reforms in developed economies may lead to sustained declines in economic growth rates and inflation rates; third, potential vulnerabilities, such as high corporate debt, declining profits, weak bank balance sheets, and weak policy buffers still exist in some emerging economies, which may influenced by the balance sheet effect generated from the tightening financial environment, capital-flow reversals, and the sharp depreciation of the currency; fourth, geopolitical risks and other non-economic factors continue to affect the economic prospects in all regions. For advanced economies, the risk of continued low inflation still exists so that an easy monetary policy must be maintained and unconventional strategies should be adopted depending on the circumstances. But easy monetary policy alone is not enough to fully boost the demand. As a result, financial support is of critical importance to the growth momentum of developed economies. Emerging markets and developing economies should strengthen their financial resilience and risk management to effectively cope with the challenges brought by the tightening global financial environment, sharp fluctuation of exchange rate and capital-flow reversals.

   Economic globalization refers to the globalization of trade, investment, finance, production, and other activities, that is, the optimal allocation of production factors on a global scale. Fundamentally, it is the high development of productivity and international division of labor and the outcome which requires further crossing the boundaries of nations and countries. Economic globalization is one of the important features of the contemporary world economy and an inevitable trend of global economic development. In the new world economic environment, market opening has become an important impetus for world development. Economic globalization has not only provided a broader space for enterprises development but also presented great challenges. For enterprises, quality building in line with the international level will boost the development of enterprises, and at the same time, facilitate customers' recognition and sharing of products and services in various countries and regions. Faced with the new normal of medium and low-speed growth in the world economy, features like accelerating economic globalization, trade liberalization, internationalization of production, financial globalization, and globalization of science and technology have become more prominent.

(2) Quality building is a realistic need for iron and steel enterprises to adapt to the economic globalization and improve the level of international management level. Brand internationalization is an important means to implement the strategy of "going out", and quality is the most direct embodiment of the enterprises core competitiveness. With the acceleration of economic globalization, possession of internationally renowned brands, and high-level product quality has become an important means to lead the global resource allocation and market expansion and is the magic weapon for enterprises to win the market and customers. Against the backdrop of world economic globalization, the domestic steel market is facing severe challenges. Multinational steel giants such as Europe, Japan, and South Korea have rushed to occupy the Chinese market, putting tremendous pressure on the Chinese market while more and more Chinese iron and steel enterprises have moved to the global market due to the increasingly fierce competition. Competition in China's iron and steel industry has extended to the global market. The problem of "being big but not strong" in China's iron and steel industry has always existed, especially reflected by the lack of globally well-known brand. The competition among modern iron and steel enterprise is not only the competition of products and technology, but also the competition of quality. Implementing a quality focused strategy has become a sharp weapon for the marketing of domestic iron and steel enterprises. Faced with the radical changes in the domestic and international steel market environment, the survival and development of domestic iron and steel enterprises are under huge pressure. Implementing the quality focused strategy and creating world-class quality and brand will not only help the development of enterprises, but also facilitate the recognition and the sharing of products and services of customers in various countries and regions. It is an inevitable choice for enterprises to become bigger and stronger and to face up to the challenges of economic globalization.

2. Quality building is an important strategy for the iron and steel industry to respond to the new market atmosphere

   (1) The new market atmosphere facing the industry.

1) Overcapacity in the iron and steel industry will intensify the competition. In recent years, with the large-scale construction of China's iron and steel projects and the release of production capacity, China faces severe overcapacity in the iron and steel industry as the demand growth slows down. China's crude steel capacity utilization rate has been 66% by the end of 2016.
2) Intensified trade friction has increased the difficulty in steel exporting. Due to the quality improvement of China's steel products and the low-cost advantage in competition, steel exports have increased significantly since 2004. In 2015, China's steel exports reached a record high with a total volume of 112.399 million tons, an increase of 19.9% on the previous year. However, with the growth of China's steel exports, trade frictions have also increased, and trade protection has become an important factor hindering China's steel exports. In 2015, there were 37 cases of trade friction in the iron and steel industry, involving a total amount of 4.7 billion US dollars; in 2016, 21 countries or regions initiated 49 investigations, involving a total amount of 7.9 billion US dollars, a year-on-year increase in the number and amount of cases. Under such influence, China exported 108.49 million tons of steel in 2016, down 3.5% from the previous year.
3) Enterprises have generally optimized their product structure, and the competition in the high-quality and special steel field become fiercer. In recent years, with the intensification of overcapacity in China's iron and steel industry, China's steel products have also entered an era of overall surplus, especially the excessive surplus in products such as re-bars, wire rods, and medium and heavy plates, and even in the high-grade steel varieties like oriented silicon steel, tire cord steel, stainless steel, etc. As a result, China's iron and steel enterprises have proposed to optimize and upgrade their product structure and develop competitive varieties that fit their realities. Many enterprises with a focus on plain steel have extended their businesses to the field of high-grade and special steel, making the competition in this sector more severe.
4) Customers put new and higher requirements on quality. As the downstream customers are committed to transformation and upgrading, there some new changes in the demand for steel materials: more common use of high-strength steel, and increasing requirements on steel's comprehensive performances. For example, in the machinery industry, steel for engineering machinery is transforming to high-strength and high-wearing type with special performances; in the shipbuilding industry, demand has been raised for the corrosion resistant, ultra-low-temperature, high-strength plates with ultra length, width and thinness or of special shape in building the liquefied natural gas (LNG) ships. At the same time, customers have more urgent requirements for the stability, consistency, and applicability of steel products and lay more emphasis on the integrated pre-sale, in-sale, and after-sale individualized service.

On the whole, the demand of downstream customers for iron and steel enterprise has evolved from pure quantity to quality focused requirement integrating quality, variety, and service.

(2) Quality building is an important strategy for the iron and steel industry to respond to the new market atmosphere and grasp the new competitive edge. Confronting severe market environment with weak domestic and external demand, serious overcapacity, and fierce competition of product homogeneity in the steel market, for the iron and steel enterprises to reverse the passive situation and achieve stable operation and sustainable development, they must actively adjust their development and management strategies, produce brand products with high quality, high credibility, high market share, and high economic benefit through market segmentation, specialized production, individualized service, and enhance the core competitiveness with quality and brand advantages to win market recognition.

### *7.2.2 Policy Orientation*

Facing the complicated domestic and international economic situation and severe market environment, China put forward a series of guiding opinions on quality building from the perspective of policies, which is of great significance to the quality focused development of China's iron and steel enterprises.

1. *Guiding Opinions on Accelerating Brand Building of Industrial Enterprises in China* ([2011] No. 347, Ministry of Industry and Information Technology).

In July 2011, the Ministry of Industry and Information Technology, the National Development and Reform Commission, and other departments jointly issued the *Guiding Opinions on Accelerating Brand Building of Industrial Enterprises in China*. Its main purpose is to implement the guidelines of "promoting the building of independent brands, enhancing brand value and effects, and pushing forward the development of the large-scale enterprises owning globally well-known brands and international competitiveness" pointed out in the *Outline of the 12th Five-Year Plan for National Economic and Social Development of the People's Republic of China*. "Accelerating brand building of China's industrial enterprises is an inevitable requirement for promoting economic restructuring, transforming development models, and taking a new road of industrialization with Chinese characteristics; an objective need to adhere to the strategy of expanding domestic demand, release consumption potential, and enhance international competitiveness; a support to promote the development of industrial innovation and the transformation of scientific and technological achievements into practical productivity; a solid foundation for establishing and maintaining the quality reputation and creating an international image and influence of Made in China" are proposed in the guiding opinions. The overall goal of brand building in the opinions is: "By 2015, China's industrial enterprises' innovation ability and brand cultivation ability will be significantly enhanced, and the market environment for the growth of industrial enterprise brand will be significantly improved. More than 50% of the large and medium-sized industrial enterprises will have formulated and implemented their brand strategies and their market share of brand products and brand added value will increase significantly. Focus will be attached on cultivating a number of self-owned brands with international influence".

This guiding opinion has raised the brand building to a strategic level that affects the overall development of the national economy and is of great significance for understanding the brand building of industrial enterprises. More and more iron and steel enterprises have incorporated brand building into their enterprise development program, and the international influence of Baosteel, Shougang, TISCO, and so on has been continuously enhanced. According to the *Proposal of the Central Committee of the Communist Party of China on Formulating the Thirteenth Five-Year Plan for National Economic and Social Development*, during the "13th Five-Year Plan" period, quality building is still the priority of development of manufacturing enterprises, and the main ideological line of the *Guiding Opinions on Accelerating Brand Building of Industrial Enterprises in China* will be maintained.

2. *Quality Development Outline* (2011–2020) (No. 9 [2012] of the State Council)

In February 2012, the State Council issued the *Quality Development Outline (2011–2020)*. Development goals put forward in the Outline are: "By 2020, the construction of a strong country in quality will achieve remarkable results, quality foundation will be further consolidated, the overall quality level will be significantly improved, and the achievements of quality focused development will benefit all the people. A group of competitive enterprises with internationally renowned brands and core competitiveness will be cultivated, a batch of modern enterprises and industrial clusters with outstanding brand image, complete service platform, and first-class quality will be fostered, the inspection system for product quality safety and key products' quality will be established basically. Therefore, a solid foundation will be laid for building a moderately prosperous society in an all-round way and basically achieving socialist modernization in the middle of this century".

The Outline puts forward the work policy of "winning by quality" and requires to "comprehensively improve the quality management level, promote the construction of a strong country in quality, and boost the sound and rapid development of the economy and society". This has played a guiding role for the iron and steel industry to improve product quality awareness.

3. *Guiding Opinions on Accelerating Brand Building of Central State Owned Enterprises in China* (No. 266 [2013] of SASAC)

In December 2013, the State-Owned Assets Supervision and Administration Commission issued the *Guiding Opinions on Accelerating Brand Building of Central Government Enterprises in China*. The main purpose of this document is to help to improve the brand building level of central government enterprises, facilitate the transformation and upgrading of central government enterprises, and achieve the goals of strengthening and perfecting central government enterprises and cultivating world-class enterprises with international competitiveness. It points out that "central government enterprises, as a major force in the international competition, must lead China's mature products, technologies and standards to go global, compete or cooperate with multinational companies in a wider field and at a higher level and strive to build a brand strength that matches the economic strength through positively building international famous brands". The guiding thought is to "gradually establish a sound institutional mechanism for brand cultivation, protection, and development of central government enterprises and achieve the goal of 'being stronger and better with world-class strength' by adhering to the scientific outlook on development as the guide, transformation of economic development mode as the main principle, independent innovation as the core, high quality as the cornerstone, refined management as the guarantee, and honesty as the lifeline". The main objectives are: "By the end of 2020, a number of enterprises with clear brand strategies, sound brand management systems, and outstanding brand building achievements will emerge; a number of well-known brands with high-quality products, excellent service, and wide influence will be formed; a batch of self-owned brands with proprietary intellectual property rights and international competitiveness will be fostered".

This document puts forward the main contents and implementation measures for the brand building of the central government enterprises, which charts the course for cultivating world-class central government enterprises, improving the level of international operations, winning new competitive advantages, and realizing the value maintenance and appreciation of state-owned assets. At the same time, it also provides guidance and reference for the state-owned and private iron and steel enterprises.

4. *Notice on Building Industrial Quality Brands in 2016* of the General Office of the Ministry of Industry and Information Technology (No. 104 [2016])

In February 2016, the General Office of the Ministry of Industry and Information Technology issued the *Notice on Building Industrial Quality Brands in 2016*. The notice is intended to require that "the local industry and information authorities and relevant industry associations should implement the *Proposal of the Central Committee of the Communist Party of China on Formulating the Thirteenth Five-Year Plan for National Economic and Social Development* and *Made in China 2025* based on the regional and industrial requirements to strengthen the planning of quality and brand building, formulate regional and industrial quality brand planning (action plan), or put quality improvement and brand cultivation at the prior place in relevant planning (action plan), as well as determine the work objectives and work content".

Quality and brand building will be included in the key tasks during the "13th Five-Year Plan" period to provide necessary conditions for a good start of the "13th Five-Year Plan". When iron and steel enterprises formulate their own "13th Five-Year Plan" for enterprise development, they should also incorporate quality building into the important development strategies by clarifying the direction and construction content of the enterprise's own quality building and carrying out works related to quality building in an orderly, efficient, and scientific manner.

## 7.3 Case Analysis

From the growth and development experience of the first-class iron and steel enterprises at home and abroad, product quality and service quality as well as building a brand system integrating quality, service, innovation, and culture are crucial to create a strong enterprise with international renowned reputation and comprehensive competitiveness.

### *7.3.1 Pohang Iron and Steel Co. Ltd. (POSCO)*

Founded in 1968, Pohang Iron and Steel Co. Ltd. (hereinafter referred to as POSCO) is the largest steel complex in Korea and one of the most competitive iron and steel enterprises in the world.

The late 1990s was the second half of the rapid development of POSCO. During this period, POSCO was committed to improve customers' satisfaction and had become a world-class iron and steel enterprise through high-quality products and competitive technologies. POSCO successfully combined the traditional production process of blast furnace with new technologies like thin slabs to ensure an environmentally friendly and competitive process to the greatest extent, and to achieve greater competitive edge through high value-added products, and appropriate product combinations.

The auto sheet is POSCO' s flagship product and the key to keep its optimal competitiveness. Although the output of auto sheets accounts for only about 24% of POSCO' s total volume of products, the operating profit account for more than half of its total profit. Vehicle manufacturers across the world are using the auto sheets made by POSCO due to its excellent brand and quality. At present, the development direction of POSCO' s auto sheets is mainly making high value-added products such as TWIP steel and hot-forming steel. At the same time, it is actively promoting the application of magnesium sheets in automobiles. POSCO' s future development strategy is to steadily increase the sales of high value-added steel plates to improve economic performance.

In recent years, POSCO has maintained a strong competitive advantage by vigorously developing World Premium (WP) products and providing solution-based marketing service. Despite the sluggish global steel market, POSCO's 2016 financial report showed that its steel sales volume fell 5.0% to 24.32 trillion won due to the average sales price of carbon steel decreased 6.6% from KRW 606,000/ton in 2015 to KRW 566,000/ton in 2016. However, by expanding the sales of high value-added WP products and reducing costs while increasing efficiency, its operating profit increased by 17.7% to 2.64 trillion won with an profit margin of 10.8%, up 2.1% from the previous year. This was the first time since 2011 that the operating profit margin has reached double digits; the net profit in 2016 was 1.79 trillion won, with a significant year-on-year increase of 35.4% and the debt ratio fell to 17.4%, which is the lowest level since it was founded. POSCO aims to improve its sales volume of WP products to 16.86 million tons in 2017, with an increased sales proportion to 52% and increase the number of products sold through solution-based marketing to above 4.5 million tons.

Thanks to its high-quality steel products and efficient marketing services, POSCO has earned considerable profits and achieved sustainable development, which well demonstrates what a world-class enterprise with quality focused development is.

## *7.3.2 Nippon Steel & Sumitomo Metal Corporation (NSSMC)*

Nippon Steel & Sumitomo Metal Corporation (NSSMC) was formed through a merger between Nippon Steel Corporation and Sumitomo Metal Industries Co., Ltd. in 2012. For many years, high-strength auto sheets, high-corrosion resistant ship plates, high-power electrical steel sheets, and special material for steel cords are the four leading cutting-edge products of Nippon Steel. The company's focus is on materials for the high value-added sheets, pipes, and metal products in machinery, electronics, automobiles, home appliances, light industry, etc. These high-grade auto sheets and high-grade long products have brought enormous economic benefits to Nippon Steel.

With strong technical strength, Nippon Steel & Sumitomo Metal has long been committed to the development of high-end steel products and it can produce most of the special steels for world's supply. At present, Japanese iron and steel enterprises, including Nippon Steel, top the world in terms of output and export volume of high-quality and high-value-added steel.

Nippon Steel & Sumitomo Metal has been at the leading level in such products as the ultra-high-end seamless steel tubes, special engineering and machinery steel, and high-end auto spare parts. Based on the close cooperation with the automobile and energy industries, it has established a long-term common development mechanism with the customers. Meanwhile, after the integration of the manufacturing technology, commodity technology, and R&D capability of Nippon Steel Corporation and Sumitomo Metal Industries Co., Ltd., the company has further consolidate its strong market status by continuously improving the productivity and the comprehensive competitiveness of high value-added products via process innovations.

Nippon Steel & Sumitomo Metal uses its advanced technology and brand advantages to implement the globalization strategy so as to seize the emerging high-end steel consumption market. The main featured product of Nippon Steel & Sumitomo Metal is the high-grade auto sheets. The company has established joint ventures with ArcelorMittal and Baosteel, a leading domestic enterprise, thus increasing the production capacity of high-grade auto sheets on a global scale and becoming a technological leader in auto sheets manufacturing. High-grade auto sheets production bases of Nippon Steel & Sumitomo Metal locate in Mexico (TENIGAL, with annual production of 400,000 tons), Thailand (NSGT, 360,000 tons), India (JCAPCPL,

600,000 tons), etc., while the production bases of high-grade steel for energy locate in the USA (NSBS, 1.4 million tons), Vietnam (CSVC, 1.2 million tons), etc. In the future, it will continue to expand overseas production lines and seize the global market.

Nippon Steel & Sumitomo Metal aims to become the world's first iron and steel enterprise in comprehensive strength. It has been constantly pursuing the world's most advanced technologies and manufacturing capabilities to provide customers with quality products and services. Through promoting global development, strengthening cost competitiveness, and leveraging the advantages of advanced technologies, it will build an optimal production system and enhance the system of an iron and steel group company. The development path of quality-assisted internationalization is well reflected by Nippon Steel & Sumitomo Metal Corporation. The company has been vigorously optimizing its production technology, continuously increasing the market share of its flagship products while enhancing innovation, improving services, and expanding the downstream industry chain of high value-added products. As a result, it ultimately achieves sustainable development.

### *7.3.3 ArcelorMittal*

ArcelorMittal is the world's largest iron and steel enterprise with businesses covering more than 60 countries and regions. ArcelorMittal is also the world's leading iron and steel enterprise, with its downstream markets, including automobiles, construction, household appliances, packaging, etc. Boasting the world's leading research and development capabilities and technologies, the company's businesses cover the world's major iron and steel markets, from the emerging markets to mature markets.

ArcelorMittal's becoming the world's best steel company is inseparable from its strong technological innovation system centering on product R&D. The R&D work of ArcelorMittal is carried out by the members of the Group's Management Board and the Senior Executive Vice President. The funds are all raised internally from the enterprise and directly allocated by the President for the product and process, quality assurance as well as long-term strategic development and research.

Auto business is one of ArcelorMittal's main businesses. In the global automotive steel sector, ArcelorMittal ranks the first among all suppliers of flat products. It also provides tailor-welded blanks, long products, and pipe products. In the field of advanced high-strength steel, coated steel, and patented products, ArcelorMittal is also a forerunner in technology. The company has an automobile department, which specializes in the production, R&D, and service of auto steels and the related products. It has established dozens of service centers related to the automobile industry in Europe in which hundreds of researchers have long been engaged in the design and research of the steel varieties for auto sheets and customer application technologies, such as stamping simulation, component performance, welding, etc. ArcelorMittal also built its own R&D laboratory in many countries.

Arcelor Automotive Steel R&D Center has established a common "design and partner" partnership with automobile manufacturers. Such partnership has ensured a close strategic collaboration with customers throughout the entire process of automobiles' development, design, and production. The center develops a full range of new automotive steel varieties, and researchers have conducted collaborative researches in hundreds of projects with car manufacturers, equipment manufacturers, and distributors. Currently, the group company has set up its new technology centers, respectively, in Detroit, the USA and Tokyo, Japan to jointly develop new series of steels for automobiles and strengthen cooperation with automobile manufacturers and equipment manufacturers.

Automotive steels of ArcelorMittal are outstanding in performance, including coating and surface treatment. Its technologies for external parts' galvanized steel and hot-stamped coated steel are all at the forefront status. By striking the best balance between light weighting and formability and assisted by its own international network and the co-processing centers, tailor-welded blank manufacturers and stamping manufacturers, the company provides products and technical services on a global scale.

ArcelorMittal's emphasis on the quality focused development by relying on innovation and its development model featuring innovation-driven brand building are worth learning by large domestic iron and steel enterprises.

### *7.3.4 China Baowu Steel Group Corporation Ltd. (China Baowu)*

China Baowu Steel Group Corporation Ltd. (hereinafter referred to as "China Baowu") is the most powerful and modernized iron and steel enterprise in China. Apart from its unique product quality, its product service concept and service level are also far ahead among all domestic iron and steel enterprises.

China Baowu is the first one in China to carry out product R&D and customer technology research with the "early involvement" model, thus creating a strategic alliance of "Industry-University-Research Application". Creating value for customers is the long-standing service concept that China Baowu has been adhering to. Targeting at the needs and characteristics of China's auto industry, China Baowu carries out the "early involvement" to participate in the early stage of automobile design, provide technical solutions for steel materials application, conduct researches on advanced materials as well as provide technical support during application so as to meet the needs of automobile manufacturers. As a result, it has promoted the common progress of China Baowu and the automobile plants. For example, China Baowu has signed long-term technical cooperation agreements with many domestic automobile plants. It has carried out researches on auto steel sheets application with SAIC Volkswagen, built cooperation on auto body development and auto steel technologies with Hainan Automobile Company, and established a "Joint Laboratory of Auto Steel" with FAW; together with the power generation equipment manufacturing enterprises, research institutes, and higher-education institutions in Shanghai, a "Joint Entity for

Key Problems-Tackling of Equipment Materials and Its Heavy Castings and Forgings of the Power Station in Shanghai" was established to accelerate the progress of key materials for advanced power station equipment and heavy casting and forging and enhance Shanghai's major equipment manufacturing capability; it also established a "Joint Laboratory for Application Technologies" with the research center of Tsingtao Brewery Co., Ltd. to promote the application of tinplate. China Metal Packaging Research and Development Center was also founded to promote the overall technical level of China's metal packaging industry.

In order to enhance the opportunities and capabilities of early involvement and to provide customers with a "package solution", China Baowu also carried out researches on the extension technologies of related products. In the field of auto sheets, researches on the forming technologies of automobiles' spare parts such as hot stamping and hydroforming have been carried out, and industrialization of related technologies have been achieved, therefore, promoting the products application of China Baowu (hot-stamped steel sheets and high-strength steel) in domestic vehicles, stabilizing, and increasing the market share and facilitating the development of the iron and steel industry.

China Baowu also explores the "early involvement" that suits the development of the downstream industries. In the future, based on the automobile industry, it will expand into industries such as construction, home appliances, shipbuilding, and energy.

### *7.3.5 CITIC Pacific Special Steel Group Co. Ltd.*

CITIC Pacific Special Steel Group Co. Ltd. is a specialized group company controlled by CITIC Pacific Limited, a Hong Kong-listed company subordinated to CITIC Group. At present, CITIC Pacific owns such core enterprises as Xingcheng Special Steel, Hubei Xinyegang Steel, Tongling Pacific Special Materials, and Jiangdu Pacific Special Materials. The group now has an annual production capacity of 9 million tons of quality special steel and is one of the global professional special steel enterprises with the most comprehensive production process equipment, varieties, and specifications.

Special steel is the key material for major equipment manufacturing and national key projects. With high technological requirements, special steel's production and application represent the level of industrial development of a country. Quality requirements for special steel products are very strict, including narrow chemical composition, high purity, good surface quality, high shape and dimensional accuracy, compact and uniform macrostructure, and desirable metallographic structure. Therefore, the proportion of special steel in total steel volume and the product structure, quality, and application of special steel are important indicators reflecting the development level of a country's iron and steel industry.

CITIC Pacific Special Steel Group Co. Ltd. has been implementing the strategy of creating famous brands. Xingcheng Special Steel has obtained well-known trademarks and famous provincial trademarks in terms of domestic brand building and enjoys a high reputation among domestic and foreign counterparts. Special steels with high quality, high grade, high surface quality, and high added value in line with international quality standards have been developed by Xingcheng Special Steel. Some products have even occupied the high ground of domestic and international markets and met the needs of major domestic customers. "Xingcheng" special steel products mainly include high-quality bearing steel, high-pressure tube billet, spring steel, high-performance gear steel, free-cutting non-tempered steel, mooring chain steel, automotive steel, etc. With relatively a high domestic market share, these products have well satisfied the demand of the downstream market, earning a high reputation in the special steel industry at home and abroad. Among them, "Xingcheng" high-standard bearing steel is the only special steel enterprise in China that has passed the certification of Sweden's SKF, German FAG, French SNR, Japanese NSK, etc. "Xingcheng" anchor-chain steel has passed the certification of classification societies in the USA, Britain, Germany, Japan, ROK, China, and Norway; "Xingcheng" engineering machinery steel has been the qualified supplier for the world's largest engineering machinery manufacturer Caterpillar Inc. for many years.

CITIC Pacific Special Steel takes the professional, special, and refined route of quality focused development. Built on its high-quality special steel products, it has been continuously enhancing its brand influence at home and abroad, and the goal of it in the future is to become the most competitive special steel enterprise in the world.

## 7.4 Prospects and Path Analysis of Quality Focused Trend

### *7.4.1 Prospects of Quality Focused Trend*

Under the new normal, China's economic growth has transformed from high speed to medium-high speed. With the transformation of economic structure, development impetus, and development modes, the consumption intensity of steel per unit of GDP has decreased significantly and steel consumption has passed the peak value and begun going down. The "double reduction" of production and consumption indicates that China's iron and steel industry has entered a new era of development with decreasing quantity. During this period, iron and steel enterprises should attach great importance to brand building and product quality improvement to take the road of quality focused development.

Quality building covers both the brand building and product quality. The trend of quality focused development is to create a brand enterprise with strong comprehensive competitiveness by laying emphasis on brand building and promoting product quality.

Brand Building. Brand building is an integral part of an enterprise's decision-making and operation. It determines the image of the enterprise itself and the brand image of the enterprise's products and also directly affects the survival and development of the enterprise so that every enterprise should attach great importance to it. Quality is the cornerstone of brand building, service is the guarantee of brand promotion, innovation is the source of brand continuation, and culture is the essence of brand promotion.

Product quality. Product quality is the foundation for the survival and development of an enterprise. Stable quality and excellent products are the basis for doing a good job, and the key to winning the market and achieving success. Through technological innovation and cultural construction, a "four-in-one" quality system of standardization, inspection, traceability, and informatization shall be built, so as to comprehensively improve product quality and service quality, and fully build a brand system that integrates quality, service, innovation, and culture, therefore, creating a brand enterprise with strong comprehensive competitiveness.

### 7.4.2 Path Analysis

1. Brand Building

Doing well in brand building to support the sustainable development of an enterprises mainly centers on brand building in product quality, service, innovation, culture, etc., and improving the enterprise's comprehensive competitiveness through these quality focused development measures.

(1) Quality culture construction. Construction of corporate quality culture should be promoted to continuously improve the quality management level. The first is to establish and improve the quality management system. Enterprises must establish and improve a management system and a consistent product quality model with quality as its core. The second is to implement the quality responsibility system. The quality responsibility system, production condition confirmation system, and process compliance system should be fully implemented. The third is to focus on process control. Enterprises should intensify the internal control management mode of product quality, lean quality management activities in production process and information platform building for quality process. The fourth is to promote the construction of quality culture. Enterprises should comprehensively promote the construction of quality culture featuring "integrity quality".

(2) Construction of the service system. A secondary customer service system for the company and the workshop should be established. The company should be responsible for the service guarantee of the core product customers and strategic customers, promoting the connection with the downstream customers' industrial chain and guarantee of product research and development, promoting EVI service model, enhancing the interdependency and tightness of cooperation between the two parties, and maintaining the market status of core products. Each production plant is responsible for the quality assurance and special requirements of all products, forming a rapid resolution method for quality

objections, and assigning quality tracking to key customers. The workshops are responsible for quality assurance in the manufacturing process.

(3) Construction of technological innovation. Technological innovation and new product development efforts should be encouraged. In accordance with the idea of "a generation of promotion, trial production, research, and planning", iron and steel enterprises should continue to invest in R&D and talent teams, vigorously develop new products, and implement quality brand strategy. Relying on the integration mechanism combining internal Industry-Sale-Research and external Industry-University-Research, enterprises should adapt to market changes and tap potential market demand, strengthen the development, and promotion of new products and new technologies, make innovations for the customer demand-oriented products, and take the initiative in market competition.

(4) Cultural construction.

1) Building culturally advanced company. In the critical period of accelerating transformation and upgrading of the iron and steel industry, enterprises should start from the four dimensions of "cultural progress", "value creation", "production safety", and "green development" to carry out the building of culturally advanced companies internally. Such effort is of essential and practical meaning to guide the management to pay attention to the basic management work, enhance the ideological and moral cultivation of civilization of all employees, further reinforce the overall cohesiveness and effectiveness of the team, promote stable, and sound development, and improve the positive image of the enterprise. Formulating enterprises' documents needs to be integrated with the system construction. By making integration into the culture of employee responsibility, it can be more standardized on the foundation of laws so as to push ahead the implementation of building culturally advanced companies.

2) Perfection of Enterprises Culture Handbook. Cultivation of cultural atmosphere requires standardized management, guidance, and publicity. Therefore, it is very important to establish an effective publicity platform to maximize and sustain the positive cultural impact. Enterprise Culture Handbook is an effective formal publicity platform that can exert lasting influence.

   Enterprises should further optimize the Enterprise Culture Handbook with the goal of establishing a cultural handbook for internal learning and external communication. It should be ensured that the mission, vision, core values, related concepts guided by core values, and the enterprise spirit are complete and expounded clearly combined with the enterprise's key emphasis in the handbook. Content in the handbook should clearly answer the questions of "why", "what", and "how to do", and provide targeted descriptions at the organizational, individual, and departmental levels. Preparation of the handbook should be conducted from the perspectives of awareness, capability, behavior, effect assessment, mechanism improvement, and establishment of system process. In-depth enrichment of the concept and content of the culture handbook can be made in which the concepts of talent, quality, marketing, etc., corporate spirit and corporate image could be added. Meanwhile, consistency between the content and form in the handbook should be strengthened to make designing based on the overall style and ensure the color and picture are matched in each part and each page. Via the cultural handbook, the purpose of enhancing internal quality while building external image can be achieved.

2. Quality Improvement

(1) Quality system requirements should be improved, and product research and development production process should be optimized. Enterprises should implement full process management control by adhering to the links of contract review, quality planning, and process control, etc., and set up an inspection point to conduct review, analysis, and summary for critical technologies, quality, delivery, service, and others, improve the technical level and the ability in stabilizing the quality of key processes so as to ensure a stable product quality.

(2) Strengthen quality management and improve the stability of product quality. Enterprises should strengthen process quality control, take effective corrective or preventive measures to solve problems in-process quality control, continuously improve the capability in-process quality control, and improve process management level; they should establish process quality control points and adopt effective control methods to have control on key processes and ensure the normal processes are under control; meanwhile, delicacy management should be strengthened and operating procedures should be strictly performed to reduce product quality problems caused by human and improve the overall product quality.

(3) Strengthen the Industry-University-Research Application cooperation study, accelerate the R&D of new products and improve quality. Enterprises should strengthen the Industry-University-Research Cooperation, carry out technical cooperation with relevant universities and research institutes, actively draw from advanced product development and production technologies at home and abroad, and realize the R&D, production, and sales services for high-end varieties through cooperative joint venture development and patent technologies transfer so that product quality and market share can be improved gradually.

While making cooperation with research institutes, they should attach great importance to the collaboration with downstream customers. Special new products for customers can be developed jointly with the customers through “early involvement” to extend the cooperative relationship from simple sales to comprehensive services, and the product quality can be improved via the first-hand feedback to enhance product competitiveness.

## 7.5 Industrial Practices of Quality Building

In the past several years, the concept of quality focused development has been actively implemented by China Metallurgical Industry Planning and Research Institute (hereinafter referred to as MPI) in its planning and consulting processes, which has guided the industry and enterprises to embark on a path of quality focused development.

As a staff department for the central government and local authorities at all levels, MPI participates in a large number of the researches and preparation for the policies in the iron and steel industry as well as in the development planning for some major steel provinces and regional industries. Among these efforts, requirements for quality focused development run through the whole process of policy-making and industry development planning. For example, the institute has participated in preparation of the *Planning for Adjustment and Upgrading in Iron and Steel Industry (2016–2020)*, in which chapters on "Improving Quality Level" and "Strengthening Brand Building" are illustrated and the importance of improving product quality is repeatedly emphasized through the text.

During making planning for iron and steel enterprises, MPI is also guiding them to head toward quality focused development. When doing consulting research of market analysis and product positioning for enterprises, it will have an in-depth understanding of downstream customers' demand for steel varieties, specifications, and quality, as well as market demand trend for product quality and service so as to chart the course for enterprises' product quality improvement, adjustment in variety, and structure and brand establishment.

As an state-level authoritative consulting institution for iron and steel, MPI has been continuously innovating the concept of consulting services, playing a role in innovation and transformation and leading the development of the industry, which includes the steel brand evaluation and iron and steel enterprise competitiveness rating. MPI has formulated the steel brand evaluation system, which evaluates the steel brands of domestic iron and steel enterprises from five aspects: quality, market, service, innovation, and brand building. By focusing on brand strategy, brand recognition, brand communication, brand crisis, brand assets, brand applications, etc., this system helps steel companies to cultivate their brands, expand brand influence, and enhance internal quality. In addition, MPI has carried out and issued the "Competitiveness Rating for Iron and Steel Enterprises" for seven consecutive years since 2011. The rating is based on three aspects, namely the enterprise's basic competitiveness, development competitiveness, and business performance competitiveness, 13 elements and 25 indicators. The result is evaluated combined with experts' study and assessment, exerting a strong influence inside and outside the iron and steel industry. Above two tasks have effectively aroused the attention of a great number of enterprises on the role of brand and quality in enhancing the overall competitiveness of enterprises, and have strongly promoted the quality focused development of the whole industry. In the future, China Metallurgical Industry Planning and Research

**Table 7.2** Practice of China Metallurgical Industry Planning and Research Institute in promoting industrial quality focused development

| No. | Type | Main content | Typical case |
|---|---|---|---|
| 1 | Policy research and formulating | Including the basic research and preparation of the national iron and steel industry development plan, a series of research projects in the major steel provinces and regional industrial planning, etc. | *Planning for Adjustment and Upgrading in Iron and Steel Industry (2016–2020)* *Study on Major Issues in China's Iron and Steel Industry in 2020—Variety* |
| 2 | Planning and consulting for enterprises | Including steel varieties, quality, brand, and other aspects | Development plan in the "13th Five-Year Plan" period for a batch of large and medium-sized iron and steel enterprises such as Wuhan Steel, HBIS, Shandong Steel and Ma'anshan Steel |
| 3 | Evaluation and rating of brand, competitiveness, etc. | Incorporating indicators like quality, innovation, service, and brand building into the evaluation of steel product brand and enterprise competitiveness | "Competitiveness Rating for Iron and Steel Enterprises" has been released for 7 consecutive years since 2011 |

Institute will continue to fulfill its role in promoting the quality building of the industry while being a good consultant for the government, an industry leader, and a think tank for enterprises.

The specific practices of China Metallurgical Industry Planning and Research Institute in promoting quality focused development of the industry is shown in Table 7.2.

## References

1. Annual Report of Baoshan Iron and Steel Co., Ltd., 2006–2016.
2. Memoir of Baoshan Iron and Steel Co., Ltd., 2006–2016.

# Chapter 8
# Standardization

## 8.1 History Review and Status Analysis

The metallurgical industry has always been at the forefront of industrial standardization. In the period from 1950s to 1970s, the standard system was initially established by referring to or adopting the standards of the former Soviet Union. In the period from 1980s to 1990s, the Japanese and American standards were adopted, and the products were based on Japanese standards and the methods were based on American standards. In the period from the 1990s to the beginning of the twenty-first century, the strategy of adopting "International Standards and Foreign Advanced Standards" was implemented, encouraging active adoption of international standards and European standards. Since 2000, new standards have been added and existing standards have been revised with focus on technological innovation, products upgrading, and customers' satisfaction. At present, the iron and steel industry involves a variety of standards. In materials sector of the industry, it includes standards for irons and steels, refractory materials, metallic ores, non-metallic minerals, cokes, carbons, pig irons, ferroalloys, metallurgical electromechanical equipment, and other products. It also includes standards for energy conservation and comprehensive utilization (metallurgical solid wastes, metallurgical energy conservation, metallurgical water saving, and clean production), safety production, and standard samples.

With the development of standardization, the current standard system and management measures cannot fully meet the actual needs. First, the scope of the standard system is too narrow, mainly limited to industrial products, engineering construction, and environmental protection requirements; second, the mandatory standard formulators are multiple, leading to authority overlapping; third, the government plays a dominant role in standards formulation, while the main market player, such as associations and enterprises, has less share in independently formulating standards that quickly reflect the market demand, resulting in insufficient standard supply; fourth, supervision measures for standard formulation, implementation, evaluation, standardization, etc., are insufficient and this is not conducive to strengthening in-process and post-event supervision.



X. C. Li, *The Road Map of China's Steel Industry*,
https://doi.org/10.1007/978-981-15-2074-7_8

At present, the *Standardization Law* is under revision. The General Administration of Quality Supervision, Inspection and Quarantine of the People's Republic of China, and the Standardization Administration jointly issued the *Guiding Opinions on Cultivating and Developing Non-Governmental Organization's Standards* and *Non-Governmental Organization Standardization Part I: Good Practices* (GB/T 20004/1-2016). The document pointed out that the unified management of compulsory standards shall be strengthened, and the mandatory national standards shall be subject to unified administration of the standardization administrative departments of the State Council, which will be responsible for project approval, numbering, publication, and external notification; non-governmental organization's standards shall be introduced and it stipulates that social organizations established according to the law can formulate non-governmental standards for voluntary adoption; the relevant provisions on corporate standards are improved, requiring the establishment of an open system for self-declaration of enterprise products or service standards and replacing the filing mechanism of enterprise product standards.

### *8.1.1 National Standards*

1. Standards for Raw Materials and Fuels
   (1) Development Status. At present, there are 102 national standards related to iron ores, and most of the standards are the ones for iron ore elemental analysis and measurement except two, which are the *Iron Ore Grade Division* (GB/T 32545-2016) and *Terminology of Iron Ores and Direct Reduced Irons* (GB/T 20565-2006).
   (2) Main Existing Problems. The existing iron ore standards are currently focused on analytical testing, and the standards of iron ore quality, especially the mandatory standards for harmful impurities such as S and P in iron ores, are yet to develop. As a result, the quality of imported iron ores in China in the past decade has been varying and there were even shoddy iron ores in disguise of good ones.
2. Technological Equipment Standards
   (1) Development Status. At present, the national standards for technological equipment in the iron and steel industry mainly include *Design Standard for Sintering Plants* (GB 50408-2015), *Code for Design of Iron Ore Pelletizing Works* (GB 50491-2009), *Design Code for Blast Furnace Ironmaking Works* (GB 50427-2015), *Design Code for Steelmaking Works* (GB 50439-2015), *Design Code for Section Steel Rolling Works* (GB 50410-2014), *Design Code for Sheet and Plate Rolling Process* (GB 50629-2010), *Design Code for Wire Rod Rolling Process* (GB 50436-2007), *Design Code for Cold Rolled Strip Lines* (GB 50930-2013), *Design Code for Cold Rolled Electrical Steel* (GB 50997-2014), etc. The standards are mainly

published by the Ministry of Housing and Urban-Rural Development of the People's Republic of China and the General Administration of Quality Supervision, Inspection and Quarantine of the People's Republic of China. The drafting enterprises are metallurgical design providers such as Capital Engineering & Research Incorporation Ltd. (CERI), CISDI Group Co., Ltd. (CISDI), WISDRI Engineering & Research Incorporation Limited (WISDRI), BERIS Engineering and Research Corporation (BERIS), Zhongye Changtian International Engineering Co., Ltd. (CIE), and Northern Engineering & Technology Corporation (NETIC).

(2) Main Existing Problems. The main problem existing in the standardization of technological equipment is the lack of talents involved in the formulation of the standards. In the process of standard implementation, an effective problem feedback mechanism has not been established, which is not conducive to further updating and perfecting the standards. In addition, the supervision of the implementation of technological equipment standards is yet to be improved.

3. Product Standards

(1) Development Status. After the adoption of the standards of the former Soviet Union from 1950s to 1970s and the international standards and advanced foreign standards from 1980s to 1990s, the formulation and revision of product-related standards of China's iron and steel industry mainly centered on technological innovation, product upgrades, and satisfaction of customers' needs since 2000. The product standardization has transformed from production orientation to trade orientation and has played a supporting role in the transformation of the iron and steel industry to the market economy. Since the 11th Five-Year Plan, the standardization level of steel products has been continuously improved, the applicability of standards has been continuously enhanced, and the standard age has been gradually shortened. At the same time, China has developed a number of new product standards that are urgently needed in the market, filling in gaps in domestic-related product standards. In recent years, the national relevant metallurgical product standards have not only played a role in eliminating backward products, but also promoted the structural adjustment and upgrading of the iron and steel industry by actively promoting the standardization of high-end metal structural materials and special metal functional materials.

(2) Main Existing Problems. The national standard formulation and revision process for metallurgical products is still carried out according to the traditional model for many years. The formulation and revision of the relevant standards are mainly planned and approved by government departments, and there are restrictions on standardized management systems and mechanisms, resulting in a series of problems such as slow response of product standards to the market, delayed formulation and revision of standards in connection with emerging fields, and the difficulty of cooperating with the steel-consuming industries.

4. Energy Standards

    (1) Development Status. The energy conservation standard is the basis of the national energy conservation system. Since the 12th Five-Year Plan, the National Standards Committee and the National Development and Reform Commission have approved a total of 206 basic national standards for energy efficiency, energy consumption limits, and energy conservation. Up to now, China has issued 73 mandatory energy efficiency standards, 104 mandatory energy consumption limitation standards, and more than 150 recommended national energy conservation standards, which have played an important role in cutting overcapacity, optimizing industrial structure and achieving energy conservation goals. The energy conservation-related standards committee includes the National Primary Energy Management Technical Standards Committee (TC20) and the National Energy System Technical Standards Committee (TC459), and the committee secretariat is set up at the China National Institute of Standardization. The Metallurgical Industry Planning and Research Institute undertook the routine work of the secretariat of the TC20/Metallurgical Energy Foundation and Management Standardization Working Committee.
    (2) Existing Problems. Compared with the current urgent need to cut overcapacity, promote industrial structure adjustment and optimization, and carry forward the ecological civilization construction, the energy conservation standard system is still not complete, the basis for implementation is still weak, the management mechanism is still not perfect, and some important energy conservation standards are missing, and some of the standards are technically backward and not timely updated. This is particularly true for the standards of energy efficiency of energy-using products, the energy consumption limit of energy-intensive industries, and the energy efficiency of buildings, which are in urgent need of updating.

5. Environmental Protection Standards

    (1) Development Status. On October 1, 2012, the new environmental protection standards for the iron and steel industry were officially implemented, covering eight emission standards such as "*Pollutant Discharge Standards for Iron Ore Mining and Beneficiation*" and "*Air Pollutant Emission Standards for Sintering and Pelletizing Industry*", which were significantly tighter than the old standards. See Fig. 8.1 for more. Through the installation of online monitoring facilities in the key waste generation and discharge processes, the enterprise emission data is connected to the system of the environmental protection departments on real-time basis for strict control on illegal emission. At present, the improvement of environmental protection standards promotes the industrial application of a number of innovative governance technologies.

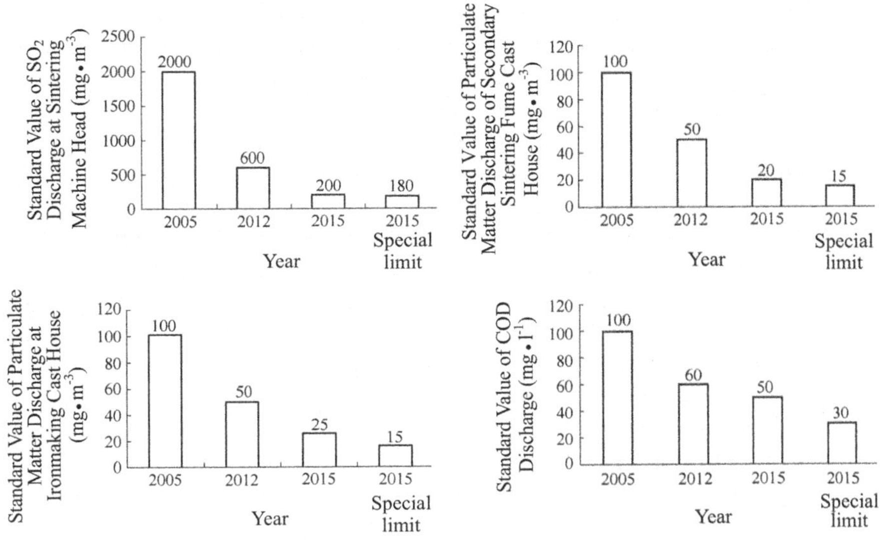


**Fig. 8.1** Comparison of new and old environmental protection standards in iron and steel industry

(2) Main Existing Problems. The emission standards of gas pollutants in the sintering process of the iron and steel industry will be upgraded with the successful commissioning of the integrated elimination technology for the pollutants from sintering process, especially for the emission concentration of NO*x* and dioxins. Formulation of the standards shall fully solicit advice from environmental experts in the iron and steel industry. At the supervisory level, it is urgent to promote the dioxin online monitoring system in the iron and steel industry through technological innovation, and to monitor the emission concentration in real time to ensure that the supervision work of the environmental protection department is well grounded.

6. Water Standards

(1) Development Status. The State Council's *Opinion on the Implementation of the Most Stringent Water Resources Management System* ([2012] No. 3 of the State Council) proposes the formulation of mandatory water-saving standards and accelerates the formulation of national standards for water consumption limit in high-water consumption industries and service industries. The government attaches great importance to industrial water-saving standardization work, and the standardization work of high-water consumption industries is entering a stage of rapid standard development and extensive publications. In recent years, in order to meet the demand for industrial water and adapt to the new water-saving situation, the government has accelerated the pace of water-saving standard system establishment. At present, a series of standards have been formulated and issued, including

water-consuming and water-saving basic standards, water-consuming and water-saving assessment standards, water quality standards, water-saving facilities and product standards, water treatment chemicals and materials standards, water-saving design specifications, etc.

Since *General Rules for the Preparation of Water Intake Norm of Industrial Enterprises* was issued, a scientific and reasonable standard system for water intake quotas of industrial enterprises was established. The terms, definitions, and calculation methods for water consumption per unit of product were established. According to the General Rules, the national standards for water intake quotas for high-water consumption industries such as thermal power generation, iron and steel, petroleum refining, cotton printing and dyeing, paper making, beer brewing, alcohol making, monosodium glutamates making, synthetic ammonia making, medicines, electrolytic aluminums, aluminum oxides, and coal beneficiation. have been published. In order to further promote the establishment of water-saving enterprises in the industrial sector, the government has successively formulated the following standards: *Water-saving Enterprise—Thermal Power Generation Industry, Water-saving Enterprise—Iron and Steel Industry, Water-saving Enterprise—Textile Dyeing and Finishing Industry* and *Water-saving Enterprise—Paper Making Industry* and *Water-saving Enterprise—Petroleum Refining Industry*. Among them, *Water Consumption Quota Part II: Steel Complex* (GB/T 18916.2) was released in 2002 and revised in 2012, which has contributed to the reduction of the water consumption per ton of steel in China's iron and steel industry from 40 to 3.5 $m^3$.

(2) Main Existing Problems. China's water-saving effort is still in a relatively extensive management mode. One of the important reasons is that the water-saving standard system is not complete. At present, there are too many blank areas of water-saving standards, which is difficult to adapt to the needs of the rapid socio-economic development of new technologies in the new phase of the new era. The severe backwardness of water-saving standardization restricts the orderly development of water-saving work and the effective improvement of water-saving precision management.

The publicity and training on water-saving standards are not in place, leading to the inability to accurately understand the scope and statement of the standards in the water-saving management level, causing various disputes in water-saving management, and even slacking the implementation requirements for water-saving standards.

7. Circular Economy Standards

(1) Development Status. The recycling utilization standard system includes green design and manufacturing, clean production, comprehensive utilization, industrial linkage, and circulated revamping. The pilot program for standardization of circular economy was launched in 2007. Up to now, the circular economy standard system has been substantially established and

improved, and a number of circular economy standards have been formulated to fill in the gaps. As of the beginning of 2017, the metallurgical industry's circular economy standardization has seen the formulation of 23 national-level standards. A set of typical models of circular economy standardization that can be replicated and promoted has been created, a number of versatile talents for circular economy standardization have been cultivated, and significant economic, environmental, and social benefits have been achieved. The standardization efficiency has been improved by using "Internet + Standards". The current resource recycling standard system is established on the basis of the theoretical connotation and framework of resource recycling, complies with classification of resource recycling, comprehensively considers life cycle stages of production, circulation and consumption and means of waste recycling, as well as emphasizes industrial symbiosis and linkage. There are currently five national-level standards under preparation.

(2) Main Existing Problems. The main problems existing in the standardization work of cyclic economy are that the standard system is still not complete, the quantity of standards is insufficient, and there are still many vacancies, failing to cover the full production and application fields; there is a deep gap to national policy requirements; the standards committees of each specialty are decentralized and difficult to form synergy, and their substantive participation in international standardization work is insufficient; the current standards are incomplete, outdated, backward, overlapped or in conflict, the standard system is not reasonable enough, and the standard coordination promotion mechanism is imperfect.

8. Logistics Standards

(1) Development Status. At the national level, China has not yet had a complete set of modern logistics standards system that is closely integrated with the supply chain logistics and steel product circulation logistics of the iron and steel industry. In the newly released "*Logistics Standard Catalogue Manual*" by the National Logistics Standardization Technical Committee (Logistics Standards Committee), there is no self-contained logistics standard for iron and steel industry in its specialty category. Five national standards pertaining to the metallurgical category are now in the wait list of Logistics Standards Committee and nothing more in this regard. Therefore, the standardization work in the logistics industry for iron and steel industry is just in its infancy.

(2) Main Existing Problems. There is a lack of relevant research institutions for steel product logistics standards at the national level. The responsible domestic governing authority for standardization is the National Logistics Standardization Technical Committee (TC269). Previously, there was no working group for the research and preparation of steel logistics standardization (TC 269/Steel Logistics Working Group was established on November 15, 2016, and its secretariat is located in the Metallurgical Industry Planning and Research Institute), which has left the formulation and revision

of steel logistics standards stagnate at a relatively backward stage. Under this circumstance, the implementation of interrelated standards in various industries often leads to a situation of "fragmented management", which makes the standard practicability and feasibility difficult.

Top-level design for steel logistics standards is absent. So far, China does not have a complete set of modern logistics standards covering the entire supply chain of iron and steel industry. From the perspective of top-level design, the development of specific work for the formulation and revision of steel logistics standards shall be effectively guided.

9. Informatization Standards

(1) Development Status. Comprehensively promoting the national economy and social informatization is a strategic measure concerning the overall situation of modernization. Informatization is an inevitable choice for China to accelerate industrialization and modernization. The premise of informatization is the sharing of information resources. The basis of information resources sharing is the standardization of informatization.

The standardization of informatization means the formulation, revision, management, and implementation of various information technology standards in various practical activities of developing and utilizing information technology adhering to the principle of staying in line with international standards, publicity of national standards and industry standards, development of local standards and standard system of practicality, and promotion of initiation, implementation, acceptance, the conformity testing, and certification of informatization application projects in order to obtain the best economic and social benefits.

The standardization of informatization, in terms of standardization, includes many links such as the formulation, revision, management, publicity, implementation, and review of various information technology standards. In terms of information technology, it includes the generation, identification, collection, classification, encoding, storage, processing, transmission, and other aspects. At present, informatization standards mainly include informatization standard systems, information technology basic standards, network infrastructure standards, and information security standards.

(2) Main Existing Problems. China's standardization efforts have provided effective support for promoting the sound and rapid development of China's economy and society. However, there are many problems in the standardization of informatization. Compared with the national strategic deployment and the demand of industrial development and market, there is still a big gap to fill, mainly in the following aspects:

The ability to participate in international standardization events is still far from the developed countries;

The adaptability and effectiveness of standards need to be further improved;

The implementation of standardization work needs to be further strengthened;

The scientificity of some standards, especially those related to life and health, needs to be further improved;

Testing methods of some standards need to be further improved.

### *8.1.2 Industry Standards*

1. Standards for Raw Materials and Fuels
   (1) Development Status. At present, there are 102 national standards related to iron ores, and most of the standards are the ones for iron ore elemental analysis and measurement except two that are the *Iron Ore Grade Classification* (GB/T 32545-2016) and *Terminology of Iron Ores and Direct Reduced Irons* (GB/T 20565-2006). Relevant national authorities have issued many industry standards in terms of coal component testing technology, coal mining equipment, and safety technologies.
   (2) Main Existing Problems. The existing iron ore standards are currently focused on analytical testing standards, and the standards of iron ore quality, especially the mandatory standards for harmful impurities such as S and P in iron ores, are yet to be developed. As a result, the quality of imported iron ores in China in the past decade has been varying and there were even shoddy iron ores in disguise of good ones. There are few industry standards available for the coal and coke, and most of the available standards are national standards.
2. Technological Equipment Standards
   (1) Development Status. The industry standards of China's metallurgical technology equipment have formed a relatively complete system, involving more than 100 industry standards and covering the full process from sintering to ironmaking, steelmaking, and steel rolling process. It provides a technical basis for the design, manufacture, and application of large-scale packaged equipment, and the level of manufacture and application of metallurgical equipment in China has been greatly improved as a result. The industry standards have played their role in optimizing the industrial structure, eliminating the backward technology and high energy consumption equipment in the industry, and developing China's ability of manufacturing technology and independent system integration of large-scale equipment in China's iron and steel industry, thus creating conditions for the rapid development of China's industry.
   (2) Main Existing Problems. China's iron and steel industry is characterized of large scale and uneven level of technological equipment. And most of the industry standards are recommended standards. The overall level of equipment for small and medium-sized private enterprises is poor, and they tend to use standardized process equipment optionally. In addition, steel

smelting and steel rolling equipment are mostly large-scale, packaged, and production line equipment. It is difficult to formulate a uniform standard for a minority of products according to the changing conditions of raw and fuel materials. The manufacturing standards for single machines in packaged equipment cannot meet the requirements of wide applicability, high precision, high speed, and customized production, quantitatively or technically.

3. Product Standards

   (1) Development Status. The total number of national standards and industry standards in China's iron and steel industry has reached more than 2370, of which more than 1100 are products and methodology standards, accounting for less than 50%. The rate of key standards adopting international standards and foreign advanced standards has reached more than 75%. A product standardization system that is oriented to market demand, gradually improved in level, and rational in structure has been substantially established.
   (2) Main Existing Problems. There are still many gaps in product standard system of China's iron and steel industry compared to the standards of the developed countries and the international standards. The existing problems are reflected as: less variety coverage, low indicators, incomplete category, and unable to meet the requirements of the international market; moreover, product standards are still dominated by production standards; technical content lacks freedom and adaptability; in addition, national standards and industry standards are not complete enough to cover corporate standards, failing to meet the individualized need of customers.

4. Energy Standards

   (1) Development Status. At present, the main energy-saving standards available in the industry include energy consumption limitation standards such as *Energy-Saving Design Specifications for Iron and Steel Enterprises*, *Specific Energy Consumption Limit for Crude Steel Production*, *Specific Energy Consumption Limit for Electric Arc Furnace Smelting*, and *Specific Energy Consumption Limit for Coke Production,* and they have played a role in promoting and standardizing energy-saving work across the industry.
   (2) Main Existing Problems. However, there are still two problems: First, the number of standards is less. The iron and steel industry is a highly correlative industry, and however, there are only a dozen of standards for energy conservation available in the industry. More standards are demanded to regulate the industry. Second, the formulation and revision periods of standards are long. Taking the energy consumption limitation standard for crude steel production as an example, the revision time interval of the last two editions has been six years, which cannot adapt to the rapid development of the industry.

5. Environmental Protection Standards

(1) Development Status. As a major stationary industrial pollution source, the iron and steel industry has achieved relatively outstanding achievements in the fields of waste gases, wastewater, and solid wastes treatment in recent years. Among them, the iron and steel industry was once the main source of sulfur dioxide emissions second to power plants, and through the full implementation of sintering flue gas desulfurization and recovery of surplus gas technologies, and the transformation of oil to gas, coal to gas, and other transformation projects in a decade, the quantity of sulfur dioxide emissions has been greatly reduced. In the past ten years, the sulfur dioxide emissions per ton of steel have been reduced from nearly 3 kg to about 0.85 kg, and about two-thirds of the emissions have been reduced. Smoke dust is the most "intuitive" air pollutant in the iron and steel industry. Through continuous revamping of dust removal facilities and increasing dust removal capacity and by using advanced technologies such as the bag filter, the emission of smoke dust per ton of steel has been reduced from 2 to 0.81 kg at the end of 2015. The discharge of wastewater per ton of steel has achieved a decline of about 75%. The total amount and intensity of wastewater discharge have been greatly reduced during the development period of this decade. The total discharge of wastewater from the iron and steel industry also fell to about 400 million cubic meters at the end of the 12th Five-Year Plan from the 1200 million cubic meters at the beginning of the 11th Five-Year Plan. The discharge of wastewater per ton of steel dropped from 3.8 to 0.8 $m^3$, and enterprises in many consumption limitation areas realized "nearly zero emissions" of wastewater. The comprehensive utilization rate of solid wastes also increased from 94.8% in 2005 to 97.5% in 2015. The amount of solid wastes generated per ton of steel decreased from 628 kg/ton in 2005 to 585 kg/ton in 2015. The environmental protection achievements of the iron and steel industry over the past years are attributed to the increasingly strict environmental emission standards and the implementation of the *New Environmental Protection Law*. On the other hand, the other inseparable impetus is the official release of the *Clean Production Standard—Iron and Steel Industry* (HJ/T 189-2006). Thanks to the leading and mandatory role of the standards, the iron and steel industry has changed from a high-consumption, high-emission, and extensive management growth pattern to the clean production featuring intensive, efficient, and low emission.

In recent years, the research work on the standardization of atmospheric pollutant discharge mainly focuses on the following eight emission standards, for example, the *Emission Standards for Pollutants from the Mining and Beneficiation Process of Iron Ores* (GB 28661-2012), *Emission Standards for Air Pollutants from Metallurgical Sintering and Pelletizing Processes* (GB 28662-2012), *Emission Standard for Pollutants from Iron Making Process* (GB 28663-2012), and the *Pollutant Discharge Standard for Steel-Making Process* (GB 28664-2012), which are stricter compared with the old standards. In particular, the special emission limits implemented since January 1, 2015, are called the "most stringent" emission standards. The emission concentration limits of various pollution factors required therein are significantly tightened compared with the old standards. Some of the limit values stipulated in the new standards are as low as one-tenth of the old ones. Following emission standards, relevant technical and equipment standards of the metallurgical industry were issued, such as *Technical Specifications for Flue Gas Purification and Recycling Facilities for Silicon-Based Ferro-Alloy Electric Furnaces* (YB/T 4166-2007), *Composite Filter Bag* (GB/T 27869-2011), *Performance Test Method of Electrostatic Precipitator* (GB/T 13931-2002), *Evaluation Technical Requirements for High-Efficiency Environmental Protection Equipment—Electrostatic Precipitator*, and other standards in a series. This promoted the application of advanced environmental protection technology and advanced equipment and ensured the stability and performance guarantee of the operation of environmental protection facilities in the production process.

(2) Main Existing Problems. In recent years, as the tough stance on environmental protection in the iron and steel industry keeps heating up, the market of environmental protection expands and the potential is huge. With the successive introduction of "The 10-Chapter Water Pollution Prevention Action Plan" (The Action Plan for Prevention and Treatment of Water Pollution), "The 10-Chapter Air Pollution Prevention Action Plan" (Air Pollution Prevention and Control Action Plan), and "The 10-Chapter Soil Pollution Prevention Action Plan" (the Action Plan for Soil Pollution Prevention and Control), the demand for assessment of technical equipment, products, and facilities performance is more urgent than ever. However, the current environmental protection market in the metallurgical industry still presents obvious discrete characteristics. The upstream and downstream of the industry chain need to be standardized and guided. The technical regulations and energy efficiency evaluation methods of such environmental protection equipment as desulfurization, denitration, dedusting, water treatment, and slag treatment critically required by raw materials systems, coking, sintering, pelletizing, ironmaking, steelmaking, and steel rolling processes in the metallurgical industry are still insufficient. Even if a small number of environmental protection technology equipment own evaluation index system, they are yet to be tested for practicability by production plants.

Therefore, in combination with the development characteristics of the industry's environmental protection field, a series of green product standards and evaluation regulations, standard methods for business to business environmental protection cost accounting, and standards for evaluating performance of existing environmental protection facilities of enterprises that respond to the needs of the industry shall be carried out, against the standardized applications of some advanced and industrial environmental protection technologies. A standard system that is compatible with the environmental protection in the industry shall be established and gradually improved.

6. Water Standards

    (1) Development Status. The iron and steel industry is a high-water-consumption and high-pollution industry, with water consumption accounting for more than 10% of industrial water consumption, ranking third in the industrial sector. In recent years, China's iron and steel industry has achieved remarkable results in water-saving work. The water consumption of key steel enterprises in the industry has been declining year by year. By 2015, the water consumption per ton of steel has dropped to 3.53 $m^3$. The iron and steel industry has seen establishment of a number of standards, mainly for water supply hydrogeology, seawater desalination technology, and metallurgical wastewater desalination process.

        The industry standards for water saving in iron and steel enterprises mainly include *Water Supply Hydrogeological Survey and Water Supply Pipeline and Well Works* (YB/T 9033-1998) for inspection, acceptance, and quality assessment of water supply hydrogeological survey and water supply pipeline and well works; *Technical Specification for Seawater Desalination in Iron and Steel Industry Part I: Low-temperature Multi-effect Distillation* (YB/T 4256.1-2012) for iron and steel enterprises that use low-parameter steam to produce freshwater through low-temperature multi-effect seawater desalination system; *Technical Specification for Metallurgical Production Wastewater Desalination Part I: Reverse Osmosis* (YB/T 4257.1-2012) for the membrane desalination system of metallurgical production wastewater.

    (2) Main Existing Problems. Since China's iron and steel industry differs in the production structures, availability of regional water resources, and degree of enthusiasm for water saving, a situation mixed with advanced and backward indicators among enterprises is the fact and is reflected in the difference between the north and the south. Due to the abundant water resources, the southern iron and steel enterprises generally have higher water consumption per unit of product. With the deep processing of China's steel products, the continuous extension of the industry chain and the demand for water is increasing, and the contradiction between water shortage in some regions and the development of the iron and steel industry has been highlighted.

        The water-saving standard system in the iron and steel industry is still not perfect. On the one hand, the basic standards are lacking, which is not conducive to guiding and standardizing the formulation and revision of water-saving standards. On the other hand, the water consumption in the iron and steel industry is a complicated one, and the standard shall be refined to each production process. There is still a long way to go for standardization work.

7. Circular Economy Standards

    (1) Development Status. China's iron and steel industry has not yet seen the establishment of a systematic circular economy standard system. Although

the National Metallurgical Standards Committee (SAC/TC 183) has specifically carried out the metallurgical solid waste standardization, it only covers the comprehensive utilization standards of part of metallurgical slag and ferrous dust, while the comprehensive utilization standards of tailings, waste rocks, ferroalloy slag, and non-ferrous slag are blank. In addition, the secondary resources comprehensive utilization technology and product standards such as wastewater, waste gas, solid waste, residual heat, and residual energy generated in the metallurgical production process are missing. The standardization of circular economies such as metallurgical large-scale equipment, machinery remanufacturing, and metallurgical industrial parks is still in its infancy. As of the beginning of 2017, the metallurgical industry has seen the establishment of 38 industry-level standards regarding circular economy, including three aspects: basic standards (terms, classification, sample preparation, stacking, and packaging signs), product standards (metallurgical slag and dust) and methodology standards (physical and chemical), and there are currently 11 standards under development.

(2) Main Existing Problems. The main problems existing in the standardization work of circular economy are that the standard system is still not complete, the quantity of standards is insufficient, and there are still many vacancies, failing to cover the full production and application fields; there is a deep gap to national policy requirements; the standards committees of each specialty are decentralized and difficult to form synergy, and their substantive participation in international standardization work is insufficient; the current standards are incomplete, outdated, backward, overlapped or in conflict, the standard system is not reasonable enough, and the standard coordination promotion mechanism is imperfect.

8. Logistics Standards

(1) Development Status. At the industry level, in the absence of a logistics standard system for iron and steel industry, the formulation and revision of industry standards are also lagging behind and mainly focus on the relevant content of intermediate links. Till now, there have been only seven steel logistics industry standards published, and there are still a lot of gaps in the relevant logistics standards as far as the overall supply chain system for iron and steel industry is concerned.

(2) Main Existing Problems. The basis of standardization of steel logistics is weak. First, the steel logistics industry started late and the basis is relatively weak. Due to long-term extensive operation of the logistics links in the iron and steel industry, the management system, management level, personnel quality, and equipment and infrastructures are considered non-core business and subject to long-term insufficiency. At the same time, relevant practitioners are ill-informed about modern logistics concepts, logistics informatization, and logistics standardization and even lack a clear understanding of the connotation and extension of the "Logistics". Second, the market base

of steel logistics standardization is relatively weak, which directly affects the implementation of logistics standardization of metallurgical links.

Service quality and management level are low. Although the government has paid more and more attention to the logistics link in recent years, and has successively issued a number of related policies, the service and management in the steel logistics link have been at a low level. For example, the inventory management of some iron and steel enterprises is relatively extensive, and the backlog of raw materials and fuels is serious; the transportation department delays in transition and poorly controls it, leading to delays in the logistics links and serious losses on the way; the storage management of circulation links is poor, resulting in frequent internal reclaiming and high cost; the connection between loading/unloading and transportation does not match, resulting in long waiting time of transportation vehicles and ships, and low efficiency.

There is no uniform standard for steel logistics standards. At present, the development of China's steel logistics system is still in a low-level stage. There are many problems in the various links of steel logistics, mainly in the following aspects. First of all, production enterprises and circulation enterprises in iron and steel industry mainly follow their own "conventional custom" process and methods that took form in a long run, which are backward in terms of equipment conditions and management and control methods, and could not keep up with the development requirements of the national overall logistics industry. The technical level and structure of various steel logistics facilities and equipment are not reasonable, the control process is relatively extensive, and the degree of standardization is low, which cannot achieve the effect of high-efficiency operation of modern logistics. Secondly, the standard of steel logistics equipment is not matched, and a standard system of logistics equipment covering raw materials and fuels, auxiliary materials, semi-finished products and finished products from supply, production, and sales to circulation is still absent. Finally, there is no effective connection among the steel logistics process flow, the steel logistics equipment standards, and the steel logistics management and control standards, which leads to an increase in ineffective logistics operations and logistics costs, and a backward logistics service quality, thus seriously restricting the improvement of logistics efficiency.

9. Informatization Standards

(1) Development Status. Before 2016, there were few informatization standards for the iron and steel industry. The standards issued then by the electronics industry and the communication industry will contain some informatization standards, but they were also common ones. In 2016, the Ministry of Industry and Information Technology issued industry standard formulation and revision plan to be carried out in three batches, of which there were 108 standards for metallurgy, but only one was the standard for informatization

of the iron and steel industry. There was only one standard for manufacturing execution system (MES) among the standards released in 2016 by the electronics industry, namely "*Specification for the Manufacturing Execution System (MES) Part II: Software Function of Manufacturing Execution System for Metallurgical Industry*", which is an information standard for the iron and steel industry.

(2) Main Existing Problems. The iron and steel industry is relatively lacking in building information technology standards. In order to implement the strategic plan of "*Made in China 2025*" accelerate the development of intelligent manufacture in the iron and steel industry, and give full play to the regulating and guiding role of standards, measures shall be taken to speed up the progress of setting up information technology standards of iron and steel industry according to the guidance of National Intelligent Manufacturing.

### *8.1.3 Corporate Standards*

1. Product Standards

   (1) Development Status. Product quality must comply with national and industry standards. This is the minimum quality standard for iron and steel enterprises to meet the needs of general customers. The corporate standards are the spire of the standard "Pyramid", which are more inclusive and more competitive. Under the current excess market environment, iron and steel enterprises have developed their own corporate standards to improve product competitiveness, while some of them have developed customized products to meet the individual requirements of high-end customers. "The national standard is the threshold, and the corporate standard shall be the guide". This conception is gradually being recognized and accepted by iron and steel enterprises.

   (2) Main Existing Problems. At present, the guiding role of corporate product standards in the industry is not prominent, partly because of the immature construction of the standardization talents in most iron and steel enterprises in China and the lack of institutions, positions, and trainings established for standardization work. Therefore, the iron and steel enterprises shall speed up the efforts in training their own standardization experts and let them play the role of "the Spokesperson of Corporate Interests".

2. Energy Standards

   (1) Development Status. Energy standards are an important basis for enterprises to implement energy-saving management and are also the technical basis for the government to implement energy-saving policies and strengthen energy-saving supervision. Corporate standardization is an important part of the standardization work and is a comprehensive basic task of the enterprise,

which shall run through the whole process of the production technology and management activities of the enterprise.

With the continuous improvement of the market economy system and the standards being the main basis for market regulation and operation, more and more enterprises have established a standardization system within the enterprise, and energy standardization is an important part of the system. In metallurgical enterprises, the reliable energy security is of great significance for the safe operation of the energy system and the safety of employees. Standardization job is a long-term, arduous, and complex system construction project, which requires the long-term and arduous efforts of management personnel of all levels and kinds and demands persistent attention.

(2) Main Existing Problems. The energy department is the key auxiliary unit of metallurgical enterprises. Generally, it is responsible for the production and supply of energy products (media) such as water, electricity, air, and gas (steam) of metallurgical enterprises. The uninterrupted and non-storage nature of energy product (media) production determines that its output is not only a tangible physical product but a utility. Since energy is not the main product of metallurgical enterprises and each energy medium has common standards available in its respective fields, most of the metallurgical enterprises are not aware of the importance of energy standards, and they have not invested much in this area. Quite a number of enterprises have still not yet set up full-time standardization staff and standardization management institutions; this is particularly true with energy standardization professionals.

3. Environmental Protection Standards

(1) Development Status. With the promulgation of the *New Environmental Protection Law* and the government's increasingly strict requirements of environmental standards of the iron and steel industry, iron and steel enterprises continue to strengthen their efforts in the implementation of clean production, legal discharge and emissions, and the stable operation of the entire plant environmental protection facilities. On the one hand, we must resolutely implement relevant national standards, such as post-evaluation of environmental protection facilities for enterprises, and special restrictions on the emission concentration of local enterprises. On the other hand, with its own enterprise as the frame of reference, the implementation of standardization work shall lead the enterprise's green development transformation through formulating the *Environmental Management Regulations and Regulations for Iron and Steel Enterprise*, the certification of the *Environmental Management System*, and the audit of enterprises' promotion of clean production policy. With the gradual enhancement of corporate environmental protection awareness, corporate environmental management has been gradually standardized and institutionalized. By the end of 2015, according to

the list of enterprises in conformity with the *Regulations on the Standardization of the Iron and Steel Industry (2012 Revision)* published by the Ministry of Industry and Information Technology, 305 iron and steel enterprises have obtained the ISO 14001 environmental quality management system certification, accounting for 90% of the national steel production capacity.

(2) Main Existing Problems. The rapid development of China's iron and steel industry has led to the rapid promotion and application of advanced environmental protection technology, laying a solid foundation for improving the environmental protection level of iron and steel enterprises. However, the corresponding technical specifications, the evaluation of environmental protection facilities' energy efficiency and pollutant removal effect, performance evaluation of environmental emission, corporate environmental cost accounting, and many other standardization works are still in blank or relatively lag period. Despite Baosteel, TISCO, Tangshan Steel, and other benchmarking enterprises that have reached the international advanced level have continued practicing the environmental management and green development concepts, insisting in implementation, learning from internationally advanced counterparts, and creating an internal standardization evaluation system of the enterprise. However, as far as the whole industry is concerned, due to regional diversity, differences in technical equipment levels, and differences in environmental protection concepts, there is still disparity in the level of environmental protection among Chinese iron and steel enterprises. According to the "Statistics of Environmental Protection of Iron and Steel Industry", the per capita emission of backward enterprises is ten times higher than that of the advanced enterprises. The cause of this situation is precisely that the above-mentioned environmental protection standards of the iron and steel industry have not been uniformly introduced in the industry. In addition, environmental protection technology providers with low credit standing and poor environmental protection facilities deteriorated the situation, resulting in the current poor environmental protection facilities in the iron and steel industry and significant differences in the standard conformity effect. In future, enterprises in the industry shall be constrained by a unified standard so that it can comply with the standards of resource-saving and environment-friendly enterprises.

4. Water Standards

(1) Development Status. Water demands of iron and steel enterprises are diversified. There are different processes that require water such as coking, sintering, pelletizing, ironmaking, steelmaking, and steel rolling. It is also reflected in the coexistence of long process and short process, ordinary steel enterprises coexisting with special steel enterprises, primary steel products coexisting with secondary and tertiary products, and coexisting of plants with and without captive power plant. Therefore, due to the different production conditions of enterprises, there will be certain differences in the water consumption structure and water utilization efficiency, which will cause

difficulty in the benchmarking work in terms of enterprise's water-saving analysis and compromise the development of water-saving endeavor.

Iron and steel enterprises usually have internal rules and regulations that suit themselves, for example, *Safety Regulations, Technical Operation Procedures, Equipment Operation and Maintenance Regulations,* and *Regulation for Shifting of Duty*. The relevant corporate standards are also formulated, mainly including the inspection standards for raw materials and fuel procurement, the test standards for product quality and the performance index standards of steel products, and they are used to guide production practices. However, the water-saving standards are still a blank area and corporate water-saving standards are yet to be formulated.

(2) Main Existing Problems. The current water consumption status in iron and steel enterprises is yet to perfect. Water circulation systems are inadequate and there is direct cooling water system still in operation. The water quality stabilization measures such as increasing the concentration multiple are not in place. The energy consumption of system equipment such as pumps is high. Therefore, a set of water-saving standards which have feasibility and adapt to the production and operation of the enterprise itself is badly needed to help enterprises in their efforts of carrying out water-saving work and improving water efficiency.

The water supply and drainage facilities are auxiliary to iron and steel enterprises. There are many problems in its management, such as defective water metering system, low coverage of water quality online monitoring system, poor water-saving awareness of iron and steel enterprises, imperfect management institutions of water supply and drainage, unreasonable water management mechanism, and less attention to the technical transformation of water facilities. In the aspect of enterprise's water-saving management, it is necessary to standardize enterprise water management and strengthen standardization work.

5. Circular Economy Standards

(1) Development Status. The establishment of corporate standard system of circular economy provides a means for the sustainable development of enterprises and can comprehensively guide enterprises to carry out standardization work of circular economy. At the same time, it will promote the enterprise to systematically sort out the existing standards and recognize deficiencies. The corporate standard system of circular economy shall be developed from several aspects such as comprehensive utilization of resources, heat energy recovery and utilization, power saving, water-saving, and waste gas recycling. In response to the development needs of circular economy, a series of corporate standards suitable for the circular economy model shall be formulated to improve and perfect the corporate standards system. For example, Baosteel has successfully developed the slag into microgrits with good corrosion resistance, micro-shrinkage resistance, and high strength and converted this scientific research result into the corporate

standards, the "*Baosteel Microgrits of Blast Furnace Slag*". Based on the standard, it opened the market door and promoted the wide application of the microgrits in large-scale projects such as the cross-river tunnels, which not only save energy but also realize comprehensive utilization of wastes, and can also generate a profit of 50 million yuan per year.

(2) Main Existing Problems. Although enterprises have established a sound and numerous management systems pertaining to the standardization management, they have not really been "Implemented into management and embodied in practice" in the daily production and operation. In the process of standardization management, some enterprises have no specific analysis of problems and have not followed the objective laws. The management standards established tend to lack scientificity, practicality, and feasibility, ending up aiming at the moon. Some enterprises bypass the standard management system under the excuses of exceptions handling, which hinders the implementation and promotion of the same.

6. Logistics Standards

(1) Development Status. The logistics at the iron and steel enterprise level has been treated as the supporting service for quite a long time, and the priority is always given to the "Guarantee production". In recent years, with the decline of the iron and steel industry, logistics has just become the focus of the enterprise's "Cost reduction". However, because it has always been in the auxiliary position with less attention, the enterprise has been adopting the "conventional thinking" of the logistics standardization, and there have been no systematic regulation and formulation of the steel enterprise logistics standards.

The enterprises in the field of market and circulation include a variety of enterprises such as warehousing, transportation, comprehensive service, freight forwarding, ship agencies, processing and distribution, and logistics parks. These enterprises have always followed relevant standards in the industries of logistics, transportation, and trade.

(2) Main Existing Problems. In the downturn background of the iron and steel industry, the logistics link that has been in the role of ancillary services in the past has just been valued by steel enterprises. The cost reduction through logistics has become a fashionable term, but the method of reducing costs and starting points has not been figured out. The cost reduction always takes the form of administrative orders on how much cost shall be reduced, passed down to logistics departments arbitrarily. In the absence of logistics history statistics, information technology, modern logistics centralized control concepts and normative standards, cost reduction has ended up being a digital game. Due to the lack of normative guidelines, incomplete cost reduction that one cost item is reduced only to find another rising becomes a common occurrence, going against the original intention of overall cost reduction.

The logistics in circulation link of iron and steel industry. Due to the incompatibility of the standards, in the fields of warehousing, transportation, loading, and unloading, with various links of steel logistics, the relevant standards encounter problems with their usability and popularity in the industry and the situation is even worsened by the small-scale and disorderly situation of logistics service providers in the industry.

7. Informatization Standards

(1) Development Status. Enterprise informatization refers to the process in which enterprises use computer and network technology to continuously improve the management level of production and management, improve the efficiency and quality of enterprise decision-making through deep development and extensive use of information resources, thereby improving the economic efficiency and core competitiveness of enterprises.

Corporate standardization is a comprehensive basic work of an enterprise, which runs through the entire process of production, technology, and management activities. Standardization is the most basic supporting element for enterprises to implement informatization construction and is the basic insurance of realizing the globalization of information systems and data sharing.

In recent years, the informatization of China's iron and steel industry has developed rapidly. More and more iron and steel enterprises regard informatization as an important means to improve their core competitiveness. Enterprises have invested a large amount of funds in information construction. However, under the traditional management mode, the operation of corporate informatization system is less effective in Chinese enterprises than in world leading enterprises. Domestic iron and steel enterprises generally have large organizational structure and complicated business processes, resulting in long development cycle, high cost, and low operational efficiency of information technology projects. In general, the lack of enterprise information standards, the crude management, and the lag of standardization have seriously hindered the effectiveness of enterprise information operation.

(2) Main Existing Problems. The information system does not match the enterprise management model. Many enterprises do not recognize the relationship between informatization and management standardization. For the same enterprise management information system, its effect will be greatly different from enterprises with different management system. At present, many enterprises have problems in their daily management system that does not match with the software standards.

Lack of Basic Data Standardization. Data standardization is the basic guarantee for enterprises to carry out information construction. Enterprise informatization is about digital design, implementation, application, and management of enterprise data acquisition. Ensuring the standardization of data acquisition is a key factor in the construction of enterprise information projects. At present, many enterprises fail to standardize data management during the implementation of information technology projects.

Standardization of information system development has not been achieved. The standardization of information system development mainly refers to the compliance with unified system design specifications, program development specifications, and project management specifications in the development of system. At present, many enterprises fail to comply with the design specifications of software engineering when constructing information projects, resulting in repeated changes required by customers, unsatisfactory system operation, and difficulty in system update iteration.

## 8.2 Development Environment and Policy Orientation

As the comprehensive deepening of reform and opening-up has pushed China's economy to the middle and high-end level, the government's attention to standardization has never been higher. The General Secretary Xi Jinping pointed out that China will actively implement the standardization strategy, give play to the basic, strategic and leading role of standardization, and promote the five developments that are innovation-driven development, coordinated development, green development, development for global progress, and development for the benefit of all through standardization. Premier Li Keqiang stressed that it is necessary to promote industrial upgrading by comprehensively improving standards, develop a new competitive advantage, and promote high-speed economic growth and lead the way to the middle and high-end ranks. The ideas of letting standards lead the improvement of China's manufacturing quality, and promoting the integration of international and domestic standards has been mentioned many times in the papers issued by the State Council such as *Made in China 2025* [1], *Opinions on Giving Play to the Leading Role of Brand Building and Promoting Upgrading of Supply and Demand Structure*, *Standardization and Quality Improvement Plan for Equipment Manufacturing Industry* [2], and *Consumer Goods Standardization and Quality Improvement Plan (2016–2020)*.

Standardization plays an indispensable role in promoting the transformation and upgrading of the iron and steel industry. More and more standards are included in the relevant documents of industrial policies. *Opinions on Cutting Overcapacity of Iron and Steel Industry to Achieve Development by Solving Difficulties* issued by the State Council clearly stated that standards shall play a restrictive and guiding role to reduce the structural overcapacity. The following standards of environmental protection, safety protection, and energy consumption must be followed, and any production facilities failing to do so shall be closed down according to legal procedures: Standards of environment protection: *Water Pollutant Emission Standards for Iron and Steel Industry*, *Air Pollutant Emission Standards for Sintering and Pelletizing Industry*, *Air Pollutant Emission Standards for Iron Making Industry*, *Air Pollutant*

*Emission Standards for Steel Making Industry*, *Air Pollutant Emission Standards for Steel Rolling Industry*, etc.; energy consumption must meet the mandatory standards such as *Limitation on Specific Energy Consumption of Major Crude Steel Production Processes*; in safety aspects, the following standards must be complied with the *Safety Regulations for Iron-Making*, *Safety Regulations for Steel-Making,* and *Gas Safety Regulations Pellet for Industrial Enterprises*.

### 8.2.1 Policies and Regulations

In 1957, in order to strengthen the standardization work, China formulated a number of national and ministerial standards according to national conditions, and since then, China's standardization work has embarked on a stage of independent development. After 1958, due to the influence of the leftism, the standardization work suffered severe setbacks. Until the national economic adjustment period in 1962, the standardization work was restored and strengthened. The State Council issued the *Management Measures for Technical Standards for Industrial and Agricultural Products and Engineering Construction*, marking the beginning of a new development era. During the Ten Years of Turmoil, the standardization work was once at a standstill, the legislature was destroyed, and the legal system was seriously devastated.

After the Third Plenary Session of the 11th Central Committee of the Party, with the comprehensive restoration of economic development, the standardization work has received the attention of the government. In order to strengthen the efforts in standardization, in 1979, the State Council promulgated the *Regulations on Standardization Management of the People's Republic of China*. Standardization work has entered a new period of development national wide. The operation mechanism of China's standardization management system has been gradually improved, and the standard system has taken its initial shape. In 1989, the government promulgated the *Standardization Law of the People's Republic of China*, which further defined the framework of China's standard system, standardization management system, and operational mechanism. Subsequently, the State Council promulgated the *Regulations on the Implementation of the Standardization Law of the People's Republic of China* in 1990, putting forward the specific regulations for the implementation of the *Standardization Law*. In order to implement the *Standardization Law* and the *Regulations on the Implementation of the Standardization Law*, as the governing body of standardization work, the former State Bureau of Technical Supervision promulgated a series of more detailed regulations covering the formulation and revision, publishing and standard management of various standards in sectors such as industrial engineering, agriculture, energy, and transportation. During this period, China's standardization laws and regulations were initially established.

Since the *18th National Congress of the Communist Party of China*, the Party Central Committee and the State Council have made a series of decision-making arrangements for strengthening standardization work and promoting standardization reform and development. The Second Plenary Session of the 18th CPC Central Committee proposed to strengthen the construction of technical standards system;

the Third Plenary Session requested the government to strengthen the formulation and implementation of development strategies, plans, policies, and standards; the Fourth Plenary Session regarded standards as an important means of governing the country according to law; the Fifth Plenary Session added the five new development concepts of "Innovation, coordination, green, openness and sharing" into the management of standards; the Leading Group for Overall Reform listed the standardization reform as a key task in 2015.

In March 2015, the State Council issued *The Plan for Furthering the Standardization Reforms* [3] to vigorously promote the reform of China's standardization deployment. The target is to transform the current government-dominated standard formulation system to a new standard system consisting of standards provided by the government and the market players. The standards formulated by the government are reduced from six to four categories, which are mandatory national standards, recommended national standards, recommended industry standards, and recommended local standards. The standards independently formulated by market players are divided into non-governmental organization standards and corporate standards. The government-led standards focus on the basics, and the standards originated from the market are focused on improving competitiveness. At the same time, a standardization management system compatible with the new standard system shall be established and improved.

In May 2015, the State Council issued the *Inter-Ministerial Joint Conference System for Promoting Standardization of the State Council*. The General Administration of Quality Supervision, Inspection and Quarantine (National Standards Committee) is the leader to coordinate the work among 39 ministries. The national standardization work is coordinated under the leadership of the State Council. The working contents include the study and formulation of major policies and guidelines to promote the reform and development of standardization, coordination and cope with major problems in the development process of standardization reform, coordination of the formulation and implementation of cross-sectoral cross-disciplinary and major disputed standards, and review and finalization of the standards that need to be submitted to the State Council for approval.

In December 2015, the *National Standardization System Construction and Development Plan (2016–2020)* [4] issued by the State Council stipulated that by 2020, the international influence and contribution of "China Standards" will be greatly enhanced, and China will enter the ranks of the world standard power. The standard system shall be more complete. The standards provided by the government and the standards independently developed by the market shall coexist in a coordinated manner. The mandatory standards are the bottom line, the recommended standards are the basic frame, and the corporate standards shall be quality enhancer. A set of world influential standards shall be fostered and developed in fields where technology is developing rapidly and the market is innovation-driven. The benefits of standardization have taken full effect. The internationalization level of local standards has been greatly improved. And the foundation of the standardization has been continuously consolidated.

In February 2016, the *Guiding Opinions on Cultivating and Developing Non-Governmental Organization Standards* was jointly issued by the General Administration of Quality Supervision, Inspection and Quarantine and the National Standards Committee to motivate social groups to develop and apply standards, standardize the work of developing non-governmental organization standards, and increase the effective supply of standards. It requires that, by 2020, the development of non-governmental standards independently formulated by the market shall be mature, the number of non-governmental organization standards and their competitiveness shall be steadily improved, the organization standardization achievements shall be widely recognized by the society, and a group of world-known and influential organization standards shall come into effect.

In May 2016, the *Outline of National Innovation Driven Development Strategy* jointly issued by the Central Committee of the Communist Party of China and the State Council dedicated one chapter for the statement of "Implementing the strategy of intellectual property rights, standards, quality and brand" from the height of the Party and the Government, demanding the "Improvement of China's standardization". Therefore, standardization has become an important part of implementing an innovation-driven development strategy.

In February 2017, Premier Li Keqiang presided over executive meeting of the State Council. The meeting saw the approval of the *Standardization Law of the People's Republic of China (Revised Draft)* and decided to submit the draft to the Standing Committee of the National People's Congress for deliberation. This is an important milestone in the revision of the Standard Law, which means that the *Standardization Law of the People's Republic of China*, which has been in force since 1989, will be revised for the first time. The current *Standardization Law of the People's Republic of China* [5] only deals with the standards of industrial production and industrial products. The problems such as "Existence of missing, aging and lagging standards", "Overlapping, redundancy and contradiction", and "Delayed publication of important standards" plagued the development of standardization. The new *Standardization Law* will focus on improving the standards of industrial products, strive to make greater efforts in the formulation and revision of service standards, encourage enterprises to promote industry standards through standard innovation, and promote industrial transformation and economic structural adjustment. The draft also emphasized that it is forbidden to use the standards to engage in setting up industrial barriers, imposing regional blockades, and pursuing unfair competition.

With the attention of the Party Central Committee, the State Council, and the whole society to the standards work, a series of laws and regulations have been formulated and revised, and China has stood at a new historical starting point for the development of China's standardization industry.

### *8.2.2 Industrial Development*

China's iron and steel industry has entered a stage of reduced development, but the demand for iron ore is still at a high level. China Metallurgical Industry Planning and Research Institute predicts that China's iron ore dependence will remain above 85% by 2030; global high-grade resources will gradually shrink and the average grade is showing a downward trend. Under the background of China's environmental protection and energy consumption policies, it is one of the important means to regulate the order of importing iron ores through tighter standards.

There are dozens of various design institutes engaging in China's metallurgical equipment research, and a strong engineering design team focusing on metallurgical geology, mining, smelting, processing, and other specialties has been formed. Under the strategic deployment of the Belt and Road Initiative, China's metallurgical engineering design institutes are actively "Going abroad" and contracting a large number of international capacity cooperation projects in iron and steel making. Relevant technological equipment standards must keep up the pace and carry out standardization of metallurgical process equipment with the aim of achieving the strategic goal of improving the quality and efficiency of the iron and steel industry, transforming, and upgrading. Highlights of follow-up efforts: first is to strengthen the top-level design and management work of standardization. From the perspective of government, it shall support and guide the manufacturing standard system with priority given to major metallurgical equipment standard system, based on *Made in China 2025* and in accordance with the *China Standardization System Construction and Development Plan (2016–2020)*. The second is to speed up the pace of formulation and revision of process equipment standards, responding to the development needs of the market and industry. According to the market requirements, a number of urgently needed standards shall be formulated and revised—for example, the standards of equipment for blast furnaces of 4000 $m^3$ and above. The third is to increase the efforts on the development and standards formulating of components and supporting parts that are key to packaged metallurgical equipment, as well as formulating standards pertaining to the safety, environmental protection, energy saving, and emission reduction in the production process with packaged metallurgical equipment, in order to improve the technical level and supporting level of metallurgical equipment.

To implement the most stringent management and assessment system of water resources by the government, carry out the *National Standard System Construction and Development Plan*, perform major standardization projects for energy conservation and emission reduction, fulfill the *Opinions on Promoting Contracted Water Saving Management to Boost Water Saving Service Industry Development*, and establish technical standard system of the contracted water-saving management, it is necessary to prepare a series of metallurgical water-saving standards to accelerate the implementation. China's water resources are in short supply and the contradictions between industrial water supply and demand are prominent. The iron and steel industry is a high-water-consumption industry, with huge water consumption. Water resources are an important constraint to the sustainable development of the iron and steel industry.

It is necessary to effectively promote the water-saving standardization work in the iron and steel industry as soon as possible, giving play to the leading role of standards. The existing water-saving standards in China's iron and steel industry are less in quantity, low in quality and always poor in operability, water-saving standardization work is still inadequate, and the water-saving standard system is still not perfect. With the gradual deepening of water-saving work in the iron and steel industry, the development of the water-saving standard system lags behind. It is difficult to adapt to the needs of the industry's water-saving management. It is urgent for professional institutions to improve, innovate, develop and update the water-saving standards system of the iron and steel industry, standardize and promote the industry water-saving standardization work. The water-saving management in the iron and steel industry is backward. The awareness of water-saving is not enough, and it cannot meet the requirements of the country's most stringent water resources management system. It is necessary to introduce a series of water-saving standards for the iron and steel industry as soon as possible, comprehensively covering all aspects of the industry's water-saving management, and strengthen efforts in developing and publicizing standards in order to standardize and guide water conservation management in the industry.

Circular economy is a new technical paradigm and a new way of increasing productivity, providing new ideas and new mechanisms for industrial restructuring, corporate incentives, and the social pursuit of new model of sustainable development. At present, the contradiction between economic development and resource shortage and environmental damage is becoming more and fiercer. Therefore, how to deal with the balance between resources and development and environment and development is a major issue in China's pursuit of the coordinated development of the people and the nature. Standardization is the basis for carrying out circular economy activities and has a fundamental position in the entire circular economy system. To construct a standard system that is oriented to circular economy is not only a top-level design but also a foundation work of standardization. Therefore, the establishment of an industrial circular economy standard system has far-reaching significance for promoting economic, social, and environmental sustainable development.

As a professional logistics standard, steel logistics standardization is an important part of the logistics standard system. The steel logistics standard system shall cover all walks of the sector, including logistics safety, logistics credibility, green logistics, logistics information and advanced facilities and equipment in terms of composition, and drop and pull transport, urban joint distribution, and multimodal transportation. Through formulating the standard system for this specialty, we shall aim to link up all parts of steel logistics industry with all kinds of logistics equipment and information, while ensuring the integrity and safe operation of logistics. Finally, through the implementation of the standards, the goal of promoting efficient, safe, smooth, and green operation of the steel logistics service system is realized.

The level of basic automation and informatization differs much among iron and steel enterprises in China, and they are at different development stages of Industry 1.0, Industry 2.0, and Industry 3.0. The standards of setting up information system are less in quantity, low in quality and poor in operability, and they need to be improved.

In order to improve the integration level between informatization and industrialization and achieve transformation and upgrading of the iron and steel industry, it is necessary to improve the construction of the corresponding information standard system. Due to the semi-continuous and semi-discrete production mode of the iron and steel industry, the informatization construction presents certain complexity. Standard system in iron and steel industry covers a wide range, consumes longer time, faces more resistance, and is hard to yield. And categories of information in the industry are various. To address the imperfect corporate information standard system, it is necessary to formulate a series of standards for the informatization of iron and steel enterprises to constrain and standardize the information construction of them and improve the success rate of information system implementation. In May 2015, the State Council officially issued *Made in China 2025*, the second item of which was to promote the in-depth integration of informatization and industrialization. In May 2016, the State Council issued the *Guiding Opinions on Deepening the Integration of Manufacturing and Internet Development* to deploy the integration of manufacturing and Internet. In order to speed up the implementation of national policies and truly improve the level of integration of informatization and industrialization of iron and steel enterprises, it is necessary to compile a series of standards for metallurgical informatization construction, highlight the leading role of standards, and achieve intelligent upgrading.

### *8.2.3 Econological Environment*

1. Environmental constraints continue to increase and the quality of the ecological environment still needs improvement.

In recent years, China has introduced a series of environmental protection laws and regulations, and the control of pollutants in the iron and steel industry has reached an unprecedented historical time. On January 1, 2015, the newly revised *Environmental Protection Law* was officially implemented and its supporting regulations: the *Interim Measures for the Implementation of Continuous Penalty by Environmental Protection Departments* and the *Interim Measures for Environmental Protection Administrations to Restrict and Stop Production*, were put into effect at the same time. Environmental violations such as excessive discharges will be severely punished by means such as "Daily punishment without upper limit" and "Security detention and criminal responsibility". Subsequently, the *Opinions of the CPC Central Committee and the State Council on Accelerating the Construction of Ecological Civilization* on April 25 clearly stated the specific objectives of comprehensively promoting pollution prevention and control, requested the local governments to hold the bottom line of environmental protection, supervise polluting enterprises to strictly abide by the quality of environmental protection, and encouraged the persistent construction of ecological civilization in an in-depth and sustained manner. In August of the same year, the *Law of the People's Republic of China on the Prevention and Control of*

*Atmospheric Pollution* was revised. It was called the "Most stringent" air pollution prevention and control law in the history. It constructively proposed to strengthen the comprehensive prevention and control of air pollution sources from coal burning and industrial plants, conduct coordinated management of gas pollutants such as air-born particles, sulfur dioxide, and nitrogen oxides. In parallel with the full enforcement of the *Action Plan of Water Pollution Prevention* and the *Action Plan of Soil Pollution Prevention*, the environmental constraints become increasingly tighter and the governance pressure is unprecedentedly mounted.

At present, the environmental carrying capacity of $SO_2$, $NO_x$, primary PM2.5, and $NH_3$ in the country is about 13.6 million tons, 12.6 million tons, 6.2 million tons, and 6.3 million tons, respectively. In 2010, the actual emissions of $SO_2$, $NO_x$, primary PM2.5, and $NH_3$ in the country exceeded the environmental carrying capacity respectively by 66%, 81%, 96%, and 52%. The emissions of major air pollutants far exceeded the regional environmental carrying capacity. Among the key emission-controlled areas, the Beijing-Tianjin-Hebei region has become a serious over-limit area. The over-limit ratios of $SO_2$, $NO_x$, and primary PM2.5 in Beijing, Tianjin, and Hebei are all greater than 150%, and the $NH_3$ over-limit rate is over 100%. Among the 31 provinces and municipalities, the emissions of the four pollutants in the highly polluted six provinces and cities, that are, Henan, Hebei, Tianjin, Anhui, Shandong, and Beijing, exceeded the environmental carrying capacity by more than one time. In the Yangtze River Delta region, the average days of fine air quality among the 25 cities in the first quarter reached 71.0%, with an increase of 7.6 percentage points year on year. The PM2.5 concentration was 61 $\mu g/m^3$, down by 10.3% year on year; the PM10 concentration was 86 $\mu g/m^3$, decreased by 15.7% year on year. In February 2017, the proportion of average days of fine air quality in the 25 cities in the Yangtze River Delta region was 67.6%, decreased by 1.7 percentage points year on year. The PM2.5 concentration was 63 $\mu g/m^3$, flat year on year; the PM10 concentration was 87 $\mu g/m^3$, decreased by 11.2% year on year. In the Pearl River Delta region, the average days of fine air quality among the nine cities in the first quarter reached 87.4%, with a decrease of 10.9 percentage points year on year. The PM2.5 concentration was 46 $\mu g/m^3$, increased by 43.8% year on year; the PM10 concentration was 62 $\mu g/m^3$, increased by 31.9% year on year. In February 2017, the proportion of average days of fine air quality in the nine cities in the Pearl River Delta region was 90.9%, decreased by 8.3 percentage points year on year. The PM2.5 concentration was 41 $\mu g/m^3$, increased by 28.1% year on year; the PM10 concentration was 55 $\mu g/m^3$, increased by 17.0% year on year. Although the overall quality of the environment has improved year by year, the situation is still not optimistic. In particular, the time for improving the quality of the atmospheric environment is urgent and the task is arduous. With the objective demand for maintaining a relatively fast economic development rate, the quality of the ecological environment is still a grim issue. It is necessary to adhere to pollution control according to law, lift up the pollutant discharge threshold in overloaded areas, and introduce standardized supervision and priority monitoring methods for key pollution-producing sources, comprehensively strengthen environmental supervision and law enforcement, and strive to continuously improve environmental quality.

2. Green development is the only way to successful transformation and upgrading of iron and steel enterprises.

As the pillar industry of China's national economy, the iron and steel industry has experienced rapid development since the twenty-first century. However, the gradual accumulation of disorderly capacity expansion and long-term extensive development has caused excessive consumption of resources and energy, seriously affected the ecological environment, and China's environmental carrying capacity has reached the upper limit. With the increasing excess capacity and deterioration of industrial-originated polluting sources, the resources and environment have reached its limits and unable to support the sustainable development of the economy. In recent years, the smog pollution incidents have further aggravated the public's concerns about the pollution of the iron and steel industry. Among the major cities with their air quality announced by the Ministry of Environmental Protection, most of the 20 most polluted cities have steel companies. Therefore, the iron and steel industry's transition to green development is an urgent task. The Chinese iron and steel enterprises represented by Baosteel, TISCO, and HBIS Group have begun to actively explore green management, green development and green transformation, and have achieved remarkable results. On March 25, 2015, at the meeting of the Political Bureau of the Central Committee, General Secretary Xi Jinping proposed "Greenization" for the first time on the top of the "New Industrialization, Urbanization, Informatization and Agricultural Modernization". Green development was mentioned at an unprecedented height. Then, the green development concept has become one of the five development concepts of China's 13th Five-Year Plan. Iron and steel enterprises must adapt to the realistic requirements of China's economic development and green transformation in the new era, vigorously implement the green development model, and adhere to the six-in-one development strategy of "Green Mines, Green Procurement, Green Logistics, Green Manufacturing, Green Products and Green Industries" to enhance comprehensive competitiveness, do better job in the mutualism and intergrowth of steel plants and cities, and achieve comprehensive, coordinated and sustainable development.

At present, the operation effect of the terminal environmental protection facilities in the iron and steel industry is not optimistic. Based on the results of government-led spot inspections to 207 sintering machines in 82 iron and steel enterprises located in 18 provinces, autonomous regions, and municipalities, it was found that the wet desulfurization process, already adopted by 70% of the CISA members, maintains a relatively high desulfurization efficiency, of which the lime-gypsum process, Mg-desulfurization method, and ammonia-sulfur method have real efficiency more than 95%, while the desulfurization efficiency of CFB circulating fluidized bed, SDA rotary spray method, and dense phase semi-dry method can be up to 85% or more. The $SO_2$ emission concentration in the flue gas can be stably maintained below 200 mg/m$^3$, and most enterprises can meet the special discharge limit below 180 mg/m$^3$. Synchronous operation rate of the CISA members can reach 90%. However, through interviews and investigations, it was also found that a small number of private iron and steel enterprises lacked the overall judgment on desulfurization technology at the beginning of the construction of desulfurization facility and handed over the project to some environmental protection companies with poor creditability, resulting in bad quality in equipment and material selection and high failure rate. The actual operation effect of some desulfurization facilities is less than 40% of the design capability. In addition, the operating cost of the desulfurization device of sintering machines is high and at the expenses of the enterprise, and the government supervision cannot cover each one of the sintering machines. With high law-abiding cost on the one hand and low unlawful cost on the other hand, some enterprises also choose to manipulate monitored data and play false. The active coke dry desulfurization process applied by TISCO, Jiangsu Yonggang, and Baosteel has achieved high desulfurization efficiency while achieving an operating rate of over 95%. The $SO_2$ emission concentration in the flue gas is controlled below 100 mg/m$^3$.

The root cause of the huge difference in environmental protection between the above enterprises is that the relevant technical norms and post-evaluation standards have not been issued for regulating relevant environmental protection technologies, technological parameters, and equipment levels. Tightened emission standards and the lag of engineering and technical standards have jointly led to a market of chaos in which environment protection technology providers with poor credit ratings "Swarm in seek of profits only". At this stage, authorities at the national and industry levels have been aware of such problems and taken measures to promote the introduction of relevant restrictive standards, tighten the market access and the approval threshold of environmental protection projects, maximally eliminating such abnormal market practices as the lowest price bid law and the Gresham's law. Only through tighter regulation on the end-of-pipe technology and setting overall standards for environmental protection facilities and the unified accounting standards for environmental protection cost calculation can China's environmental protection work in the iron and steel industry be on the right track, the fair competition and healthy development among all enterprises in the industry be promoted, and the road of transformation and upgrading of green development be embarked on. The purpose is to use the standardization of deep governance measures to lead the sustainable development trend of ecological environment where people and nature coexist harmoniously.

## 8.3 Case Analysis

### *8.3.1 Baoshan Iron & Steel Co. Ltd. (Baosteel)*

1. Overview of Baosteel's Standardization

At the beginning of establishing the plant, Baosteel also introduced packaged advanced management technologies while purchasing packaged foreign advanced equipment and production technology. On the basis of learning, digesting, and mastering foreign advanced management systems and scientific management experience, Baosteel has gradually formed a systematic management system of its own with social professional collaboration as conditions, centralized management as the core and grassroots management as basis, and system optimization in consideration of national conditions and factory conditions through persistent reform, development and innovation. Management system based on grassroots management and. It strengthened the grassroots management through the "authority decentralization" and letting down the gravity center of management. It developed the basic management model of "setting target values, giving play to operation head leadership, following operation standards, highlighting the check-based repair mechanism and the self-management".

2. Contents of Baosteel's Standardization

Baosteel's standardization includes eight aspects: standardization of criterion and standards, standardization of management methods, standardization of behavioral actions, standardization of time and series, standardization of work procedures, standardization of safety work, standardization of etiquette, and standardization of clothing and logos. Main contents are shown in Fig. 8.2.

(1) Standardization of Criterion and Standards. It includes criterion standards and technical standards, and their standardization is the premise and core content of Baosteel's standardization operations. The criterion standards are classified into two categories: general criterion and special criterion. The technical standards include product standards, raw material standards, process standards, equipment standards, safety and environmental protection standards and standards for product quality acceptance, packaging, and marking.
(2) Standardization of Management Methods. It includes standardized management of management levels, management businesses, management methods and management tools, and the standardized management over the management system itself by means of standardized principles and methods.
(3) Standardization of Behavioral Actions. It includes standardized methods such as methodology research, job measurement, operation actions, and communication languages.

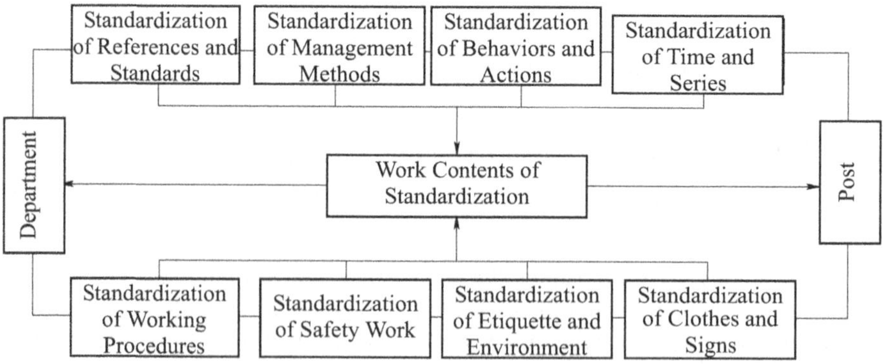


**Fig. 8.2** Main contents of Baosteel's standardization operation

(4) Standardization of Time and Series. For example, employees are required to strictly abide by the work schedule; attention is paid to the control accuracy of planned work timing; pursue the standardization of working hours.
(5) Standardization of Working Procedures. It includes process standardization, standardization of work processes, and standardization of procedures in the fields of enterprise technology, businesses and managements, and is the basis and prerequisite for developing job work (job) standards.
(6) Standardization of Safety Work. This includes the development of safety or labor protection standards (or systems) as well as the implementation of safety standards, etc.
(7) Standardization of Etiquette. Including factory environment standardization, work environment standardization, conference management standardization, and social etiquette standardization.
(8) Standardization of Clothing and Logos. Including dress standardization, standardization of logo symbols, and standardization of corporate visual identity systems.

3. Basic Promotion System of Baosteel's Standardization

The company has set up a standardization promotion committee and standardization operation promotion offices under it. Each secondary unit has a departmental standardization operation promotion group and corresponding coordinator.

The company's standardization promotion committee is responsible for defining the objectives of promotion work of standardization and making the summary. According to the company's on-site production situation, it hosts meetings and conferences to promote the company's standardization and holds on-site inspections. It approves the incentives for advanced collectives and individuals who perform well in standardization operations.

The company's standardization promotion office is responsible for the formulation and post-analysis of the company's standardization promotion work plan, as well as preparing meetings and conducting on-site inspections.

The departmental standardization operation promotion team is responsible for formulating the management measures for the standardization operation of the department, defining the milestones and conducting the on-site inspection of the department.

The departmental standardization operation promotion coordinator is responsible for establishing the network system for promoting the standardization operation tasks of this department and is responsible for the daily standardization operation promotion activities.

4. Knowledge and Practices of Baosteel's Standardization

After nearly two decades of promotion, Baosteel's standardized contents have been continuously perfected. The requirements, standards, and practices of promotion have been continuously improved. This has played an important role in improving the on-site management level of the enterprise, enhancing the quality of the workforce and ensuring safe, continuous and smooth on-site production operations, and achieving business goals.

At the same time as standardization operations are carried out within the company, Baosteel continues to explore and has achieved positive results in terms of standard formulation, product certification, scientific research standardization management, and technical regulations. In recent years, Baosteel has compiled and completed a series of internationally advanced national standards such as *Cold-Rolled Steel Sheets and Strips for Deep Drawing*, *Steel Sheets and Steel Strips for Containers*, *Hot-Rolled Wide Strips for Petroleum and Gas Transfer Pipelines*, *Color Coated Steel Sheets and Strips*, and the *Test Methods of Color Coated Steel Sheets and Strips*. At the same time, Baosteel sent its staff relevant to standardization work to participate in meetings of international organizations such as ISO, IEC, API, European Coil Coating Association, and American Coil Coating Association; organized academic exchange activities with American Society for Testing Materials (ASTM); worked together in the GM Steel Products Working Group; and kept itself updated with the latest LCA and environmental reports to prepare for the certification of green signs.

### *8.3.2 Ansteel Group Corporation (Ansteel)*

1. Development Strategy of Ansteel on Standardization

Ansteel Group Corporation (hereinafter referred to as Ansteel) is one of the first 26 standardization demonstration enterprises in Liaoning Province and a member among more than 20 domestic technical committees or technical subcommittees. It has developed standards applicable for productions and managements such as standards for the market research, product development, manufacturing, marketing, and after-sales service. It has undertaken and completed the formulation and revision of over 80 national (industrial) standards and the revision and reorganization of more than 300 corporate standards, contributing to regulating iron and steel enterprises becoming standardized, generalized, and serialized. Ansteel has incorporated the standardization development strategy into the company's overall development strategy. Its guiding lines for standardization work are to thoroughly implement the Outlook of Scientific Development, serve the company's business development, support the company's technological innovation, and ultimately achieve leading position in technology. It formulated its development goals of standardization work in the 12th Five-Year Plan period:

(1) Establishment and Strengthening of Corporate Standards. Optimize the company's standard system and structure and formulate and improve the corporate standards to cover the whole process; 85% of the corporate standards in parallel with international standards shall meet or precede their international counterparts; promote the effective implementation of standards; and significantly enhance the implementation benefits. At present, Ansteel has established a corporate standard system covering the whole process of production and has formed a standard system including procurement standards of raw and fuel materials, intermediate product, finished products, and inspection method standard system, which provides a strong guarantee for the production and operation of the enterprise.

Among the existing corporate standards of Ansteel, more than 50% of them reach and exceed international standard level. Among them, the corporate standards for hot rolling and cold rolling and coated products have accounted for above 90%. The percentage of implementing standards is far higher than the industry level. By promoting the implementation of corporate standards, customers have widely recognized and accepted the Ansteel standards. So far, all of the hot-rolled and cold-rolled products have followed the Ansteel standards.

(2) Multi-Point Coverage to National Standards. Ansteel actively participated in the activities organized by the Standardization Technical Committee, such as Steel Standards Committee, fully involved in the formulation and revision of national standards, and gradually turned the core technology of Ansteel into a national standard in key areas, making Ansteel a leader in national standards work.

By virtue of the opportunities that come with the government's recent advocacy to improve and perfect national standard system and strive to develop the national standards in important sectors, the company has fully engaged in national standardization work and increased its efforts and input in the formulation and revision of national standards. It fully connected to the National Steel Standards Committee and the Standard Committee of Iron Ores and Direct Reduced Iron, Pig Iron and Ferroalloys, became the committee members of the three major committees and its 22 sub-committees, and joined hands with them to apply, submit for approval of and develop national standards. Ansteel has improved the coordinated mechanism of the company's scientific research and standardization work to establish a close cooperation between these two tasks in order to carry out research and formulation of the national standards.

(3) Active Participation in Formulating International Standards. As a leading enterprise in the domestic iron and steel industry, Ansteel has reached a consensus on "Technology must be patented, patents must be embodied in standards and standards must be internationally accepted". Formulating international standards has become part of the company's strategic goals. With leadership support, the company has established a long-term mechanism for international standardization work. The development of international standards has been included in the company's scientific research management. New breakthroughs have been made in participating in international standardization work. Since Anshan Steel became the chair company to the secretariat of the ISO/TC17/SC17 subcommittee of "International Committee of Wire Rod and Steel Wire" in 1998, Anshan Steel has formulated five international standards as a domestic convener or participant. Among them, ISO 7989-1:2006 and another international standard, which were complied mainly by Anshan Steel, won the third prize of 2009 China Standard Innovation Contribution. The two international standards (ISO 22034-1:2007 and ISO 22034-2:2007) co-complied by Anshan Steel received the first prize of 2010 China Standard Innovation Contribution. The goal of Ansteel's international standardization has made a great step forward, and the role of Anshan Steel

has gradually transformed from a follower of the international standardization to a substantive participant.

2. Practical Effects of Ansteel Standardization Work

The standardization work of Ansteel has brought good economic benefits and huge social benefits to Ansteel. Through the adoption of European standards, international standards, and foreign advanced standards, Ansteel's heavy rails, shipbuilding steel plates, and container steel plates have won the title of China's famous brand products. The standardization work has promoted the improvement of product quality through the optimization of products, which triggered the brand name effect of enterprises. The national standard for low-power detection method of dendritic structure in continuous casting billets developed by Ansteel can save the company's inspection cost by 500,000 yuan per year. If it is promoted to the whole country, it will produce huge direct benefits. And what's more, it provides relevant parameters that are valuable to continuous casting production, technological parameter setting, and equipment adjustment and have a very significant effect on the improvement of product yield.

The standardization management work has also brought great benefits to Ansteel. The company has studied the application of the "GB/T 2828 sampling inspection procedure in the inspection of steel products". Some steel products were changed from the previous 100% inspection to sampling inspection or exemption from inspection. The inspection cost was saved up to 7 million yuan per year, and the method is still improving through continuous research.

3. Objectives of Standardization Work

(1) Continue to develop the corporate standard system and improve the advancement and applicability of the standards. During the 13th Five-Year Plan period, more than 80% of the corporate standards of Ansteel shall reach or exceed international standard level and the percentage for hot rolling, cold rolling, and coating products shall reach 100%. The structure of the corporate standards shall be improved so that all products be classified according to market demand and corporate standards be developed, in order to achieve green product standards and better meet market demand.
(2) Further improve the competitiveness of Ansteel in the national standardization work and play a leading role in terms of the number of national standards provided by Ansteel in the domestic industry, and, among the standards, 50% were listed in the 12th Five-Year Plan.
(3) Full Participation in International Standardization Work. The international standards formulated or revised by Ansteel shall not be less than one for each year; at the same time, professional international talents who understand both international standardization rules and can independently draft and participate in international standards exchanges shall be cultivated as soon as possible.

### 8.3.3 *Nippon Steel & Sumitomo Metal Corporation (NSSMC)*

Nippon Steel & Sumitomo Metal Co., Ltd. (hereinafter referred to as Nippon Steel) is one of the world's largest steel companies and Japan's largest multinational steel company. In addition to the Japanese headquarter, it has manufacturing bases in Asia, Europe, America, the Middle East, and Africa with products covering thick plates, thin sheets, rods and wires, steel pipes, and so on.

Japan has a complete set of industry standards, namely the JIS standard. The JIS standard is the most important and authoritative standard in Japan's national standards and is formulated by the Japanese Industrial Standards Committee (JISC). The JIS standard is divided into 19 items including automobiles, railways, ships, steel products, and fibers. Among them, steel standards include carbon structural steel, alloy structural steel, stainless steel and heat-resistant steel, spring steel, chromium-containing bearing steel, tool steel, and silicon steel for electrical grade.

In addition to the above national standards, the standard used in the Japanese iron and steel industry is the JFS standard made by Japan Iron and Steel Federation. The Japan Iron and Steel Federation is a juridical association among non-profit corporations. One of its main functions is to promote the standardization of the iron and steel industry. In conjunction with the situation of the downstream iron and steel industry, the standardization of steel specifications is carried out. Many Japanese technical regulations and standards are proposed by Japanese companies, reported to the competent government departments through industry associations, and finally issued by the government. The current president of the Japan Iron and Steel Federation is Kosei Shindo, the president of Nippon Steel & Sumitomo.

The Japan Iron and Steel Federation has always attached great importance to the integration of Japanese national standards with international standards to enhance the competitiveness of Japanese companies in the international market. The basic standards or methodology standards shall be directly adopted, while the product standards shall be directly adopted as much as possible. If not, they shall be equivalently adopted as much as possible. If not equivalently adopted, keep the inherent standards. This principle provides a convenient production standard reference for the global businesses of Nippon Steel & Sumitomo Metal.

As early as the twentieth century, when Nippon Steel accepted an order, JIS standards, or international standard +a was written in the technical conditions. In order to meet the customer requirements and win out in the market competition, Nippon Steel developed standard a, namely product standard or supplement technical conditions. This not only reflects the customer's requirements, but also reflects the company's level of technological innovation.

In the steel manufacturing process, Nippon Steel & Sumitomo Metal Co., Ltd., adopts various standards and specifications according to market demand. For example, the steel pipe products such as oil and gas transmission pipes produced by Nippon Steel & Sumitomo Metal Co., Ltd., adopt the standards of the global American Petroleum Institute (API standard), and the bars and wire rods for automotive steel adopt the standards of the Japanese Automobile Standards Organization (JASO standard) or the standards of the Society of Automotive Engineers (SAE standard); mechanical structural steel, spring steel, bearing steel, and other bar and wire and stainless steel pipes generally follow the JIS standard.

As the Nippon Steel & Sumitomo Metal has continuously developed product research and improved its technical content, it has also formed a corporate standard more superior than the national standards. The "ABREX" series of thick plates produced by Nippon Steel & Sumitomo Metal Co., Ltd. is a wear-resistant component material used for various industrial machines such as construction machinery. It has effectively extended the life of machinery and components, contributing to weight reduction and achieving economy. The "ABREX" series of thick plates are available in four standard types and three high toughness types, which are suitable for a wide range of applications. The environment-friendly steel sheet for electronic parts produced by Nippon Steel & Sumitomo Metal Co., Ltd., which is named the "ECOTRIO" series steel sheet, is used in the electrical and electronic components industry. It is a Ni–Sn–Zn alloy-coated product by alloying the steel sheets with three metal electroplating of Ni, Sn, and Zn.

### *8.3.4 Pohang Iron & Steel Co. Ltd. (POSCO)*

1. Overview of Corporate Standardization

The development of the standardization system of POSCO started from the 1980s, and it established the technical standards and operating standards of the company mainly based on the requirements of the Korean KS standards. Since 1986, the standardization system has been promoted throughout the company. In 1993, the ISO 9002 standard was introduced and the detailed contents of duties, operation procedures, and authority were standardized. In 1995, the standard management system was launched. From 1997 to 1999, QS9000, ISO 9002, ISO 14001, and other system were successively certified. Since 2000, with the official launch of the "Customer-Centered" PI (Process Innovation) project, POSCO has made appropriate adjustments to the original business procedures, simplified the contents of the original standard documents, cleared up the redundancy of standards among departments, set up standard documents conforming to the IT form, and further modified and improved the standard system. In 2012, the POSDMS standard management system was built to set up a networked management system. The functions of standard registration and inquiry were greatly improved.

2. Introduction to Typical Corporate Standards

In addition to actively adjusting the internal standard system, POSCO has also developed a series of standards related to advanced steel products and held a leading position in Korea.

(1) Steel HSA800 for Super High-Rise Architecture. The HSA800 (High Performance Steel for Architecture) developed by POSCO and the Research Institute

of Industrial Science and Technology (RIST) is made through TMCP. Its tensile strength is 800–950 MPa and yield strength is 650–770 MPa. Compared with the existing SM570 for building structure, its minimum tensile strength is increased by more than 40%, which ensures uniform strength and quality. The steel yield ratio is limited to less than 0.85, thus ensuring stability and shock resistance. In addition to the ultra-high strength, if used as a very thick plate material with a thickness of 100 mm, it can also present good deformation properties and is suitable for building structural systems. The cantilever beams assembled with HSA800 steel are not welded. Compared with SM490 steel, the load-bearing structure made of HSA800 steel can reduce the weight by 30% and shorten the construction period by 10%. The steel grade was adopted by the Korea Agency for Technology and Standards (KATS) on October 26, 2011, as a new standard "KSD 5994" (High Performance Steel for Architecture).

(2) High Manganese Steel Technology for LNG Storage Tanks. In March 2014, POSCO and the Korea Advanced Institute of Science and Technology (KAIST) jointly developed a large-capacity LNG storage tank with a storage capacity of 20,000 $m^3$, which is a 20 times increase compared with the storage capacity of 1000 $m^3$ of existing storage tanks. The material is high manganese austenitic steel developed by POSCO that is an energy steel that can withstand ultra-low temperature. It is especially suitable for storing LNG at minus 162 °C. Compared with the current stainless steel, it is not only superior in the welding performance but also easier to manufacture the tank body. In January 2015, in cooperation with the Ministry of Trade, Industry and Energy, POSCO formulated a series of new standards such as the *Standard for High Manganese Steel Sheets for Low Temperature Pressure Vessels (KSD 30131)*, *Arc Welding Rods for High Manganese Steel (KSD 7142)*, *Welded Flux Cored Wire for High Manganese Steel (KSD 7143)*, and *Submerged Arc Welding Wire and Flux for High Manganese Steel (KSD 7144)*. The development of these standards can reduce the number of welding, not only increasing production efficiency for customers but also saving manufacturing costs.

3. Remarks

By visiting the plant area, POSCO has mastered the actual situation and incorporated the optimal operation practices in formulating standards. It uses the charts to accurately distinguish the standard content and clean up unnecessary and formal standards. According to the PI project, POSCO revised the system operation manual, deleted duplicative standards existing among departments, unified the business guidelines, improved operation mechanism of the business guidelines, and increased efficiency. Through external cooperation, POSCO developed industry standards, comprehensively improved the practicality of technical standards and operation standards at the same time, vigorously promoted new technologies and new products, and guided the rapid development of the industry.

## 8.4 Prospect and Path Analysis of Standardization Trend

During the 13th Five-Year Plan period, China's iron and steel industry entered a stage of comprehensively advancing the supply-side structural reform. *The Adjustment and Upgrading Plan of Iron and Steel Industry (2016–2020)* proposes that, given that the iron and steel industry faces problems such as overcapacity, low level of independent innovation, intensifying constraints on resource and environmental protection and poor management of business operations, many of these issues require standards to lead and regulate; cutting excess capacity requires the use of environmental, energy, quality, safety, and technology standards to eliminate backward production capacity; iron and steel industry management and enterprise production and operation activities also require standards to standardize and guide.

### *8.4.1 Standards of Raw Materials and Fuels*

1. Standardization Trend

The standardization work on the raw materials and fuels in China's iron and steel industry started relatively early, and the standardization system is relatively complete. In China, raw materials and fuels such as iron ores, cokes, coal, refractory materials, lime, and ferroalloy carbons have been defined with corresponding quality, classification, testing, and other standards at the national level and industry level. Some product quality standards have been classified very finely; for example, the standards of cokes are classified into metallurgical cokes, foundry cokes, gasification cokes, semi-cokes, and other varieties of quality standards. In addition, some enterprises have also established corporate standards that are suitable for the development needs of enterprises and are used to ensure the implementation of concentrate input strategy of the steel production.

At present, the quality, classification, and testing standards based on single source of raw material and fuel are basically complete. In the future, the standardization of raw materials and fuels will focus on the acceptance, efficient utilization, product grade breakdown and related organizational standards, such as acceptance technical specifications of coking coal, guidelines for cokes graded supply evaluation, and guidelines for evaluation of scrap utilization efficiency.

2. Path Analysis

   (1) Continue to improve the standardization system of raw materials and fuels of China's iron and steel industry, learn from previous work experience in national, industry, enterprise and local standardization, and improve the organizational standard system of raw materials and fuels.
   (2) Focus on the standardization of raw materials and fuel grade subdivision, evaluation of utilization efficiency, green supply chain and technical guide

of resource security in order to promote the stable supply of raw materials and fuels, and meet the transformation and upgrading needs of the iron and steel industry.

### 8.4.2 *Process Equipment Standards*

1. Standardization Trend

China has formed a relatively complete standard system for process equipment in iron and steel industry, and with the continuous improvement of the level of process equipment, the standard system has been continuously improved. It has been embodied in national standards, industry standards, and corporate standards. Before the twenty-first century, the innovation of process equipment in China's iron and steel industry is mainly reflected in upsizing and increase of efficiency in single equipment, such as blast furnace replacing cupola, converter replacing open hearth furnace, continuous casting process replacing die casting, high-speed wire mill replacing double duo mill, mechanized coke ovens replacing primitive ovens, and the like. Since the twenty-first century, the progress of process equipment made by China's iron and steel industry is mainly reflected in the application of synergistic high-efficiency technology between processes, clean steel production platform technology, and intelligent manufacturing technology, such as blast furnace expert system, one hot metal ladle technology, converter one-button steelmaking, thin slab continuous casting and rolling, endless hot rolling, and intelligent manufacturing system supporting mass customization steel. Therefore, in the future, the standardization work of process equipment will focus on new processes and intelligent production, such as technical specifications of ESP endless strip rolling technology and technical specifications of intelligent 7-m and larger large-scale coke oven.

2. Path Analysis

   (1) Continue to improve the standardization system of process equipment in China's iron and steel industry, learn from the previous work experience in national, industry, enterprise and local standardization, and improve the organizational standard system of process equipment.
   (2) Standardization work in the process equipment of the iron and steel industry shall focus on new technologies and intelligent production aspects. In addition to the standards of the entire equipment, standards for certain specific components can be formulated.

### 8.4.3 *Product Standards*

1. Standardization Trend

    (1) Make the selection of standard of metallurgical products more autonomous. In March 1989, China promulgated the *Standardization Law of the People's Republic of China* for the first time, which classified the national standards and industry standards into mandatory standards and recommended standards. It stipulated: The standards and laws that serve to safeguard human health, personal and property safety and enforceable standards stipulated by regulations and laws are mandatory standards while the other standards are recommended standards. Mandatory standards must be observed. The recommended standards are the ones that the government encourages enterprises to adopt voluntarily, and once the company adopts the recommended standards in the contract or agreement, it is subject to the law and becomes mandatory. Measures shall be taken to gradually release the degree of voluntary adoption of recommended standards, giving customers more choices and greater autonomy to reflect the flexibility, timeliness, and economy of standards in the market economy. At present, the national and industrial standards for metallurgical products have all been changed from mandatory ones to the recommended ones.

    (2) Make metallurgical product standards more suitable for market needs. Due to the transformation and upgrading optimization adjustment of China's iron and steel industry, the demand for steels has undergone great changes and the consumption structure becomes more multi-level and diverse. Eliminating backwardness, improving product quality and efficiency, and accelerating the pace of industrial restructuring and product upgrading through scientific and technological progress, it is the fundamental guarantee for China's metallurgical industry to meet market demand and improve competitiveness. It is also the center and focus of metallurgical standardization work.

    (3) Make metallurgical product standards more international. Standards are an important means of competition in the current international market. The differentiated competition between individual products gradually evolves into the competition of industry standards. Controlling and influencing the process of standard formulation and revision become the new focus of market competition. The competitive advantage in terms of standards is an important basis for a country (region) to get more shares of interests in the international market competition. More and more countries and regions recognize the leadership in the formulation of international standards as an important means to promote industrial upgrading, enhance market competitiveness, and resort to advanced technical standards to control the dominant position in international market competition. By adopting international standards and advanced foreign standards, China's iron and steel industry standards can accelerate its pace to meet international standards and narrow the gap to international counterparts. Promoting Chinese standards to "Go global"

can speed up the internationalization of local standards and the output of standards can drive the export of steel products and capacity projects.

2. Path Analysis

(1) Adhere to the combination of advanced and practical standards with priority on the advancement of standards. In today's market competition, standards have become the most powerful means of market competition. Whoever obtains the right to formulate standards and embody its technologies in standards will have the initiative in the market. Technical standards have become the commanding height of competition in an industry, especially high-tech industry. In the mode of conventional large-scale industrial production, products come first before standards. In the era of rapid development of the knowledge economy and science and technology, standards often take the lead. In a certain sense, the competition of the future products is, first of all, the competition of the advanced and practical standards of products. It depends on who can better meet the needs of customers, guide the future development, and create more values for the development of customers.

(2) Actively participate in international standardization activities to promote and accelerate the integration with the international community. Standards are the most important discourse system in current international competition. The winners of standard competition can control the development direction of related technologies and market innovation direction for a long period of time and enjoy extensive control and industry leadership in the international market. Attention shall be paid to actively participating in international standardization activities and actively adopting international standards. In key areas, it is necessary to follow up the whole process of drafting international standards or directly participate in it by sending delegates. We shall involve more in the routine businesses of the secretariat and work groups and control the formulation and revision projects of specific international standards, promote and accelerate the integration of China's product standards with international standards, expand China's influence in the field of international standardization, and safeguard the trade interests of the industry.

### *8.4.4 Energy Standards*

1. Standardization Trend

The iron and steel industry is a key industry for energy conservation and emission reduction, and it is also the industry with the most potential and the most mature conditions to carry out energy conservation standards. At present, the iron and steel industry has energy consumption limit standards only for the main production processes, and the industry standards for basic energy utilization such as energy conservation and energy utilization are scarce, which is not conducive to standardizing enterprise's basic energy management and developing the industry-based energy conservation work. In view of the current industry development situation, the industry urgently needs enterprises to focus on energy conservation and emission reduction, take initiatives to change its development mode, improve the management level and competitiveness of enterprises through energy-saving standards in order to promote the green transformation of the industry, and take a new industrialized road featuring high technology content, good economic benefits, low resource consumption, and low environmental pollution.

2. Development Path

   (1) Establish a pyramid-shaped model structure, breaking the pattern of small correlation and poor mobility between the previous standards. The national standard is a mandatory standard and the basic standard that all enterprises must strictly abide by in their production activities. It takes position at the bottom of the pyramid. The industry standard is the entry threshold for the industry. It is among the second tier standard system that combines the energy consumption characteristics of the industry with the development trend of the industry on the basis of national standards. The organizational standards shall represent the vanguard of energy-saving technologies in the industry and have three-tier standard systems consisting of national standards, industry standards, and local standards.
   (2) Strengthen the formulation and revision of energy conservation standards in key areas. Accelerate the formulation and revision of energy-saving standards for industries such as steel and non-ferrous metals and establish a standard system covering energy conservation, energy conservation monitoring and management, energy management and auditing of production equipment.
   (3) Implement demonstration projects for energy conservation standardization. Build metallurgical parks or key steel enterprises with demonstration and radiation effects into demonstration projects of energy-saving standardization, promote advanced energy-saving technologies and equipment such as low-temperature waste heat power generation, absorption heat pump heating, ice thermal storage, high-efficiency motors and motor systems, and enhance energy utilization efficiency of enterprises.
   (4) Promote the internationalization of energy conservation standards, keep pace with the development of international standards in the field of metallurgical energy conservation, substantively participate in or lead the formulation of a number of international standards for metallurgical energy conservation, and expand the international market share of metallurgical energy-saving technologies, products, and services. Strengthen bilateral and multilateral international cooperation in formulating metallurgical energy conservation standards and promote the establishment of mutual recognition mechanisms for energy conservation standards with major trading countries.

### *8.4.5 Environmental Protection Standards*

1. Standardization Trend

With the frequent occurrence of air pollution and the increasingly tight water resources, the prevention and control of air pollution in the iron and steel industry and water-saving standardization have drawn more and more attention from environmental protection and related departments. In the future, government-led national-level standards in the field of environmental protection will be formulated and the conflict between regional environmental capacity and production capacity scale in the region will loom large. Under this background, differentiated pollutant discharge standards for different iron and steel enterprises will be put into force and local environmental protection facilities and technical application will be standardized to foster a development mode in which new standards coexist with existing national environmental protection standards and all standards are strictly implemented. The mandatory nature of the standardization work will be in full play to meet the annual reduction of air pollution in the region and continuously upgrade waste gas and wastewater treatment facilities. The aim is to let iron and steel enterprises have laws and standards to follow so as to finally cure the problems such as air pollution across the country, low efficiency of using water resources, difficulty in accounting environmental protection costs, low efficiency of end-of-pipe treatment facilities of steel enterprises, and short equipment service life.

As the competition in the metallurgical market become increasingly fierce, the benefits from standardization will be fully demonstrated. In the competition of steel products, the standardization work for green design products will lay sound basis for controlling the energy consumption of producing steel products and the design of the whole life cycle. Through the evaluation on steel enterprises themselves and its green production involvement, it is intended to encourage a positive market in which enterprises tap their potentials by adhering to standards and enhance utilization efficiency of energy resources in true accordance with national and local environmental protection requirements. Through the leading and restraint role of standards, the undesirable enterprises that sacrifice environment in exchange for its cost advantage of production shall be forced to quit the industry so as to improve the market competitiveness of steel enterprises that comply with environmental protection regulations and follow the green design concept and let them grow bigger and stronger. By doing so, we gradually realize the "high-end quality" transformation requirements of *Made in China 2025* and achieve the comprehensive and sustainable development of iron and steel industry featuring green and orderly growth, higher quality, and efficiency.

It is true that, by looking at the emission standards and water resources utilization indicators in the industry, some indicators have reached or even exceeded the international advanced level. On the surface, the international level of environmental protection standards in China's iron and steel industry has increased significantly, but in fact the operational effects and technical indicators of atmospheric and water pollution control facilities are relatively lagging behind, which greatly restricts the function of evaluation mechanism for the actual effects of environmental protection facilities in all enterprises. Such problems will be expected to be improved through standardization in future. In addition, China's talent pool in the environmental protection field of the iron and steel industry needs to be expanded. Only through in-depth understanding of all processes and current problems existing in the iron and steel industry can we continuously deepen the understanding of standardization work while improving environmental protection standards in the relevant fields of the iron and steel industry and put forward feasible and forward-looking advice for standardization work. In the future, by cultivating more professionals oriented to metallurgical environmental protection standardization and relying on them to participate in the environmental protection standardization activities of international metallurgical community, China's steel environmental protection standards will rival the world counterparts or even make international standards, leading the environmental standardization work globally.

2. Path Analysis

China has successively issued a number of macro-level laws and regulations such as the *Air Pollution Prevention Action Plan*, *Water Pollution Prevention Action Plan,* and *Soil Pollution Prevention Action Plan* involving many sensitive issues such as air pollution control and total water consumption, water utilization efficiency, and prevention of water pollution. However, it is still necessary to attribute metallurgical nature and features to the general environment protection standards in order to achieve the environmental expectations. *The CPC Central Committee Recommendations for the 13th Five-Year Plan for Economic and Social Development* clearly stated that restrictive targets shall be imposed on the emission of atmospheric pollutants and the most stringent water resources management system in which water resources determine how much and where to produce shall be implemented to build a water-saving society. In addition, the *National Standard System Construction and Development Plan* put the implementation of major projects for energy conservation and emission reduction standard in a prominent position. The revision of the pollutant discharge standards in the iron and steel industry during the 13th Five-Year Plan period will shock the market of the environmental protection transformation at a certain level. The major green renovation projects such as the desulfurization and denitrification of coke oven flue gas and the denitrification of sintering flue gas mentioned in the "*Adjustment and Upgrade Plan for Iron and Steel Industry (2016–2020)*" will also be gradually implemented. *The Opinions on Promoting Contract Water Saving Management to Boost Water Saving Service Industry Development* stated to establish a technical standard system of contract water-saving management. To implement the above-mentioned policies, we must have a "Breakthrough point" and take feasible monitoring and treatment measures to fully play the core role of the standardization work in environment protection and to reduce the standardization work into various steps.

In view that the current environmental protection standards system in China's iron and steel industry is still not perfect, the environmental protection standardization work requires the technical support from a third-party authoritative consulting organization with a strong industry background to comprehensively combine the research and development of environmental protection standards and the enforcement regulations and conduct post-assessment on implemented standards. It shall promptly offer advice to the technical committee of SAC/TC207/metallurgical environmental management standardization and the technical committee of SAC/TC275/metallurgical environmental protection industry standardization on the validity, revision, or abolition of the relevant standards. At the same time, it organizes the presentation, interpretation, training, discussion, and justification of environmental protection standards for the iron and steel industry, and organizes or participates in the formulation, revision, justification, and review of environmental protection standards in the industry.

The Working Group of the Metallurgical Environmental Protection Industry and Environmental Management Standardization will carry out the formulation and revision of standards for the key equipment of environmental protection in the iron and steel industry such as the fully enclosed raw material yard, flue gas desulfurization and denitrification, dust removal, wet electrostatic precipitator, advanced treatment of coking wastewater, and the treatment of cold rolling wastewater; the formulation and revision of products standards; the formulation and revision of standards for flue gas desulfurization equipment from steel slag treatment, equipment for synergistic governance of active coke flue gas; all-in-one equipment for coke oven desulfurization and denitrification, coking wastewater treatment equipment, life cycle evaluation of circulating water pumps, high-efficiency evaluation and energy efficiency evaluation; the formulation and revision of energy efficiency evaluation standards; the standardization and revision of products such as coking and sintering flue gas denitration catalysts, water treatment agents, and high voltage power supply of electrostatic precipitators. It organizes publicity and interpretation of standards of environmental protection equipment and products in metallurgical industry, investigates the implementation of enforced standards for evaluation analysis and information feedback, and writes reports; it submits projects with encouraging achievements in professional fields to the National Standardization Administration Committee and relevant competent authorities for awarding purpose; it organizes the formulation and revision of standards pertaining to environment management and crucial evaluation elements in terms of environmental performance evaluation on China's iron and steel enterprises and on each production unit, the environmental information disclosure of iron and steel enterprises and the environmental cost accounting of iron and steel enterprises; it develops detailed regulations for the technical specifications of the upcoming "Pollutant discharge licensing system" and extends its application in key links and submits assessment specifications for the green products; it aims to comprehensively improve the coverage and precision implementation of environmental protection standards in the iron and steel industry and guide the enterprises in their efforts of standardized the green, standardized, systematic transformation, and upgrading.

### 8.4.6 Water-Saving Standards

1. Standardization Trend

With the increasingly tight water resources situation, the water-saving standardization work in the iron and steel industry will gradually attract more attention. There will be a situation in which the standards set by the government and the standards independently developed according to the market situation by the iron and steel enterprises develop in parallel and supplement each other in effect. This gives full play to both the mandatory standards and recommended standards and meets the requirements of steel enterprises on saving regional water resources, reclaiming wastewater, reducing the cost of water, and extending the service life of water treatment equipment in steel companies.

With the continuous development of water-saving standardization work, the foundation of water-saving standards in steel industry will continue to be consolidated. Relying on the water-saving work platform, the water-saving work in the iron and steel industry will be more standardized and normalized, promoting the establishment of water-saving standards in the iron and steel industry, developing and cultivating a number of advanced enterprises who lead the way in water-saving standardization in the industry, and creating driving force of water-saving standardization work in the iron and steel industry.

As the competition in the metallurgical market become increasingly fierce, the benefits from standardization will be fully demonstrated. Among the competitors of steel products, standardization work will make the energy consumption of steel products traceable and enable steel enterprises to improve the efficiency of energy resource utilization, meet local environmental protection requirements, enhance product competitiveness, and achieve sustainable development.

China's water consumption indicators in iron and steel industry have been at the international advanced level, and the internationalization of Chinese water-saving standards in the iron and steel industry has increased significantly. In the future, in the process of improving the water-saving standards in the iron and steel industry, China will cultivate more talents with expertise in metallurgical water-saving standardization, participate more in international steel water-saving standardization activities, spread the metallurgical water-saving standards to the world, and lead the global metallurgical water-saving standardization.

2. Path Analysis

China has successively issued the *Measures for the Implementation of the Most Strict Water Resources Management System* and *Water Pollution Prevention Action Plan* and other systems, and put forward requirements from the aspects of total water consumption, water use efficiency, and prevention of water pollution. *The CPC Central Committee Recommendations for the 13th Five-Year Plan for Economic and Social Development* clearly stated that "the most stringent water resources management system in which water resources determine how much and where to produce, shall be implemented to build a water-saving society". *National Standard System Construction and Development Plan* encouraged to implement major projects for energy conservation and emission reduction standardization; *Opinions on Promoting Contract Water Saving Management to Boost Water Saving Service Industry Development* established a technical standard system of contract water-saving management. To implement these policy documents, we must have a "Breakthrough point" and feasible measures. Standardization work will be the core of water-saving endeavor and one of the important ways to reduce water-saving work into various fields.

It shall be noted that the water-saving standard system in China's iron and steel industry is still not complete. China's steel water-saving standardization work requires a special platform organization who shall investigate and study the implementation of water-saving standards in the iron and steel industry, conduct post-evaluation on the enforced standards, and give advice to the National Standardization Administration Committee and the National Water Conservation Standardization Technical Committee on the validity, revision, and abolish of the relevant standards. At the same time, it shall organize the propaganda, interpretation, training, discussion, and demonstration of water-saving standards in the iron and steel industry, and organize or participate in the revision, formulation, demonstration, and review of water-saving standards in the iron and steel industry.

Through the research on water system in the iron and steel industry, combining with the characteristics of water use in the iron and steel industry, the water-saving standards in the iron and steel industry mainly involve water-saving terminology classification and definition, water-saving management and statistics, water quality and consumption monitoring and measurement, water quotas and water efficiency standards, and calculations and assessments to contract water conservation and management, water balance test methods and assessments, water-saving assessments and appraisal, and other standards.

The Metallurgical Water Saving Standard Working Group will carry out research work on the water-saving standard system of the iron and steel industry, and sort out, establish, and improve the water-saving standard system framework of the iron and steel industry. The focus will be placed on six aspects to develop water-saving standards for the iron and steel industry: basic standards, water quotas, contract water-saving management, wastewater zero-discharge, calculation methods, and assessment and evaluation system.

### 8.4.7 Circular Economy Standards

1. Standardization Trend

Vigorously promote the comprehensive utilization of metallurgical resources and the standardization of circular economy; accelerate the promotion of standardized management of metallurgical secondary resources recycling and the promotion and application of advanced energy conservation, environmental protection, and comprehensive utilization of resources. Action is taken mainly from the following three aspects.

Technical product standards: standards for comprehensive utilization technologies, products and equipment of metallurgical tailings, waste rocks, smelter wastes, dust and slurry, waste materials, and domestic wastes; standards for comprehensive utilization technologies, products, and equipment of secondary resources such as metallurgical wastewater, waste liquid, waste gas, residual heat, and residual energy.

Iron and steel industry park standards: standards for park spatial layout and industrial structure standards, standards for park industry chain and recycling production, standards for park resource utilization, standards for park pollution prevention and information technology and management.

Key common standards: standards for basic management over statistics, inspection and tests, recovery, storage, and transportation of secondary resources like metallurgical wastewater, waste gas, solid waste, and residual heat and residual energy; standards for comprehensive utilization technologies of existing resources, evaluation standards, and indicator system of cyclic economy of iron and steel enterprises.

2. Path Analysis

Based on the theory of circular economy, the model of circular economy implementation is deeply analyzed. The life cycle analysis method is adopted to build a circular economy system framework based on the feasibility of technology and the development management requirements. Construct a technical standard system, management standard system, and job standard system framework covering all aspects of processes, equipment, inspection, measurement, energy, safety, environmental protection, and quality management. In addition, organize the revision and formulation of relevant standards through standardization means, proceeding from the actual development of circular economy and in combination with the characteristics of the industry. Standardization practitioners shall continue to explore and develop absent standards, and turn the advanced technological achievements and experiences in production management operations into standards. For example, efforts in metallurgical slag treatment shall focus on strengthening the combination of resource utilization technology and standardization, and incorporating new technologies and processes into the standards. In treatment of iron-containing dust sludge, it is urgent to develop corresponding standards to promote clean production in the industry, promote the transformation of scientific research results into productivity, and reduce the dumping of dust pollutants.

### 8.4.8 Logistics Standards

1. Standardization Trend

Under the current rapid development of "Internet +" drive, the traditional logistics industry, under the influence of Internet technology and new development models, is improving operational efficiency and transforming the industry ecology. Internet technology will integrate logistics resources and bring revolutionary changes to the logistics industry. The main development directions include the logistics system integration and intelligent development, the integration of enterprise logistics system and production system, the diversified development of logistics system by integrating individual technologies, the massive application of IoT technology, and the low-carbon and environment-friendly construction of logistics systems.

With the rapid development of the logistics industry and the active embrace of the Internet technology, the pace of logistics standardization will be further accelerated. Standardization of steel logistics system must base itself on the standardization of hardware facilities while software standardization is distillation. On the basis of the unification of information standards, facilities, and equipment standards, the improvement of operational standards, management standards, and service standards shall be the focus and represent the developing trend of logistics industry in this professional field.

2. Path Analysis

(1) Establish a steel logistics standard system with steel enterprises as the core of the supply chain. The logistics standard system is an organic integration of various standards within the scope of logistics standardization framework. The steel logistics standard system is an institutional framework involving logistics-related standards in the iron and steel industry. The establishment of this standard system is the basis for organizing a scientific and rational structure of standards within the scope of steel logistics standardization work. It is a blueprint that covers the existing practices and future development standards in the field of steel logistics, and is a guiding and fundamental work for the formulation and revision of steel logistics standards.

Due to the core position of the iron and steel enterprises in the metallurgical supply chain system and their significant agglomeration effect on flow of goods, logistics services and logistics information, establishing a steel logistics standard system that treats iron and steel enterprises as core of supply chain and covers the metallurgical circulation field is an effective measure to build logistics standards covering the whole metallurgical supply chain (a complete framework). The scope of the standard system shall help the effective connection between social logistics and enterprise logistics, improve enterprises' dependency on social logistics, indirectly promote the socialization and industrialization of enterprise logistics, and effectively improve the efficiency of enterprise logistics and reduce the logistics of cost.

(2) Guided by the national logistics standard system, a steel logistics standard system framework will be built. The metallurgical supply chain system is segmented into links according to the scope of the entire supply chain covering the supply, production, sales, and circulation of the industry. At the same time, according to the framework of the national logistics standard system, the above-mentioned links are further segmented in accordance with logistics technical standards and logistics services standards, logistics information standards, logistics management standards, etc., establishing a metallurgical supply chain logistics standard framework that covers the whole industry chain. The framework focuses on the content of relevant logistics standards in the micro-logistics sector of the iron and steel industry, including the standards for logistics technology and work-related standards involved in the metallurgical supply chain system. After the main framework, the "Tree trunk" of the steel logistics standard system, is established, the "Branches and leaves" can "Grow" therefrom according to the existing steel logistics standards and the expected ones, and finally, a complete steel logistics standard system can come into being.

(3) Taking the formulation of key standards as a start point, initiate the formulation and revision of steel logistics standards in an orderly manner. The development and revision of steel logistics-related standards are still in its infancy and there are still some gaps in the steel logistics standards. However, as a principle, the formulation and revision of relevant standards shall prioritize the formulation of key standards and unfold orderly the other steel logistics standards. Key standards shall have strong market demand, can immediately solve the problems existing in the current steel logistics operation process, and can improve the logistics operation efficiency and reduce the logistics operation cost for the industry and enterprises. Only through the significance of the key standards can the important role of the formulation and revision of steel logistics standards be highlighted, thus initiating the preparation of other steel logistics standards.

(4) Paying attention to the standardization of production logistics in iron and steel enterprises. Iron and steel enterprises have huge flow of production logistics and complex logistics processes. However, because they belong to the internal logistics of enterprises, they have always been kept in a closed development mode, independent of the social logistics system, with a low degree of socialization. Since the production logistics in steel enterprises are all operated by usage due to its closed model, there are serious shortages of logistics standards, the logistics operation efficiency is low, the connection of logistics links and the logistics operation cost is unsatisfactory. Being confronted with the grim situation of the overall "Cold winter" of the iron and steel industry, the standardization of production logistics in iron and steel enterprises is imminent. At the same time, the production logistics standards of iron and steel enterprises shall be effectively combined with the relevant standards of social logistics and the channels between internal logistics and social logistics must be opened. Internal logistics must

gradually move its development focus to non-steel businesses, expand the degree of enterprise logistics outsourcing, reduce the number of enterprise personnel, and improve the labor productivity of enterprises.

### 8.4.9 Informatization Standards

1. Standardization Trend

*The Adjustment and Upgrade Plan for Iron and Steel Industry (2016–2020)* issued by the Ministry of Industry and Information Technology pointed out that the smart manufacturing base must be consolidated. It is necessary to comprehensively carry out the work of standardization and assessment for the integration of information technology and industrialization management system in iron and steel enterprises and promote the standardization of steel intelligent manufacturing.

*The Guidelines for the Construction of National Intelligent Manufacturing Standards System* (2015 Edition) jointly issued by the Ministry of Industry and Information Technology and the National Standardization Administration Committee pointed out that we will strive to establish a relatively complete system of intelligent manufacturing standards by 2020. More than 500 intelligent manufacturing standards will be formulated and revised so that basic common standards and key technical standards are fully covered. Intelligent manufacturing standards will be widely verified in enterprises and applied in all areas of manufacturing, thus promoting the improvement of China's intelligent manufacturing level and increasing the international competitiveness of Chinese manufacturing standards significantly.

According to the *Construction Guidelines*, the development of informatization standards in the iron and steel industry will also be carried out from three dimensions: life cycle, system level, and intelligent function. The content will contain basic commonality, key technologies, and iron and steel industry standards. There is no affiliation between the intelligent manufacturing standard system and the iron and steel industry standard system, and the contents of the two are intersected. The intersection lies on the part of iron and steel industry application standards in the intelligent manufacturing standard system—for example, the iron and steel industry standards that are used to guide steel-related production, design, management, etc., and the key standards of the iron and steel industry in the intelligent manufacturing standard system that involve the related aspects of intelligent manufacturing such as interconnection and connectivity in the iron and steel manufacturing process.

2. Path Analysis

Intelligent manufacturing is the most effective way to achieve transformation and upgrading of China's iron and steel industry. Intelligent manufacturing has a strong comprehensive nature. It is not only meant to the breakthrough and application of single technology and equipment, but the deep integration and innovative integration of manufacturing technology and information technology. Therefore, the informatization standards of the iron and steel industry will focus on smart manufacturing in the future and developed from the following aspects:

(1) *The National Intelligent Manufacturing Standard System Construction Guide* will be used as basis for approving projects. *The Construction Guidelines* will be used as the basis for guiding the establishment and revision of national standards and industry standards for intelligent manufacturing in the coming period.
(2) In accordance with the principle of "Prioritizing Common and General Standards and Standards in Urgent Need", the focus will be on establishing standards of cross-domain and cross-industry system integration. By coordinating standard resources and optimizing standard structure, it focuses on solving the bottlenecks in promoting intelligent manufacturing like data integration and interconnection and connectivity.
(3) Being Based on National Conditions and Clearing Obstacles for Cooperation. Based on the characteristics of China's intelligent manufacturing standards such as weak foundation and the imbalance of industry development, we must fully consider the applicability of standards, strengthen the standardization and industrialization of independent intellectual property rights, strengthen communication with advanced manufacturing countries and international standardization organizations, and duly include independent intellectual property rights standards as international standards. At the same time, the international standards that are suitable for the development needs of China's manufacturing industry must be transformed into national standards in a timely manner. Our goal is to set up an intelligent manufacturing standard system with good compatibility and openness.
(4) Advance with the Times and Make Sustained Efforts. Intelligent manufacturing is a huge system of dynamic development. The industry's understanding of intelligent manufacturing will be a gradually deepening process. With the development of intelligent manufacturing technology and industry, new modes and new forms of businesses continue to emerge, and the intelligent manufacturing standard system will be dynamically adjusted and improved.
(5) Establish Application Standards for Intelligent Manufacturing in the iron and steel industry. On the basis of the common standards related to intelligent manufacturing and according to the characteristics of the iron and steel industry, we must accelerate the formulation of application standards for the iron and steel industry such as smart factories, intelligent logistics, network collaborative manufacturing, large-scale personalized customization, and remote operation and maintenance in the iron and steel industry, aiming at a breakthrough in the field of intelligent manufacturing.
(6) Establish a new standard system to develop and coordinate with the standards independently formulated by the market under the leadership of the government.

## 8.5 Industrial Practices of Standardization

China Metallurgical Industry Planning and Research Institute (hereinafter referred to as MPI) has undertaken secretariat work of six standardization working groups including groups for TC20/metallurgical energy foundation and management, TC207/metallurgical environmental management, TC275/metallurgical environmental protection industry, TC442/water saving in iron and steel industry, TC415/metallurgical resources comprehensive utilization, and TC269/steel logistics. MPI has undertaken the work of the secretariat of five standardization working committees including China Special Steel Enterprise Association, China Steel Structure Association, China Circular Economy Association, Henan Iron and Steel Association, and China Metallurgical Mining Enterprise Association. The China Metallurgical Industry Planning and Research Institute can provide enterprises with comprehensive and systematic standardization consulting services, including formulation and revision of national, industrial, association's, corporate standards, corporate standard system construction and strategic planning, corporate standard-related training, corporate standards public statement, evaluation and improvement, domestic and international standard analytic research and publicity, etc., which are shown in Table 8.1 for detail.

**Table 8.1** Practices of MPI in promoting industry standardization

| No. | Types | Main content | Typical cases |
|---|---|---|---|
| 1 | National standards | Be committed to establishing standard systems for metallurgical energy conservation and environmental protection, comprehensive utilization of resources and steel logistics, and strive to promote green, recycling and low-carbon transformation of metallurgical industry, and promote efficient, safe and optimized operation of steel logistics. Relying on six standards working group as platforms, MPI provides services for the establishment, revision, organization and coordination, standard-related training and publicity of relevant national standards for enterprises in need | MPI provides technical services for the formulation and revision of national standards such as the *Methods for Calculating Water Savings in the Iron and Steel Industry* and *Energy Audit Methods for Iron and Steel Enterprises* |

(continued)

**Table 8.1** (continued)

| No. | Types | Main content | Typical cases |
|---|---|---|---|
| 2 | Industrial standards | We will carry out research and development and conduct pilot work of industry standards in the fields of iron and steel industry products, energy conservation and comprehensive utilization, environmental protection, logistics, green manufacturing, and integration of informatization and industrialization, and promote the transformation and upgrading of China's metallurgical industry and sustainable green development. MPI provides services for the establishment, revision, organization and coordination, standard-related training and publicity of relevant industrial standards for enterprises in need | We have provided technical services for formulation and revision of the following industrial standards: *Guidelines for the Evaluation of Green Manufacturing Plants in Iron and Steel Enterprises*, *Scope and Calculation of Logistics Costs of Steel Enterprises*, *Technical Specifications for Power Demand-Side Management Platforms in the Iron and Steel Industry*, *Technical Requirements for the Charged Raw Materials and Fuels to 4000* $m^3$ *Blast Furnaces and Above*, *Code for the High Level Production Scheduling System (APS) of Iron and Steel Industry*, *Technical Requirements for Waste Heat Recovery of Blast Furnace Slag Granulating Water*, *Technical Specification for Treatment of Chromium-Containing Heavy Metal Wastes by Blast Furnace Method*, *Green Design Product Evaluation Specification for Stainless Steel Kitchenware*, *Technical Specification for Purification Through Sintering/Pellet Wet Desulfurization Flue Gas Cyclone Tube Electrostatic Precipitator* |

(continued)

**Table 8.1** (continued)

| No. | Types | Main content | Typical cases |
|---|---|---|---|
| 3 | Associational standards | MPI is committed to the formation of a series of standards that adapt to market demand, lead the industry development, and fill gaps. It promotes the supply-side reform of the iron and steel industry through associational standards and meets the customized needs of the upstream and downstream industries of the industry chain for steel products. Relying on five associational standards working group as platforms, MPI provides services for the establishment, revision, organization and coordination, standard-related training and publicity of relevant associational standards | MPI has provided technical services for formulation and revision of the following associational standards: *Sorbite High-Strength Stainless Structural Steel Hot-Rolled Ribbed Steel Bar*, *Sorbite High-Strength Stainless Structural Steel Hot-Rolled Wire Rod*, *Sorbite High-Strength Stainless Structural Steel Hot-Rolled Steel Sheet And Strip*, *Sorbite High-Strength Stainless Cold-Rolled Steel Sheet And Strip*, *Sintered Products Made from Industrial Waste Slag for Stainless Steel Smelting*, *Alloy Square Steel*, *Steel and Alloy Silver Bright Bars for Internal Combustion Engine Valves*, *Stainless Steel Wire Rod for Cold Heading*, *Large Diameter Thick Wall Forging Pipes*, and *High Alloy Hot Die Forging Bars* |
| 4 | Corporate standards | The Metallurgical Planning Institute can provide enterprises with comprehensive and systematic standardization consulting services, including formulation and revision of corporate standards, corporate standard system building and strategic planning, corporate standard-related training, corporate standards public statement, evaluation and improvement, domestic and international standard analytic research and publicity, etc. | We have conducted consulting services such as *Product Standard System Construction for Zhengzhou Yongtong Special Steel Co., Ltd.*, *Product Standard System Construction for Jiangsu Shenyuan Group Co., Ltd.*, *Standardization Promotion Strategy for Tianjin Yuantai Derun Steel Pipe Manufacturing Group Co., Ltd.* |

## References

1. China Government Network. Notice of the state council on issuing and distributing "*Made in China 2025*" [2015-05-19]. http://www.gov.cn/zhengce/content/2015-05/19/content_9784.htm.
2. Official website of the Ministry of industry and information technology. Equipment manufacturing standardization and quality improvement planning [2016-08-02]. http://www.miit.gov.cn/n1146285/n1146352/n3054355/n3057497/n3057498/c5180111/content.html.
3. State Council. Deepening the reform of standardization work [2015-03-26]. http://www.gov.cn/zhengce/content/2015-03/26/content_9557.htm.
4. General Office of the State Council, National Standardization System Construction and Development Plan (2016–2020) [2015-12-30].http://www.gov.cn/zhengce/content/2015-12/30/content_10523.htm.
5. China Government Network. The Standardization Law of People's Republic of China [2017-11-05]. http://www.gov.cn/xinwen/2017-11/05/content_5237328.htm.

# Chapter 9
# Differentiation

Adopting a differentiation strategy, providing distinctive products and services, and winning the trust of customers are the important development directions for iron and steel enterprises in the future, especially for small- and medium-sized ones. Specifically, the first is the differentiation of development strategy, namely strengthening the research on market demand and competitors and identifying the enterprise positioning in the process of "becoming bigger and stronger", "doing the finer", "becoming moderately diversified" and "relatively diversified", and developing corresponding strategy to achieve industrial chain differentiation and regional differentiation. The second is the differentiation of products, namely optimizing product structure, increasing the share of competitive products, focusing on product serialization, and improving the "one-stop" supply capacity. The third is the differentiation of production lines, namely achieving specialized production, greatly improving production efficiency, and reducing production costs through the technical transformation of equipment. The fourth is the differentiation of services, namely improving pertinence and recognition by adopting different service models for different customers and different regions. The fifth is the differentiation of sales, namely common products are mainly sold by e-commerce platform and direct sales to large customers, so as to reduce intermediate links and sales expenses. The high-quality special steel products mainly focus on the modes of technology sales, strategic customers, early intervention, and continuous tracking, keeping up with customers' demands and gradually expanding the market. The sixth is the differentiation of control, namely vigorously implementing low-cost production strategies for common products to improve their competitiveness. The high-quality special steel products are oriented by strengthening product quality to open up the market and improve customer recognition.



X. C. Li, *The Road Map of China's Steel Industry*,
https://doi.org/10.1007/978-981-15-2074-7_9

## 9.1 History Review and Status Analysis

Since the reform and opening-up, Chinese iron and steel industry has been rapidly developing with the support of strong cost competitiveness and which has also objectively caused the extensive growth mode of the iron and steel industry dominated with quantitative expansion. In recent years, the state has focused on the industrial restructuring for the development of the iron and steel industry, which means that the scale expansion in the traditional way is no longer feasible. In addition to the continuous implementation of energy conservation and emission reduction policies, the growth in iron and steel consumption is constrained to some extent. With the continuous adjustment of the product structure and technological advancement of the enterprise, the homogenization of products has become more and more serious and the competition has become increasingly fierce.

At present, the disorderly competition of low-end products in Chinese steel market is fierce, mainly in three aspects: First, the price is disorderly. Chinese iron and steel industry shows overcapacity and sluggish downstream demand, the low-end products of all enterprises are at the same level, and there is no obvious gap in varieties, specifications, and quality among their products. Consequently, iron and steel enterprises are pressing each other's prices to seize the market, so their profits are meager or even minus, causing disorder competition. Second, the export is disorderly. The steel products exported from China are mainly low-end products with low added value, and the average export price is significantly lower than the average import price. In 2016, the average price difference between import and export was USD 493.1/t. The steel products exported from China still have problems such as the concentration of export destinations. The disorderly export of such low-end products has intensified trade frictions among countries, resulting in multi-faceted anti-dumping against Chinese steel products by European countries and the USA. Third, existence of shoddy products. The existence of "shoddy" products such as counterfeit hot-rolled rebars by substandard steel and counterfeit micro-alloyed or ultra-fine grain rebars with water-cooled rebars is also an important reason for the "Malignant Competition" in the iron and steel market in some regions of China. Substandard steel products are produced in many small steel plants in China. Although their quality is not high, they are sold in a low-cost dumping operation mode because of their low cost and low burden on enterprises, which further deteriorates the market environment.

Iron and steel enterprises are eager to break away from the homogeneous competition of low-end products. Consequently, many high-end production lines have been launched in recent years, and their equipment levels are very close. However, many enterprises have similar product mix and lack of scientific and rational market positioning, which has led to the increasingly homogeneous competition of high-value-added products. For example, as of the end of 2016, 15 sets of hot rolling wide-strip mills over 2000 mm had been completed in China (Table 9.1), with a total output of 71.5 million tons. They have high equipment levels and are generally positioned to produce automobile panels and home appliance panels, resulting in repeated construction of a large number of high-end equipment with similar product

**Table 9.1** Status of hot rolling wide-strip mills over 2000 mm in China

| No. | Enterprise | Mill specification/mm | Output/× $10^4$ t | Mill type | Date of put into operation | Equipment level |
|---|---|---|---|---|---|---|
| 1 | Baosteel | 2050 | 400 | 3/4 tandem | 1989 | Domestic advanced level in China |
| 2 | WISCO | 2250 | 450 | Semi-tandem | 2003 | Domestic advanced level in China |
| 3 | Liuzhou Steel | 2032 | 350 | Semi-tandem rolling | 2005 | Domestic advanced level in China |
| 4 | Anshan Steel | 2150 | 500 | ASP | 2005 | International advanced level |
| 5 | TISCO | 2250 | 450 | Semi-tandem | 2006 | International advanced level |
| 6 | Shougang Qiangang | 2160 | 450 | Semi-tandem | 2006 | International advanced level |
| 7 | Masteel | 2250 | 550 | Hot tandem rolling | 2007 | International advanced level |
| 8 | Shougang Jingtang | 2250 | 550 | Semi-tandem rolling | 2008 | International advanced level |
| 9 | Hanbao Steel | 2250 | 450 | Semi-tandem rolling | 2008 | International advanced level |
| 10 | Rizhao | 2150 | 500 | Semi-tandem | 2008 | International advanced level |
| 11 | Benxi Steel | 2300 | 550 | Semi-tandem rolling | 2009 | International advanced level |
| 12 | Valin Steel | 2250 | 450 | Semi-tandem rolling | 2009 | International advanced level |

(continued)

**Table 9.1** (continued)

| No. | Enterprise | Mill specification/mm | Output/× $10^4$ t | Mill type | Date of put into operation | Equipment level |
|---|---|---|---|---|---|---|
| 13 | Pangang Group Xichang Steel | 2050 | 450 | Semi-tandem rolling | 2011 | International advanced level |
| 14 | Baotou Steel | 2250 | 500 | Semi-tandem rolling | 2014 | International advanced level |
| 15 | Zhanjiang Base of Baosteel | 2250 | 550 | Semi-tandem rolling | 2015 | International advanced level |
| | Total | | 7150 | | | |

positioning. Taking pipeline steel as an example, all the hot rolling wide-strip mills of WISCO, Baosteel, Ansteel, TISCO, Shougang, Masteel, Benxi Steel, and other large iron and steel enterprises in China are able to produce pipeline steel products, resulting in serious product homogenization and obvious overcapacity and outstanding contradiction between supply and demand; consequently, it directly leads to a sharp drop in the price of pipeline steel products, which ultimately causes the profit space of all enterprises' pipeline steel to shrink sharply.

With the increasingly tight constraints on the environment and resources faced by the iron and steel industry, the pressure on iron and steel enterprises is enormous due to the cyclical adjustment of upstream and downstream industries, the unreasonable industrial layout, and the overcapacity caused by disorderly competition in various regions. "Homogeneous competition" is one of the main culprits for the current chaos and drawbacks of the iron and steel industry. The homogenization of products causes vicious competition among enterprises; the homogenization competition of enterprises makes the development of the iron and steel industry uncoordinated, unbalanced, and unsustainable. To this end, if China's iron and steel industry wants to complete the transformation and upgrading, its way out is to vigorously promote the differentiation strategy.

Michael Porter, the master of strategic management in the USA and the "father of competitive strategy", pointed out that the economic significance of differentiation is to manufacture scarce products, which means the enterprise shall manufacture the scarce product that is different from competitors in a certain aspect or a certain link of the business process under the market structure of supply and demand balance or oversupply, namely "partial short supply", so that the enterprise will have their own competitive advantage and gain excess value of innovation. As Michael Porter pointed out, the significance of company's advantages or disadvantages ultimately depends on the extent to which the company can respond to the market with relatively low cost and differentiation. By implementing differentiation strategies and by relying on innovative products, innovative brands, and innovative markets, opening-up their own exclusive market space, maximizing value, and winning lucrative market returns should be the most important goals of each iron and steel enterprise for transformation and development.

During the 12th Five-Year Plan period, most iron and steel enterprises in China began to consider implementing differentiated development strategies. While adhering to moderate scale operations, they constantly adjusted their operating variety structure and expanded their differentiated competitive advantages. However, it is still in the infancy for Chinese iron and steel enterprises to adopt differentiation strategy, and is necessary for them to strengthen understanding and implementating of the same. At present, there are two points worthy of attention in the competitive environment of the iron and steel market. The first is the pressure from similar marketing: Since most steel products are standardized, homogeneous products and the differences among products of different enterprises are less, and the marketing strategies and means are similar. Secondly, technical competition is homogenized: Advanced technology is widely used by large iron and steel enterprises, which is one of the reasons for similar products.

## 9.2 Development Environment and Policy Orientation

### *9.2.1 Development Environment*

1. Economic Environment

At present, the global economy is still in a stage of deep adjustment, the recovery momentum is insufficient, instability, and uncertainties are numerous and varied, the economic situation is complicated, the recovery base of developed economies is relatively fragile, and the major emerging economies are facing pressures such as economic slowdown and structural adjustment. From the perspective of the domestic environment, China will further promote the transformation of development mode and economic restructuring, and economic growth will maintain a new normal in the stage of medium-high speed. On the other hand, with the further advancement of China's new industrialization, informationization, urbanization, agricultural modernization, and greening, domestic demand will gradually be released, consumption will become the backbone of economic growth, and the proportion of the added value of the service industry has surpassed that of the secondary industry and ranks first. At the same time, with the implementation of China's "the Belt and Road Initiative", coordinated development of Beijing-Tianjin-Hebei and the Yangtze River Economic Belt, new regional economic growth pole will gradually take shape. The domestic economic situation is nurturing opportunities in difficulties; especially, "the Belt and Road Initiative" has brought unprecedented opportunities for international cooperation between iron and steel enterprises. On the whole, the macroeconomic environment at home and abroad is still tightening in the coming period.

2. Environment of the Industry

The competition pattern of global iron and steel industry is undergoing profound changes and adjustments. After the international financial crisis, developed countries have implemented the "re-industrialization" strategy to reshape the new competitive advantage in manufacturing; especially as Germany, the USA, Japan, Britain, France, and other countries announced their plans or strategies to revitalize the manufacturing industry, the new technological revolution characterized by the deep integration of the new generation of information technology and manufacturing is triggering far-reaching industrial changes and forming new production methods, industrial forms, and business models, which will greatly enhance the competitiveness of the iron and steel industry in the established powers. Some developing countries have also seized the opportunity of global industry re-division to actively participate in undertaking industry and capital transfer to expand international market space by taking advantages of accelerating growth of local steel demand, labor, and resources. It can be said that China's iron and steel industry is facing "double-side pressure" from developed countries and other developing countries.

China is in the decisive stage of building a well-off society in an all-round way. As an important basic industry of the national economy, the iron and steel industry has entered into the second half of production and consumption peak period, the period of profound adjustment of the market structure, the key decision period for the establishment of an iron and steel powerhouse, and the historical opportunity period for innovation and development. In the future, the risks and challenges faced by China's iron and steel industry are mainly reflected in the fact that international trade protection is not optimistic, the task of resolving overcapacity is arduous, the risk of business chain failure is increasing, and the risk of relocation of iron and steel enterprises in core areas is intensified. Overall, the environment of China's iron and steel industry is not optimistic.

3. Market Environment

The demand of the international iron and steel market will remain generally in a state of mildness, trade remedy measures are prevalent, resource nationalism has risen, and international market competition among iron and steel enterprises will be further intensified. The pattern of China's economic growth depending on the investment is changing, the steel consumption intensity of specific economic aggregate is further reduced, steel production and consumption show a downward trend from the peak arc region, and the industry has entered into a difficult stage of reduction development. With the upgrading of the demand structure and the adjustment of the industrial structure, the market will put forward higher requirements on the iron and steel enterprises in terms of variety, quality, and service. In addition, environmental resources and human resource bottlenecks will further intensify, so the pressure on steel manufacturing costs will increase. Various measures on comprehensively deepening the reform will stimulate the vitality of various economic factors and create a fair competitive market environment. At the same time, it also requires iron and steel enterprises to accelerate reform, innovation, and transformation development to adapt to the rapidly changing external market environment. Market competition is gradually shifting to quality and differentiated-based competition from quality expansion and price competition.

In the current and in the long run, the most obvious characteristic of the iron and steel market is still serious oversupply, and the situation of brutal market competition is difficult to change. Under the new normal, the increases of the quality of consumption and the proportion of personalized demand in China's iron and steel market will become more obvious. During the 13th Five-Year Plan period, with the adjustment to emphasized consumption and service in the adjustment and transformation of the domestic economic structure and the weakening of downstream demands, it is expected that the peak arc zone of consumption quantity of China's iron and steel market will be extended. From the perspective of consumption structure, with the implementation of *Made in China 2025* [1], and the establishment of the steel structure promotion group by the Ministry of Industry and Information Technology and the Ministry of Construction to promote green buildings, the equipment manufacturing industry and steel structure construction will maintain a relatively fast growth rate in the future, which will drive the demands for strip products, and China's market share of strip will expand in the future.

4. Environmental Protection Constraints

The iron and steel industry is a key industry for the prevention and control of air pollution. Since 2015, the new environmental protection law and the new series standards for iron and steel industry have been fully implemented. Known as the most stringent new environmental protection law in history, it adds more measures, such as government accountability, public interest litigation, no-limit punishment on a daily basis, and criminal detention of responsible persons. The high-pressure situation formed by the new environmental protection law will deter illegal activities for a long time. The Beijing-Tianjin-Hebei region is the region with the most serious air pollution in China. According to the monitoring results of the environmental protection department, none of the 13 key cities in the Beijing-Tianjin-Hebei region can conform to comprehensive standards, and the situation of air pollution prevention and control in the Beijing-Tianjin-Hebei region is very serious.

The new series of environmental emission standards in the iron and steel industry not only increase the assessment indicators, but also significantly tighten emissions. Among them, the discharge standard of wastewater from iron and steel industry is supplemented with total nitrogen and other 13 water pollutant indicators, the pollutant emission standards are supplemented with pollutant indicators such as dioxins and nitrogen oxides, and the pollutant discharge standards of coking and chemical industries are supplemented with PAH and other 14 indicators of pollutants with industry characteristics. The emission indicators stipulated in the new standards obviously become stricter, and the environmental protection costs of enterprises will increase by a large margin. Especially in the regions of Beijing-Tianjin-Hebei, Yangtze River Delta and Pearl River Delta that are implemented with special emission limits for atmospheric pollutants, the environmental protection supervision faced by iron and steel enterprises will be more stringent, and some iron and steel enterprises will be forced to shut down due to non-compliance of environmental protection.

### 9.2.2 Policy Orientation

*The Adjustment and Upgrade Plan of Iron and Steel Industry (2016–2020)* issued by the Ministry of Industry and Information Technology proposes that China's iron and steel industry will no longer be in the period of large-scale development and will enter into the development stage of structural adjustment, transformation, and upgrading in the next five years, and it will insist on structural adjustment, structural adjustment, innovation drive, green development, quality first, and open development to accelerate the realization of adjustment and upgrading and improve the development quality and efficiency of China's iron and steel industry. It can be seen that China's iron and steel industry will implement "differentiated competition" in the future, and policy orientation will change from blindly pursuing high value-added in the past to encouraging enterprises to purse differentiated and characteristic development according to their own development situation and market demands.

On January 27, 2016, the State Council held a third meeting in the month to guide financial institutions to adjust their credit policies and support industrial enterprises to upgrade their quality and efficiency. It was clearly stated for the first time for the iron and steel industry that "for the enterprises and backward production capacity suffering long-term losses or losing their liquidity or failing to meet environmental protection and safety production standards or unable to be rectified, their relevant loans should be resolutely compressed and withdrawn and the dissolution of overcapacity will be supported"; "the enterprises under standard operation and able to repay debt are encouraged to issue their corporate credit bonds in order to adjust their debt structure". From the perspective of policy influence, with the supply-side reform entering into the practical stage, the financing channels and financing costs of iron and steel enterprises will accelerate differentiation. The iron and steel enterprises with backward production capacity, nonstandard operation, weak profitability, poor solvency, and high reliance on external financial support may face difficulties in turnover due to tightening of financing environment and thus facing risk of default or bankruptcy; eligible corporate financing channels (including equity and bond financing, accounts receivable financing, etc.) are expected to be broadened, and the proportion of direct financing will increase.

*Guidance Catalogue for Foreign Investment Industries* (2015 Edition) announced by National Development and Reform Commission points out that the common manufacturing industry will be further liberalized and foreign capital will be allowed to take holdings over domestic iron and steel enterprises. *Adjustment Policy for Iron and Steel Industry* (2015 Edition) (Draft for Comment) proposes to encourage foreign capital to participate in the merger and reorganization of domestic iron and steel enterprises, to build a new open economic system, to liberalize the restrictions of foreign investment on the domestic iron and steel sector, and to enjoy the same investment policy of domestic and foreign enterprises. This reflects that China's market will be gradually opened in the future, China's iron and steel industry chain is highly mature with perfect infrastructure and the good development environment will attract the steel plants with foreign investment to seek differentiated construction, and it will be more conducive to Chinese enterprises to participate in international competition, especially opening-up emerging markets in South Asia and Southeast Asia such as India and Vietnam.

## 9.3 Case Analysis

### *9.3.1 Differentiation of Development Strategy—Baosteel Ltd.*

The development path of Baosteel in the past 40 years can be highly summarized as follows: taking "striving for first-class" as the strategic orientation, implementing "excellent product strategy", and realizing "import substitution". The essence of Baosteel's excellent product strategy is the differentiation strategy around elaborate works. For a long time, the differentiation strategy has always been the leading strategy for Baosteel's development, leading Baosteel to become a world-class steel company with international competitiveness.

Since its establishment in 1978, Baosteel has been positioned to produce high-value-added, high-tech, and high-end steel products that are in short supply in the domestic market and cannot be produced by other domestic steel plants and can replace import products, has been fully promoting market-oriented product upgrades, and has focused on the development of high-end steel products represented by automobile panels, electrical steel, stainless steel, oil and gas steel, and special high-temperature alloy steels to form a competitive strategic product group; in terms of quality and technology, through the further development of processes, equipment, and products, Baosteel has developed high-quality, high-tech, and high-value-added steel products; through the implementation of the excellent product strategy, Baosteel's high-standard and high-quality brand image has been firmly established in the market.

Baosteel has been adhering to the principle of "customer-oriented and always right" for a long time. Since 1990s, Baosteel began to promote customer satisfaction (CS) project and implement total satisfaction management (TSM) strategy and proposed the new concept of "standard + α" in 1997 to realize the customization of product quality from satisfying the enterprise standard to satisfying the customer's use requirements and taking sanctification of customer's demand as the standard for the effective operation of the inspection quality system. "Standard + α" is a concentrated expression of Baosteel's business philosophy of "customer-centered and customer-satisfied" and is also a concrete embodiment of serving the people by serving customers.

When discussing the products positioning of Project Phase-III in the early 1990s, the Baosteel leadership has resolutely and decisively listed the automobile panels as the first strategic product in virtue of anticipation of huge market demand and development potential of automobile industry in China under a far-sighted and unique strategic version. And they also decided to change the main product of completed Project Phase-II from mechanical steel into automobile panels. Through continuous learning and innovation, Baosteel achieved a success in developing automobile panels ahead of the industry. Today, Baosteel's automobile panels have firmly occupied half of the domestic market of the same.

Following the automotive panels, in 2003 Baosteel listed silicon steel as the second dominant strategic product. In addition to the strong market demand, the more important point of choosing automotive panels and silicon steel as the strategic leading products is their development and production which have a strong driving force for Baosteel's product upgrade and a revolutionary and comprehensive improvement of Baosteel's management, innovation, production, and technology level. In particular, oriented silicon steel is a handicraft in steel boutiques, and its product quality is often considered as an important indicator of a country's iron and steel production technology and represents the top level of today's iron and steel production technology. Internationally, only a few companies, such as Nippon Steel, are able to produce oriented silicon steel, and they at the time imposed a technical blockade on the fast-growing Baosteel. Therefore, Baosteel officially decided at the end of 2004 that silicon steel technology would be independently developed and silicon steel engineering would be independently integrated. In May 2008, the oriented silicon steel production line of Baosteel was successfully put into operation. It can be said that the strategic vision and the decision-making qualification on bravely climbing the innovating peak of Baosteel's leadership have played a key leading role in the successful development of automotive panels and silicon steel.

### *9.3.2 Product Differentiation—Shiheng Special Steel*

In recent years, in the face of the complex and ever-changing market environment, Shiheng Special Steel has fully utilized the decision-making mechanism of rapid response, firmly grasped the "three lifelines" of market, efficiency, and innovation, constantly optimized the product mix with the reform as the driving force, adhered to the low-cost strategy, and continuously explored new profit growth points, so its profit per ton of steel and its tax per ton of steel always rank among the top three in the industry. In 2016, the company realized a profit of 1.42 billion yuan and a treasury tax of 620 million yuan and achieved a profit of 401 yuan per ton of steel, ranking first in the industry.

In the increasingly competitive development environment of the iron and steel industry, Shiheng Special Steel has explored their own successful enterprise transformation experience. Shiheng Special Steel firmly grasps the ability of the company's products to adapt to the market, profitability, and innovation ability. With the aim of "doing better and stronger", it has achieved the overall goal of high operational efficiency, market share, strong resource control, low overall cost, and good economic returns. The main experience is as follows:

The first is product differentiation and specialization. Shiheng Special Steel fully utilizes the advantages of the existing equipment level, is committed to product mix optimization, and has established a differentiation strategy and "five-oriented" product policy, namely small-scale, high-strength, specialization, micro-alloying, and high-quality, so as to create their own competitive advantage. Its market share of anchor steel in the country is more than 50%; ultra-fine grain steel accounts for more than 50% of the total steel products, and the production technology has reached an international leading level; the proportion of high-strength steel rebars reaches 100%, and the proportion of high-strength steel rebars of 500 MPa is far more than the industry average.

The second is to rely on technological innovation to enhance the competitiveness of the enterprise. Shiheng Special Steel has established a standard process for project approval, implementation, process management, achievement evaluation, assessment, and incentives in technological innovation and formed a virtuous cycle model of information collection, subject research, and project transformation. Shiheng Special Steel is the only enterprise in China that has mastered the rolling technology of ultra-fine grain steel bar. Continuous equipment innovation and technological breakthroughs make its rebar wire rod products that have two competitive advantages: brand quality and cost.

The third is the innovative management model to enhance corporate efficiency. The first is to strengthen the system construction for management innovation. The management organization shall be improved, the innovation of all employees shall be deepened, and a scientific project establishment, management, review, and reward system shall be established. The second is to establish a management model of "efficient, long-term, and breakthrough". The value chain is divided into more than 50 core businesses, the management model is innovated by taking each core business as the subject, and it is implemented continuously. The third is to create an economic and technical indicator system with unique characteristics. Sixty-seven key economic and technical indicators were determined, among which indicators such as product mix, burden structure, capital operation, cost performance, and purchasing index were creatively included, which played a significant role in improving the ability of enterprises to adapt to the market.

### 9.3.3 *Differentiation of Production Line—Fangda Special Steel*

Since the completion of reform and reorganization, Fangda Special Steel has taken a series of effective measures in the areas of technological transformation, product development, and information technology, focusing on the development of producing better and stronger spring flat steel and industry chain of "spring flat steel—plate spring". After the transformation of the original spring flat steel production line, a new high-end special steel production line was built to increase the annual production capacity of the spring flat steel from 0.3 million tons to 1 million tons and the proportion of high-end special steel to be more than 45%, so the product mix has become more reasonable.

In the aspect of industry chain construction, it has merged and holds shares of some domestic enterprises producing automobile plate springs and suspensions and has formed a characteristic industry chain of "mine-smelting-spring flat steel plate spring". Its domestic market share of automobile plate springs has been significantly improved.

In terms of product development and construction of laboratory facilities, a plate spring laboratory has been built and successively provided with a series of advanced inspection and experimental equipment worth more than 30 million yuan, which lays a foundation for product quality testing and research and development of new spring flat steel products. New process and new technology are adopted, and research activities are carried out to improve the quality of spring flat steel. The research and development of new products were strengthened, and 17 varieties of rectangular spring flat steel with rounded corner and thick section spring and the new spring flat steel products of more than 300 sizes were successfully developed.

In the aspect of automation control, the information transformation of the spring flat steel production, manufacture, and quality inspection data was carried out, and the whole spring flat steel production process from blanking, production, and manufacturing to the product delivery under dynamic computer monitoring and management was realized, so the production efficiency of the spring flat steel is significantly improved.

By means of those effective measures, significant results have achieved. After the production line is perfected, it has the characteristics of high efficiency, stability, and high control level and has formed differentiated competition with other domestic enterprises. The differentiation of the production line is finally reflected in the product. The spring flat steel products of Fangda Special Steel cover GB standard, Japanese standard, German standard, American standard, and more than 20 designations specially developed for customers with the width of 44–150 mm and the thickness of 5–50 mm. Fangda Special Steel is the most comprehensive and most productive spring flat steel production enterprise in China, and its spring flat steel production ability is constantly getting better, stronger, and higher, and spring flat steel is the product which has always provided Shiheng Special Steel with profit after the financial crisis [2].

### *9.3.4 Service Differentiation—Baosteel Ltd.*

In the implementation of transformation strategy of "from manufacturing to service", Baosteel took the lead in channeling resources to both ends of the smile curve—R&D

and marketing network, and formed a R&D service system featuring "Early Vendor Involvement" as the business card, and the processing and distribution service system all over the world, thus enhancing the original product advantages of Baosteel's automotive plates into a package solution featuring comprehensive competitive advantages from product development, material selection, and deep processing of steel products until JIT distribution.

With the continuous development of the steel market, Baosteel's original unique leading products have also become homogenized products. In order to enhance the differentiated competitiveness of those products, Baosteel took its own advantages to start with service. After years of operation, it has realized the service creating value for Baosteel. At present, Baosteel's services have been carried out in the entire process of pre-sale, in-sale, and after-sales, and its technical content is also constantly improving. Technical service and technical support have become the main features of Baosteel's customer service system, such as on-site instruction and guidance to users on Baosteel's product application technology, helping customers to select materials reasonably, providing product consultation, up to participate in the production, operation, and even development process of customers, and establishing a customer application technology research center to help them improve productivity, assist them to develop new products, form a set of leading efficient service system, and cultivate and form their core capability with the overall advantages of production, sales, and research. Benefited from the successful application of the differentiation strategy, Baosteel has achieved outstanding performance, and its comprehensive competitiveness ranks among the best in the global iron and steel industry.

Early Vendor Involvement (EVI) means the early involvement of vendors. It is a customer-centered and customer-oriented value creation activity of the supply chain comprehensively participated by the suppliers. For Baosteel, EVI is not only a technical marketing activity but also an innovative business model that aims to satisfy the demands of customers, promotes the development of products and technologies of Baosteel, and accelerates the marketing promotion and business cooperation. Through EVI, Baosteel focuses on customer's demands, fully involves all aspects of the customer from R&D to mass production, and gives full play to the capabilities and advantages of Baowu Steel Group to create customer value, thus conquering the highest position in market competition.

At present, the EVI of Baosteel has formed an EVI cooperation type, which covers the whole process of automobile development and manufacturing from advanced engineering design, early (concept) and body design, mold (tool) design and development, vehicle type production and mass production, and satisfies different demand levels of customers. Baowu Steel Group has the ability to provide customers with a "target cost-oriented" and "lightweight-oriented" package solution and has established in-depth cooperation with all major domestic automobile manufacturers. At the same time, the EVI work is no longer limited to automotive customers, but also expanded to customers in different industries such as home appliances, electrical steel, and metal packaging. The EVI work of Baosteel is an effective exploration and practice in the process of transforming from manufacturing to services.

At present, China's automobile industry is facing the pressure from the constraints of petroleum, environment, and transportation, and the whole industry will enter a stable developing period with low growth. At the same time, domestic competitors in the iron and steel industry have entered or target to the demands of the automobile industry and launched fierce competition with Baosteel through strategies such as new units, new product development, improved production technology level, and "low price penetration". Particularly influencing by the factor of the price, the Baosteel is currently facing tremendous pressure on the fixed-point acquisition of new models and the share maintaining of the old models; especially, the price competitiveness of hot-dip galvanized products has declined significantly. Facing with such a brutal market competition situation, Baosteel needs to comprehensively enhance its competitiveness in an all-round manner from the perspectives of products, technology, services, marketing, etc. to maintain and consolidate its current market share. The EVI work of Baosteel is a highlight of Baosteel's concept of "competitive products + service". While creating value for customers, it also greatly enhances the added value and competitiveness of Baosteel's products. Today, with the increasingly fierce competition for product homogenization, Baosteel fully demonstrates the value of the company's technological advantages and service capabilities through EVI work, avoiding falling into simple and low-level price competition, thus maintaining the market share of the company's products and higher ability of premium, and creating a competitive path for differentiated services [3].

### *9.3.5 Sales Differentiation—POSCO*

In recent years, the situation in the international iron and steel market has deteriorated rapidly. In order to get rid of the crisis, various iron and steel enterprises have also chosen to take diverse measures for improving quality. POSCO will improve the competitiveness of the iron and steel industry as one of its core strategic directions. That is to say, through the combination of product technology and marketing strategy, POSCO will provide differentiated value to customers and maintain a profit gap of more than 5 percentage points with competitors, thereby improving its competitiveness.

Even if the high-quality steel products are produced, if they are not convenient for customers or with poor economic nature, they will not be favored by customers. In view of that, POSCO has built a new business model "technology-based platform business" that integrates "highest performance steel products" (hardware) with "customer application technology" and "commercial support" (software) into the entire marketing campaign, facilitating customers to use POSCO' s products, which is the original intention of "solution-based" marketing. POSCO not only participates in the research and development process of customer products, but also provides customers with product application technology and even necessary financial support. This kind of comprehensive, customer-oriented, and "solution-based" marketing is now fruitful and has greatly boosted the growth of sales of high-value-added products.

1. Solution-Based Marketing of "Product Technology Services"

Since 2014, as the market competition has become increasingly fierce, in order to ensure its own competitiveness, POSCO has officially proposed a "solution-based marketing" concept that is more advanced than EVI activities, aiming to meet the potential needs of customers and provide customers with differentiated value.

After Kwon Oh-joon became the chairman of the board, he established a "solution center" at the headquarters of iron and steel business, which aggregates the company's internal distributed solution-based marketing functions to provide customers with systematic solutions, starting a new business model. Through the implementation of solution-based marketing activities that are differentiated from competitors, POSCO has established strong partnerships with Korean enterprises such as LG and Samsung Electronics, as well as other overseas well-known enterprises such as Volkswagen and Mahindra.

POSCO has been supplying side plates to Volkswagen for more than 10 years, and its Motive Power is the strategy of solution-based marketing. The outer side plate requires to be processed by advanced mold technology such as crack processing and wrinkle correction, and strict quality management is required after the steel plates are supplied. In the automobile markets of global emerging country, POSCO not only provides the major automotive manufacturers with the necessary mold technology for the outer side plates, but also continuously strengthens the locking effect on automotive materials. Through continuous development of "solution-based marketing" of mold technology, POSCO has supplied the German Volkswagen with outer side plates for more than 10 years.

In March 2014, POSCO successfully won the bid for construction of the Dongdaegu Compound Transfer Center under the control of Korean Shinsegae Department Store, becoming the sole designated supplier of H-beam and medium and heavy plate products for the project. The material designed for the project is RH section steel, which is not produced by POSCO, so POSCO does not meet the conditions for the tender. However, the POSCO Research Institute of Science and Technology (RIST) worked closely with the construction steel product sales office to design and develop the high-strength HSA800 steel product, which can completely replace RH section steel, and proposed a cost-saving budget plan to the construction contractor, and finally won the bid. At the same time, POSCO also received the order of medium and heavy plate products for the project, which is fully contributed to the strategies of "Total Solution Provider" developed by POSCO P&S and the steel processing center.

In April 2014, POSCO and DK Austech in Korea signed a memorandum of understanding on "supporting overseas export of auto parts materials", providing solutions in forming analysis and mold technology, and promoting export sales deploitation of Korean parts manufacturing industry.

DK Austech has been doing business with POSCO since 2004. It mainly produces auto suspensions, chassis, and other parts and supplies them to domestic and foreign auto manufacturers. The company participated in the "common development" activities of POSCO and introduced QSS, security diagnosis, and other activities. Benefiting from POSCO's "solution-based marketing" campaign, its cost of products and export competitiveness have got different degrees of improvement, resulting in obtaining orders for chassis parts from Japanese auto enterprises. Last year, it began to expand the proportion of exports to Japanese auto enterprises.

2. Solution-Based Marketing of "Fund Support"

POSCO's "solution-based" marketing not only provides technical support for the development or application of new steel grades, but also provides financial solutions for customers with capital operation difficulties.

NSB is a design company for building girders. Despite its technical strength is strong, its financial strength is obviously insufficient, making it difficult to carry out post-business expansion and effective project management. Previously, the company successfully cooperated with POSCO to develop a special "Sbarch composite girders" used inside the arched steel box, which can save more than 30% of the steel consumption. Meanwhile, due to the internal filling of the concrete, not only the anti-settling and seismic performance of the building are improved, but also the beauty of the arch shape can be reflected. Based on this technology, the company undertook the design and construction project of the Bridge on the Busan Outer Ring Expressway, with the construction cost of 22 billion won and the steel product consumption of 8000 tons. Hence, the company needs to invest 5.3 billion won for steel product procurement to manufacture girders.

At the beginning of the project, the company only received 10% of the contract amount, and its credit transaction with POSCO also required a two-month cycle, so its initial steel procurement funds were obviously insufficient. To this end, NSB was planned to outsource steel structure manufacturing enterprises, which generally adopts low-cost imported steel products. In response to this problem, POSCO also provided NSB not only with the information on steel prices and market conditions but also a cost-saving solution, which made NSB aware that it is very beneficial to directly purchase steel products to manufacture girders. In order to solve the shortage of initial procurement funds, POSCO provided procurement funds supporting service and studied related credit programs with the assistance of POSTECH Venture Capital Corporation.

Although NSB is a small- and medium-sized construction company with certain industry risks, after many studies, POSTECH Venture Capital Corporation still decided to provide 1 billion won in credit. In the end, NSB reduced the cost by adopting the cost reduction plan provided by POSCO under the background of initial capital shortage. Up to now, 1200t steel products have been supplied directly by POSCO, and the rest are supplied by POSCO's processing center—Toyo S-TEC.

In exchange for providing the necessary support to the customers with capital turnover difficulties, the customers will give priority to the steel products provided by POSCO, instead of importing or outsourcing steel products, thus achieving a win-

win situation. There are still a lot of small- and medium-sized customers like NSB. Therefore, POSCO will continue to provide solution-based marketing with "funding support".

3. Expansion of Sales and Transfer of High-End Products and Technology

At present, POSCO mainly improves its profitability by increasing the proportion of high-value-added products. Its research mainly focuses on high-strength automotive plates, extremely thick plates for energy service, and steel products for extremely low-temperature LNG storage tanks. Under unfavorable conditions such as global steel products oversupply and unsatisfactory demand, POSCO has continuously improved its market power through the differentiated strategy of WP products (World Premium).

POSCO selected 220 items of WP products in 2014, and the sales volume of WP products maintained a good growth trend, laying a good foundation for the long-term development of customers and helping to enhance the competitiveness of customers. In 2014, the sales volume of WP products was 10.207 million tons with a year-on-year growth of 13%, and the proportion of WP products increased from 30.9% in 2013 to 33.3%. In 2015, the sales target of WP products was 12.15 million tons, and the proportion of the same increased to 36.0%. As of 2016, POSCO is planned to increase it to 41%.

With the rapid development of the automotive industry, automobile production has increased significantly, and automobile varieties have also become more diversified. The demand of lightweight automotive vehicles for high-strength automotive plates is also increasing. To this end, POSCO selected automotive plates such as AHSS, MAFE, and HPF as WP products and strengthened the marketing efforts on it. POSCO has continuously set up technical service centers at home and abroad to build "solution-base" around the world. Through close cooperation among headquarters of iron and steel business, Pohang/Gwangyang Steel Plants, and Technical Research Institute, POSCO has established a response mechanism for production, marketing, research and development, and technical services throughout the company and vigorously implemented "solution-based" marketing to customers. The A&C Construction Company of POSCO is also promoting various high-performance steel products such as PosMAC, POS-ALUWALL, and POS-BH beam from the beginning of the building engineering to the procurement of raw materials to the completion of construction. At present, the service effect of WP products has been widely recognized by the industry, which indicates that POSCO's technical strength has also been upgraded by leaps and bounds, thereby further consolidating the competitive advantage of its advanced product production and sales systems.

In addition, in order to improve the competitiveness of its iron and steel industry, POSCO is planning to sell its own innovative technology to overseas and provide technical solutions to create new benefits for it. POSCO's goal this year is to build the POIST business globally. The so-called POSCO's Innovation Steel Technology (POIST) means POSCO's exclusive steelmaking technology. As an innovative, compact, integrated steelmaking process, this technology integrates three processes, namely FINEX for ironmaking, PS-BOP for steelmaking, and CEM for continuous casting and rolling, so as to maximize competitiveness compared with traditional processes and there is an epoch-making improvement in profitability and environmental friendliness.

In order to improve production efficiency, POSCO is also promoting the sales of CEM technology (Compact Endless Casting and Rolling Mill) downstream to FINEX. CEM is a short-process steel rolling technology that integrates continuous casting and continuous rolling. Its core process technologies include high-speed continuous casting, headless hot strip rolling, and batch rolling technology.

At the end of May 2014, POSCO signed a basic agreement on "Transfer License and Joint Marketing of CEM Technology" with SMS Group. According to this Agreement, POSCO will transfer the innovative CEM Technology self-developed by himself to a globally well-known equipment and engineering company. POSCO will implement the necessary management supervision and education and training for CEM technology transfer, and charge technology transfer fees from SMS Group. Group subsidiaries such as POSCO Construction and POSCO ICT will also participate in the project construction. The cooperation indicates the market competitiveness of the SMS Group, and the technical strength of POSCO will be seamlessly connected. In order to fully promote CEM technology, two enterprises will continue to carry out in-depth technical cooperation.

In order to build a new technology sales platform, POSCO has classified the existing 221 technologies according to different levels to distinguish between strategic sales and common sales. There are three levels of these related technologies: strategic sales (grade S) include 50 items of technologies such as Slim and FINEX, strategic sales (grade A) include 76 items of technologies such as continuous ultra-thin sheet hot rolling technology, and common sales (grade B) include 95 items of technologies such as efficient sintering operation technology.

POSCO has always attached importance to significantly increase revenue by improving the market competitiveness of high-end products and high-value-added products while implementing "solution-based marketing" activities. Pohang not only participates in the research and development process of products, but also provides customers with product application technology and even necessary financial support. This all-round, customer-oriented, and "solution-based" marketing campaign has achieved remarkable results and greatly promoted the growth of sales of high-value-added products. Despite the decline in the average selling price of steel products, the profitability has improved significantly as a result of increasing the proportion of sales of high-value-added products. In recent years, POSCO has accelerated the pace of exporting FINEX/CEM technology. It has successively signed cooperation agreements with enterprises in China, India, and Germany to obtain a new source of revenue through the transfer of high-value-added technology.

### *9.3.6 Control Differentiation—POSCO*

In 2017, World Steel Dynamics (WSD) released the list of "world's most competitive iron and steel enterprises", and POSCO once again ranked on the list, with high scores in terms of technological innovation, cost savings, labor skills, structural restructuring, and investment environment, reflecting a high core competitiveness. Since 2010, POSCO has been continuously ranking the world's most competitive iron and steel enterprises for eight years.

In order to achieve the goal of "becoming a leading iron and steel enterprise in the world", POSCO pays great attention to product technology research and development and boldly innovates existing production processes and operating procedures to enhance the competitiveness of high-value-added products and save production costs. In recent years, POSCO has identified cost reduction, manufacturing competitiveness improvement, and the creation of new industrial demands ("3C" strategy) as the basic development direction, launching a new round of reform and development.

1. Cost Control

The per capita steel output of POSCO is more than twice that of China's iron and steel enterprises. Although the wage level in China is low, the total labor cost is very close to that of POSCO. POSCO's steel plants in South Korea have a relatively simple layout, only including two steel plants in Pohang and Gwangyang. Among them, the annual crude steel output of the steel plant in Pohang is 21 million tons, which consists of BF, two FINEX furnaces, two stainless steel plants, and one electric furnace steelmaking plant; the annual crude steel output of Gwangyang steel plant is 19 million tons, which consists of five BFs. This simple layout structure allows POSCO to reduce logistics costs, which generates significant economic benefits.

From the perspective of cost indicators, POSCO has advantages in terms of equipment, energy, and fixed costs. The utilization rate of CHP facilities in POSCO is 70%, and its energy cost is only USD 23/ton due to its long-term LNG procurement contract with Tangguh Indonesia at a low price. The new technology developed by POSCO will allow the application of low-quality coal with a percentage up to 50%, which will make the large-size coke oven run more economical. The increased use of cheap iron ore and coal and combined with the use of FINEX-economic ironmaking technology give POSCO a significant advantage in the competitiveness of production cost.

In terms of entire cost of the company, POSCO is more competitive than Japanese enterprises, except for the raw materials, labor, and depreciation cost. In recent years, in order to save costs, POSCO has also been implementing the "Simplification Plan". In 2014, the company adopted effective measures such as low-cost raw materials and fuels, energy recovery, equipment efficiency, and by-product gas and set a cost reduction target of 603 billion won, with an increase of about 20% year-on-year. Among targets, the raw material cost is 355 billion won, the maintenance cost is 57 billion won, the energy cost is 78 billion won, and the material cost is 113 billion won.

2. Product Competitiveness Control

POSCO is pursuing high-value-added products and seeking survival and development by entering emerging markets. Its high-value-added steel products have a profit margin of 15–20%. In 2015, POSCO's high-value-added products accounted for

38.4%. The automotive plate is representative of POSCO's high-value-added products. The company has been committed to manufacturing the world's best automotive plates and expanding sales in emerging markets, including 450,000 tons in India, 450,000 tons in China, and 500,000 tons in Mexico. At the same time, the company's sales of other high-value-added products as energy service steel continued to increase.

POSCO's R&D system on iron and steel business is strong, forming the "Industry-University-Research Cooperation System over POSCO-Involving Field" consisting of Technical Management Committee, Technology Strategy Committee, Pohang Engineering University, Pohang Industrial Technology Research Institute, Pohang Technology Research Institute, Global R&D Center, overseas research institutions, and their cooperative R&D institutions. POSCO's investment in research and development continues to increase, taking the leading positions in iron and steel enterprises all over the world, and its share of total sales is expected to increase to 2% by 2020. The company's core iron and steel department has a total of 1085 researchers, and its research and development direction is mainly focused on automotive plate, TMCP steel, electrical steel, etc. The recent research and development results mainly include ultra-high-strength tensile steel plate for automobiles and advanced high-strength automotive plate, strength control technology, as well as welding technology for polymer polyester coated automotive plate and high-strength TMCP super thick plate.

3. Market Control

Since 2010, in order to reduce the impact of Hyundai Steel's entry into the Korean HRC market and improve its profitability, POSCO has adjusted its production and sales pattern in a timely manner, gradually reduced its proportion of HRC products, and increased its proportion of downstream high-value-added products.

While getting its competitiveness in the domestic market continues to increase, POSCO has been paying close attention to overseas markets. Its exports have achieved rapid growth in the past five years, and its exports accounted for more than 40% of total freight. The primary exporting countries and regions of POSCO are Japan and Southeast Asia, accounting for 60% of its total exports, and the rest 40% are mainly flowing to North America, Europe, and the Middle East. In order to establish close contact with its terminal customers, POSCO has established a series of production bases in its overseas markets around the world, including one steel complex, 10 downstream plants, and 47 machining centers. 20% of POSCO's automotive plates are supplied to POSCO's overseas downstream plants as raw materials, and this proportion will further increase as POSCO's large-scale expansion in overseas markets.

## 9.4 Prospects and Path Analysis of Differentiation Trend

### *9.4.1 Prospects of Differentiation Trend*

1. Market Needs Differentiated Competition

Differentiation strategy means that the strategies taken by the iron and steel suppliers to provide customers with more product added value or to meet the unique functional needs of customers in a certain aspect, relying on some of their own advantages, in order to make the company's products, services, technology, image, marketing, channels, brands and other aspects, have obvious differences with competitors, thus obtaining a competitive advantage. The goal is to get a "premium" that is higher profit by investing additional costs to make the product unique.

After stopping to pursue pure production expansion, differentiated competition will be an important strategy for iron and steel enterprises adapting to the market environment under the new situation. In terms of differentiation strategy, iron and steel enterprises should consider differences in technology, region, service, industry chain, and specialization. It shall be enhanced that the non-repeatability of high-value-added products to create core competitiveness by means of core technology. The regional characteristic demand of iron and steel shall be found out, and regional differences are the easiest way to form the core competitiveness of enterprises. By providing solutions to customers and improving service competitiveness by means of service differences, the transition from material suppliers to comprehensive service providers is completed. Differentiation strategies can be found out from all aspects of the industry chain. The professional and distinctive strategic orientation shall be put into all aspects of product development, standards, and customer specifications. The enterprises will be able to gain great benefit from production of non-mainstream products and produce high efficiency by low-end equipments if the research on differentiation has been implemented well.

There is an iron and steel enterprise in Germany that has been in operation for hundreds of years, whose product positioning is to produce only super thick steel plates. More than 100 years ago, its annual production output was 200,000t. Up to now, its annual production capacity has only expanded to 300,000t. In the economic tide of global integration, the company has created two miracles: First, the company has not been merged in the face of numerous iron and steel enterprises that have expanded, and second, since the outbreak of the global financial crisis in 2008, the entire iron and steel industry has generally suffered losses and reduced output, but the company still operates in good status. The secret of the development of this company lies in its globally differentiated market strategy.

First, its differentiation strategy follows the basic rules of "survival of the fittest". Any business operator will definitely choose the industry and products that he believes to be the most promising for his development. However, the rationality of the enterprise itself is limited, so it often cannot make correct judgments on itself and the market and it is easy to blindly enter the industries that appear to be lucrative and behave as a bee colony effect in the competition. As a result, the market quickly reached saturation, profits fell, and thus enterprises and even the entire industry suffered losses. This shows that the company is doomed to fail in the fierce market competition if it does not have the advantage of differentiation. If one company wants to gain a foothold in the market, it shall play its own advantages in an appropriate environment by making its own differentiation comparison, thus clarifying its own advantages.

Second, the differentiation strategy fits into the sustainable development strategy. From the perspective of world development trends, the booming of the new technological revolution has brought about tremendous changes in the living environment of enterprises. In the industrial society that enterprises will sustain with unique technologies and unique products as its core competitiveness, the enterprise survival competition mode is gradually changed, which means that under the new technology ecology, new technologies and new products are constantly emerging, any product and technology can be imitated and surpassed, and no enterprise can rely on some specific technologies and specific products to dominate the competition. Therefore, grasping the differentiated strategic thinking and marketing means is an important path for current enterprises to get rid of the fierce competition of product homogenization and win the future market. With the changes in the external environment and internal conditions, enterprises should adjust their business strategies at any time, find differentiated advantages, and create differentiated advantages, so as to achieve sustainable development and survive in disorderly market competition.

2. Correct Understanding of Differentiation Strategy

In recent years, the constraints on the environment and resources faced by the iron and steel industry have become tighter. Moreover, because of the cyclical adjustment of upstream and downstream industries, the unreasonable industrial layout, and the disorderly competition due to the overcapacity, all kinds of chaos and drawbacks are concentrated in the iron and steel industry. "Homogeneous competition" is the culprit of the situation. Not only at the micro-level, product homogenization has caused vicious competition among enterprises and made them suffered losses, but also at the macro-level, enterprise homogenization competition has made the development of the iron and steel industry increasingly uncoordinated, unbalanced, and unsustainable. To this end, if China's iron and steel industry wants to complete the transformation and upgrading, its way out is to vigorously promote the differentiation strategy. At present, the understanding of iron and steel enterprises on the differentiation strategy needs to be further strengthened.

First, the differentiation strategy is not the same as the diversification strategy. Diversification strategy is also called diversified operation and cross-industry operation in China and means the business strategy of expanding the scale and obtaining market profits in a number of related or unrelated industrial fields after the company has developed to a certain extent. According to the degree of association between the existing business field and the new business field, the diversification strategy can be classified as two types: related diversification and unrelated diversification. The diversified practice of iron and steel enterprises is essentially the development of non-steel-related business, but the diversified development solves the problems between main business and auxiliary industry or in the multi-business of the enterprise. It is the expansion of the business field or the adjustment of the business direction of the enterprise.

Second, the differentiation strategy is not necessarily a low-cost strategy. Achieving product differentiation sometimes contradicts the struggle to occupy a larger market share. It often requires enterprise to be mentally prepared for the exclusiveness of this strategy, that is, this strategy and the increase in market share cannot be balanced. The more common situation is that costs will increase in the process of establishing differentiation, such as extensive research, updating product design, developing high-quality materials, or providing thorough customer service. But even if customers across the industry field understand the unique strengths of the enterprise, not all customers are willing or able to pay the higher prices that the enterprise requires. Therefore, the essence of the differentiation strategy is not the oddity or novelty of the form, the key of that is to provide more value to customers, or to create comparable value at a lower cost.

The differentiation put forward by Michael Porter, who is the US strategic management master and the "father of competitive strategy", is to manufacture scarce products from the aspect of economic significance, which means the enterprise shall manufacture the scarce product that is different from competitors in a certain aspect or a certain link of the business process under the market structure where supply and demand are balanced or oversupply, namely "partial short supply", so that the enterprise will have their own competitive advantage and gain excess value of innovation. The enterprise only can stand out from the competition by having strong differentiation capabilities.

3. Vigorous Implementation of Differentiated Competition Strategy

To eliminate the increasingly deteriorating homogenization competition, the main responsible body is the enterprise with the method of innovation. A variety of differentiation, including brand differentiation, quality differentiation, and scale differentiation, are the core and foundation for achieving differentiated strategies.

Brand Differentiation. The cultural traditions and orientations in the brand are important factors that evoke people's psychological identity, sometimes even as a symbol going deep into the consumers' mind, and thus being able to win consumers. At present, the brand awareness of China's iron and steel enterprises is still very weak, and most of them are identified by the company name lacking individuality. In building a competitive advantage in brand image, first is to create differences and establish a prominent image for the brand in the homogenized market, which will help manufacturers to obtain higher market share and more profits in the market. Second is to establish individuality, namely the creation of brand individuality must be based on the overall master and careful application of a variety of factors, including the factors related or unrelated to products, so as to get prominent and distinct brand individuality and enhance customers' perception of brand individuality. Third, it must be consistent with the long-term goals of the enterprise. Brand image is a long-term strategy. Therefore, it is necessary to make long-term investment in brand image so that the image can continue to grow and become full, thus accumulating brand assets.

Quality Differentiation. To create differentiated quality, first is to establish a correct quality concept. The company must pay attention to both the inherent quality of products and the external quality of products as well as to the development of service resources and the simultaneous launch of brand services and brand products. Second is to create high-quality products, develop new products according to the actual needs of customers, create quality characteristics of products, continuously introduce new products with high technological content, establish a new brand image, and create their own brand advantages. "High quality is sourced from the production, not from the testing". The highest standard of Baosteel's product quality is not the national standard or the enterprise standard, or even the technical parameters specified on the contract, but whether the product is suitable for actual use of customer, and whether it can create good benefits for customers [4].

Scale Differentiation. In the fierce market competition, it is not enough for enterprises to create brands only. It is also necessary to increase their market share on the basis of high quality. It is necessary to selectively and specifically expand the competitive brands according to the situation. At the same time, enterprises shall continue to develop high-tech and high-value-added products and continuously extend to related products and industries, by taking brand advantage as the core and adopting merging and reorganization and other means, thus stimulating the expansion of tangible assets with the advantage of intangible brand assets. Through reorganization, the large-scale iron and steel enterprises can concentrate more production capacity, facilitate unified planning within the enterprise, coordinate industrial layout and product structure, and avoid repeated competition, thus achieving coordinated development.

The significance of a company's advantages or disadvantages ultimately depends on the extent to which the company can respond to the market with relatively low cost and differentiation. By implementing differentiated strategies, with innovative products, innovative brands, and innovative markets, opening-up their own exclusive market space, maximizing value and winning lucrative market returns should be the goals of each iron and steel enterprise for transformation and development.

### *9.4.2 Path Analysis of Differentiation*

1. Differentiation of Development Strategy

There is a saying that "not rushing in case of short supply, and not losing in case of recession", which means that when a certain product is popular, it shall not blindly follow the trend; when it is unsalable, it shall not abandon it because of its low price. The scientific connotation reflected by that saying actually lies in the fact that it shall not follow the trend and the road of differentiation shall be taken. There is a business strategy terminology in China: "providing the product that others do not have and providing the better one if others have" and it also reflects the differentiation strategy.

The same is true for the iron and steel industry. While adhering to the operation in a moderate scale, the enterprise shall constantly adjust the product mix, expand the differentiated competitive advantage, and manufacture the scarce product that is different from competitors in a certain aspect or a certain link of the business process under the market structure of supply and demand balance or oversupply, namely "partial short supply", so that the enterprise will have their own competitive advantage and gain excess value of innovation.

Grasping the differentiated strategic thinking is an important path for the current enterprises to get rid of the fierce competition of product homogenization and win the future market. With the changes in the external environment and internal conditions, enterprises should adjust their business strategies at any time, find and create differentiated advantages in order to achieve sustainable development. Differentiation strategies are not the same as diversification strategies, and differentiation strategies are not necessarily low-cost strategies. The key for differentiation strategy is to provide more value to customers or to create comparable value at a lower cost.

The essence of vigorously implementing differentiated competitive strategies is to create products and services with outstanding personalities. First, it shall continue to innovate in technology, create an atmosphere of innovation for all employees, and adopt "technical differences" to create core competitiveness; second, it shall find out the characteristic demands of steel products in the region by studying market and form core competitiveness of enterprises based on "regional differences"; third, it shall establish a customer-oriented business operation mechanism, provide complete solutions for customers, improve competitiveness with "service differences", and complete the transition from material suppliers to integrated service providers; fourth, it shall seek differentiation strategies from all aspects of the industry chain, jump out of the steel production sector, and through the improvement of product development, raw material procurement, warehousing and transportation, processing and distribution, customer services, and other links in the industry chain to obtain higher added value; fifth, it shall implement the professional and distinctive strategic orientation in all aspects of product development, standards, and customer specifications, explore the operation mode of steel and customer's industry from product development to large-scale application, and promote competition with "professional differences".

2. Product Differentiation

Product differentiation means the incompleteness of alternatives between products produced by competitive enterprises in the industry. Specifically, it refers to the conditions by which a company provides products to the market or sells products, and its distinguishable characteristics compared with other companies in the same industry. The existence of product differentiation determines the unequal market segmentation between enterprises with competitive relationships, and the market share of enterprises in differentiated markets is different regardless of the same price or different prices. The degree of differentiation also has an impact on the content and the degree of competition among enterprises. The greater the difference in products is, the enterprises with differentiated advantages can sell products at high prices and obtain excess profits, so the non-price competition among enterprises is fierce.

If the enterprise wants to make product differentiation, it needs to put in more effort on technological innovation, product quality, product serialization, brand image building, and other aspects.

(1) Enhancing technological innovation. Product differentiation is a manifestation of technological innovation. Therefore, enterprises should increase their investment in research and development, actively track the development trend of world technology and the technology in the same industry, study the latest technological development trends of equipment and raw materials required by themselves, correctly make technical decisions, product decisions, and determine type of new product to be developed.
(2) Paying attention to the realization of product serialization and improving the "one-stop" supply capacity. Product serialization means the supply of serialized products with different performances according to different consumption requirements of customers. For example, the products will become luxury products (or high-end products) by adding some functions and will become middle or low-end products by disabling some functions. Consumers can choose the products with corresponding functions according to their habits and affordability.
(3) Improving product quality. Quality means not only the narrowly defined quality of natural attributes such as product suitability, durability, reliability, safety, and economy, but also its social attributes, such as the subjective feelings of consumers, the gap between the ability to meet specific needs and expectations, and the social attributes of quality, which play a very important role in product differentiation.
(4) Optimizing brand image. Although the brand is at the level of product form, its meaning to the product has surpassed the simple mark that distinguishes the product from others. It more represents the product image, and it is the external appearance of product differentiation. If some products want to attract consumers' attention and awareness out of many similar ones, thus attracting them to purchase, enterprises are required to enhance and shape the brand image through CI design and brand strategy, thus highlighting its personalities and creating its brand image differentiation advantages.

3. Production Line Differentiation

Since the twenty-first century, China's iron and steel industry has witnessed the phenomenon of homogeneous construction of technology and equipment and repeated introduction of foreign technology and equipment, resulting in the similar technical structure and product structure among enterprises. It is necessary to achieve the production line differentiation through technical transformation and improve the level of smart manufacturing to greatly improve production efficiency and reduce production costs, thus achieving specialized production.

(1) Improving production efficiency. The production efficiency of iron and steel enterprises in China is still very low compared with that of foreign advanced enterprises such as POSCO. The per capita steel output of enterprises is generally less than 500 tons per year. It is necessary to combine the development opportunities of "Made in China 2025" to vigorously develop smart manufacturing and greatly improve the production efficiency of the production line in the meantime.

Therefore, how to improve the production efficiency by the production lines differentiation, the key is on the development of smart manufacturing technology. Promoting the smart manufacturing in the iron and steel enterprises meets the overall requirements of "Made in China 2025", and the integration of informatization and industrialization is the basic support for realizing the smart manufacturing of the same. At present, the integration of informatization and industrialization in iron and steel enterprises such as enterprise resource planning (ERP) and manufacturing execution system (MES) has been applied on a large scale and from individual applications to integrative applications. Baosteel and other enterprises are gradually transforming into intelligent enterprises. The *13th Five-Year Plan* will be the period for rapid development of smart manufacturing, which shall be paid attention by all the enterprises. The key tasks include continuing to improve and integrate ERP and MES systems and develop the product lifecycle management (PLM) technology, intelligent control system of main process, real-time production management system based on network platform, and intelligent decision-making system.

(2) Striving to reduce production costs. The implementation of low-cost strategy is a strategy that all enterprises should attach great importance to during the difficult period of industry operation. In a certain sense, low-cost manufacturing depends on the efficient operation of the production line. Therefore, to reduce production cost by the upgrading of the production line, at least the following aspects should be prepared: First, a more reasonable, more efficient, and low-cost production process operation system shall be developed by tapping potential from the whole process; second, energy and resource utilization shall be diagnosed to maximize energy and resource utilization benefits based on the entire process; and third, the diagnosis and optimization for the whole process shall be carried out, combined with the brand and market competitiveness, to decisively abandon uncompetitive products, processes, and equipment.

(3) Striving to achieve specialized production. Achieving specialized production has become the consensus of iron and steel enterprises, especially special steel enterprises. How to optimize existing production lines to help enterprises achieve specialized production: First, by taking necessary technological innovations or small-scale technological transformation measures, the production line can fully adapt to the production needs of specialized products and can improve the product quality to a certain extent, especially to improve the stability of product quality, and second, by combining with the development strategy of

enterprises and investment shall be made for construct and develop new products for professional development, thus greatly increasing the added value of the product.

4. Service Differentiation

The traditional steel product service is petal-style, providing only basic services. Evocatively, pre-sales service is "self-boasting", and after-sales service is "firefighting by firefighters". The core idea of traditional services is: service ends if the produce can be used. "The objection about quality shall be immediately dealt with in case of any product problem" is an incomplete understanding of after-sales service; no quality objection or less quality objection means good product. This service, which is limited to dealing with quality objection, has no intrinsic value connection with the product itself. It is only a remedy for the loss caused by the application problem of the product. This service is indispensable, but does not have any added value.

The advanced model and future model of services provided by the iron and steel enterprises lie in the continuously providing the product to meet the new needs of customers. How to make good use of steel products and bring into play the potential of steel products to enhance the competitiveness of customer is what customers really care about and need. The advanced service model is to help customers use steel products well, improve their production technology, and further increase the added value of products, so that customers can consume the minimum resources to create the maximum value. The vision for advanced and future service models is: "achieving high efficiency, benefit, and energy efficiency of steel products, which is associated with the services characterized by soft and hard integration, networking, and full lifecycle". Service is not only for the sale of products. The value of the service is to make the customer transferring from the loyalty of the product brand to that of the company's brand. Products can only capture customer's sight, but good service can capture customer's heart.

5. Sales Differentiation

In general, the differentiated sales strategies for steel products are mainly classified into three types, including steel product differentiation, price differentiation, and payment differentiation.

(1) Product differentiation. Product differentiation mainly means the performance, and quality of the steel products sold by the enterprise is significantly better than that of competitors. In order to be able to differentiate from competitors and provide customers with differentiated products, the enterprises shall change the previous single product sales and provide customers with different products according to their needs, thus meeting the diverse needs of customers. For the appearance of products, corresponding adjustment shall be made according to the different requirements of customers, in order to expand the sales of steel products. Finally, for the delivery of products, the traditional product delivery only means transporting the original products to the customers and the customers have to process them before use after they got the products. In order to meet the requirements of differentiation, manufacturers now will process the products

to facilitate customers before delivering the products. Such product sales will meet the needs of customers and will be greatly in favor of the steel product sales.

(2) Price differentiation. Different price strategies should be adopted for different customers and at different stages of sales. For example, for the customers with long-term cooperation relationship and relatively large demand, appropriate price reduction measures can be adopted to make product price lower than market price, which can consolidate long-term cooperation with customers and strengthen the trust between them; for the customers with less demand and importance, the product prices can be increased based on product quality. Reasonable price differences can greatly increase the sales of steel products.
(3) Payment differentiation. Payment is a very important link during the sales of steel products. Whether an enterprise will have a long-term cooperation with customers, right payment method is the key. When the long-term customers face difficulties in capital turnover, the enterprise can allow customers to pay installments or even credits, which is based on the trust to quality customers, and that is the prerequisite for cooperation between two parties. But other types of customers are not allowed to pay installments or credits in advance, because the products cannot be easily handed over to the customer without a thorough understanding to it, and the customer is required to pay in cash. Because the funds are the key to a company's normal operation, when the company's funds cannot normally turn over, the company will fall into a huge crisis. Therefore, different payment methods should be adopted for different customers.

6. Control Differentiation

Control differentiation is involved in two aspects, namely low-cost production strategy shall be vigorously implemented for ordinary products to improve its competitiveness; while for high-quality special steel, its product quality shall be strengthened to open up the market and improve the recognition by customers. The implementation path of differentiated control can also be achieved from two aspects: "Scale + Cost" and "Variety + Quality + Service".

"Scale + Cost" control can be implemented for ordinary products. Under the homogenization competition, the low-cost competition strategy is a conventional option for enterprises to expand market share and improve their competitiveness. According to the formula, total profit = scale × specific product profit rate, it can be seen that the only way for enterprises to become stronger and bigger is to expand its scale and tap the potential of its internal costs. In the fierce market competition, according to their own circumstances, enterprises can selectively and targetedly expand their competitive products and reduce the proportion of costs invested in products through mass production, thus getting "scale effect"; by taking the measures of mergers and reorganization, the large-scale iron and steel enterprises can concentrate more production capacity, thus facilitating unified planning within the enterprise, coordinating industrial layout and product structure, avoiding duplication of competition, and achieving coordinated development. For the iron and steel industry, as a major task to enhance the competitiveness of enterprises, cost reduction can be implemented by reducing external costs such as transaction costs, tax burden, financing costs, and social costs, as well as improving labor productivity, further tapping potential and increasing efficiency and reducing internal costs.

"Variety + Quality + Service" control can be implemented for high-quality special steel products. In response to the customer's personalized and differentiated needs, the marketing philosophy of common steel product shall be weakened, while that of high-quality special steel products shall be strengthened, and more resources shall be put into research and development, marketing, and service, thus improving product profitability. On the one hand, technicians develop the market against the production process and participate in from pre-sales, sales to after-sales, so as to provide customers with "one-on-one" personalized and differentiated technical services; on the other hand, enterprises and customers jointly build research and development platform to deeply embedded research and development chain into the research, development, and manufacturing process for strategic customers, thus continuously creating value for customers, creating a service-oriented business model against channels, transferring from producers to comprehensive service providers, and becoming a community of true interests with customers.

## 9.5 Industrial Practices of Differentiation

Over the years, the China Metallurgical Industry Planning and Research Institute (hereinafter referred to as MPI) has seized the differentiated development direction of the iron and steel industry and adopted different differentiation strategic path such as development strategy differentiation, product differentiation, production line differentiation, service differentiation, sales differentiation, and control differentiation according to the characteristics of enterprises, so as to provide customers with customized solutions, including providing diversified development strategy consulting, market research and product structure optimization consultancies, production operation diagnosis and technical transformation scheme research, and management consulting. The details are shown in Table 9.2.

**Table 9.2** Practices of MPI in promoting industrial differentiation

| No. | Type | Typical practice case of MPI |
|---|---|---|
| 1 | Differentiation of development strategy | It means strengthening the research on market demand and competitors and identifying the enterprise positioning in the "bigger and stronger", "doing the finer", "moderately diversified" and "relatively diversified", and developing corresponding strategy to achieve industry chain differentiation and regional differentiation. Its main work performances include: *13th Five-Year Plan*, development strategy planning, diversified development planning, overall development planning, and industry chain extension planning for iron and steel enterprises. For example: *13th Five-Year Plan for WISCO*, *13th Five-Year Plan for HBIS Group*, *13th Five-Year Plan for CITIC Pacific Group*, *13th Five-Year Plan for Baogang Group*, *13th Five-Year Plan for Ma Steel Group*, *13th Five-Year Plan for SD Steel Group*, and *13th Five-Year Plan for Hebei Jingye Group* |
| 2 | Product differentiation | It means optimizing product structure, increasing the share of competitive products, focusing on product serialization, and improving the "one-stop" supply capacity. Its main work performances include: product positioning research, product structure adjustment and market analysis report, product upgrade and adjustment planning. For example: *Product Positioning Research of HBIS Group ShiSteel Company, Product Positioning Research on HBIS Group ChengSteel Company, and Product Upgrade and Adjustment Plan of Angang Group* |
| 3 | Production line differentiation | It means achieving specialized production, greatly improving production efficiency, and reducing production costs through the technical transformation of equipment. Its main work performances include: capacity balance and product structure research for main production lines, configuration and product positioning research for production lines, balance research for production lines, etc. For example: *Capacity Balance and Product Structure Research for Main Production Lines of Handan Steel during 13th Five-Year Plan*, *Configuration and Product Positioning Research for EAF Production Lines of Angang Group*, and *Balance Research for Bar Production Lines of Delong Group* |
| 4 | Service differentiation | It means improving pertinence and recognition by adopting different service models against different customers and different regions. Its main work performances include: providing iron and steel enterprises with special consultancy regarding merger and reorganization, special technical demonstration, green factory establishment, standardized information service, data information report, and equipment level assessment. For example: *Merger and Reorganization Program for Iron and Steel Enterprises in Hebei Province, Green Factory Establishment of Shanxi Taigang Stainless Steel Co., Ltd., and Standardized Service of Jiangsu Shenyuan Special Steel Co., Ltd.* |

(continued)

**Table 9.2** (continued)

| No. | Type | Typical practice case of MPI |
|---|---|---|
| 5 | Sales differentiation | Common products are mainly sold by e-commerce platform and direct sales to large customers, so as to reduce intermediate links and sales expenses. The high-quality special steel products are mainly sold in the modes of technology sales, strategic customers, early intervention and continuous tracking, keeping up with customer demands, and gradually expanding the market. Its main work performances include: Market Analysis of Steel Products, Analysis and Research of Product Sales by E-Commerce Platform, Regular Analysis of E-Commerce Platform for Iron and Steel Industry, Innovation Research on Sales Models of Some Iron and Steel Enterprises, Advanced Technology Promotion Services for Energy Conservation and Environmental Protection, and Product Agency Conversion Services. For example: *Product Structure Adjustment and Market Analysis Report for Ma Steel* and *Report on E-commerce Development for Iron and Steel Industry* |
| 6 | Control differentiation | For ordinary products, low-cost production strategies shall be vigorously implemented to improve their competitiveness. The high-quality special steel products are oriented by strengthening product quality to open up the market and improve customer recognition. Its main work performances include: optimization for cost system of iron and steel enterprises, analysis and evaluation of competitiveness, upgrading and implementation of smart manufacturing systems, upgrading of job and staff design and human resources management, and creation of product quality standards system. For example: *Optimization for Cost System of Reafon steel*, *Optimization for Cost System of Weifang Special steel Group*, *Evaluation of Competitiveness of Shanxi Jianbang Group*, *Evaluation of Comprehensive Competitiveness of Jianlong Group*, and *Standardization Improvement Strategy for Yuantai Derun Pipe Manufacturing Group Co., Ltd.* |

# References


1. China Government Network. Notice of the State Council on printing and distributing "Made in China 2025" [2015-05-19]. http://www.gov.cn/zhengce/content/2015-05/19/content_9784.htm.
2. Fangda Special Steel Technology Co., Ltd. (2012). Transforming the development mode and insisting on differentiated development to achieve a huge leap from backwardness to advanced. *Iron and Steel Industry of China, 8*, 18–20.
3. Pan, J.H. (2014). Differentiation is the inevitable choice for the iron and steel industry to get breakthrough—Baosteel's differentiation strategy application practice. *Metall Econ Manag, 2*, 45–48.
4. Zuo, K.J. (2005). Differentiation: The key to the internationalization of China's Iron and Steel Enterprises. *Journal of Huazhong Agricultural University: Social Sciences, 3*, 1–4.

# Chapter 10
# Servitization

## 10.1 History Review and Status Analysis

### *10.1.1 Connotation of Servitization in Iron and Steel Industry*

1. Definition of Servitization

As early as 1972, Levitt pointed out that manufacturing enterprises should integrate services into the goods they provide, focus on the development of relationships with customers, and propose a service in mode of production line. In 1988, Vandermerwe and Rada first proposed the definition of servitization: Enterprises of manufacturing industry are switching from providing only goods to providing goods—service packages. Complete goods—services include goods, services, support, self-service and knowledge, and the services—dominate the package and are the main sources of added value. Since then, many scholars have begun research on the field of servitization in manufacturing industry. For example, White and other researchers stated in 1999 that servitization is the transformation of manufacturing enterprises from product producers to service providers, and it is a dynamic transformation. In 2000, Reiskin defined servitization as the transformation of enterprises from taking production of goods as the center to taking service as the center; Szalavetz (2003) thought that "business servitization" does not fully reflect the trend of servitization in manufacturing industry, the servitization of manufacturing industry at least has two meanings: One is that the efficiency of internal services is becoming increasingly important for the competitiveness of manufacturing enterprises, including product development, design, training, accounting, legal, and financial services; the other is that external services related to goods are becoming more and more complex and important to customers. In addition, other scholars have proposed similar concepts from a strategic perspective, such as "Service-oriented Manufacturing Industry" by Drucker (1990), "Service Enhancement" by Berger and Lester (1997), and "Service-based Value Innovation" by Kim and Mauborgne (1997) [1].



X. C. Li, *The Road Map of China's Steel Industry*,
https://doi.org/10.1007/978-981-15-2074-7_10

Compared with foreign countries, domestic academic circles have started late in the study of servitization of manufacturing industry. However, after entering the twenty-first century, research in this field has attracted more and more attention from domestic academic circles. Despite the current definition of servitization in manufacturing industry, the management community has not reached a consensus. However, from the documents obtained so far, the definition of Szalavetz has been recognized by many domestic scholars [1], that is, servitization has two meanings: One is input servitization, that is, service elements (such as market knowledge, R&D design, internal logistics, management efficiency, marketing, law, and finance) occupy an increasingly important position in the total investment of manufacturing industry. Industrial enterprises tend to input less in physical factors and more in service factors in production and operation; the other is output servitization, that is, service products occupy an increasingly important position in the overall output of the manufacturing industry. Industrial enterprises shift their focus to the service sector, causing the proportion of physical factors in their output be decreased, and the proportion of service factors be increased.

Combining the domestic and foreign scholars' researches on servitization, the servitization of the iron and steel industry, from the perspective of enterprise value chain, is the ultimate goal of iron and steel enterprises to better meet the needs of customers, to achieve enterprise value and gain competitive advantage. The enterprise value chain is a dynamic transformation process from taking manufacturing as the center to taking service as the center. From the perspective of the industrial value chain, it is a dynamically changing process during industrial operation in which the proportion of services in the input and output of iron and steel enterprises continues to increase, the service factors are increasingly becoming the dominant factor in the creation of iron and steel manufacturing value, and the industrial added value and brand benefits are continuously improved.

2. Necessity of Servitization in Iron and Steel Industry

The external environment has forced the iron and steel industry to shift from taking manufacturing as the center to taking service as the center. The first is that China's demographic dividend is coming to an end, and the core competitiveness of iron and steel manufacturing, i.e., low labor costs, will no longer exist. The second is that the high load of the environment is becoming more and more strict on the iron and steel industry, and the environmental cost of iron and steel manufacturing is becoming more and more difficult to be externalized. The third is that the transformation of macroeconomic growth mode will reduce the intensity of iron and steel demand, and the demand for services will increase.

The internal environment of the industry also requires service transformation. From the inside of the industry: The first is that overcapacity has become the normal state, which means that more surplus production capacity can be shifted to meeting individual demands and high-level demands. The second is that the competition for homogeneous products in the iron and steel industry is intensifying day by day. The era when any enterprise attempts to dominate the world with its products has ended, and enterprises need to take the lead in innovating the competition. The third is that the product structure and technical level of China's iron and steel industry have been greatly improved after years of development. The industry has initially met the conditions for service transformation. The fourth is that the "hard strength" of Chinese iron and steel industry transformation and upgrading has basically been exhausted, but the "soft" strength required for service transformation has larger potential.

The challenge of iron and steel industry to emerging industries also requires service transformation. First of all, the rise of strategic emerging industries will fundamentally change the service targets of iron and steel enterprises. The difference between strategic emerging industries and traditional industries lies not only in the difference in industrial forms but also in the difference in service demand. The second is the new challenges from the new industrial revolution on iron and steel materials. Intelligent manufacturing, interconnected manufacturing, green manufacturing, customized manufacturing, etc. are considered to be the symbols of the third industrial revolution. Iron and steel industry as a basic material industry must adapt to the subversive changes in material demand of the new industrial revolution, which not only requires Chinese iron and steel enterprises to speed up the transformation from iron and steel to materials but also requires Chinese iron and steel enterprises to accelerate the service transformation of manufacturing because the new industrial revolution will end the era of homogeneous large-scale batch production, and services will infiltrate every link of the iron and steel industry chain.

Internationalization is a road that Chinese iron and steel industry must take in the future. However, Chinese iron and steel industry must first strengthen its service capabilities by taking the international development path. Servitization of manufacturing industry is the precondition for Chinese iron and steel industry to go global. It is difficult to find a manufacturing-exclusive enterprise in the camp of the world's outstanding multinational companies. On the contrary, the proportion of integrated service-oriented enterprises in this camp is getting larger and larger. To become an international enterprise, Chinese iron and steel enterprises shall not either simply sell steel products overseas, or simply build overseas iron and steel manufacturing bases, what is more important is to internationalize their service capabilities. Only when the service transformation is realized in China, can it become an excellent multinational iron and steel company.

3. Significance of Servitization in Iron and Steel Industry

Regardless of whether it is input servitization or output servitization, the iron and steel servitization is in terms of the overall development trend of the industry. From the industrial level, it is the development trend of the iron and steel industry, while from the micro-level, it is the development strategy of the iron and steel enterprises. Under the current conditions of China's economic development, the servitization in the iron and steel industry is not "abandoning the manufacturing industry", but a rational choice for iron and steel enterprises to enhance their competitiveness according to the actual situation of the enterprise and the development environment of the industry. The fundamental goal is to expand the value chain of the enterprise and improve the industrial added value and brand benefits. Servitization is an important direction and way for the transformation and upgrading of Chinese iron and steel industry. It is of great significance to the transformation of China's future economic development mode and strategic adjustment of economic structure [2].

(1) Satisfy the demands of customers. With the development of economy, most steel product customers are no longer satisfied with the steel itself, but need more services that are accompanied with the steel, such as logistics, inventory, deep processing, material design, and model calculation. Redefining the product as a product-service package meets the customer's expectations and helps to satisfy the demands of customers. In this way, the traditional practice of satisfying customer demand by simply manufacturing products is no longer applicable, and the focus of iron and steel enterprises is increasingly put on establishing and maintaining relationships with customers. Iron and steel enterprises are increasingly paying attention to end customers, actively seeking opportunities to understand their own problems, and providing services to enhance credibility and create demand.

(2) Improve competitiveness of enterprises. The most important role and driving force of servitization is that it can increase the competitive advantage of enterprises. Managers in mature industrial sectors regard service as a tool of differentiation to extend the life cycle of products and prevent enterprises from being eliminated. Similarly, for iron and steel industry which is traditionally a cost-oriented industry, the strategy of servitization is an important means of creating differentiated advantages. Service has low visibility, high labor dependency, and is difficult to be imitated, which make it a sustainable source of competitive advantage. Providing better services than competitors can increase a company's competitive advantage, make its products more attractive and conducive to differentiate itself from competitors.

    Although China has stepped up its efforts to close down backward capacity, the problem of cyclical overcapacity and structural overcapacity remains outstanding. Overcapacity makes the competition between iron and steel enterprises increasingly fierce, the income of enterprises is diluted, and the whole industry is in a state of low profit or loss. In response to this situation, all iron and steel enterprises must find another way to ensure survival and development. Management guru Peter Drucker said: "The competition among future enterprises is the competition of business models". Realizing the servitization development of enterprises and implementing service-oriented manufacturing are major innovations in the business model of iron and steel enterprises. Through the construction of servitization, we can better satisfy the demands of customers, enhance customer satisfaction and loyalty, and thus create differentiated competitive advantages.

(3) Increase economic income. Product-related services can increase the income while also reducing the fragility and variability of cash flow, helping to increase shareholder value. The economic reason for manufacturing enterprises to integrate services into their core products is that the considerable income of the enterprises comes from the customer group in the entire life cycle of the products; services usually generate higher profits than products; and services provide

a more stable source of income. From a realistic perspective, services do bring lucrative profits to some enterprises.

In the fierce market competition, the iron and steel enterprises are faced with the choice of survival of the fittest. The competition between enterprises has infiltrated into all links and all aspects. It is difficult to overcome the competitors through only one certain advantage. The proprietary technology of production is converging among different iron and steel plants, the same type of iron and steel products tend to be homogeneous, and the difference of profits generated from the manufacturing process is very small, and thus, the competitive advantage of the enterprises is difficult to be reflected from the quality and price of the products. Deeply implementing the servitization and obtaining service profits can become a new profit growth point and an important source of profit in the future for iron and steel enterprises.

(4) Improve management level of enterprises. Iron and steel enterprises have been immersed in the "planned economy" and "seller's market" environment for a long time. The relaxed internal and external environment has caused many problems for iron and steel enterprises in terms of operation philosophy, organizational structure, management mode, work style, etc. Contradictorily, today's customers' demands are diversified and differentiated. Single-specification orders with large quantity are becoming less and less, and customers often need a combination of products with multiple specifications but in small quantities. Under this background, by grasping the trend of service-oriented development, the iron and steel enterprises can fundamentally realize the "customer-centered" transformation, achieve tremendous leap forward in operation philosophy, management mode and management method, greatly enhance the service capabilities of enterprises, and comprehensively promote the improvement of enterprise management level.

4. Manifestation of Servitization in Iron and Steel Industry

Related studies show that manufacturing enterprises have a variety of different forms of services available: consulting services, design and development services, financial services, installation and implementation services, leasing services, maintenance and support services, outsourcing and operation services, procurement services, intellectual property and real estate, retail and distribution services, systems and solutions, passenger transport and freight transport services, etc. However, it is not limited to these services. In fact, the connotation of servitization in manufacturing industry is very rich, including not only all the services provided by the output of manufacturing enterprises but also the support required by the development of the manufacturing industry, i.e., input servitization.

The services provided by iron and steel industry to the downstream customers are not only just simple materials but also all-round services from the supply of materials, quality assurance, technical services, semi-finished product processing, parts manufacturing, logistics, etc., and even technical support and consulting services can be provided for links prone to problems during use of the material, such as welding, strength calculation, forming, and mold design. By analyzing the path of service-oriented transformation of global manufacturing enterprises, there are mainly four manifestations for servitization in the iron and steel industry:

(1) Value-added services based on improvement of product efficiency. Intensified market competition has enabled enterprises to continuously enhance product value, satisfy the demands of customers, better reflect differentiated competition, create profits, and make customers loyal through services to stand out in the new round of competition. The service modes for improving the product efficiency are diversified. No matter at the early stage of product R&D design or in the operation after the product is delivered to customers, the value of product can be added through service-oriented means and new business models will be innovated continuously.
(2) Value-added services based on convenient product transactions. Based on the wide application of information technology, iron and steel enterprises can improve the transaction efficiency and convenience of iron and steel products through diversified financial services, precise management of supply chain and convenient e-commerce, thus enhancing the core competitiveness of enterprises. From the perspective of the development history of foreign manufacturing industry, more and more manufacturing enterprises have taken improvement of product transaction efficiency and convenience as an important means and approach to enhance their competitiveness.
(3) Integration of products and services. With the upgrading of customer demand from a single product to a comprehensive aspect of products and related services, providing customers with product integration and comprehensive solutions has become an important means and approach for manufacturing enterprises to enhance their core competitiveness and win the market. The integrated professional services and systematic product integration from production to materials and from manufacturing to services in the iron and steel industry are becoming an important mode for iron and steel enterprises to expand their business and a commanding height for industrial competition.
(4) From product-based services to demand-based services. No matter the improvement based on product efficiency, convenience based on product transactions, or value-added services based on product integration, the starting point of service is based on R&D design, product transaction and functional enhancement of tangible products, but this is still the initial stage for service-oriented transformation of manufacturing industry. The higher stage for service-oriented transformation of manufacturing enterprises is that the enterprise development strategy has realized the transition from product-based services to demand-based services of customers, and the service model is no longer bound to its original products, but to extend their operation capacities as services in terms of market R&D, supply chain, sales, etc. The enterprises are no longer the product providers, but become the solution providers, who take advantage of their operation in the value chain to provide professional services not relying on the products. The enterprises will focus on exploring and discovering the potential demands

of customers, using a strong service system to help customers solve complex and tough problems and provide the customers with "packaged" and "one-stop" solutions and implementation results, so as to create more values for customers [3, 4].

### *10.1.2 Evolution of Servitization in Iron and Steel Industry*

Due to the different understanding on concept of servitization in manufacturing industry, there are also different perspectives in the evolution stage of servitization of manufacturing industry. The evolution of servitization in the iron and steel industry generally consists of four stages: stage of goods, stage of goods and additional services, stage of value-added services, and stage of overall solution.

1. Stage of Goods

Enterprises still belong to the traditional manufacturing industry, and the business center still stays at the stage of production and sales of products. The identity of the enterprises is only the producers and sellers of products, and almost no one is playing the role of a service provider. Most Chinese iron and steel enterprises are still exploring in this stage.

2. Stage of Goods and Additional Services

This stage is an extension of weak-service stage, in which services exist as a form of additional product, and the production and sales still hold the leading position. The services at this time are only a tool for producers to increase sales, helping products to increase market competitiveness and share, and generally free of charge.

3. Stage of Value-Added Services

Enterprises bind a series of intangible services to tangible goods and provide it for consumers. From simple transportation, installation, repair, maintenance, etc. to complex credit insurance, process support, etc., they are all connotations of services. At this moment, the service can be priced and sold by the enterprises as if it were a product. The previous assistant and attached effect to the product have been changed by the service, and it has become a new way to make the value of the product to be innovated and added. At this stage, many manufacturing enterprises have greatly increased their operating income by increasing services.

4. Stage of Overall Solution

At this stage, the core competitiveness of the enterprise is not the production or sales of products. As part of the services, the product can only serve as a carrier for the enterprise to attract consumers and only a link to the overall solution of the enterprise. Services and products exist at the same time, but the position has changed. That means the product is not a major part of the enterprises' production, but a part of service. Enterprises began to move closer to the service-oriented model, and its core competitiveness also began to shift to services.

### 10.1.3 Status of Servitization in Iron and Steel Industry

1. Status of Servitization in Foreign Iron and Steel Industry

Servitization has become an important trend in the development of global manufacturing industry. In the process of economic servitization in developed countries, the manufacturing industry takes the lead in the servitization and becomes the industrial support for the overall economic servitization. Since the 1980s and 1990s, with the widespread application of information and communication technologies, the economic globalization, and the changes in the market environment, the development of the service industry and the manufacturing industry is from division of work and differentiation to interaction and integration, and the servitization has become an important trend for the development and upgrading of the global manufacturing industry. In Japan, France, Australia, and the UK, the dependency degree of manufacturing industry on service industry in the mid-1990s increased by about 10 percentage points compared with the early 1970s. In the mid-1990s, that in Germany also increased by nearly 10 percentage points compared with the early 1980s; the degree of dependency also greatly increased in Canada, the USA, Denmark, and the Netherlands, if the data in the mid-1990s is considered. Although the growth rate of input in services by manufacturing industry of different countries is different, in general, the intermediate input in the services by manufacturing industry is showing a clear upward trend; the output of manufacturing industry is showing a trend of servitization, and more and more manufacturing enterprises take service as an important means of differentiated competition. According to the report of Deloitte in 2010, the share of services in the output of manufacturing industry in countries of the Organization for Economic Co-operation and Development (OECD) rose from 15% in the 1970s to 30% in 2000. Among the 80 manufacturing companies surveyed, service revenue accounts for more than 25% of the total sales revenue on average, of which, the service revenue of 19% of manufacturing companies accounts for more than 50% of the total revenue [5, 6].

From the perspective of the international iron and steel industry, in recent years, the relevant prices that constitute the main cost of iron and steel products have increased, making more and more profits for the upstream enterprises in the entire iron and steel industry, while bringing lower and lower profit margins for the iron and steel enterprises. In order to compete with the downstream steel customers and the upstream monopolized raw material suppliers, the iron and steel enterprises around the world have begun a super-large-scale merger to establish large iron and steel groups. In order to enhance the profitability, the world-class iron and steel producers began to transform into providing specialized services by adopting services such as design, distribution, and product "customization" to support the production. Some enterprises began to provide all-around logistics, engineering, distribution, and other services and participate in the design services during the entire process, such as joint development of high-strength steel sheets for automobiles with customers; besides, they provide product extension services such as laser tailor-welded blanks and components; they also provide customized services such as steel solutions

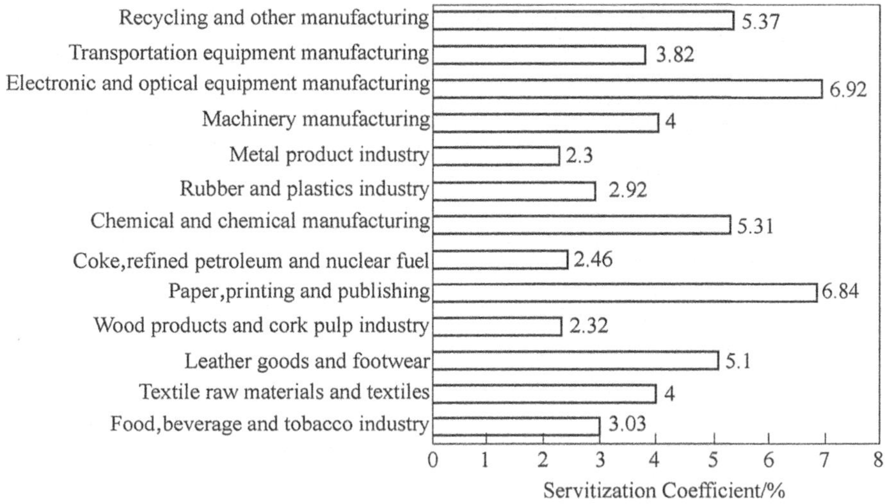


**Fig. 10.1** Comparison of servitization coefficients of different manufacturing industries in 20 representative countries (2009, unit: %)

for the manufacturing of an entire ship and services with multiple varieties but in small quantities, etc. By deepening the services, the profitability of the enterprise is effectively improved.

By comparing the servitization coefficients of different manufacturing industries in 20 representative countries (Fig. 10.1), relative to paper/printing and publishing industry (6.84%), leather products and shoemaking industry (5.1%), chemicals and chemical product manufacturing industry (5.31%), electronics and optical equipment manufacturing industry (6.92%), and recycling and other manufacturing industry (5.37%), the servitization coefficient of metal products industry is relatively low, only accounts for 2.3%.

Among the 20 representative countries, the servitization coefficient of the metal products industry in the Netherlands, Poland, Australia, Canada, the UK, and Germany is relatively high (Fig. 10.2).

2. Status of Servitization in China's Iron and Steel Industry

The servitization of China's iron and steel industry is still in its infancy, and most companies still do not aware enough about this, and they are still in the stage of goods or additional services. However, no matter in terms of input servitization or output servitization, the service-oriented development of China's iron and steel industry has already begun to take shape.

(1) Increase investment in R&D and accelerate the cultivation of innovation capabilities. The servitization of manufacturing is not to weaken the role of R&D innovation, but to highlight the importance of R&D innovation. In recent years, many iron and steel enterprises have continued to increase investment in R&D and accelerate the cultivation of innovation capabilities. Among the A-share

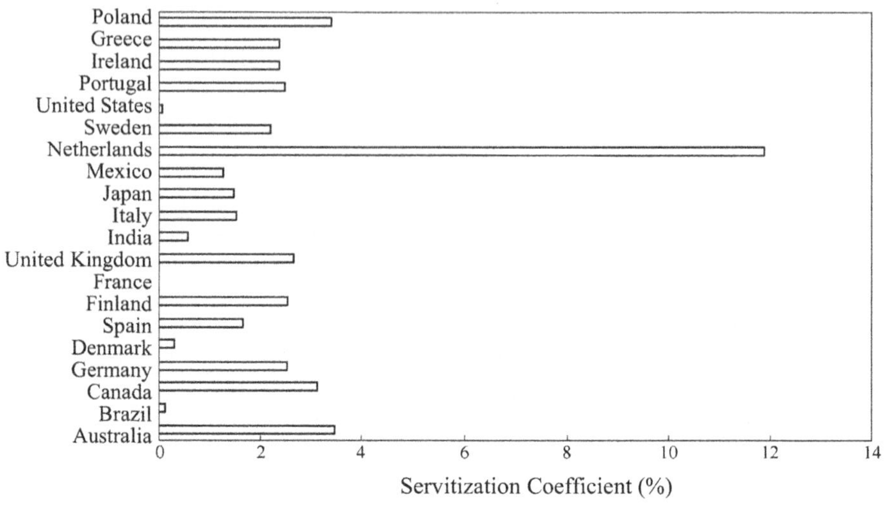


**Fig. 10.2** Comparison of servitization coefficient of metal product industry in 20 representative countries (2009, unit: %)

**Table 10.1** R&D expenditures of some A-share iron and steel enterprises in 2016

| Ranking | Stock name | Total R&D expenses/yuan |
|---|---|---|
| 1 | Baosteel | 3,662,000,000 |
| 2 | TISCO Stainless Steel Co., Ltd. | 2,017,030,802 |
| 3 | Benxi Steel Plate Co., Ltd. | 1,437,398,000 |
| 4 | Angang Steel | 1,002,000,000 |
| 5 | HBIS | 1,413,191,379 |
| 6 | JISCO Hongxing | 1,040,726,490 |
| 7 | Nanjing Iron and Steel Co., Ltd. | 801,780,000 |
| 8 | Ma'anshan Iron and Steel Co., Ltd. | 752,570,000 |
| 9 | Daye Special Steel | 238,114,262 |
| 10 | Lingyuan Iron and Steel Co., Ltd. | 227,715,650 |

*Source of data* Annual report of listed companies in 2016

iron and steel enterprises in 2016, Baosteel was ranked first with R&D expenditure of 3.66 billion yuan, while TISCO Stainless Steel Co., Ltd. and Benxi Steel Plate Co., Ltd. were ranked second and third, respectively (Table 10.1).

Taking China Baowu as an example, the same has maintained a high-level R&D investment, and the R&D investment rates from 2014 to 2016 were 2.0%, 2.1%, and 2.0%, respectively. China Baowu attaches great importance to the cultivation of scientific and technological talents and the construction of incentive mechanisms. On the basis of forming a series of talent training and incentive policies, it takes the lead in introducing equity incentive scheme in the industry to carry out equity incentives for core management personnel and technical talents and further encourage the innovation enthusiasm of core talents. In 2016, Baosteel increased its investment in technological innovation and made continuous progress in product development and technological progress. The R&D input rate was 2.0%, the invention patent application accounted for more than 70%, the proportion for new test alone accounted for more than 71%, and the high magnetic induction grain-oriented silicon steel B27R080, B30R090, and coiled tubing CT110 are the world's first products. China Baowu has built a R&D system with the Central Research Institute as the core by referring to the model of international advanced enterprises setting their professional technology development organizations. The Central Research Institute has established large categories of product development module oriented to market competition, a common technology development module oriented to improvement of manufacturing technology, a cutting-edge technology development module oriented to future development, and a project team oriented to R&D of new material. In 2013, the National Key Laboratory for Automotive Steel Development and Application was established. China Baowu continuously improves its ability to acquire external technical resources by cooperating with colleges and universities, establishing strategic alliances for industry-college-institute cooperation, carrying out strategic cooperation between the industry chain and peers, and building international exchange and cooperation platforms, etc. In terms of technological innovation infrastructure construction, China Baowu has built an experimental platform for the whole metallurgical process from smelting → hot rolling → cold rolling → galvanizing → post-processing and has the ability to guide large-scale production through pilot production tests.

(2) Transform to the "customer-centered" marketing organization model, enhance the ability to capture and create high-quality customers, and improve customer level. The evident goal of being "customer-centered" is to "stick to the customer", i.e., to establish and maintain a highly stable customer relationship, so as to prevent the loss of customer due to the intrusion of competitors. Whether the service transformation of manufacturing enterprises is successful is closely related to their customer groups. The high-quality and high-end customer base is the basic guarantee for the industry to realize service transformation.

Among China's large-scale iron and steel enterprises, a number of enterprises, such as HBIS Group, China Baowu, and Jinan Iron and Steel Group, are implementing a major customer manager system to enhance the ability to capture and create high-quality customers and upgrade the customer level.

In 2015, the operation plan of the major Customer Service Center of HBIS Group was officially implemented, and the confirmation of the Group's first batch of major customers and customer teams, the construction of CRM management, and control information platform and other specific works have been developed one by one.

After several years of trial operation, China Baowu's major customer manager system has played a good role in promoting the establishment of the company's "customer-centered" operation philosophy, enhancing the relationship between China Baowu and its customers, solving customer problems, and improving the market share of the company's products. Dozens of potential and strategic customers such as China FAW Group Corporation, Dongfeng Automobile, and CIMC have been developed to become the major customers, which are mainly focusing on six major industries such as automobiles, home appliances, petroleum, and transportation.

(3) Improve ability to manage big data and strive for accurate services. Service-oriented iron and steel enterprises can provide better services only by more accurately grasping the subdivided characteristics and individual demands of customers and employees. In today's information age, the operational activities of market and enterprises have generated massive amounts of data. The maturity of data mining technology based on big data in the new processing mode in the Cloud Era will help enterprises to achieve accurate services through analysis of big data.

Some domestic iron and steel enterprises have begun to build platforms, collect data, build big data, research, and develop services such as technical services, technology industrialization and industrial organization, and collaboration based on platform data analysis. For example, product technical services and engineering services with the combination of online and offline: Provide product operating guides, performance databases, standard comparisons, technical trend analysis of applied materials for supporting database of material and application technology solutions; provide recommended material solutions for innovative products and complex operation environment; provide engineering analysis services required by customers, etc. In addition, technology R&D trends and sources can be obtained through data analysis in technical services, so as to provide steel plants with customized application schemes of steel products corresponding to analysis on characteristics of the customer's processing and use; recommend quality objection tracking feedback and solutions to problems; and provide a variety of technical services such as complete technical solution for transformation of iron and steel products into the customer's products. It can also provide technical services to processing and distribution service providers such as material reliability database, processing method database and processing scheme design in terms of modification, reshaping, surface changing, etc. of the steel products based on analysis of the customers' demands. In the field of product R&D and production technology services, technical services can be provided to iron and steel plants.

(4) Improve logistics optimization capability and develop iron and steel e-commerce. The e-commerce in the iron and steel industry is a promising business. According to statistics, the total number of steel e-commerce companies

established through iron and steel enterprises, steel trade enterprises and the third parties has exceeded 200, accounting for 27.6% of the national bulk commodity e-commerce enterprises. There are about 20 entrepreneurial iron and steel e-commerce companies built through the mode of transactions rather than the mode of information. Since 2011, the proportion of sales through the iron and steel e-commerce has gradually increased. In 2016, the online iron and steel trading volume on the platform of national iron and steel e-commerce reached 240 million tons, accounting for 25% of the annual output of steel products. At present, the overall level of e-commerce development of large-scale iron and steel production enterprises is still low, and there is a very obvious differentiation.

However, one of the biggest bottlenecks in the development of iron and steel e-commerce is iron and steel logistics, just like the great development of e-commerce in the consumer goods sector requires strong support from the express industry. The logistics distribution of iron and steel products is obviously different from the general express logistics, which not only requires more rigorous organization and optimization capabilities, but also the supporting of logistics infrastructure. Iron and steel enterprises such as China Baowu, as well as steel trade enterprises such as Tewoo and Minmetals, are reshaping the logistics system, so as to cope with the needs of platform economic development brought by iron and steel e-commerce.

(5) Establish and improve the service system of early vender involvement (EVI). EVI (Early Vender Involvement) means early involvement of the vender, i.e., the vender takes the customer as the main body and orientation and fully intervenes in all aspects of the customer from R&D to mass production, and transmits the customer's demand to the internal of the enterprise in an instant and forward-looking manner, so as to prompt enterprises to continuously improve their capabilities and integrate with customers' demands earlier and deeper, thus further expanding the space for creating value between the two parties.

In recent years, the introduction of the EVI technology service model has become an important strategy for many iron and steel enterprises in China to accelerate their transformation from manufacturers to service providers. Especially in the field of automotive sheet, the target-oriented automotive EVI service model of Baowu Steel Group has formed a powerful automotive EVI service system to provide whole process design and technical support of the automotive steel with the characteristics of China Baowu. Initiate advanced engineering design—early (conceptual) and body design—mold (tools) design and development—vehicle model production—mass production at various stages, implementation of corresponding market development—design support—tool support—production support—production monitoring and other whole process systematic and individualized supports according to the procedures of automobile plants. For the new models in automobile plants, China Baowu proposed a detailed EVI layering principle and EVI working mechanism for the new models of self-owned brands, joint venture brands, and foreign design. By implementation of the automotive EVI service model, China Baowu realized cost reduction through the EVI service-supported technology in terms of cost optimization, production support, mold design, body design and advanced engineering design, etc. of the automotive sheet EVI service and achieved a good win-win situation for both automobile plants and iron and steel plants.

In 2016, Ma'anshan Steel was taking the technology center as the core, launching the EVI service strategy in an all-round way and creating a new path of differentiated competition. Recently, after elaborate planning and repeated demonstration, the technology center of Ma'anshan Steel has launched the first batch of ten EVI technical service action plans covering eight categories of products of the company.

(6) Develop deep processing and extend the iron and steel industrial chain. Deep processing of steel products is one of the ways for service-oriented transformation of enterprises. Some iron and steel enterprises in China have done a lot of work in serving customers and deep processing of iron and steel products. When the steel market is booming, service for customer shall be placed in the first place. In today's iron and steel market with downturn and serious overcapacity, the development of deep steel processing will help iron and steel enterprises to realize the transformation from simple material production to technical services and material services.

China's iron and steel enterprises have made many breakthroughs in both marketing-driven deep processing and industrial deep processing. As for marketing-driven deep processing, in order to make customers more convenient, intermediate products are provided for customers to add value to the marketing. For example, according to the customer's requirements, by establishing a steel cutting, processing, and distribution center, the metal products will be directly used by the end customers after straightening, cleaning, cutting, stamping, color printing and other processes, as well as through the logistics links such as warehousing and transportation. Marketing-driven deep processing means a series of services to the customers, which is a value-added chain connected to the end customers. Due to the advantages of high efficiency and low cost generated by socialized and modernized services, it can bring huge benefits for both the supply and the demand parties of metallurgical products and the society.

In industrial deep processing, steel is used as raw material and processed into products that customers can directly use. At present, the field which China's metallurgical enterprises are engaged in industrial deep processing includes construction steel structures, metal packaging, fasteners, hardware furniture, auto parts, steel cords, bridge cables, and so on.

(7) Scale customized services. Digital, network, and intelligent technology integrates the scale manufacturing with customization, which not only fully reflects customer's demand in all aspects of product design, production, and application,

forming a new industrial model, but also applies the Industrial Internet to the manufacturing industry, so as to achieve the transformation of industrial form from production-oriented manufacturing to service-oriented manufacturing. More importantly, under the condition of overcapacity and low profit operation in the iron and steel industry, the integrated manufacturing and management system of customer-driven modern metallurgical enterprise shall be built based on the demands of customers, forming a large-scale customized production system in terms of quality, production, cost, and logistics management to realize the individualized service management innovation under large-scale customization, the dynamic cost control of the whole process of production under lean manufacturing, and the agility in the manufacturing process.

China Baowu has made beneficial explorations in the field of large-scale customization and achieved initial success. As a large-scale iron and steel enterprise with most advanced technology in China, Baowu Steel Group has independently developed an optimization system for production organization of steelmaking–rolling process based on ERP and MES systems against the characteristics of large-scale and long process flow and developed a virtual manufacturing and discrete event simulation platform, realized the adaptive optimization of human–machine interaction, which shortens the order-to-order scheduling time from 4 h to 10 min, thus greatly improving the efficiency of planning system and production adaptability.

HBIS Tang Steel promotes "customized" services, responds to market changes with the rapid transformation of production and management concepts, produces customized products for customers, and fosters customer loyalty with special services. Tangshan Steel can provide costumers with non-standard cut-to-length deformed steel bar, hard wire rod, boron-added angle steel, angle steel with non-standard thickness, steel used for low-sulfur submerged arc welding wire, boron-added gas-shielded welding wire, and other products through customized production. Tang steel also explored a set of "customized" precision R&D program and established a new model of high-end product technology R&D and market development.

In addition to the production of regular 9 and 12 m high-strength deformed steel bars, Hebei Jingye Group provides customized production service of "specification + material + packaging" for high-strength deformed steel bars, with its specifications, materials, and packages being selected by customers, through which, about 11% of steel product costs can be saved for customers, and the order can be placed starting from 500 tons.

## 10.2 Development Environment and Policy Orientation

### *10.2.1 Development Environment*

1. Macro-External Environment

    (1) China's economy has entered a new normal state, in which economic growth has slowed down and consumption has become the main driving force. In 2016, the growth rate of gross domestic product (GDP) was 6.7% which was 0.2% lower than that of the previous year. In the "troika" which stimulates the GDP growth, consumption continued to be the main engine for economic growth, while the driving force of investment and export was weakened. In 2016, the contribution rate of consumption to social economic growth reached 64.6%, with an increase of 4.9% over 2015, and the consumption power continued to increase.

    Since stepping into the new economic normal, China has made efforts to adjust its economic growth model, reduced its dependence on investment, and stimulated personal consumption. In the future, China's consumption will play a more important role in driving the economy, the contribution rate of consumption to China's GDP will increase steadily, individualized, diversified and customized consumption will gradually become the mainstream, and the proportion of service consumption will increase.

    (2) The international economy recovers slowly and the development prospects are not optimistic. Due to the varying degrees of sluggish in most countries who have emerging markets, and coupled with factors such as low commodity prices and slowing trade and capital flows, the global economy will recover slowly in the coming years. The slowdown of global economic growth has forced enterprises to focus more on the improvement of product quality and services.

2. Internal Environment of Industry

    (1) The steel demand of China has entered the arc top of a peak, and the demand will decline with fluctuation. At present, the steel consumption of China has entered the arc top of a peak. In 2015, the actual consumption of steel in China was 664 million tons, declined by 5.4% on a year-on-year basis, which was the first decline since 1996. In 2016, the actual consumption of steel in China rebounded to 673 million tons, with a year-on-year growth of 1.36%. According to the economic development situation and economic structure changes in China in the future and combined with the characteristics of steel consumption in China and the development trend of downstream industries, it is predicted that during the 13th Five-Year Plan period, the steel production and consumption in China will show a downward trend. The further decline in steel demand poses a severe challenge for iron and steel enterprises, making them to continuously improve service levels, extend service

chains, and provide individualized and customized product services while improving product quality and performance.

(2) Services are valued by enterprises universally, and the market competition will become more intense in the future. In recent years, with the intensified overcapacity in China's iron and steel industry, the steel products in China have entered an era of overall surplus, especially the excessive surplus in products such as rebars, wire rods and medium and heavy plates, and even in the "high-grade" steel varieties such as grain-oriented silicon steel, tire cord steel and stainless steel, etc. Therefore, the iron and steel enterprises in China have adjusted and optimized the product structure, developed competitive varieties that are suitable for themselves and expanded the scope of services and service varieties. At the same time, each iron and steel enterprise has paid more attention to the improvement of service level, improved the service model and continuously promoted the scope, level and depth of services.

3. Challenges of Emerging Industries

(1) Development of steel alternative products. At present and for a long time from now on, steel will be the basic raw material to support the development of the economy and downstream industries in China, but its development and application are facing the threat of other alternative metal materials. For example, with the lightweight development of automobile, the application of aluminum alloy in automobiles has been expanding, and some parts such as automobile wheels have been replaced by aluminum alloy materials in a certain proportion; in the future, with the continuously stressing requirements on energy saving and emission reduction, the replacement of steel by aluminum alloy, magnesium alloys and titanium alloys in the fields such as automobile and high-speed trains will be accelerated.

(2) Emergence of new materials. In recent years, with the continuous progress of science and technology, a number of advanced new materials such as carbon fiber, carbon nanotubes, nanoceramics, and graphene have emerged. These materials generally have higher strength and lower weight and are the main alternatives for iron and steel materials in the future. Carbon fiber has the advantages of low density, high strength and modulus, good corrosion resistance, and fatigue performance and is widely used in fields including construction, medical and health, automobile, aerospace, etc. The density of carbon nanotubes is only one-sixth of that of steel, but its strength is 100 times higher than that of steel, and it has been used in high-end fields such as military and aerospace in the USA.

4. Internationalization Development Path

Since 2000, steel exports in China have shown an overall upward trend. In 2016, after a six-year rapid growth of steel exports, steel exports declined for the first time, with an export quantity of 108 million tons and a year-on-year decrease of 3.5%. The service of iron and steel enterprises in China is also accompanied by the export of steel products to the world. The service level also turned from simply distributing products for local distributors of the importing countries into developing deep-level cooperation with international traders, then further developed to providing direct supply service for large customers in the importing countries, and ultimately promoted to establish cutting, processing, and distribution centers in foreign countries to provide product–supplementary services and accurate services.

### *10.2.2 Policy Orientation*

1. China's Macro-Policy Pays More Attention to Services

With the continuous improvement of China's industrial manufacturing capacity and product performance and quality, in China's policies, more and more attention is paid to services. *The 13th Five-Year Plan for National Economic and Social Development* of China clearly proposes to promote the transformation of manufacturing industry from production-oriented mode to production service-oriented mode and guides manufacturing enterprises to extend the service chain and promote value-added services.

*Made in China 2025* (No. 28 [2015] of the State Council) proposes to guide and support manufacturing enterprises to extend the service chain and transform from the main provision of product manufacturing to the provision of products and services; encourage manufacturing enterprises to increase service input and develop individualized and customized services, full life-cycle management, network precision marketing and online support services; support competent enterprises to transform from the provision of equipment to the provision of system integrated general contracting services, and from the provision of products to the provision of overall solutions; encourage competitive manufacturing enterprises to disintegrate the professional advantages, and provide socialized and specialized services to the industry through rebuilding of the business process.

2. Service-Oriented Manufacturing Leads Development of Manufacturing in China

Service-oriented manufacturing is a new industrial form for integrative development of manufacturing and service development in the process of industrialization and is an important direction for the transformation and upgrading of manufacturing industry. Developing service-oriented manufacturing and reshaping the manufacturing value chain are the inevitable requirements for enhancing industrial competitiveness and promoting the transformation of the manufacturing industry from being big to strong. It is an active choice to adapt to the new round of scientific and technological revolution and industrial transformation and is an important measure to effectively improve the supply system and adapt to the upgrading of consumption structure.

In order to implement *Made in China 2025*, the Ministry of Industry and Information Technology and the National Development and Reform Commission and the

Chinese Academy of Engineering formulated and issued the *Special Action Guidelines for the Development of Service-Oriented Manufacturing* (No. 231 (2016) of the MIIT and CFIE, referred to as the Action Guidelines), which is a guiding document for promoting the development of service-oriented manufacturing.

*The Action Guidelines* points out to integrate manufacturing with services in an all-around, wide-ranging and in-depth way. Through three years of development, the level of service-oriented manufacturing has been significantly improved, which further strengthens its role in improving the quality, increasing the efficiency, and promoting the transformation and upgrading of enterprises. *The Action Guidelines* proposes to promote innovative design development, expand customized services, optimize the management of supply chain, promote networked collaborative manufacturing services, and provide system solutions.

3. Policies Related to Iron and Steel Industry Constantly Bring Forth Services

In recent years, most of the iron and steel-related policies and plans issued by China propose to support the iron and steel industry to change service concept, extend service range, and improve service level. *Several Opinions of the General Office of the State Council on Further Strengthening Energy Saving and Emission Reduction to Accelerate the Structural Adjustment of the Iron and Steel Industry* (No. 34 [2010] of the General Office of the State Council) proposes to actively guide iron and steel enterprises to focus on brand, standards, services, and benefits and comprehensively improve product quality, and enhance international competitiveness.

The Key Task (IV) of improving the effective supply level of iron and steel industry specified in the *Plan for Adjustment and Upgrading of Iron and Steel Industry (2016–2020)* is to promote service-oriented manufacturing. Fully establish customer-centered product concept and service awareness, and promote the transformation of iron and steel enterprises from manufacturers to service providers. Encourage iron and steel enterprises to actively connect with the downstream steel-consuming enterprises, focus on the demands of customers, and combine with the mode of early R&D involvement and late continuous tracking and improvement (EVI), to innovate technical support and after-sales service, improve logistics distribution system, and provide a series of extended services such as material recommendation and follow-up processing and application solutions, so as to create and lead the high-end demands.

## 10.3 Case Analysis

### *10.3.1 China Baowu Steel Group Corporation Ltd. (China Baowu)*

After more than 30 years of development, China Baowu Steel Corporation Ltd. (hereinafter referred to as China Baowu) has become the most modernized and competitive iron and steel complex. The mission of the Group is to "become a world-class supplier of steel products, technologies and services", and the development path is to continue to promote the strategic transformation "from steel to materials, from manufacturing to services, and from China to the world". China Baowu upgrades its services to the strategic level of the Group and places its services in a critical position, which shows that China Baowu attaches great importance to services. Baosteel practices the Group's strategy. Through early services intervention, director responsibility system for major customers, e-commerce trading platform, rapid response to a sound marketing service system, and with a marketing network all over the world, Baosteel provides the customers with first-class products, technologies, and services. While meeting the demands of the domestic market, the products are exported to more than 70 countries and regions such as Japan, South Korea, Europe, and America.

1. Application of EVI in China Baowu

EVI (Early Vendor Involvement) means the early involvement of vendors. It is a customer-centered and customer-oriented value creation activity in the supply chain comprehensively participated by the suppliers. For China Baowu, EVI is not only a technical marketing activity that aims to satisfy the demands of customers, promote the development of products and technologies of China Baowu, and accelerate the marketing promotion and business cooperation, but also an innovative business model. Through EVI, China Baowu focuses on customer's demands, fully involves in all aspects of the customers from R&D to mass production, and gives full play to the capabilities and advantages of China Baowu to create customer value, thus occupying the dominant position in market competition.

At present, the EVI of China Baowu has formed an EVI cooperation mode, which covers the whole process of automobile development and manufacturing from advanced engineering design, early (conceptual) and body design, mold (tool) design and development, vehicle model production, and mass production and satisfies different levels of customers' demands. China Baowu has the ability to provide customers with a "target cost-oriented" and "lightweight-oriented" package solution and has established in-depth cooperation with all major domestic automobile manufacturers. At the same time, the EVI work is no longer limited to automotive customers, but also expanded to customers in different industries such as home appliances, electrical steel, and metal packaging. The EVI work of China Baowu is an effective exploration and practice in the process of transforming from a manufacturer to a service provider.

At present, China's automobile industry is facing the pressure from the constraints of petroleum, environment, and transportation, and the whole industry will enter a stable but slow developing period. At the same time, domestic competitors in the iron and steel industry have entered or targeted to the demands of the automobile industry and launched fierce competition with China Baowu through strategies such as new units, new product development, improved production technological level and "low price penetration". Particularly on the factor of price, China Baowu is currently facing tremendous pressure on acquiring fixed-point acquisition of new vehicle models and maintaining the share of old models. Especially, the price competitiveness of hot-dip galvanized products has declined significantly. Facing with such a brutal market competition situation, China Baowu needs to comprehensively enhance its competitiveness in an all-round manner from the perspectives of products, technology, services, marketing, etc. to maintain and consolidate its current market share. The EVI work of China Baowu is a highlight of Baowu Group's philosophy of "competitive products + service". While creating value for customers, it also greatly enhances the added value and competitiveness of Baowu Group's products. Today, with the increasingly fierce competition in product homogenization, China Baowu, through EVI work, fully demonstrates the value of the company's technological advantages and service capabilities and avoid simple and low-level price competition, thus maintaining the market share of the company's products and higher ability of premium, and creating a competitive path for differentiated services.

In recent years, the trend and demand for lightweight development in the automobile industry have become obvious, which has further promoted the advance of materials and manufacturing process of automobile sheet. "Environmentally friendliness, safety, long service life, and low price" has become the lasting driving force for the development of automotive industry. According to the development trend of automobile materials in Europe, the USA, Japan, and other countries for over the past 20 years, the consumption of steel plates is decreasing, while the consumption of lightweight metals (aluminum alloys and magnesium alloys) and plastics is increasing. However, steel sheets are still the leading material as automobile sheets in the future, accounting for about 55–70% of the weight of vehicle, but the connotation of the steel sheets has been changed greatly. In the future, automobile sheets will be developed in the direction of high strength, lightweight, high corrosion resistance, and environmental protection, and at the same time, it is necessary to provide technical solutions to customers. In the future, high-strength steel, ultra-high-strength steel, and automobile outer sheets will still be the products which will be highly concerned and promoted by Baowu Group. The innovative development of harmless and environmentally friendly automobile sheet products and the application technology is the development direction in the future.

China Baowu is currently the only iron and steel supplier of all types of automobile sheets such as hot-rolled and pickled (HR-PO) steel applied for passenger-car chassis and structural parts, cold-rolled (CR) sheet, hot-dip galvanized pure zinc (GI) sheet, ZnFe alloy (GA) sheet, and electro-galvanizing (EG) sheet for vehicle body. It can not only meet the needs for automobile sheet products of proprietary brands but also meet the needs for domestic and localized automobile sheet products of joint venture brand. At the same time, China Baowu is also the only iron and steel manufacturer in China with the ability to develop and supply the first, the second, and the third generation of ultra-high-strength steels and has established a complete material database and a application knowledge base, which have the ability to cover all kinds of materials with all strength demands used for a whole vehicle. In July 2013, China Baowu's Q&P980 steel grade was successfully launched in the world, achieving a major breakthrough in the development and application of Baowu Group's automobile sheet products from the mode of "going after" to "taking the lead". In October 2015, 1700MS steel (martensitic steel with tensile strength of 1,700 MPa) was successfully produced in China Baowu. This is another breakthrough for China Baowu in the R&D and production of high-strength steel. 1700MS steel is the high-strength steel grade used for automobiles which has the highest tensile strength in the world, it is mainly used for manufacturing bumpers and door anti-collision bars, and it can greatly improve the anti-collision performance and reduce the weight when applied in front and rear bumpers. Under the premise of ensuring collision safety, 1700MS steel has a potential to lose weight by 10–20% compared with the previous high-strength steel, which can guarantee the safety of driving to the maximum extent.

Product development and technological advancement are the foundation of enterprise development. EVI of China Baowu is the booster for the development and technology upgrade of the Group's iron and steel products. In the days to come, China Baowu will adhere to the EVI work concept, insist on the customer-oriented principle, exert the synergistic effect of production, sales, and research, accelerate the R&D and breakthrough of key products and key technologies, strengthen the R&D investment in equipment technology and manufacturing technology of automobile sheet, accelerate the industrialization of R&D results, and constantly meet the demands of customers. China Baowu should further explore the demand for materials in the development of automobiles, refine the essence of technology, and develop in the direction of the automobile sheets with high strength, high corrosion resistance, and high surface technology in the field of lightweight technology.

The EVI work of China Baowu is an effective exploration and practice of the company in the process of transforming from manufacturing to providing service. The team members have fully implemented the requirements on the culture of China Baowu and closely integrated it with the business characteristics, forming an EVI culture with distinctive characteristics. The EVI culture is gradually formed by the EVI team of China Baowu in the long-term work exploration and practice. Its core concepts mainly include:

Customer thinking—from customers, serve customers, accomplish customers
Synergistic thinking—the same goal, the net-type work, the same voice
Enterprising thinking—skilled in profession, dedicate to achievement, do more than expectations.

These concepts have been infiltrated into all aspects of EVI activities. Besides, on this basis, the common guidelines of the EVI team of China Baowu and some vivid stories and typical figures are formed. These are also the key to the success of the Group's EVI activities. Through the practice and communication of the EVI culture, China Baowu has enriched its brand connotation of "Building Shared Value", enhanced the brand value of its automobile sheets, and embodied the soft power of servitization.

2. Ouyeel Expands its Service Scope

   (1) Development history. Under the development trend of centralized manufacturing, standardization and quantitative, and personalized demands of customers in the iron and steel industry, Baowu Group integrates the relevant resources of the existing iron and steel electronic transaction, thinks about

and reconstructs each link of the business in the whole process, and builds an iron and steel service platform of Ouyeel in a brand-new mode to tackle with problems such as long-term information asymmetry, high circulation cost, low trade efficiency, and collapse of steel trade credit environment in the iron and steel circulation field.

Ouyeel Co., Ltd. (hereinafter referred to as Ouyeel) was established on February 13, 2015, with a registered capital of 2 billion yuan. The goal of it is to build an ecological iron and steel service platform in the iron and steel circulation field based on new technical methods such as the Internet, the Internet of Things, big data, and mobile Internet, which integrate information, settlement of transactions, logistics and warehousing, processing and distribution, investment and financing, financial intermediaries, technology, special services in the industry, and other functions and follows the principle of symbiosis and win-win among iron and steel manufacturing enterprises, iron and steel trading companies, logistics processing service providers, steel consumers and other major parties, and with a service scope covering the whole country.

(2) Business scope. Ouyeel includes five business platforms, i.e., Ouyeel E-commerce, Ouyeel Logistics, Ouyeel Finance, Ouyeel Materials, and Ouyeel Data. These five business platforms are both collaborative and relatively independent, each of which can be expanded outward.

Ouyeel E-commerce is not only an iron and steel trading platform but also the gateway interface to the entire iron and steel service platform. It integrates intelligent trading and professional services, supports various trading modes such as self-operation, consignment sales and matchmaking, and has a dedicated third-party payment channel, i.e., Easternpay. According to six major functional orientations, i.e., iron and steel trading, information, logistics, finance, R&D, and renewable resources, Ouyeel E-commerce breaks the boundary of ownership and realizes the powerful combination among central state-owned enterprises, private enterprises, and local enterprises in leading enterprises of the cross-ownership industry to build an international advanced online electronic service platform in iron and steel industry and form a steel supply chain service system characterized by large production, large circulation and large market, and enhance the pricing power and discourse power of China's iron and steel industry in the international market. Through the innovative trading mode, Ouyeel E-commerce promotes the digital operation of the iron and steel industry chain; promotes the standardized operation of the iron and steel logistics industry by establishing a standard operation system of logistics and warehousing; promotes the sound development of the iron and steel industry and the financing environment of small- and medium-sized enterprises through the combination of industry and finance; based on the large-scale real-time transactions, Ouyeel E-commerce forms "Shanghai Price" and "Shanghai Standard" for steel trading and services, and a modern service system of iron and steel supply chain, so as to lead the transformation of steel circulation, and build a leading and sustainable B2B (inter-company trading) service platform in China.

Ouyeel Logistics establishes a specialized online logistics service platform for the bulk material circulation field. The business scope of the platform includes warehousing services, transportation services, processing services, logistics transactions, financing supervision, etc. in the process of bulk materials logistics. In terms of transportation capacity, it forms a fourth party logistics platform for transportation operations, which provides a safe and convenient one-stop logistics service. Logistics supervision is realized through multi-dimensional means such as warehouse property right mortgage, information system monitoring, and personnel dispatch management, to provide escort for iron and steel product trading and management and control of financial and entity risk.

Ouyeel Finance is committed to providing professional services of Internet finance for the whole industry chain of bulk commodities, providing customers with Internet financial products such as financing services, investment and wealth management, online payment, and asset management, and creating an ecosystem of Internet financial services for the bulk commodity industry. As one of the most competitive financial service platforms of Internet + industry chain in the country, Ouyeel Finance can provide one-stop service for customers and realize "All in One Net" service in the financial field.

Ouyeel Materials integrates Ouyeel Standards to provide customers with technical and deep processing services, trade business support services, and overall solutions for steel applications.

Ouyeel Data collects the data of trillion-scale transaction volume in the entire industry chain, providing customers with industrial big data services such as data storage, data mining, demand forecasting, precision marketing, price forecasting, and logistics optimization, as well as e-commerce platform services such as platform design, system construction, and system hosting.

(3) Operation mode. Ouyeel is committed to becoming a third-party iron and steel service platform that truly belongs to the entire industry. It adheres to the business model of "Service-oriented Production System" and advocates the value concept of "Co-building, Sharing, and Trustworthiness", and creates a brand-new mode of open equity with the utmost sincerity.

Based on the differences in development depth and the future development orientation of e-commerce business of various iron and steel plants at present, the e-commerce platform of Ouyeel provides a variety of cooperation modes for iron and steel plants to participate in the cooperation.

1) Business cooperation modes

Hypermarket mode: It is the most direct and convenient cooperation model for iron and steel plants looking to rapidly expand their spot sales channels. Namely, it is to put the spot products to the e-commerce platform of Ouyeel, so as to realize the low-cost and high-efficiency network sales with the help of the mature fixed-price sales, logistics and warehousing, payment and settlement, and other functions of Ouyeel E-commerce, as well as a huge group of customers.

Exclusive store model: it is designed for rapid establishment of a relatively independent e-commerce sales platform (futures and spot goods), and Ouyeel provides support services by utilizing mature technical solutions for platform construction. Taking Bsteel of Baosteel, a subsidiary of Baowu Steel Group, as an example, which closely connects marketing services with production control, and improves operational efficiency and optimizes customer experience by adopting Internet means. Under this mode, the iron and steel plants not only satisfy the independence of its own brand management and enjoy the private cloud database, but also take advantage of the support of Ouyeel in terms of consumer introducing, financial services, and warehousing and logistics.

E-commerce alliance mode: It is designed for the iron and steel plants that have developed their own e-commerce platform (or plan to establish), sharing the win-win effect of $1 + 1 > 2$ in the form of alliance. That is to say, in this mode, Ouyeel E-commerce and other e-commerce platforms conduct data exchange and immediately realize the sharing of various resources such as the number of customers, the amount of listed resources, the warehousing and logistics network, and selectable financial services. If customers concerns about the openness of information, Ouyeel can promise the one-way data transparency, i.e., customers of Ouyeel platform can access the alliance platform for procurement, while customers of the alliance platform cannot access the Ouyeel platform for procurement.

2) E-commerce platform equity cooperation mode. Ouyeel plans to go public in about three years through several rounds of investment invitations. As the first listed operating platform, Ouyeel E-commerce will introduce multiple investment entities such as iron and steel plants, steel traders, and other strategic investors to create a truly third-party public iron and steel service platform.

   The cooperative iron and steel plants can participate in the shareholding platform specially built in the form of limited partnerships, take the accumulated transaction volume contributed on the platform before listing as the basis of share allocation, and lock the shares of the e-commerce platform at low prices, so as not only to enjoy the benefits of transformation and upgrading of business mode brought by the platform of Ouyeel E-commerce, but also maximize the value of assets through the capital market.

(4) Development trend. Ouyeel E-commerce is not only an iron and steel trading platform but also the gateway interface to the entire iron and steel service platform. It currently has more than 50,000 registered customers. Ouyeel E-commerce can support a variety of trading modes such as self-operation, consignment sales, and matchmaking and has a dedicated third-party payment channel, i.e., the Easternpay. It has business transactions covering

all provinces and cities in China (including Taiwan of China). Besides, it launched cross-border e-commerce transaction in 2015 and trading interface in English. As the Taobao of the iron and steel industry, the platform mainly covers a wide range of steel grades such as hot-rolled steel, pickled steel, cold-rolled steel, and galvanized steel. In terms of suppliers, the Shanghai Iron and Steel Exchange Center maintains stable contact with 152 social iron and steel manufacturing enterprises. The major suppliers that have joined the platform to open stores are Baowu Group, Angang Steel, Shougang, Shagang, Ma'anshan Steel, Baotou Steel, TISCO, Ningbo Steel, Meishan Steel, and other large-scale iron and steel plants in China. In terms of traders, there is also a broad basis for cooperation with social steel traders. At present, there are 428 long-term spot trading objects in the society including Zhejiang Materials Industry, Minmetals Group, Nanjing Huaneng, etc. Since it is established, the trading volume of the Ouyeel E-commerce platform has been continuously increasing. The trading volume was 10.18 million tons in 2015 and 38.76 million tons in 2016. In the future, Ouyeel E-commerce will develop into one of the world's largest third-party iron and steel trading platforms.

At present, Ouyeel Logistics has more than 60 capital joined warehousing enterprises and more than 200 agreement joined warehousing enterprises in the whole country, with its management and control covering more than 500 warehousing enterprises and having an annual warehousing turnover capacity of 50 million tons.

Ouyeel Finance has three financial licenses for payment, pawn and factoring, and a bank credit of nearly 200 billion yuan. It can provide customers with various Internet financial services such as pledge financing, credit loans, trade financing, third-party payment, and online financial management. Within one month after its launch, it has provided more than 40 enterprises with the financing services of more than 100 million yuan and has more than 4300 financial management customers and a cumulative amount of more than 100 million yuan.

### *10.3.2 HBIS Group Co. Ltd. (HBIS)*

HBIS Group Co. Ltd. (hereinafter referred to as HBIS) works actively to make the transformation from manufacturing to providing services, so as to meet the demands of individualized steel consumption of enterprises. Continuous exploration and innovation are made in terms of marketing mode, iron and steel e-commerce, customized production, EVI services, extension of industrial chain, etc. Measures taken by HBIS for developing customer-oriented servitization are as follows.

1. Implement major customer marketing services and build a relationship network and service mechanism for high-quality customers

"The height of the customer base determines the height of an enterprise's product" and "the attitude toward the customer and the product determines the value of the product" are the ideas and focal points for the transformation of marketing mode of HBIS. All companies of HBIS focus on the demands of high-end customers, further strengthen the connection with high-end customers, strive to build a high-quality customer relationship network and service mechanism by differentiated and individualized services, and accelerate the transformation from iron and steel manufacturers to iron and steel service providers.

Based on the marketing department, Tangsteel establishes a major customer service center, builds a comprehensive customer service system, strongly promotes the major customer manager system, and plans, organizes, implements, and manages the development and service of strategic customers, so as to realize the situation of "sales maintain customers, and production supports sales" and accelerate the establishment of strategic customer groups. Major Customer Service Center of Tangsteel is the core management organization for developing and providing services to Tangsteel's strategic and directly supplied customers. Its construction scheme includes two core contents, i.e., the establishment of a comprehensive customer service system and the promotion of the major customer manager system. The main responsibilities include designing and setting up major customer service system, planning and formulating development strategy for major customers, establishing and improving major customer manager marketing system, establishing customer VIP management mechanism, creating customer relationship network, managing strategic customers, promoting transmitting of customer's demand to the production line, and pushing the reform of internal mechanism and so on. Tangsteel takes the major customer manager system as the "bridge" to effectively connect customers and himself, and deeply develop the market and customers, promotes the production, sales and research operation internally, so as to practically realize the production and operation mode of "selling for customers and producing for sales". At present, Tangsteel has set up 14 "major customer manager" service teams according to key products and key customers, including major customer team of Lingyun Technology, major customer teams of Great Wall Automobile and in Baoding Region, and major customer team of container sheet, etc. to accurately target and develop strategic customers, innovate business solutions and improve customer satisfaction. In order to ensure the smooth progress of the major customer manager system, Tangsteel establishes a comprehensive customer service system, mobilizes and integrates all available resources, establishes a complete process for acquisition, transmission and implementation of customer's demand, and carries out professionalized customer classification and hierarchical management internally, so as to provide corresponding professionalized service resources for customers of different classifications and different levels, and ensure that all customers get satisfactory services while achieving the rational allocation of the company's limited resources.

Chengde Steel actively promotes the marketing transformation and establishes the "Major Customer Service Center of Chengde Steel" as an independent unit in the Market Management Department. Clearly orientation is made for the Center, and by taking the comprehensive advantages such as platform resources, channels and technologies of the "Major Customer Service Center" of HBIS Group, major customer managers, customer managers, customer engineers, and marketing systems of Chengde Steel are seamlessly connected, so as to develop and stabilize high-end product demand channels, high-end customer base, and achieve comprehensive increase in terms of product quality, quantity and price. The newly appointed nine major customer managers are mainly the leaders for sales, technology, and R&D and are in one-to-one correspondence with the categories of strategic varieties. They provide tailor-made services for high-end customers and are mainly responsible for coordinating, resolving and handling of feedback problems during pre-sales, in-sales, and after-sales, supervising the management of customer managers and customer engineers, making regular return visits of customers, and implementing the individualized requirements of customers. At present, Chengde Steel is fully committed to building a "one-stop" service hall, aiming to further improve the efficiency of marketing services. The large customer service center is stepping up the development of strict performance incentive and assessment system, which fully reflects the linkage between performance and achievements, so as to enhance the enthusiasm of major customer managers and their teams, ensure steady sales of high-end products, continuously expand market share, and realize continuous increase in the prices of vanadium-titanium featured products.

Handan Steel appoints 20 major customer managers and opens a series of customer service platforms such as customer service hotline and inquiry system of electronic warranty; Xuanhua Steel implements customer manager system, builds a rapid response mechanism which takes customer managers as the core, improves the service system and realizes a hierarchical responsible system by relying on the "customer manager" mechanism; the Hengstrip further clarifies the customer cooperation direction and the target customer base, classifies the customer's popularity and cooperation stability in levels, establishes VIP of strategically close level, core cooperation close level, and general cooperative customers, and makes every effort to create a brand-new marketing mode; Xuanhua Steel further strengthens the management and assessment of service agents, improves the service response speed by optimizing the consolidated storage of parts and completing the inventory variety structure, sets up customer service hotline, etc., so as to establish an unobstructed communication channel with customers.

2. HBISTC aims to create the most influential service platform of iron and steel supply chain in North China

HBISTC (formerly known as Hebei Iron and Steel Trading Center) is an enterprise owned by the whole people and operated and managed by HBIS, which was approved by People's Government of Hebei Province and was registered and established by the state-owned Assets Supervision and Administration Commission of Hebei Province in May 2012.

Relying on modern e-commerce means and integrating various high-quality resources such as commodities, customers, logistics, and finance in the industrial chain, HBISTC takes trading services as the basis, logistics, and financial services as the support, and with the supplementation of data information services, to help the iron and steel plants to innovate the customer-centered marketing service mode, continuously improve marketing service capabilities and levels, help downstream dealers and end customers to obtain directly supplied resources and high-quality products from iron and steel plants conveniently and efficiently, and gradually form a specialized and large-scale online trading market by the way that resources drive channel and channel promotes resources.

HBISTC is launched online for trial operation on October 23, 2012, and successively built and completed online direct spot sales of iron and steel plants, online agreement sales of iron and steel plants, online inquiry, price comparison and procurement of raw materials and fuel, online auction of recycled materials, online auction of chemical products, online financing of production capacity, online financing of orders, online financing of warehouse receipt, online agency ordering, and other online services. After two years of exploration and practice, HBISTC fully relies on the resources advantages and scale advantages of HBIS to deepen trading, logistics and financial services, create and improve the customer service system, innovate a diversified service pattern, and gradually develop into a network service provider in the iron and steel supply chain. HBISTC is selected by the Ministry of Industry and Information Technology as one of the "E-commerce Integrated Innovation Pilot Projects in 2013" and is listed as one of the ten major commodities electronic trading platforms in the e-commerce activity of "Ten, Hundred, and Thousand Project" in Hebei Province.

Up to now, trading of all the resources of HBIS' s deformed steel bars, some hot-rolled plates and cold-rolled plates, and section and strip products have been sold online. The bulk raw materials and fuels such as ferroalloy, coals, cokes, and iron concentrates are purchased through online bidding, and the recycled materials such as waste inferior materials, waste and old materials, and idle waste equipment are all auctioned online. In 2015, the total online trading volume of HBISTC Platform reached 40.43 million tons, achieving total trading revenue of 62.65 billion yuan with an increase of 141% and 105%, respectively, over the previous year.

3. Customized service

HBIS promotes "customized" services, responds to market changes with the rapid transformation of production and management concepts, produces customized products for customers, and fosters customer loyalty with special services. As for long products, Tangsteel can provide the customer with non-standard cut-to-length deformed steel bar, hard wire rod, boron-added angle steel, angle steel with non-standard thickness, steel used for low-sulfur submerged arc welding wire, boron-added gas-shielded welding wire, and other products through customized production. In January 2016, the total number of customized wire rods of Tangsteel Long Product Department reached 7852 tons, accounting for 23% of the total output, with a 5% increase over the same period last year, of which 7737 tons were customized products sold to key customers with a 60% year-on-year increase, showing an upward trend in an adverse market environment. Hot Rolling Department of Tangsteel actively changes its role and the old mode of self-organized production according to orders, launches "high-end customized" service, takes the initiative to hunt in high-end market, carries out in-depth technical services, customizes individualized products according to the demand of the customers, and at the same time reaches a intent price higher than the common sheet with end customers, so as to achieve the goal of mutual benefit. Tangsteel also explored a set of "customized" precision R&D program and established a new model of high-end product technology R&D and market development.

4. EVI

Vigorously promoting the EVI model (Early Vender Involvement) is an important deployment for HBIS to realize new breakthroughs in market planning and products in 2016.

Relying on the advantages of the national enterprise technology center, and through the co-construction of joint laboratories and other cooperation modes, Handan Steel enables the marketing team of major customer managers to participate in the customer's product design, mold development, on-site process, and other aspects, so as to realize technical guidance and cooperative development and provide customers with value-added services. Taking Shandong Shifeng as an example, technical personnels from Handan Steel begun to involve in the work when many products of Shandong Shifeng were in the R&D stage, so as to tackle problem together with customers through the EVI technological marketing mode. At present, Shandong Wuzheng Automobile has made sample automobiles by adopting the upgraded scheme provided by Handan Steel for materials of the ordinary low-carbon steel parts with seven specifications.

Shijiazhuang Steel has carried out the "Early Involvement" or even "Cooperative R&D" on gear steel, free-cutting non-tempered steel, spring steel, etc. "Early Involvement" has also been realized for the bearing steel with fixed grade. The EVI technology service of Shijiazhuang Steel is not limited to be domestic. In 2015, it carried out early stage technical services for a high-end commercial vehicle enterprise in India. This is the first time that Shijiazhuang Steel has promoted "Early Involvement" technical services to the international market.

5. HBIS Haier Special Steel extending the industrial chain

In 2015, HBIS and Haier Group signed an equity cooperation agreement for Haier Special Steel project. According to the agreement, HBIS holds a 70% stake in Haier Special Steel, and both parties jointly operate the business of color-coated steel plate. HBIS has joined hands with Haier Group to transform from an iron and steel manufacturer to a home appliance service provider.

Haier Special Steel is a subsidiary for R&D and production of color-coated plates of Haier Group, with an annual production capacity of 300,000 tons. It is also the enterprise with the largest market share of color-coated plates for home appliance in China. The products of Haier Special Steel are mainly sold internally in Haier Group, accounting for about 80% of the sales, and the external sales accounts for about 20%. During the order cooperation period, HBIS fully enters the supply channel of Haier and has the priority to supply products under the same conditions. This means that the relationship between HBIS and Haier Group have been changed from the trade cooperation in the past to equity cooperation. This will help HBIS to extend downstream of the industrial chain and realize the deep connection among production lines, products, and end customers.

Although the annual production scale of Haier Special Steel is small at present, it will not have a substantial impact on the scale of plates for home appliance of HBIS. However, through this cooperation, HBIS can stabilize this sales channel of Haier and further expand into the field of high-end home appliance.

### 10.3.3 Ansteel Group Corporation (Ansteel)

The overall strategy of servitization of Ansteel Group Corporation (hereinafter referred to as Ansteel) is to change the service concept, improve the service level, and enhance the product value through highly efficient, fast, accurate and extended services, and achieve "win-win" cooperation with customers.

1. Establish Dalian Ship Plate Processing and Distribution Company to provide customized services

In order to stabilize and develop the market share of the shipbuilding plate and improve service levels and benefits, in 2004, Ansteel and CSIC jointly established Angang Steel—DSIC Dalian Steel Processing and Distribution Co., Ltd., which is the first specialized processing and distribution enterprise for shipbuilding steel in China. The establishment of this joint venture company has achieved a strategic alliance between iron and steel enterprises and shipbuilding enterprises, and by building a stable supply and demand chain, the enterprises have greatly improved the market competitiveness, so as to jointly resist market risks.

The joint venture company was officially put into operation in October 2008, with a designed annual processing capacity of 600,000 tons. It is mainly engaged in the processing of steel for shipbuilding and the manufacturing of structural parts, and the production capacity was basically achieved after being put into operation in 2010. The procurement price of the raw materials of the joint venture company was negotiated by DSIC and Ansteel, both of whom have signed a long-term price agreement to protect the benefits of both parties and jointly resist market risks. By establishing steel deep processing enterprises through joint ventures with powerful downstream enterprises, Ansteel extends the iron and steel industry chain, forms a firm and reliable cooperative relationship with downstream customers, and provides a stable market for the main steel business of Ansteel.

2. Implement steel processing and distribution strategies to provide accurate, efficient, and value-added services

Ansteel actively implements the steel processing and distribution strategy and has established more than ten steel processing and distribution centers in Shenyang, Shanghai, Tianjin, Wuhan, Weifang, Dalian, Changchun, Zhengzhou, etc. Most of these steel processing and distribution centers are established by Ansteel with local powerful large-scale logistics enterprises, so as to meet the demands of the regional market through steel cutting and distribution.

The Processing and Distribution Center of Ansteel is divided into two types, one is marketing-oriented steel processing and distribution center, and the other is specialized steel processing and distribution center. The steel processing and distribution centers established by Angang Steel in Shanghai, Shenyang, Tianjin, etc. are marketing-oriented type. By setting up steel processing centers in these central cities, Ansteel serves a large number of small- and medium-sized customers in machinery, steel structure, and other industries around the city. At the same time, in order to meet the domestic demands of downstream industries such as automobiles and home appliances, Ansteel establishes specialized steel processing and distribution centers, such as the steel processing and distribution center jointly built by Changchun FAW, Ansteel and MITSY, and Ansteel ThyssenKrupp Steel Distribution Center; these processing and distribution centers provide key customers with accurate, highly efficient, value-added, and high-level services in industries such as automobiles and home appliances.

### *10.3.4 ThyssenKrupp*

In the development strategy of ThyssenKrupp Group, to improve the service capability, extend the industrial chain to the downstream steel-consuming industry, and transform from production orientation to customer orientation is one of the important contents.

1. Provider of services and solutions of iron and steel material system

In order to achieve the overall development strategy of the Group, ThyssenKrupp Group adopts the high value-added product differentiation and emphasizes that the enterprise is not only a supplier of iron and steel materials, but also a provider of services and solutions of iron and steel material system oriented to demands of customers. ThyssenKrupp provides a business model of high value-added products for specific customers (i.e., providing value-added services such as processing on the base of iron and steel business) and lays the foundation for its sustainable operations.

2. Global processing and distribution service network

Raw Material Service Division of ThyssenKrupp is principally engaged in the processing, warehousing, logistics, sales and maintenance of industrial raw materials such as steel (including steel pipes), non-ferrous metals and plastics, as well as provides technical services in terms of inventory and supply chain management, engineering systems, rails and civil engineering, etc., with more than 500 processing and distribution enterprises in more than 40 countries around the world.

Taking Germany as an example, ThyssenKrupp has 60 warehousing centers, 500 transport vehicles, and a wide range of processing services, mainly including cutting, heat treatment, leveling, bending, polishing, deburring, coating, machining, welding, flame cutting, etc.

3. Promote e-commerce in Europe and globally

Facing the impact of imported products and oversupply in the market, ThyssenKrupp also actively explores new ways of selling products, including electronic sales. In the field of e-commerce, ThyssenKrupp Materials Service Department has developed a comprehensive online solution for the European market. ThyssenKrupp indicates that the time for developing steel e-commerce in the European market is mature, the demand for electronic solutions from manufacturing enterprises and customers is increasing, and the network makes the cooperation between the two parties more convenient and efficient. Currently, ThyssenKrupp offers three online stores in Europe for existing and new customers, which can provide customized solutions according to their demands.

For the existing customers, the online store not only provides detailed product catalog (including more than 15,000 products), inventory, model contracts, and customer manual downloads but also realizes online order purchase and logistics distribution management. The more customer-friendly design is that online stores can provide online processing services according to customer's demand. As long as customers provide CAD design drawings, the cut-to-length processing can be realized automatically.

For new customers, online stores can offer small-scale retail for products with standard sizes for small enterprises and even end customers. This approach is based on years of experience of online sales in the US market (since 2007), which is beneficial for ThyssenKrupp to develop potential markets, so as to reach target customers and ultimately achieve large amount supply. There are currently 65,000 varieties of products available.

In addition to promoting online solutions in Germany and the USA, ThyssenKrupp's goal is to promote this service to every European country. In 2016, the service was promoted in Belgium, the Netherlands, Luxembourg, Denmark, Sweden, Switzerland, and other countries successively. Among these, online stores for new customers have been successful in the UK market and launched in Spain.

The core of the "one steel" revitalization strategy proposed by ThyssenKrupp previously is to strengthen the market and customer demand orientation, further improve production efficiency, optimize product structure, accelerate the transformation of innovation results to promote development, and improve efficiency of supply chain. While electronic marketing is precisely in response to this strategy, which can not only understand the changes in customer's demands timely but also improve the efficiency in order execution and reduce the production costs. At present, the number of customers served by the ThyssenKrupp online solution has exceeded 90,000, and the annual orders brought by it for ThyssenKrupp is over 150,000, creating considerable economic benefits.

### *10.3.5 Pohang Iron & Steel Co. Ltd. (POSCO)*

1. Build Global Steel Processing and Distribution Centers

In order to promote international development, Pohang Iron and Steel Co. Ltd. (hereinafter referred to as POSCO) sets up steel cutting, processing, and distribution centers around the world. At present, POSCO has invested or control a total processing and distribution capacity of over 10 million tons. The processed steel products basically cover hot-rolled strip, wide and heavy plate, cold-rolled strip, silicon steel, stainless steel, etc. There are more than 40 processing centers overseas and the production capacity exceeds 4 million tons, involving such countries as China, Singapore, Vietnam, Thailand, India, and Brazil.

Through the steel processing centers all over the world, POSCO is able to process products according to customers' demands, provide customers with timely and accurate steel cutting, processing and logistics services, and also provide customers with higher level, diversified and individualized services by collecting dynamic information of steel demands in the relevant regions.

2. Provide Comprehensive Solutions for Iron and Steel Industry

In recent years, POSCO continues to improve its service level and capabilities, tries to provide overall solutions for downstream customers, and achieves good results. POSCO believes that even if iron and steel enterprises produce profitable high-quality steel, they will not attract customers if the products are inconvenient to be used or have poor economic benefits for the customers. Therefore, POSCO provides customers with the hardware, i.e., the "best steel" as well as the software, i.e., "customer application technology" and "business support", which not only allows customers to use the steel products of POSCO more conveniently but also makes them enjoy more cost-effective solutions.

Since 2014, POSCO has successfully implemented comprehensive steel solutions and signed more than 100 technical cooperation and sales agreements with major customers in domestic automobile, household appliances, and other industries. Through the new service mode of comprehensive solutions for steel, POSCO realizes the transformation from developing high-quality steel to providing application technology and business support, achieves the differentiated competitiveness to provide comprehensive solutions, improves the company's revenue, and works with customers to achieve "win-win" situation.

## 10.4 Prospects and Path Analysis of Service-Oriented Trend

### *10.4.1 Prospects of Service-Oriented Trend*

With the development of social productivity, the advancement of iron and steel production technology, the increasingly fierce market competition, and the diversity and differentiation of customer's demands, the service development trend of the iron and steel industry has attracted people's attention. As material manufacturers, the iron and steel enterprises are focusing on service-oriented development as an important part of the core competitiveness of enterprises and an effective measure to adapt to the new situation. At the same time, the process of service-oriented development is accelerating, the servitization and service functions are becoming more and more prominent, and the transformation to servitization has become an irresistible trend. How the iron and steel enterprises transform to servitization and what countermeasures will be taken by them are related to the survival and development of the enterprises.

1. Establish a New Service Concept

Establish the idea of systematic services. The services provided by the enterprises to customers are demand services, but not only the products. What a customer wants is to obtain the required products and services immediately. The iron and steel enterprises should not only solve the quality problems of the products, serve with enthusiasm and high efficiency, and improve the customer's satisfaction, but also should practice the service concept throughout the whole process of production, operation and management of the enterprises, consider for the customers in each link from the market demand research and product design to production, sales, and after-sales services, so as to maximize the customer's interests, meet the demands of customers immediately, and take the satisfaction of the customer's demands as the foothold and start point of all activities of the enterprises and as the basic principle of all business activities of the enterprises.

Establish the idea that everyone serves. Serving for customers is not the responsibility of a certain department or certain personnel, but the responsibility of each manager and employee of the enterprise, with the only difference in their labor division. The irresponsibility or incompetence of any person may result in a lack of service. All design, R&D, production, marketing, and management personnel should take service for customer as their fundamental mission, and provide customers with ever-present service, quick service, future-oriented service, and lifetime service.

2. Transform from "Manufacturing" to "Creating"

The essence of competition among iron and steel enterprises is the competition of products, and manufacturing plays a fundamental role. If an iron and steel enterprise does not have core technology or core products, it will not be competitive at all. Therefore, the servitization of iron and steel enterprises does not mean that the iron and steel enterprises neglect the capabilities of manufacturing and R&D, but from a higher perspective, that is, from the height of all-round services for customers, to continuously strengthen and improve the manufacturing and R&D capabilities of iron and steel enterprises. On the basis of continuously improving product quality, iron and steel enterprises should unceasingly improve their R&D capabilities, develop new products, transform from "manufacturing" to "creating", and continuously expand new market. Iron and steel enterprises should continuously enhance their manufacturing, R&D, marketing, and other system capabilities during service transformation.

3. Build a New Information Network and Logistics Platform

   (1) Build a new information network platform. In the era of information economy, the business environment of the enterprises has undergone fundamental changes, and the role and status of e-commerce in modern enterprise management are becoming more and more important. Iron and steel enterprises should speed up the construction of a modern service system and make full use of the existing information network technology to realize digital customized production of low-inventory, online marketing and offline customer relationship management, create an open, fair, and just production and operation environment, and strengthen information communication with dealers and customers, so as to provide customers with more individualized, specialized, and automated services to improve sales success rate and efficiency.
   (2) Build an efficient logistics platform. Iron and steel enterprises should establish efficient, convenient and low-cost logistics and transportation systems through chain operation, processing, and distribution, so as to improve service level.

4. Adjust Organizational Structure to Reconstruct Process

Iron and steel enterprises should take the customers as the center to construct business processes of the enterprises and set up corresponding organizations. The setting of traditional linear functional management organization has faced a great test, exposing the shortcomings in the speed and effect of the reaction. It is necessary to establish a system and organizational structure that is conducive to quick response to customer's demand and quick decision-making by managers, standardize customer-centered workflow, improve the response speed and competitiveness of the organization, and enable new organization to have quick response capability, regulation capability, and updating capability, so as to make the service more precise.

5. Realize Internationalization of Services of Iron and Steel Enterprises

Iron and steel enterprises should have the vision of serving the world. For export products and services, iron and steel enterprises should raise the quality issues to the height of the country, insist that products and services can not go out of the plant if it can not be provided to the foreign country, and achieve the globalization of services. The basic requirements are the internationalization of service networks, service concepts, and service teams.

## *10.4.2 Path Analysis of Service-Oriented Trend*

1. Promote Iron and Steel Industry Chain to Extend Toward Both Upstream and Downstream

   (1) Extension direction of industrial chain

1) The construction of industrial parks is a new impetus to promote the extension of the industrial chain of iron and steel enterprises. The park-oriented industrial development of various places is gradually formed, and the supporting upstream and downstream industrial chains are built along with it, including various projects for deep processing industrial chain of iron and steel products. For example, the supporting projects such as plate warehousing, processing and distribution, laser tailor-welded blank project, standard part production project, and suspension spring production project are constructed for automobile industrial park; steel structure project, ship plate preprocessing project, plate coating project, and plate processing and distribution project are constructed in the logistics industrial park nearby plate manufacturing enterprises; various deep processing and product manufacturing projects of stainless steel plate and deep processing project of stainless steel long product are constructed in the industrial park nearby the stainless steel manufacturing enterprises.
2) Build industrial chain extension projects by multi-party cooperation. The huge market of steel deep processing industrial chain has become a hot spot for investment in various industries. Iron and steel enterprises, downstream enterprises, steel traders, logistics enterprises, and even IT enterprises are the main bodies for the construction of current industrial chain of iron and steel enterprises, which can form different types of joint ventures and cooperative enterprises to promote the benign development of the industrial chain of iron and steel enterprises. To build plants by upstream and downstream enterprises through joint investment is a common demand of both enterprises in resisting market risks, and can also establish a closer strategic partnership between the iron and steel enterprises and downstream enterprises. Through cooperation, the iron and steel industry provides a more satisfactory comprehensive solution for the downstream industrial and creates a strong upstream and downstream industrial chain, so as to jointly enhance market competitiveness.
3) Actively participate in the construction of engineering projects and expand the popularity of the enterprise's products. With the acceleration of domestic social and economic development process, the demand for infrastructure and industrial products is increasing day by day, and a number of major engineering projects are constructed and put into operation one after another. These include projects with huge demand of steel, such as the Beijing National Stadium, the Three Gorges Project, high-speed railway projects, construction of petrochemical bases, and construction of nuclear power plants. These projects require large quantities of specific steel varieties with high quality. Therefore, iron and steel enterprises can aim at the steel consumption demand from these key projects and guarantee the supply of raw materials for major national

construction projects through signing contracts with the project contractors, while obtaining opportunities to have their products applied and to gain profits.

(2) Prospects for extension of the industrial chain. Iron and steel industry in China has entered an era of high cost and low efficiency. With the high cost of raw materials such as iron ore and coking coal and the slowdown in the development of the iron and steel downstream industries, the competition in the iron and steel industry will become increasingly fierce, and the iron and steel industry will enter a critical period of structural adjustment. Extending the industrial chain and developing deep processing of steel products are important measures for iron and steel enterprises to respond to the severe operating situation of the iron and steel industry and gain benefit from the high value-added services. The domestic advanced iron and steel enterprises such as China Baowu, Angang Group, and HBIS all take iron and steel as their main business to extend toward the upstream and downstream industrial chains, so as to realize diversified development, improve the enterprises' core competitiveness and open up new profit growth points. In the future, with the continuous low profit margins of iron and steel enterprises and the increasing awareness of industrial chain construction, the construction speed of downstream industrial chain will be further accelerated and the competition will become increasingly fierce. For extension of industrial chain, the iron and steel enterprises should fully understand the market conditions of the surrounding areas of steel processing and distribution plants, as well as the status of deep processing of steel in surrounding areas, and combine the production situation and development planning of its own iron and steel products to well forge their product positioning and market orientation of deep-processed products.

2. Take Advantage of E-commerce for Transformation and Development

(1) Accelerate the integration of iron and steel e-commerce platform. Compared with the e-commerce of other bulk products, the iron and steel e-commerce is relatively mature. At present, large and mature leading enterprises of iron and steel e-commerce have emerged in this market. It is recommended to rely on the leading iron and steel e-commerce enterprises to give full play to the core competitiveness of iron and steel enterprises and third-party platforms and strengthen connectivity among platforms through developing multi-dimensional and multi-level cooperations to achieve overall upgrading of the industry. It is required to increase industrial concentration ratio in the field of iron and steel e-commerce by means of merger, acquisition, integration and withdrawal with the aim of establishing a leading industry-level iron and steel e-commerce platform. At the same time, it is necessary to give full play to the integration role of iron and steel e-commerce horizontally among iron and steel enterprises and vertically along industry chain, effectively guide the industry to improve production standards and reduce production

capacity, and promote the healthy and orderly development of e-commerce industry and iron and steel entity enterprises.

(2) Encourage specialized operation of iron and steel e-commerce platform. The orientation of the iron and steel e-commerce platform needs to be defined and transformed from inappropriately expanding to specialized operation. The development characteristics of the iron and steel industry are combined to explore a mature e-commerce profit model, which shall be transformed from simple "information fee" and "membership fee" to value-added services. It is also necessary to accelerate the specialized development of financial services, and warehousing and logistics services related to iron and steel production and sales, establish a support system for e-commerce technology, logistics, payment and credit, and form a win-win closed-loop network in terms of iron and steel production, trade, customers, warehousing, processing, logistics, banking, insurance, etc. Brand building of the iron and steel e-commerce platform shall be strengthened, which means to gradually improve the system construction, build an industry-level e-commerce platform, and provide specialized services by using the professional e-commerce team instead of abusing the "Internet +" concept to make money.

(3) Use the opportunity of iron and steel e-commerce to accelerate the transformation to service-oriented manufacturing. Iron and steel e-commerce not only helps to achieve the integration of various resources, but more importantly, to realize the connectivity of information to form big data accumulation, which will become the basis of service-oriented manufacturing. Iron and steel e-commerce can be taken as the breakthrough in service-oriented manufacturing, and the steel plants are encouraged to interact closely with the downstream customers, so as to enable different types of steel customers and non-ferrous materials customers to smoothly connect with the steel plants, promote steel plants to carry out large-scale customized production according to the demand, and accelerate the high-end development of the iron and steel industry chain. Iron and steel e-commerce can also be cooperated with the iron and steel logistics park to strengthen service functions to enhance the service value of products by building an O2O organizational structure that improves industry efficiency, reduce the intermediate participants between steel plants and customers, and promote the structural adjustment of the entire industry.

(4) Steadily and cautiously carry out cross-border iron and steel e-commerce business. According to customs data, the iron and steel export volume reached more than 100 million tons both in 2015 and 2016. It can be seen that export is becoming the "blue sea" for China's development of iron and steel e-commerce and also for capacity digestion. At present, some platforms have begun to engage with cross-border business. With the large overseas customer base, cross-border matchmaking trading has been launched. However at the same time, the cross-border e-commerce is also facing a series

of risks in aspects such as import and export trade financing, cross-border culture, process control, and anti-dumping.

3. Provide Good Logistics Services

At present, China's iron and steel logistics industry is still in an important development stage from extensive style to intensive style, and the market share of domestic iron and steel logistics is still very large. The development of the iron and steel industry is required to be closely followed for the adjustment of product mix and the transformation and upgrading of industrial development. It can be foreseen that the future development trend of iron and steel logistics industry is also inseparable from merger and acquisition, transformation and upgrading as well as improvement in industrial concentration so as to expand and broaden the development path of the enterprises, and transform them from traditional logistics enterprise to multi-functional and integrated comprehensive logistics service providers through a innovative management mode. While in this process, the large-scale iron and steel logistics enterprises undoubtedly have favorable advantages; in the process of industrial consolidation, the large-scale enterprises will possess a larger market share. To develop the logistics industry, the iron and steel enterprises should make full use of the Internet, and the "Internet of Things + Iron and Steel Logistics" will be an important development direction of the iron and steel logistics industry by using the modern Internet of Things technology.

4. Promote Major Customer Manager System

As a key part of service, marketing plays a vital role in the service system. The ultimate goal of marketing is to sell products to customers, and the marketing process itself is a series of services. The "Major Customer Manager System" promoted by HBIS Group achieves good results as an innovation in marketing mode, which is worth colleagues in the iron and steel industry to learn.

Starting with the functional orientation of the customer service center, the structure and management platform of the customer service center, the customer manager management system, the determination of the first batch of major customers and customer teams, the work objectives of the major customer service center, etc., and focusing on sales targets, the enterprise needs to carefully master and deal with the relationship between the major customer service manager system and the original sales model: first, on the basis of guaranteeing contract performance, to explore the advantages of various production lines and product value of the Group, determine the key development targets on the basis of the original direct sales, establish a connecting relationship with competitive direct supply enterprises, and improve on-site maintenance and service quality; second, to establish a customer relationship network in terms of the target of major customer development to further determine the development category of major customers; third, to implement the de-administration management of customer managers and business managers. The key customers and large enterprises in the region should be listed to determine the specific personnel responsible for new product development and key major customer development; besides, special personnel shall be arranged to follow up the development and carry out system services. The administrative level treatment should be eliminated to truly realize the market-based incentive mechanism, adjust and improve the marketing assessment mechanism and responsibility performance-linked assessment, and use the mechanism to generate impetus and stimulate vitality; fourth, to further control the overall market development and operation, strengthen communication with subsidiaries, make overall planning for the development programs of high-end major customers and famous brand enterprises in each production line of the Group, jointly determine development measures and schedules, and coordinate the overall market development of the Group.

HBIS actively promotes the transformation of marketing model by focusing on the market and products, takes comprehensive marketing services and variety structure adjustment as the main line of marketing, selects technical personnel with solid professional foundation, excellent moral character and strong development ability to strengthen the marketing team and to be appointed as major customer managers, and tries to provided customers with standardized and individualized one-to-one service with high quality. The major customer managers should actively contact with and develop high-end customers, strive to promote the upgrading of production lines with high-end customers and high-end product varieties, establish a good brand image of enterprise products in regional and international markets, and consolidate the cooperation with major customers to achieve a big leap in market sales. On the basis of elaborating the customer rating system, the Group provides major customer managers, customer managers and customer engineers to major customers, strategic core customers, and key project customers to quickly establish a three-level customer service system and build an integrated management network for marketing, production, and technical services of major customers. HBIS persistently improves the service system of major customer and drives the simultaneous operation of production, sales, and research by strengthening the coordination and unification between market management and customer service. At the same time, staying close to the customer's demands, HBIS establishes a 24-h problem response mechanism to provide pre-sales, in-sales, and after-sales services for the ordered products of major customers as well as coordination and supervision of all matters such as planning and scheduling, production process, process quality, delivery and transportation, product application, customer feedback, and issue solution and improvement, truly provide major customers with more value-added services than the product itself by first-rate services, and continuously improve marketing value-added capabilities and product profitability.

5. Promote "Customized" Services

In recent years, China's iron and steel industry has a serious overcapacity and fierce market competition. Iron and steel enterprises are carrying out corresponding individualized services according to the special requirements of different customers while improving product quality and reducing production costs. With the changes in the relation between supply and demand and the market, customers put forward higher requirements on steel products, and the service of iron and steel enterprises is changed from "individualization" to "customization". China Baowu and HBIS are in the forefront of the industry in this respect. They have participated in the design of the customer's products at the beginning, and jointly develop with the customer the steel grades to meet the demands and provide a series of complete services.

Customized service is a diversified, three-dimensional marketing model based on quality, taking reputation as guarantee and brand as support. The Groups actively carry out market service work. By conducting market promotion through door-to-door visits, telephone communication, etc., they establish a smooth contact mode with customers, timely understand the potential varieties needed by customers, and strive for more orders. In response to the individualized demands of customers, the Groups integrate the product research and development, technical innovation, and product sales together in the form of technology centers, steel grade research teams, research and sales teams, etc. to form an innovation system with research and development meeting and being promoted by the market demands, so as to develop customized mode; according to the new material, specification and size of products, and with the combination of equipment and technical level, they preliminarily judge, evaluate, develop, and design process route, establish important operation points of steelmaking and rolling, carry out necessary process and equipment upgrading and revamping, develop corresponding production plans, cooperate with the whole process of steelmaking and rolling, coordinate and organize the production, track the production throughout the process, and continuously make improvement and summarization. The market service concept is strengthened to enhance the combination of production and sales, detail the order to production, inspection and warehousing, shipping, and other aspects of steelmaking and rolling to guarantee the delivery period of the customer, and continuously improve the dependence and loyalty of the customers.

## 10.5 Industrial Practices of Servitization

With the continuous improvement of the demand of downstream customers, the iron and steel industry is gradually transforming from a manufacturing center to an integrated service center; the iron and steel enterprises have an urgent need to improve their service levels. As the leading consulting organization in iron and steel industry, the China Metallurgical Industry Planning and Research Institute (hereinafter referred to as MPI) has provided many supports in transformation of the enterprises from production-oriented mode to service-oriented mode. It has been dedicated to Wuhan Steel, Baotou Steel, HBIS, Shandong Steel, Zhongte, NISCO, Hebei Jingye, and other domestic large and medium-sized iron and steel enterprises successively to develop industrial chain extension and development planning, marketing and trade planning, iron and steel e-commerce development planning, logistics park development planning and other special plans and measures, and give advice and suggestions to promote the transformation, upgrading and scientific development of iron and steel industry and iron and steel enterprises.

**Table 10.2** Practice of MPI in promoting servitization of the industry

| No. | Type | Main content | Typical case |
|---|---|---|---|
| 1 | Planning for deep processing of steel products and extension of industrial chain | It mainly includes extension and processing solution and industrial extension planning for various steel grades such as medium and heavy plate, hot-rolled coil, cold-rolled sheet, rebar, wire rod, and bar | *Research Report on Industrial Development for Steel Deep Processing of NISCO*, *Industrial Development Plan for Deep Processing of Baotou Steel*, *The 13th Five-Year Development Plan for Deep Processing Segment of Hebei Jingye Group* |
| 2 | Development plan of logistics park | The plan includes functional positioning of the logistics park, steel storage, transportation, steel product processing and distribution, etc. | *Development Plan for Service Industry Cluster of Jiulong Iron and Steel Logistics*, *Feasibility Study on Steel Trade City Project of Minmetals at Wuxi West Railway Station*, *Feasibility Study on Lanzhou Iron and Steel Logistics Park Project of Minmetals*, *Pre-feasibility Study Report of Tangshan Logistics Park* |
| 3 | Development plan of e-commerce platform | The development status and trend of e-commerce platform, functional positioning, target and operation mode of enterprise e-commerce platform, etc. | *Development Program for E-commerce Platform of NISCO* |
| 4 | Steel marketing planning | Transform marketing service concept, improve service level, optimize marketing model, adopt EVI mode for key customers, develop customized services, etc. | *The 13th Five-Year Plan of HBIS Group*, *The 13th Five-Year Plan for Development of Iron and Steel Industry of Shandong Steel Group*, *The 13th Five-Year Plan for Development of Hebei Jingye Group*, *The 13th Five-Year Plan for Dongte Group*, etc. all contain chapters for special functional planning of marketing |

The practice of the Metallurgical Industry Planning and Research Institute in promoting servitization of the industry is detailed in Table 10.2.

## References

1. Guo, H.Y. (2013). Servitization of manufacturing industry: International trends and inspiration. Globalization, 9, 100–108.
2. Zheng, X.L., & Li, Y. (2011). Discussion on the transformation of iron and steel enterprises to service-oriented enterprises. Metallurgical Economy and Management, 6, 27–29.

3. Wang, Y.D., Tang, D., Sun, J.Q, et al. (2014). Servitization transformation and industrial chain extension of iron and steel enterprises. China Iron and Steel Industry, 31(4), 6–9.
4. Nian, M., & Liu, F. (2014). How to treat the servitization in manufacturing industry. China Opening Herald, 5, 67–70.
5. Zhou, D.P. (2013). Research on evolution mechanism and development trend of servitization in manufacturing industry. Commercial Research, 55(1), 12–21.
6. Zhou, H.Q., & Huo, J.D. (2014). Global servitization level of manufacturing industry and its influencing factors—An empirical analysis based on international input-output data. Economic Management, 1(517), 1–10.

# Chapter 11
# Intelligentization

Accelerating the development of smart manufacturing is the only way to cultivate new drives for China's economic growth and a strategic choice to seize the commanding heights of the future economic and scientific and technological development. It is of great strategic significance to promote the supply-side structural reform of China's manufacturing industry, create a new competitive advantage for China's manufacturing industry, and build China into a manufacturing power [1]. Iron ore and other raw materials used in China's iron and steel industry are controlled by others and subjected to great pressure in terms of energy and environmental protection and others. Providing competitive products at a lower cost to the market is the only choice; therefore, the profit rate should be regarded as a key index to measure the comprehensive strength of China's iron and steel industry. The smart manufacturing of iron and steel can realize the integration of large-scale production and customized manufacturing by improving the high-efficiency research and development capability of new steel grades, production capacity with stable product quality, flexible production organization capacity, and comprehensive control ability of energy efficiency in the steel manufacturing process, and promote the comprehensive competitiveness of steel enterprises which take profit margin as the key indicator, so as to make a leap from a big country to a strong country of iron and steel.

## 11.1 History Review and Status Analysis

The initial form of intelligence of iron and steel enterprise is informatization. At this stage, it is not really intelligent in the true sense. However, as the leading form of intelligence, it is an indispensable basic stage of intelligent development.

Its application in international advanced iron and steel enterprises is dated back to the early 1970s. For the purpose of reducing costs and improving efficiency, the vertical integration from production site to management decision-making is constantly improved, and the digitalized factory with business collaboration as the core fully supports the survival and development of enterprises. The intelligence building of



X. C. Li, *The Road Map of China's Steel Industry*,
https://doi.org/10.1007/978-981-15-2074-7_11

China's steel industry is 10–15 years later than that of developed countries. However, compared with other domestic industries, the overall level of intelligentization and informatization of iron and steel enterprises is at an advanced level and leading position among various domestic industrial sectors.

### *11.1.1 History Review*

1. Development Process of Informatization of Foreign Iron and Steel Enterprises

The development process of informatization of foreign iron and steel enterprise is mainly divided into three stages. In the early stage, a centralized information system based on mainframe was built. In the 1980s and 1990s, its application functions were expanded, the new application function was expanded and developed based on the core information management system, many distributed information systems were added, and the information system became an indispensable pillar of enterprise operation. In the late 1990s, due to the rise of e-commerce, the direction of information construction was adjusted according to changes in the industrial environment, which was mainly embodied in strengthening the construction of e-commerce and expanding the application of information system to the application between customers and suppliers, and even the application integration between peers. The digital factory with business collaboration as the core comprehensively supports the survival and development of enterprises.

(1) Integration of Production Management
In the 1960s and 1970s, computer technology began to be applied in the iron and steel industry, which marked the beginning of informatization of iron and steel enterprises. Subsequently, the USA, Britain, and Japan successively launched the information construction of iron and steel enterprises, such as the hot rolling informatization of American McLouth, the computer hierarchical management system of British RTB Steel, the famous production management integrated system of Japan's Sumitomo Metal Industry, and the advanced production management computer system of Japan's Nippon Steel. The development of the informatization of production integrated system in the iron and steel industry is becoming more and more mature, the production transformation of iron and steel enterprises is gradually realized, and the intelligentization is also beginning to emerge in the iron steel industry.

(2) Integration of Operation Management
With the widespread application of computer and information technology, in the 1980s, the iron and steel enterprises continued to expand their production scope, and with the help of information technology, more emphasis was placed on the integration of operation management. The information-based system of iron and steel enterprises is no longer the traditional integrated system of production management, but to expand the scope, gradually covering the original

information system of iron and steel enterprises, and building enterprise ERP system that integrates market, operation, production, sales, and logistics. The system shortens the production cycle of steel products, improves the completion rate of steel delivery, and realizes the integration of operation and management. For example, Sumitomo Metal Industry in Japan has established a management system of the whole company integrated with sales, production, and logistics, which connects the head office, sales enterprises, and production plants together for unified planning and management.

(3) Integration of Strategic Management
With the development of market economy and socialized large-scale production, iron and steel enterprises must seek for their survival and development from a long-term perspective. At the same time, the widespread application of Internet technologies and the rapid development of informatization of iron and steel enterprises prompt the vertical integration and horizontal interconnection between the information systems of iron and steel enterprises, and promote the development of strategic management system and intelligent business. In the 1990s, the international advanced iron and steel enterprises were committed to the development of strategic management system, focusing on the integration of strategic management, such as South Korea's Pohang Steel Company and Japan's Sumitomo Metal Industry. The intelligentization rudiment of the iron and steel industry began to emerge.

2. Development Process of Informatization of Chinese Iron and Steel Enterprises

The informatization development of the iron and steel industry can be roughly divided into three stages. Before 2000, it was an early exploration stage, mainly the initial stage for informatization of iron and steel enterprises. From 2000 to 2010, it was a rapid development stage, and the informatization of iron and steel enterprises turned to integration. Since 2010, it is in a mature transformation stage, and the informatization construction of iron and steel enterprises has gradually matured and transformed into intelligentization.

(1) Departmental informatization: infrastructure construction and environment construction, and application of information technology to all key business links. In the 1980s and early 1990s, it was an attempt and exploration stage. The information system was used for electronic accounting based on the original manual workflow to realize electronization of functions. In the late 1990s, by recognizing the importance of management informatization, some iron and steel enterprises (China Baowu, Ma'anshan Steel, etc.) intensified their investment in production automation, began to consider the business needs such as system integration, production, sales, and financial management, and paid attention to the payback of project investment for information system, i.e., the establishment of application functions. A few leaders began to introduce external consultation and technology transfer, and build the information system that integrates production and sales.

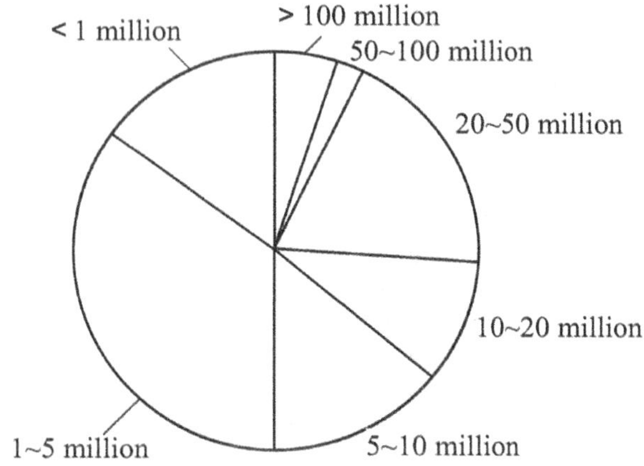


**Fig. 11.1** Proportion of investment in informatization construction of Chinese iron and steel enterprises from 1996 to 2000

According to the statistics of the China Iron and Steel Association, from 1996 to 2000, iron and steel enterprises with the investment in informatization construction reaching 100 million yuan and above accounted for 5%, such as Baotou Steel, Zhuhai Steel, and Baosteel, the enterprises with the investment reaching 50 million to 100 million yuan accounted for 2%, those reaching 20 million to 50 million yuan accounted for 10%, those reaching 10 million to 20 million yuan accounted for 10%, those reaching 5 million to 10 million yuan accounted for 14%, those reaching 1 million to 5 million accounted for 35%, and those with investment less than 1 million yuan accounted for 15%, as shown in Fig. 11.1.

The key task of the stage in the late 1990s was the combination of information technology and business processes to apply computer technology in key business links and improve the production and management efficiency of iron and steel enterprises by establishing a single application system and gradual comprehensive coverage. The development stage of the intelligentization construction of iron and steel enterprises is mainly to cut into the integrated management and control system, the computer-integrated manufacturing system (CIMS), and the enterprise resource planning (ERP), manage the expanded scope, and develop the original enterprise information-based system covering production to an enterprise resource management system that integrates market, operation, development, and distribution.

(2) Enterprise informatization: integration and improvement—develop to the direction of integrated application and realize reformation and innovation of business process and production operation mode. In the twenty-first century, the direction of informatization construction of iron and steel enterprises is adjusted with the changes in the environment of the iron and steel industry. The construction of steel e-commerce and the application of iron and steel information system are expanded to applications between customers and suppliers, and even application integration between peers. Iron and steel enterprises began to improve

the vertical integration from production site to management decision-making; new application functions based on the core information management system are expanded and developed; many distributed information systems are added; enterprises' investment in information system construction is increased, organization of information department is expanded, and personnel is increased; information management system becomes an indispensable pillar of enterprise operation.

The function expansion stage focuses on horizontal and vertical deep integration based on the production and sales system, to speed up the implementation of customer-oriented strategy inside the enterprises, promote the integration of business process and systems, further improve some functional modules in the ERP system and MES system, complete the building of equipment state management, automatic warehouse management, production planning and other functions and the transformation from integration of production management to integration of strategic management, and realize the integration of production processes, operation management, and strategic management of enterprises.

(3) Deep informatization: smart manufacturing—intelligent decision-making, innovation, and breakthrough, to realize collaboration and integration between key business links and application systems, and intelligent decision-making support. Deepening and upgrading are stages of innovation and breakthrough. Through the collaboration and integration between key business links and application systems, smart manufacturing and intelligent decision-making can be realized. Since 2010, with the policies such as smart manufacturing, integration of informatization and industrialization, and "Internet+" have been put forward, and iron and steel enterprises deeply integrate the characteristics of process industry with the modern Internet information technology and strive to build flexible production of iron and steel enterprises with the help of intelligent and customized smart manufacturing technology. The enterprise's integrated information system gradually began to transform into a smart iron and steel manufacturing system with the main goal of achieving scale manufacturing and customized production. At this stage, iron and steel enterprises no longer put scale benefit as the first priority, but pay more attention to quality and quality benefit and transformation and upgrading from the extensive operation to the intensive operation.

In 2012, the Ministry of Industry and Information Technology selected 218 enterprises as national demonstration enterprises for deep integration of informatization and industrialization from 27 industries, including iron and steel, nonferrous metals, and other industries, and ten iron and steel enterprises such as Baosteel and Jinan Steel in the iron and steel industry were successfully selected [2]. The important characteristics of the deepening and upgrading stage are manufacturing intelligence and business intelligence, which are manifested in the intelligent production and intelligent management of iron and steel enterprises, so as to create an intelligentization route that integrates management and production.

## 11.1.2 Status Analysis

1. Current Situation of Smart Manufacturing in Foreign Iron and Steel Industry

The informatization of the foreign iron and steel industry had developed earlier and is now becoming mature gradually. The investment priority of enterprises has shifted to information technology equipment, and the construction of information infrastructure has begun to take shape. More and more iron and steel enterprises have realized the scientific decision-making of production and operation through enterprise resource planning (ERP), supply chain management (SCM), business intelligence (BI), and other management systems.

At present, the international advanced iron and steel enterprises have completed the vertical integration from the production site to the management decision-making and added many distributed information systems. The direction of informatization construction is adjusted with the changes of the industrial environment, and the construction of e-commerce and the application of information system are expanded to application between customers and suppliers, and even application integration between peers.

The following is the informatization construction situation of the world-famous metallurgical enterprises.

(1) ThyssenKrupp Company of Germany is one of the world-famous iron and steel producers. In 2000, its North American branch implemented financial/management accounting, fixed asset management, materials management, production management, quality management, and sales/distribution management systems, integrating all data from the supply chain to finance.
(2) POSCO is one of the largest iron and steel enterprises in the world. Since 1974, it begun to build a management information system of the enterprise. In 1999, with the continuous expansion of equipment and the demand for market analysis, the company rebuilt a customer-centered marketing system and the ERP system, including the entire process from the comprehensive sales plan to the processing of orders, production and manufacturing management, delivery process management and after-sales management, which shortens the delivery time and improves customer satisfaction.
(3) US Steel and China Steel Corporation, have built a new generation of information system for iron and steel enterprises. The system takes the order (contract) processing as the lead, the production and sales as the main line, and the finance as the core, with the quality control throughout the whole line, so as to realize the dynamic synchronization of logistics, capital flow, and information flow. By using that system, the production is determined as per sales, the materials are determined as per production, and the allocation of resources is optimized, so that the inventory is reduced, the delivery cycle is shortened, the customer service is improved, the cost is reduced, and the operation management of the enterprise is significantly improved.

The informatization of the iron and steel industry was begun in the Britain and America, but carried forward in Japan and South Korea. The representative enterprises are Nippon Steel and POSCO. POSCO and Nippon Steel's successful informatization construction and customer-oriented manufacturing system demonstrate the rudiment and possibilities of intelligent production, and their informatization path is basically the same—first vertically integrating the manufacturing process, then horizontally integrating the supply chain information, and finally achieving cost reduction and efficiency increase. In order to adapt to the ever-changing market environment and improve customer satisfaction, the iron and steel industry in foreign developed countries has basically realized the informatization of the iron and steel manufacturing industry, and is now developing toward a highly intelligentization and networked direction.

2. Current Situation of Smart Manufacturing in China's Iron and Steel Industry

Under the policy of the integration of informatization and industrialization, during the 12th Five-Year Plan period, the application of information technology in manufacturing, enterprise management, logistics distribution, product sales, etc., has been continuously deepened, with the numerical control rate of key processes exceeding 65% and equipment rate of enterprise resource planning (ERP) exceeding 70% [3]. With the advancement of the standard certification system for integration of informatization and industrialization, China's iron and steel enterprises with advanced equipment and high equipment automation level, including several large-sized steel plants and some medium-sized steel plants which are developing rapidly, have basically realized the automatic control, processing, and data collection of the production process; the iron and steel enterprises with complete set of automatic production line and high production management level represented by China Baowu have established a relatively complete basic control and management information system, which can achieve enterprise-level integrated management of production, supply and marketing information, automatic control of the whole production process, and data collection.

(1) Current Development Situation of Smart Manufacturing in the Iron and Steel Enterprises

1) Process Control and Production Execution

At present, for the system technology of process control automation for smelting, except for the system platform of the large-sized blast furnaces, converters still depend on that of the foreign countries, and other equipment basically depends on procurement of the related systems or self-development of application software. Due to the huge system and strict requirements on control and communication speed, most enterprises introduce foreign automation systems, and some enterprises adopt independently designed and integrated process computer control system, as well as develop control model and various control functions for rolling process especially for hot and cold strip rolling.

Manufacturing execution system (MES) is an important component of the information-based structure of iron and steel enterprises, and a bridge connecting enterprise's operation management and production control. Statistics show that the proportion of enterprises that have built a workshop-level manufacturing execution management system for all or part of the production lines is about 82%, the proportion of enterprises that have realized the automatic collection of all or part of the pound scale measurement data is 87%, and the proportion of enterprises that have achieved automatic collection of all or part of the testing data is 77%. More than half of the enterprises have built energy management systems, and most of them have built environmental protection monitoring and management systems, but their support scope for business is relatively weak, and the functional scope only stays on data collection, lacking effective support in terms of analysis, optimization, execution, and intelligent model.

2) Operation Management
In terms of operation management, the internal supply chain management information system of most enterprises has been basically completed, and most enterprises have established a unified financial management system of the company. The coverage of financial management businesses such as general ledger, fixed assets, receivables, and payables is relatively good, but the informatization of enterprise budget management is basically absent. The informatization of procurement management, company-level production management, company-level quality management, sales management, collaborative office management, human resource management, comprehensive statistics, and data analysis has basically realized, while the informatization of e-commerce and engineering project management still needs to be strengthened.

Large and medium-sized key iron and steel enterprises have basically achieved a seamless connection between business and finance, but the connection between the financial system and the peripheral business systems must be further improved, such as the close connection with procurement system, sales system, logistics system, equipment system, and engineering project system.

3) System Integration
At present, some key large and medium-sized iron and steel enterprises have achieved control connection. Enterprises that have realized automatic data interaction between manufacturing execution system and enterprise resource planning system and process control system account for less than 50% of the enterprises included in the statistics. Among them, the proportion of enterprises that have achieved "two-way integration" is only 20%, and the degree of integration is low, indicating that iron and steel enterprises are still at a shallow level of application in connection with management and control.

At present, 13% of enterprises have realized the overall integration of the three systems of R&D, manufacturing and service circulation, of which the enterprises with integration of R&D and manufacturing account for 30%.

This shows that in the integration of three major systems, the enterprises are currently focusing on the integration of R&D and manufacturing.

At present, the humanized interaction of information system has been valued by enterprises. Nearly half of the enterprise portals have been integrated with major or all systems, but further efforts should be made to support the control ability of information system. In the current situation that iron and steel enterprises have made great progress in construction of hardware facilities for informatization and made initial progress in software development, only about 50% of enterprises can obtain information and issue instructions in time, while only 1/3 of those instructions can effectively executed.

4) Supply Chain Collaboration
Supply chain integration is mainly reflected in the fact that the enterprises further establish the comprehensive integration with customers and suppliers through the construction of e-commerce platform. The collaboration between enterprises and customers is relatively missing. Forty-four percentage of the enterprises have carried out collaborative informatization work of customers, mainly in ordering, production planning, and settlement and logistics distribution. It can be said that customer collaboration applications are still few, but have been greatly developed. However, there are fewer applications of supplier collaboration by enterprises, and as many as 65% of enterprises have not implemented supplier collaboration.

5) Product Life Cycle Management
The life cycle management and control of steel products are mainly reflected in the research and development of new products, the production and manufacturing of products, and the customer service of products. At present, most enterprises have adopted unified coding in the information construction system, and a good job is done in unified coding of the system. In the construction of the three major systems of R&D system, production system, and circulation system, the construction of unified production system is in a good condition. Among them, because the quality monitoring system and production, and supply and sales system of the whole plant can bring direct management benefits to the enterprises, the proportion of its construction is more than or nearly 50%. Unified customer service standards and the construction of a unified circulation service system have also been recognized by about one-third of enterprises, indicating that the enterprises constantly attach importance to customer service.

It can be said that China's iron and steel industry has achieved remarkable results in the process of industrialization and informatization, which is closely related to the introduction of China's policy on the integration of informatization and industrialization. The introduction of the national policy on integration of informatization and industrialization has greatly promoted the improvement of the integration level of iron and steel enterprises. The smart manufacturing is the advanced stage of the integration of informatization and industrialization. The improvement of the integration level of informatization and industrialization has laid a solid foundation for the development of smart manufacturing in the iron and steel industry.

In general, according to the application of informatization and automation, China's iron and steel enterprises are mainly divided into three categories: (1) Iron and steel enterprises with low level of equipment automation, long operating time, and weak capital basically do not have the conditions to implement large-scale information systems and cannot reduce costs, improve labor productivity and ensure product quality by means of automation; (2) iron and steel enterprises with advanced equipment and high level of equipment automation, including several large steel plants and some medium-sized steel plants with rapid development, represented by CITIC Pacific, have basically realized the automatic control, processing, and data collection of the production process, and realized the wide application of information technology; (3) iron and steel enterprise with complete automatic production line and high production management level is represented by Baowu Group. These enterprises have a high level of management and production automation. They have established a relatively complete basic control and management information system, which can achieve enterprise-level integrated management of production, supply and sales information, as well as automatic control of the entire production process and data collection and processing.

(2) Problems in Smart Manufacturing of Iron and Steel Industry

1) The standards for the industry's smart manufacturing are missing, and the phenomenon of disorderly market development is beginning to appear. The establishment of standards is a necessary prerequisite for smart manufacturing to achieve interconnection and information integration. The publication of the *Guidance on the Construction of the National Smart Manufacturing Standards System* marks the completion of the construction of China's smart manufacturing top-level framework, but the industry-based smart manufacturing standards have not yet been carried out in an orderly manner. The iron and steel industry is one of the industries with the highest degree of automation and the longest manufacturing process in China. The key technologies such as industrial robots and Internet of Things have been applied to different extents, but there is no clear path in the development of smart manufacturing for the future. Enterprises at home and abroad implementing the system have poor product compatibility and integration due to differences in understanding of smart manufacturing and the need to build product competition barriers. In the process of intelligent transformation, enterprises often face the risk of resource waste caused by the failure of the system integration. The lack of standards brings great challenges to the interconnectivity of smart manufacturing, and it increases the difficulty of demonstration and promotion of smart manufacturing in the industry.
2) The proportion of investment in smart manufacturing is too large, and the enterprise lacks motivation for transformation and upgrading. The iron and

steel industry belongs to the long process and asset-intensive industry. In the process of intelligent transformation, it involves the technological innovation of numerous equipments, the layout transformation of the Industrial Internet of an entire plant, the integration of process control systems, the interconnectivity of industrial software systems such as MES and ERP, the application of the optimization model and the simulation technology, etc., so the overall investment cost is too much. In terms of investment ratio, since most of the basic equipment is imported from SMS Group, SVAI Group Company, Danieli Group, and other foreign manufacturers, it is necessary to introduce foreign systems in some key intelligent technologies, which further increases the investment cost. In terms of enterprises' investment willingness, state-owned iron and steel enterprises have better infrastructure and high enthusiasm for intelligent transformation; small and medium-sized iron and steel enterprises, especially small and medium-sized private enterprises, have poor infrastructure and high proportion of manual operations, and thus they are under great financial pressure and lacking in motivation for intelligent transformation and upgrading under the background of cutting steel overcapacity and environmental protection supervision.

3) Insufficient control of core intellectual property rights and the proportion of original innovation applications are low. *The Smart Manufacturing Development Plan* (2016–2020) proposes that by 2020, a relatively complete smart manufacturing technology innovation system will be established, breakthroughs for a number of key generic technologies will be made, some technologies will reach the international advanced level, and the core support software market will meet the satisfaction rate of over 30%. China's iron and steel industry is still weak in innovation in terms of the development, management, and integration of information systems and physical systems. The comprehensive integration of product production process design and intelligent management decision-making support system and the extension of business system to the front of the industry chain lack mature industrial solutions. At present, most software and integration technologies in China are at the end of the industrial value chain, and the technical level, labor productivity, industrial value-added rate, and product added value are relatively low. In the aspect of research and development, an innovative R&D system based on industry-university-research-application has not yet been formed. The initiative of original innovation and research and development is not high, and policy support needs to be strengthened.
4) The smart manufacturing system is not well understood, and the status of industrial software needs to be strengthened. Low-end robots have low barriers to entry, and the risk of overcapacity increases. With the rise of smart manufacturing, the robot industry develops rapidly. However, the iron and steel industry belongs to a long process industry. In the introduction of robots, more consideration is given to the matching of production rhythm and the seamless connection with upstream and downstream systems. It is difficult to meet the needs of iron and steel production process by simply

transplanting robots from other industries. Some low-end robots without the attributes of the iron and steel industry will soon face market saturation and overcapacity.

The difficulty of smart manufacturing is modeling, and the focus is simulation. Currently in the "new four bases" of the iron and steel industry, namely "one hardware" (automatic control and sensing), "one software" (industrial core hardware and software), "one network" (Industrial Internet), and "one platform" (industrial cloud and intelligent service platform), the investment proportion of hardware equipment is larger, the proportion of automated robots in China Baowu, Shagang, and Rizhao Steel is significantly improved, but the proportion of investment in industrial software is relatively low and the status that only paying attention to intelligent hardware construction and ignoring intelligent integration is an outstanding problem.

(3) Evaluation of Reproducibility and Promotion of Smart Manufacturing in China's Iron and Steel Industry

1) The technical level is becoming more mature, and the barriers to entry are greatly reduced. At present, emerging technologies such as the Internet, big data, cloud computing, and Internet of Things are widely used in various industries, and barriers to entry of technologies have been greatly reduced. The iron and steel industry has a high level of automation and informatization. It has many years of industrial accumulation and implementation enterprises in the system integration and implementation routes. Therefore, it is relatively easy to copy and promote in technology.
2) The background of system implementation enterprises is complex, and the third-party implementation organization has obvious advantages. The demonstration and promotion of smart manufacturing will inevitably drive the promotion of advanced management concepts and technologies. The information system providers having the background of iron and steel enterprises, represented by Baosight and Zhongguan, have implemented system solutions such as ERP and MES for many iron and steel enterprises. However, more and more iron and steel enterprises, especially those with market competitive relations, concern about the systems implemented by them: The system technology is not shared, whether the management concepts are advanced, whether the system design meets the development strategies of the enterprises, etc. Third-party implementation agencies, such as China Metallurgical Industry Planning and Research Institute, are receiving more and more attention from enterprises. Due to their status properties such as fairness, independence, and industry authority, they consider more about the strategic objectives, output of advanced concepts, and sharing of advanced systems in the implementation process of the system. Therefore, they are more likely to promote the reproduction and advancement of smart manufacturing.

3) The return on investment is difficult to be quantified, and the development momentum of private enterprises is insufficient. The one-time investment in smart iron and steel manufacturing is large, the performance index is difficult to be quantified and mainly be presented by the improvement of service level and management level, and it can be obtained through statistical analysis after a longtime running. China Baowu has invested 400 million yuan in its 1580 intelligent workshop, the state also provided financial support, which equates smart manufacturing with huge investment in the industry and seriously affects the promotion of pilot demonstration projects, and this is especially true for most private enterprises and small and medium-sized enterprises. Private enterprises have a stricter awareness of cost control and are more cautious about one-time large-scale investment projects. Therefore, for the smart manufacturing of private enterprises, it is necessary to strengthen policy guidance and support, and the task of reproduction and promotion is difficult.

## 11.2 Development Environment and Policy Orientation

As the main direction of *Made in China 2025*, relevant national policies on smart manufacturing have been intensively introduced. Demonstration projects for smart manufacturing have been selected, and the smart manufacturing is also highlighted in the proposal of the 13th Five-Year Plan. President Xi Jinping said that China has incorporated robots and smart manufacturing into the priority areas of national science and technology innovation, while Miao Wei, Minister of the Ministry of Industry and Information Technology, said that smart manufacturing should be the main direction of *Made in China 2025*.

### *11.2.1 Policy Environment*

In April 2011, the Ministry of Industry and Information Technology, the Ministry of Science and Technology, the Ministry of Finance, the Ministry of Commerce, and the State-Owned Assets Supervision and Administration Commission issued the *Opinions on Accelerating the Deep Integration of Informatization and Industrialization* (No.160 [2011] by MIIT and CFIE) which puts forward taking science development as the theme and accelerating the transformation of economic development mode as the main line, insisting that the industrialization is driven by informatization and industrialization promotes informatization, focusing on transforming and upgrading of traditional industries, trying to promote the integrated application of information technology in manufacturing industry, making efforts to promote the development of productive service industry by using information technology, striving to improve the supporting ability of the information industry to the integration and development, accelerating the pace of new industrialization, and promoting the overall optimization and upgrading of industrial structure.

In August 2013, the Ministry of Industry and Information Technology issued the *Special Action Plan for Deep Integration of Informatization and Industrialization (2013–2018)*, which stated that it would carry out 8 major actions to promote the deep integration of informatization and industrialization, namely standard construction and promotion action of "Management System for Integration of Informatization and Industrialization of Enterprises"; ability improvement action for integration of informatization and industrialization of middle and small-sized enterprises; innovation action for integration of e-commerce and logistics informatization; intelligent-level improvement action for key areas; cultivate action for smart manufacturing production mode; innovation action for integration of Internet and industry; and ability improvement action for supporting service of information industry support.

In May 2015, the State Council issued the *Made in China 2025* and sent a notice to deploy and comprehensively promote the implementation of the strategy for becoming a powerful manufacturing country. Industrialization and informatization are the main lines of the *Made in China 2025*, and smart manufacturing is the main direction of the *Made in China 2025*. To accelerate the integration of the new generation of information technology and manufacturing technology, it is necessary to take smart manufacturing as the main direction of deep integration of informatization and industrialization, comprehensively improve the intelligence level of enterprise's R&D, production, management, and service, and promote the transformation of China's manufacturing mode from "Made in China" to "Smart Manufacturing in China" [4].

In October 2015, the Fifth Plenary Session of the 18th Central Committee of the Communist Party of China adopted the *Proposal of the Central Committee of the Communist Party of China on Formulating the 13th Five-Year Plan for National Economic and Social Development* (hereinafter referred to as the Proposal), which was officially released in November. The Proposal clarifies: implementing smart manufacturing engineering, constructing a new manufacturing system, promoting a new generation of information and communication technology, and expanding high-end CNC machine tools and robots, aerospace equipment, marine engineering equipment and high-tech ships, advanced railway transportation equipment, energy-saving and new energy vehicles, electric power equipment, agricultural machinery and equipment, new materials, biological medicine, high-performance medical equipment, and other industries.

In December 2015, the Web site of Ministry of Industry and Information Technology issued a notice on the implementation of the *Action Plan (2015–2018)* of the *Guidance of the State Council on Actively Advancing the "Internet+" Action*. The action plan takes acceleration of the deep integration between the next generation of information and communication technology and industry as the main line, focuses on implementation of "Internet+" manufacturing industry and small and micro-enterprises, and is supported by high-speed broadband network infrastructure and information technology industry, in order to detail the construction of the ecological system in an all-round way in terms of technical route, industrial mode, policy guarantee, etc., and strive to create new advantages in industrial competition under the new situation.

On May 20, 2016, the State Council issued the *Guidance on Deepening the Integrative Development of Manufacturing Industry and Internet* (No.28 [2016] of the State Council), which deploys and deepens the integrative development of manufacturing industry and Internet, jointly promotes the "Made in China 2025" and "Internet+" actions, and speeds up the construction of a powerful manufacturing country. The Proposal pointed out that major projects such as smart manufacturing shall be carried out in the whole process of production and manufacturing, the whole industry chain, the life cycle of products, to support enterprises to deepen the integration of quality management and the Internet, promote the quality control of the whole industry chain such as online measurement and online testing, and vigorously develop new production modes such as networked collaborative manufacturing.

On December 7, 2016, the Ministry of Industry and Information Technology officially announced the *13th Five-Year Plan for smart manufacturing in China* (hereinafter referred to as "Plan"), which identified two major time nodes and ten important tasks for the intelligent transformation of China's manufacturing industry. According to the *Plan*, by 2025, China will promote smart manufacturing and implement a two-step strategy: The first step is that by 2020, the development foundation and supporting capacity of smart manufacturing will be significantly enhanced, the key fields of traditional manufacturing will basically achieve digital manufacturing, and the intelligent transformation of key industries with mature capabilities and foundation will make significant progress; the second step is that by 2025, the smart manufacturing support system will be basically established, and the key industries will initially realize intelligent transformation.

In January 2017, with the approval of the State Council, the Ministry of Industry and Information Technology and the National Development and Reform Commission officially issued the *Guidelines for the Development of the Information Industry* (No.453 [2016] of MIIT and CFIE, hereinafter referred to as the Guidelines). The Guidelines proposed to enhance systematic innovation capabilities, build collaboratively optimized industrial structure, promote the deep integration and application of information technology, construct a new generation of information infrastructure, improve the management level of information and communication and radio industry, strengthen the guarantee capability of information industry, enhance the 7 tasks of development capacity for internationalization, and determine the development priorities in 9 fields, i.e., integrated circuits, basic electronics, basic software and industrial software, key application software and industry solutions, intelligent hardware and application electronics, computers and communication equipment, big data, cloud computing, and Internet of Things.

On March 5, 2017, Premier Li Keqiang pointed out in the work priorities of 2017 that it should deeply implement the *Made in China 2025*, accelerate the application of big data, cloud computing, and Internet of Things, promote the transformation of production, management, and marketing modes of traditional industries by new technologies, new forms, and new modes, take development of smart manufacturing as the main direction, and promote China's manufacturing to the mid-to-high end.

The *Made in China 2025* takes promotion of innovation and development in manufacturing industry as the theme, improvement of quality and efficiency as the center, acceleration of the integration between new generation of information technology and manufacturing industry as the main line, promotion of smart manufacturing as the main direction, and satisfaction of demands of economic and social development and national defense construction on major technical equipment as the goal, so as to strengthen industrial basic capabilities, improve the level of comprehensive integration, perfect the multi-level talent system, promote industrial transformation and upgrading, and realize the historical leap of manufacturing industry from being big to getting strong.

### *11.2.2 Technical Environment*

The essence of the new round of Industrial Revolution is the standard competition of the global new Industrial Revolution in the future. Each country is building its own smart manufacturing system, which is supported by the technical system, standard system, and industrial system. In recent years, a new generation of information technologies represented by the Internet of Things, mobile Internet, big data, and cloud computing, and new manufacturing technologies represented by 3D printing, robots, and human–machine collaboration have shown multi-point breakthroughs and cross-integration with new energy, new materials, and biotechnology, and technical innovation in smart manufacturing continues to make new breakthroughs.

In the coming year, the most notable core technologies in the field of smart manufacturing include: Industrial Internet of Things, cloud computing, industrial big data, industrial robots, 3D printing, virtual reality, artificial intelligence, and so on.

Industrial Internet: The Industrial Internet is proposed by GM, representing the connection and integration of global industrial system with intelligent sensing technology, advanced computing, big data analysis, and Internet technologies. Its three core elements include intelligent equipment, advanced data analysis tools, and human–machine interaction interface. Industrial Internet is the deep integration of smart manufacturing system and smart service system. It is the integration and extension of industry chain and value chain of the industrial system.

Cloud computing: In the framework of the virtual brain of the Internet, the central nervous system of the Internet virtual brain integrates the core hardware layer, core software layer, and Internet information layer of the Internet to provide support and services for the virtual neural systems of the Internet, and cloud computing is the central nervous system of the Internet virtual brain.

Industrial big data: It is the key to controlling the industry of the future. Industrial big data is based on the development of emerging technologies, and it expands the quantity of industrial data through technologies such as industrial sensors, radio frequency identification, bar code, industrial automatic control system, enterprise resource planning, and computer-aided design. Industrial big data runs at a high speed on the production lines of industrial enterprises and is a kind of unstructured data generated by machines.

Industrial robot: It is the best assistant of Industry 4.0. Industrial robot is a multi-joint manipulator or multi-degree-of-freedom machinery oriented to the industrial field. It can perform work automatically and is a kind of machine that realizes various functions by its own power and control ability. The industrial robot consists of three basic parts: the main body, the drive system, and the control system. It has the characteristics of being programmable, anthropomorphic, and universal. It can be commanded by humans or run in accordance with pre-programmed procedures, and modern industrial robots can also act according to the principles established by artificial intelligence technology.

3D printing: the essence of 3D printing—additive manufacturing (AM) technology, which relies on computer-aided design (CAD), big data, cloud computing, computer-aided manufacturing (CAM), Internet of Things, virtual reality, and other technical supports, to directly process the digital or computer model into 3D objects through layer-by-layer accumulating method. The mainstream processes of 3D printing include extrusion molding, photopolymerization molding, granular material molding, etc.

Virtual reality: Virtual reality technology is a kind of computer simulation system that can create and experience virtual world. It uses a computer to generate a simulation environment, and through the system simulation of interactive three-dimensional dynamic vision of multi-source information fusion and physical behavior, users are immersed in the environment. Virtual reality is a combination of multiple technologies, including real-time 3D computer graphics technology, wide-angle stereo display technology, tracking technology for observers' heads, eyes, and hands, as well as tactile and force sense feedback, stereo, network transmission, voice input and output technology, etc.

Artificial intelligence: It is a new technical science that researches and develops intelligent theories, methods, techniques, and applications for simulating, extending, and expanding humans intelligence. Artificial intelligence is a branch of computer science that attempts to understand the essence of intelligence and produce a new intelligent machine that can respond in a manner similar to human intelligence. Research in this field includes robot, speech recognition, image recognition, natural language processing, expert system, etc.

In the future, the innovative application of smart manufacturing technology will accelerate the make progress toward system integrated application, and the deep integration of "Internet+" and manufacturing industry represented by the Internet of Things will accelerate the development of smart manufacturing system platform.

## 11.3 Case Analysis

Driven by the demonstration of advanced enterprises, the iron and steel industry has made great progress in recent years in basic automation, process automation, and construction of enterprise operation and management system through independent innovation, digestion of imported technology, and re-innovation, which has laid a good foundation for smart manufacturing in the iron and steel industry. The representative enterprises include POSCO, China Baowu, Shagang, CITIC Pacific, Rockcheck, etc.

### *11.3.1 Pohang Iron & Steel Co. Ltd. (POSCO)*

Informatization of Pohang Iron & Steel Co. Ltd. (hereinafter referred to as POSCO) began in the 1970s, but with the integration of steel plants, upgrading of product structure, and changes in the market environment, the original system was unable to meet the demands of production and management. In 1999, POSCO began to reconstruct its customer-centered sales system and ERP system to achieve full process coverage from sales planning, order processing, production and manufacturing management, delivery process management to after-sales management, which improved the delivery cycle and customer satisfaction.

In 1998, the South Korean government launched the privatization process of POSCO, and POSCO began to carry out asset reorganization, structural adjustment, and industrial upgrading. However, the separation of the information system became a bottleneck for the development of POSCO, which reflected in the following aspects: First, the functions of the third-level systems are incomplete—the systems developed by Pohang and Gwangyang plants had a large time span, and the codes were not in line with the metallurgical specifications, lacking important functions such as resource optimization, scheduling optimization, and process coordination in the production of varieties of steels and cold-rolled coils; second, the fourth-level system was unable to realize information sharing—management processes and application systems for financial, personnel, materials, etc., of the two companies were not unified; thus, the information exchange was difficult, and the internal process design focused on the security lacks the flexibility to adapt to the market, which seriously affected the expansion of POSCO's overseas market.

At that time, the iron and steel industry was shifting from the seller's market to the buyer's market. Therefore, the direction of POSCO's informatization reform was to improve manufacturing execution, product competitiveness, and customer satisfaction. At the same time, the overall concept of the information system was as follows: on the basis of highly integrated information, to increase customer value as the basis of management, with a focus on building "customer-oriented business process improvement".

POSCO's process change (PI project) was divided into two major phases: In the first phase, the ERP-based company-wide management system POSPIA was launched for all departments such as sales, procurement, finance, and human resources; in the second phase, the Internet-based manufacturing execution system project was promoted. In November 2004, POSCO applied the manufacturing execution system to the ironmaking and chemical conversion process of the Pohang Steel Plant and the entire Gwangyang Steel Plant. Thus, the 81 factories of the two steel plants could be operated effectively as a whole, and the delivery cycle and customer satisfaction were further improved. By optimizing the supply chain, POSCO had achieved the goal of optimal cost for global metallurgical enterprises. At the same time, through the unified management of material codes and quality standards, the operational efficiency of the enterprise and standardization degree of management had been significantly improved.

From 2002 to 2005, POSCO had integrated internal processes and external processes through ERP/SCP and e-commerce to complete the construction of Internet-based management system of the enterprise. First of all, through the establishment of a centralized four-level management system of finance, human resources, procurement, and supply chain planning, a customer-oriented business process was established to realize integration of capital flow, logistics, and information flow as well as business intelligence. In addition, the three-level system could be re-planned, and the core production management contents such as equipment maintenance, planning, and cost management were all transplanted into ERP for unified supervision, which increased the supporting functions such as the formulation optimization for the production of varieties of steels, the optimization of continuous casting and rolling in the production of cold-rolled sheet, and processing and distribution optimization of steel coil, and realized the strategy of "optimal overall cost and optimal quality". Optimization of internal resource allocation, integrated operation of supply chain, and cost reduction and efficiency improvement via e-commerce technology enabled POSCO to surpass its international competitors at the high level of operation and management.

POSCO's informatization took place at the time of its transformation and facing crisis, which is similar to the environment faced by China's iron and steel industry. POSCO went out of the crisis through informatization, and China's iron and steel industry can also get rid of the current predicament internally by deep integration of informatization and industrialization and increasing the production efficiency. POSCO's informatization path is first vertically integrating the manufacturing process, then horizontally integrating the supply chain information, and finally achieving cost reduction and efficiency increase. After that, POSCO began to provide external solutions, relying on the technology and experience accumulated in its information construction and operation, realizing the process from liberating productivity to creating productivity.

### *11.3.2 China Baowu Steel Group Corporation Ltd. (China Baowu)*

China Baowu Steel Group Corporation Ltd. (hereinafter referred to as China Baowu) takes Baosight as the strategic execution unit for the integration of informatization and industrialization, and independent development as the main way, to implement the overall plan step by step according to the medium- and long-term plan and the annual plan. As a subordinate enterprise of China Baowu, Baosight Software has undertaken the IT close-fitting service work for the latter. It has been always adhering to the development strategy of China Baowu and cooperating with China Baowu's operation improvement system to incorporate enterprise information planning into the company's strategic planning. Based on the medium- and long-term plan and the annual plan, the integration of informatization and industrialization enables the implemented projects to focus on the strategic demands of the company and support the realization of the company's strategic goals through the integration of informatization and industrialization.

Since the construction of Baosteel project more than 30 years ago, the informatization construction has taken the road of "introducing and learning integration optimizing-independent innovation". The entire informatization has experienced four stages of development, as shown in Table 11.1.

With the continuous change of China Baowu's strategic positioning, Baosight provides a full-level and full process information service for China Baowu based on the strategic focus of different periods. In the stage of ensuring the stable and smooth production, from the basic automation to the establishment of a four-level computer system architecture in order to achieve comprehensive management of manufacturing and steel products; in the stage of adapting to the socialist market economy, to support the "finance-centered" operational management philosophy by the overall production and marketing system; in the stage of promoting the customer-oriented competitiveness, to build a customer-driven agile manufacturing management information system, so as to improve the ability of quick response to inquiries, optimized scheduling, and quick response in delivery period; in the stage of scale expansion, merger and acquisition, to realize rapid replication for management of the system through integrated sales, procurement, finance, human resources and other systems, so as to effectively support the standardized and efficient cross-regional and cross-organizational management requirements, as shown in Fig. 11.2.

After more than 30 years of unremitting efforts, from the early learning to the late independent research and development, China Baowu has become the pioneer in the informatization of domestic manufacturing enterprises in China. However, compared with foreign advanced counterpart enterprises, there is still a big gap in the intelligent application of big data, use of intelligent robots, etc., which are reflected in big data analysis, plant virtual simulation application, advanced optimization scheduling, process optimization model research, remote equipment monitoring and fault diagnosis, green environmental protection, resource utilization, etc.

Therefore, the research on smart manufacturing model based on technologies such as the Internet, Internet of Things, cloud computing, robotics, and big data analysis is of great significance for further narrowing the gap with foreign advanced enterprises in the same industry and improving the core competitiveness of China Baowu in the international iron and steel market. At present, China Baowu is advancing the strategic pattern of "one body with two wings", that is, taking iron and steel as the main body, and at the wing of manufacturing, arranging intelligent transformation and construction of intelligent plant based on the existing fine manufacturing advantages, while at the wing of service, building a "co-constructed and shared" iron and steel service ecosystem by the use of Ouyeel platform.

**Table 11.1** Development process of informatization of China Baowu

| Stages | Management development | Informatization |
|---|---|---|
| Production-centered (1978–1992, construction of Phase-I and Phase-II projects) | 1. Put forward the goal of "high quality, high efficiency, high benefit, and build a first-class iron and steel enterprise", and focused on consistent management<br>2. Supported by "five systems" to ensure stable production<br>3. Separation of main and auxiliary facilities, making production management more vigorous and efficient | Informatization construction and project construction were carried out synchronously to initiate the informatization of production line level and regional level |
| Took finance as the center (1992–2000, construction of Phase-III project) | 1. Changed from planned economy to market economy, and established the core position of finance in enterprise management<br>2. Changed from being production-oriented to being customer-oriented, and realized the importance of meeting customers' demands | The informatization construction and the management development were synchronized, and the manufacturing base-level informatization was promoted, and a four-level computer system covering first, second, and third phases of the project and vertically integrating from top to bottom was built |
| Facing the customers, and took improvement of competitiveness as the center (2000–2005, implementation of ESI project) | 1. Put forward the goal of "building the most competitive iron and steel enterprise in the world"<br>2. Established a "customer-centered" business philosophy, and implemented enterprise system innovation ESI project according to enterprise management business | Baosteel Co., Ltd. implemented the ESI project, carried out e-commerce, built a customer-oriented SCM system and data warehouse, and realized information upgrading in core business Manufacturing and management systems of The First Steel, The Fifth Steel, Meishan Steel, and Ningbo Baoxin were put into operation |

(continued)

**Table 11.1** (continued)

| Stages | Management development | Informatization |
|---|---|---|
| Soft power was improved, and leapfrog development was achieved (2005 to present, promote the integration of iron and steel industry) | 1. Baosteel Co., Ltd. has completed the new issuance and acquisition and promoted a new management system combining vertical integrated management with horizontal collaborative management 2. Realize the strategy transformation from "quality goods" to "quality goods + scale", and the expansion mode transformation from "new construction" to "combination of merging and reorganization and new construction" | Start to move toward the informatization which supports the integrated management and control mode. Basically, build an integrated operation management system to provide a system platform for integrated operation and scale expansion, and enhance the soft power of Baowu Group |

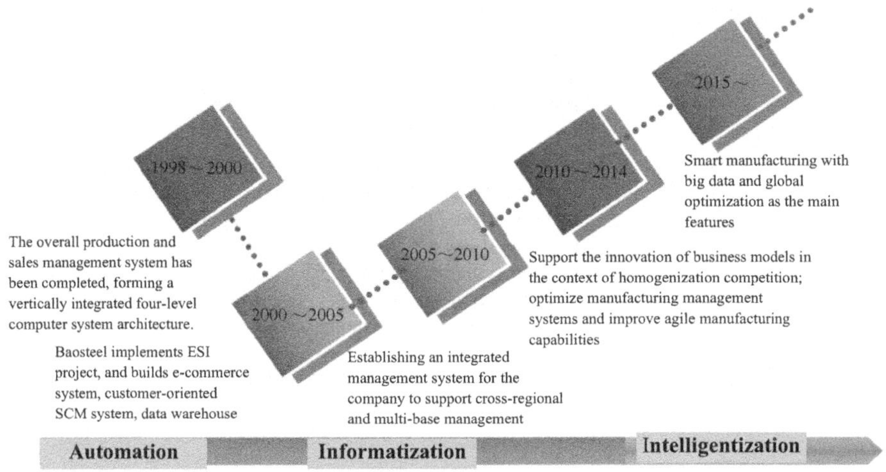


**Fig. 11.2** Development process of China Baowu's automation–informatization–intelligentization

### *11.3.3 CITIC Pacific Special Steel Group Co. Ltd.*

In September 2011, CITIC Pacific established the strategic development direction of informatization, with "building a new management model of digitization, modernization, and internationalization for CITIC Pacific Special Steel Group and establishing an information system with international advanced level of iron and steel informatization" as the construction goal, "realizing the cross-regional business coverage capability, remote real-time data transmission capability, and remote business management and control capability of the planned and constructed information system" as the construction task, "group-led and centralized management, overall planning and itemized implementation, benefit and efficiency priority and emergency first, business first and unified norms" as the basic principles, and "one network, three platforms" as the construction concept, realizing the connectivity of the Group and foreign enterprises via a special network.

In 2012, by using the second enterprise network "Pentium No.1" of CITIC Network, connectivity among Shanghai headquarters of the Group, Xingcheng Special Steel, Xinyegang Steel, Xinyaxing, Pacific Special Materials, Group Website (Hong Kong Cloud Platform), etc., was realized, which makes it possible for information system to have cross-regional service coverage capability and remote business management and control capability. At the same time, the CITIC Special Steel began to deploy data center communication project to realize the data communication between business systems of the Group's enterprises.

At the group level, a unified business platform system was established, mainly including: centralized procurement system, which integrates the original discrete "centralized procurement system" and "online bidding system" into one platform according to the needs of the Group's centralized procurement business; the foreign trade sales information system was built to cover the whole process of business management of international trade company such as "contract input, contract approval, contract tracking, production and manufacturing, finished product transshipment, terminal inventory, outward transportation and delivery, foreign trade document making, and settlement;" human resource management system, in August 2013, the Group's unified human resources management information system was implemented, and by the end of 2014, the system had covered the entire group; construction of the Group's unified financial information platform, a unified enterprise financial information platform was built to realize the integration of financial management and control of the Group's enterprises and standardization of accounting subjects, so as to meet the financial management needs of the Group and the financial management standardization needs of CITIC Pacific. The construction of the Group's unified software system platform provides a unified financial management system platform for newly established enterprises of the Group.

At the enterprise level, Xingcheng Special Steel's informatization construction began in 2000, and the total investment has exceeded 100 million yuan so far. From the initial electrification of the management to the current "integration of production, supply, and marketing", and the "simultaneous planning, implementation, and on-lining of informatization and main project", Xingcheng Special Steel has realized the integration of informatization and industrialization; after years of development of the informatization construction, Xinye Steel takes process automation control of production line as the basis and implements the plan, data acquisition, and management via various secondary and tertiary systems, and with the integrated information system of production, supply and marketing as the core, to carry out systematic and efficient integration and management of supply chain, sales, production, and finance of the enterprise. The BI system implements mining and analysis of large amount of business data, providing a solid foundation for the scientific decision-making of the enterprise management level; according to the actual development of the company and combined with the production characteristics of the coking industry and the core demands of informatization, Xinyaxing forms a three-tier architecture integration in accordance with the "enterprise operation management system-manufacturing execution system-process control system (ERP-MES-PCS)", makes overall planning, and implements it step by step, so as to promote the construction of "management and control integration", and now it has initially formed a multi-level integrated automation and informatization system which integrates human, finance, material, supply, production, and sales; Yangzhou Pacific has completed the construction of main information application systems such as port area and the pellet MES system, video monitoring system, network cabling and computer room system, collaborative office and one-card pass system, and belt scale measuring system.

In general, informatization construction of CITIC Special Steel has effectively supported the development needs of the Group's business operation. The informatization work has gradually shifted from the construction of the business platform to the construction of the data analysis platform and the information service platform, marking a new level of informatization construction.

As an outstanding representative in the special steel industry, CITIC Special Steel has made many beneficial explorations in promoting the deep integration of informatization and industrialization, and obtained remarkable achievements. The experience of CITIC Special Steel is summarized as follows.

Emphasis is made on following the advanced management concepts and development strategies. CITIC Special Steel's informatization construction focuses on improvement of production capacity, takes quality as the core, and adopts a centralized and consistent management mode, which embodies the advanced production organization mode and management concept suitable for the iron and steel industry.

Focus is made on comprehensive coverage and strengthens research and development. CITIC Special Steel adheres to the comprehensive coverage of business process by information system and incorporates production, inventory, sales, and other aspects into the control of informatization. At the same time, informatization construction is taken as an important means to support the research and development of new products. The gradual improvement of the simulation R&D system enables the strengthening of the R&D design capabilities through informatization, so as to seize the upstream of the industry chain.

Attentions are paid to independent innovation and master-independent intellectual property rights. By digesting and absorbing the key technologies of the existing information business system, CITIC Special Steel, based on its own conditions, makes exploration and innovation and conducts software secondary development closely in accordance with the production and operation management requirements of the enterprise, so as to rapidly develop new functions and new processes, inject the information system with new vitality, and highly meet the demand of enterprise's innovation management.

### 11.3.4 *Tianjin Rockcheck United Iron & Steel Group Co., Ltd.*

Tianjin Rockcheck Xiangtai Investment Holding Group Co., Ltd. (hereinafter referred to as Rockcheck Group) is located on Financial Street of Binhai New Area, Tianjing. It is a large-scale modern enterprise group with iron and steel as its main business and involved in many fields such as science technology and finance, culture and health and resource development. Rockcheck Group grasps the strategic opportunities presented by *Beijing-Tianjin-Hebei Integration* and *Made in China 2025*, takes the *"Internet+" Action Plan* as the guide, adheres to the innovation drive, and has signed a strategic cooperation with the China Metallurgical Industry Planning and Research Institute to jointly build an industry leading and well-known benchmarking "Internet + Smart Manufacturing" demonstration at home and abroad, which gives full play to the positive role of information technology in operation and decision-making optimization, improves resource utilization efficiency, integrates innovation and development, and fully promotes the transformation and upgrading of the Group.

1. Rockcheck "Internet+" Demonstration Base

The "Internet+" demonstration project of Rockcheck Group is based on the business intelligent system covering the overall situation of Rockcheck Group and crossing many different industry sectors. It is an innovative model in the industry and even in China, integrating business innovation, management innovation, technology innovation, product innovation, and service innovation. Based on the extensive extraction of enterprises' overall data, BI system carries out classification, summarization, analysis, and other operations on the enterprises' data, and provides scientific and reliable decision-making support for all kinds of enterprises' decision makers, and its core is data warehouse and data mining technology.

Rockcheck "Internet+" demonstration base mainly includes intelligent decision-making, supply chain coordination, and industrial integration.

Intelligent decision-making: As a deep application of information technology, intelligent systems can provide enterprises with various decision-making information and solutions for many business operation problems, thus improving the quality and efficiency of decision-making. The construction of Rockcheck's business intelligence system can effectively integrate the existing data in the enterprise, provide reports, and put forward decision-making basis quickly and accurately to help the enterprise make wise business operating decisions and realize the intelligentization of production and operational decision-making.

Supply chain coordination: "Internet+" can effectively eliminate information barriers and integrate upstream and downstream resources to form supply chain coordination. With the development of the 9 major e-commerce platforms, Rockcheck can provide customers with cross-sector comprehensive services including financial service, health service, product trading service, logistics service, etc. With this advantage, Rockcheck can effectively integrate the resources across the industry chain and realize the connectivity of information between enterprises, which helps the enterprises to rationally allocate resources and realize supply chain coordination.

Industry convergence: With creation of an "Internet+" demonstration base, Rockcheck can promote the continuous integration between industries, strengthen regional advantages and scale advantages, effectively promote the optimization of industrial structure, and maximize the interests of the Group.

2. Smart Manufacturing Management and Execution System of Rockcheck Steel

The overall solution for Manufacturing Management and Production Execution System (RS-MES) of Rockcheck Steel is planned and designed according to the current informatization situation of Rockcheck Steel, including Rockcheck Steel Production Management System (RS-PMS) and Rockcheck Steel Production Execution System (RS-PES), advanced planning and scheduling system, and secondary data-integrated management platform.

RS-MES serves Rockcheck Steel's entire integrated management of "research, production, and marketing". Among them, RS-PMS covers the overall production and operation management functions, including order, planning, scheduling, logistics, quality, cost, tools, reports, and other aspects. RS-PES is distributed in various production lines to realize the operation control and data collection functions of each production line, including production execution, testing, material tracking, material consumption, etc. Advanced planning and scheduling (APS) system covers procurement planning, order planning, steel rolling integration, and other aspects in the production and operation process. In APS integration and RS-PMS, the input and output of model algorithm are realized through data interaction with RS-PMS, so as to achieve the functions of intelligent optimization and decision-making. The Level-2 data-integrated management platform manages the Level-1 equipment data through unified collection, identifies the information performance feedback matching the RS-PES, and receives the production instructions issued by the RS-PES. RS-MES connects upward to the company-level management system to achieve integration with production, sales, finance, and other systems, and connects downward to various production lines to achieve integration with PCS in each operating area.

3. E-commerce Industry of Rockcheck Group

Relying on the existing resources and taking steel trading, steel deep processing, second-level product auction, finance, and logistics as the main business, Rockcheck gradually expands the business of healthy food, cultural products, timber, cross-border products, etc., finally creates an online e-commerce platform covering the whole industry chain of Rockcheck Group, and builds an Internet ecosystem.

The e-commerce industry of Rockcheck Group integrates the overall resources of Rockcheck Group, crosses multiple industry sectors, and builds a Binhai Yunshang comprehensive e-commerce platform. Rockcheck's e-commerce industry is under the overall management of Binhai Yunshang Group, and it is divided into iron and steel sector and non-steel sector. The iron and steel sector integrates resources such as materials, metallurgical exchanges, and Rongyi shopping, so as to build a unified e-commerce platform in iron and steel field, while the non-steel sector includes the rest assured shopping, the memory of the era, the wine industry of Rockcheck, the timber resources, etc., so as to create unique B2B and B2C e-commerce platforms by relying on the existing resources of Rockcheck, independent development, and open cooperation.

With the big finance as the core and based on the principle of integration of industry and financing and overall optimization, Rockcheck has realized business integration and data sharing among the e-commerce industry sectors of Rockcheck Group through unified customer management, decision-making optimization, and support system. In terms of decision-making optimization and support system, with the goal of achieving enterprise's industry integration, overall optimization, and supply chain integration, the company focuses on the construction of four aspects, i.e., product R&D support, product recommendation support, enterprise cockpit, and iron and steel sector; in terms of big finance, Rockcheck's e-commerce platforms take Rongbao Payment as the third-party payment channel and provide financial services via Ronglian Finance, Exchange, etc., so as to achieve data sharing and business integration through business process with finance running through all e-commerce platforms.

The "Internet + Smart Manufacturing" demonstration project of Rockcheck Group takes the decision-making optimization and support system (RDOSS) and the smart manufacturing management and execution system (iMES), covering the overall situation of the Rockcheck Group and crossing multiple industry sectors, as the core, and integrates business innovation, management innovation, technology innovation, product innovation, and service innovation, by integrating enterprises and partners in the supply chain, to achieve information sharing of customer resources, product design, production and operation, etc., expand the collaboration of the production process to the whole supply chain and even across the supply chain, realize networked configuration of superior resources of the whole business and superior enterprises, promote the integration of related industries, and truly realize the socialized large collaborative production.

Smart manufacturing profoundly changes the production and operation mode of enterprises, and creates new opportunities for the competitiveness reconstruction of the iron and steel industry. As a major special scientific and technological research project for smart manufacturing of iron and steel, pilot demonstration projects such as intelligent workshop, smart mine, and mass customization have played an important role in leading the development direction of smart iron and steel manufacturing. The "Internet + Smart Manufacturing" mode jointly created by the China Metallurgical Industry Planning and Research Institute and Rockcheck Group fully considers the actual demands of enterprises in the top-level design stage, focuses on the core links of production process, business operation, and enterprise decision-making, and brings real economic benefits to enterprises. Through cooperation between the two parties, a team of high-quality talents is built and a batch of core smart manufacturing technologies is reserved, which truly promote the intelligent transformation of enterprises and provide the industry with a smart manufacturing model that can be learned, operated, replicated, and promoted.

## 11.4 Prospects and Path Analysis

The general idea of *Made in China 2025* is to adhere to the path of new industrialization with Chinese characteristics, and it takes promotion of innovation and development in manufacturing industry as the theme, improvement of quality and efficiency as the center, acceleration of the integration between new generation of information technology and manufacturing industry as the main line, and promotion of smart manufacturing as the main direction. By taking the *Made in China 2025* as an opportunity, the state will promote the comprehensive integration and application of Industrial Internet, cloud computing, and big data in the whole process and whole industry chain such as enterprises' R&D and design, production and manufacturing, operation management, and sales service, so as to promote the intelligentization of manufacturing processes, and build intelligent plants/digital workshops in key fields. Iron and steel enterprises should seize the historical opportunity of the development and application of the new generation of information technology and cultivate new competitive advantages in the market.

### *11.4.1 Prospects*

Smart manufacturing is the combination of smart manufacturing technology and information technology, involving many industries. It focuses on the key aspects of smart manufacturing pilot and helps to accurately grasp the industrial investment opportunities associated with it. For iron and steel enterprises, the intelligentization of enterprises means the intelligentization of product R&D, production, procurement, sales, and service. Through the sharing of knowledge and information, the boundary between departments is broken, and rapid and effective collaboration is realized, so as to achieve intensive production and provide customers with personalized service.

1. Smart Manufacturing Comes into a Golden Period of Development

In 2015, the government and enterprises jointly made efforts to promote smart manufacturing to get into the fast track of development. The State Council successively issued *Made in China 2025* and *Guidance of the State Council on Actively Promoting the "Internet+" Action*, which clearly defines smart manufacturing as the main direction of deep integration of informatization and industrialization, and points out 11 key fields of action, such as "Internet+" collaborative manufacturing. The Ministry of Industry and Information Technology organized the "*2015 Smart Manufacturing Pilot Demonstration Special Action*" and approved the implementation of 94 smart manufacturing special projects, opening a new chapter in the development of smart manufacturing in China.

In the future, with the continuous introduction of the dividend policy for smart manufacturing, the smart manufacturing of the iron and steel industry will usher in the golden period of development. Iron and steel enterprises will accelerate the deep integration of industrialization and informatization, actively build intelligent plants and improve the level of intelligence of enterprises; a number of smart manufacturing solution providers will emerge in the market, and smart manufacturing solutions will become mature day by day; smart manufacturing standard system will be more perfect.

2. Cloud Computing, Big Data, Mobile Internet, and Other Technologies Will Achieve Large-Scale Application

At present, the wave of "new IT" with cloud computing, big data, Internet of Things, mobile applications, and intelligent control technology as the core is on the rise. During the 13th Five-Year Plan period, it will reshape the traditional information application mode.

First, the information systems will enter into the era of mobile applications. At present, the rise of mobile Internet is evolving into a subversive revolution with extensive role and far-reaching effects, and new forms of industry and new business models are constantly emerging. The innovation of mobile Internet and the constant emergence of new forms of industry and new business models have formed a forced mechanism for the development, transformation, and upgrading of traditional industries. With the rise of mobile Internet and intelligent terminals, information system will also quickly enter into the era of mobile Internet, actively embrace the concept of mobile Internet, and keep up with the development speed of the mobile Internet era, and the realization of business innovation and mobile Internet transformation is the top priority of the information construction in the future.

Second, cloud computing will jump from the bottom tier (IaaS) to the intermediate tier (PaaS). In recent years, the layout of basic resources of cloud computing has been intensified, the core technologies have been gradually improved, and the awareness of service leasing has become more and more popular, all of which have provided favorable conditions for the wide application of cloud computing technology. During the 13th Five-Year Plan period, the PaaS tier of cloud computing will gradually become mature and will truly reshape the architectural pattern of IT: The IT infrastructure is composed of a standardized bottom tier (IaaS) plus an abstracted intermediate tier (PaaS). The new IT infrastructure formed by abstracted intermediate tier integrated with the standardized physical tier supports the top-level personalized application, realizes the rapid development and deployment of applications, and ultimately supports the personalized and rapid changes of the business.

Third, the integration of big data and cloud computing will be deeply and widely applied. During the 13th Five-Year Plan period, big data technology will be integrated with cloud computing. There will be a large number of big data cloud platforms for industry applications in the market, providing government and enterprises with in-depth information analysis technology service for massive media data, which will enable the government and the enterprises to have more available resources and data services so as to enhance their information utilization and decision-making capabilities, and really make this technology enter the homes of ordinary people.

Fourth, the Internet of Things and intelligent management technology will be further applied in practice. In the next few years, the "Internet of Things" based on sensing technology and the "Internet of Everyone" based on mobile Internet as well as their integrated applications will lead to higher levels of procurement, production, sales, and logistics.

3. Data Applications Will Become the Core of Smart Manufacturing

With the continuous improvement of the level of informatization, more and more governments and enterprises have accumulated a large amount of business data. If the information system is the "blood vessel" of enterprises, the data is the "blood". With the maturity of technologies such as big data, the focus of informatization construction will gradually shift from information technology (IT) to data technology (DT), from being process-centered to being data-centered, and the focus of future information construction will be that how to conduct in-depth, multi-dimensional, real-time mining and analysis of internal and external data of the organization, so as to meet the needs of the decision-making level and promote the evolution of informatization to a higher level.

In addition, the application of industrial big data is expanding from the single marketing link of the enterprise to the whole life cycle and the whole industry chain, and becomes a key factor driving the innovation and development of the manufacturing industry. With the implementation of the national big data strategy, there will be a boom in application of industrial big data from all walks of life. A large number of iron and steel enterprises will recognize the potential value of industrial big data, explore the application of industrial big data analysis technology, mine the value of industrial data, and create data-driven enterprises to obtain market competitive advantages.

4. The Iron and Steel Industry Will Accelerate the Development of the Internet

For iron and steel industry, the introduction of the "Internet+" concept will promote the transformation of the iron and steel industry from the competition of a single enterprise to the competition of a platform economic entity and will promote the merging of manufacturers, traders, users, financial institutions, logistics distribution agencies, technical service agencies, and other resources in the iron and steel industry chain, so as to achieve joint management, sharing of results, and sharing of risks, improve efficiency, reduce costs, serve customers more effectively, and promote transformation of China from a major iron and steel manufacturing country to a powerful iron and steel manufacturing country. Therefore, under the strategy of "Internet+", the platform advantages of iron and steel e-commerce shall be made full use, so as to eliminate information barriers and optimize resource allocation, and at the same time, innovative thinking, innovative mechanism, innovative model, and innovative technology shall be taken as the core to achieve agile manufacturing, precision management, intelligent decision-making, and optimized operation of iron and steel, which is the way for transformation of iron and steel enterprises in the cold winter.

In addition, the cross-border e-commerce of iron and steel is developing rapidly, and the iron and steel industry will accelerate the journey for going abroad. On the one hand, cross-border e-commerce is becoming a disintegrator of iron and steel overcapacity, and it shall strengthen export guidance by cross-border e-commerce and transfer production capacity to Southeast Asia, South America, and other parts of the world to ease inventory pressure; on the other hand, iron and steel cross-border e-commerce breaks through the multiple restraints of traditional trade, resolves the information asymmetry of trade export circulation, and sells the products of cooperative steel plants to overseas customers, so as to improve the profits of steel plants and the turnover of goods. With the strategic guidance of the Belt and Road Initiative, cross-border e-commerce will gradually evolve into an important channel for iron and steel enterprises to "go abroad". At present, domestic iron and steel e-commerce is gradually maturing in all aspects such as channel, logistics, and payment, and it is an inevitable trend to extend the business line to overseas in the future.

### *11.4.2 Path Analysis*

Combined with the characteristics of the iron and steel industry and China's smart manufacturing, the promotion of smart manufacturing in the iron and steel industry should be carried out in stages and in a focused and orderly manner. Smart iron and steel manufacturing can simultaneously promote the implementation of four paths: First, focus on realizing the intelligentization of steel manufacturing process; second, achieve the intelligentization of enterprises' decision-making; third, integrate industry data resources and with emphasis placed on promoting the integrated application of big data; fourth, realize the interconnection and intercommunication of the steel industry chain and various nodes, and focus on promoting the cooperative application of the industry. Refer to Fig. 11.3 for details.

1. Promote the Intelligentization of Manufacturing Process

The iron and steel manufacturing process is a complex, dynamic, and holistic engineering system. It is a multifactor, multi-scale, multi-unit, and multi-level integration, with emergence rather than simple additivity [5].

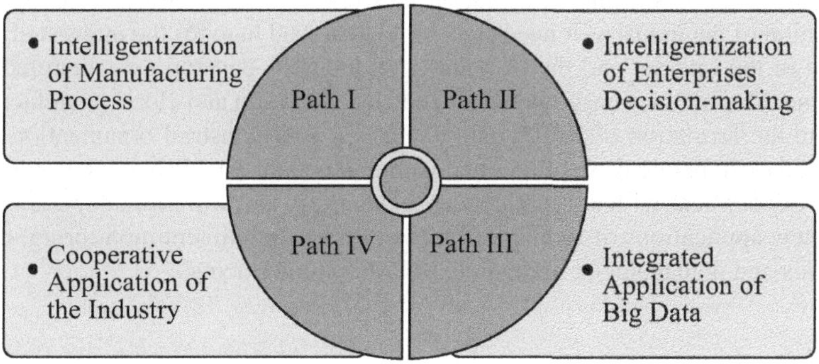


**Fig. 11.3** Four paths for the implementation of smart iron and steel manufacturing

To promote the intelligentization of the steel manufacturing process, it is necessary to pilot the construction of intelligent factories/digital workshops in key areas, accelerate the application of technologies and equipment such as human–machine intelligent interaction, industrial robots, and intelligent logistics management in the production process, and promote the simulation optimization, digital control, real-time monitoring of status information, and adaptive control of steel manufacturing process. At the same time, on that basis, the advanced planning and scheduling (APS) system is fully implemented to achieve agile manufacturing and precise delivery.

2. Promote the Intelligentization of Enterprise Decision-Making

The core of smart iron and steel manufacturing is the effective development and efficient use of information resources, the goal is to improve the overall utilization efficiency of resources, and the focus is on the intelligentization of decision-making. In order to improve the utilization efficiency of resources and energy, iron and steel enterprises should adopt the idea of system optimization to establish dual models with smelting technology and economic cost and achieve the balance between local optimization of single department and integrated overall optimization of multiple departments.

3. Promote the Integrated Application of Big Data

Big data is an extension and means of traditional database, data warehouse, and business intelligence concepts. Improve the information infrastructure of the iron and steel industry, integrate metallurgical data resources, break through the core technology of big data in the iron and steel industry, improve the ability of steel big data analysis and application, improve data security capability, cultivate versatile big data talents, organize and implement pilot projects for innovation and application of big data in manufacturing industry to promote the renovation of the manufacturing mode and the transformation and upgrading of the metallurgical industry, and foster and develop new forms of metallurgical industry.

4. Promote Cooperative Application of the Industry

Cloud manufacturing is the integration of advanced information technology, manufacturing technology, and information network [6]. To promote the cooperative application of the industry, it needs to formulate a road map for the integrated development of the Internet and the iron and steel industry, develop new manufacturing modes such as Internet-based personalized customization and cloud manufacturing, promote the formation of R&D, manufacturing, and industrial organization methods based on dynamic perception of consumer demand, establish an open-type steel industry ecosystem with complementary advantages and win-win cooperation, and foster new applications of Industrial Internet such as intelligent monitoring, remote diagnosis and management, and whole industry chain tracing.

### 11.4.3 Development Recommendation

1. Pay Adequate Attention to the Scientific and Technological Breakthroughs at Three Levels

The development of smart manufacturing in the iron and steel industry shall pay adequate attention to the scientific and technological breakthroughs at three levels. First is the major scientific and technological projects at the state level, second is the common technological breakthroughs in the industry, and third is the technological innovation at the enterprise level. Advanced iron and steel enterprises can cooperate with relevant scientific research institutes to jointly study major scientific and technological projects, and make use of the latest scientific and technological achievements to improve the level of smart manufacturing in the industry. Other enterprises can take the in-depth integration of informatization and intelligentization as a pointcut and use industry research results to gradually promote the smart manufacturing of enterprises.

2. Develop a Sound Smart Manufacturing Standard System

"Smart Manufacturing, Standards First", in order to solve the problems of lack, lagging, repetition, and overlap of standards, and give full play to the basic and guiding role of standards in promoting the development of smart manufacturing, it is recommended that the standards established by the government as the lead and the standards that are independently developed by the market shall be developed collaboratively, and the supporting new standard system shall be coordinated.

3. Establish Pilot Demonstration of Medium-Sized Enterprises

In the process of promoting smart manufacturing of iron and steel, medium-sized enterprises even more need to realize transformation and upgrading through smart manufacturing. The smart manufacturing foundation of medium-sized enterprises is similar, and only by establishing the pilot demonstration of medium-sized enterprises can the intelligent transformation of the iron and steel industry be truly promoted. At present, the "Internet + Smart Manufacturing" model jointly created by China Metallurgical Industry Planning and Research Institute and Rockcheck Group focuses on the core links such as production process, business operation, and enterprises' decision-making, cultivates professional talents, reserves core technologies, and drives intelligence transformation of enterprises with practical benefits, so as to provide the industry with a smart manufacturing paradigm that can be learned, operated, replicated, and promoted.

4. Support Third-Party Organization and Combination of Industry, University, Research, and Application

To develop smart manufacturing of iron and steel, it needs to independently research and develop solution for integrated application of smart iron and steel manufacturing, which is in line with Chinese characteristics. The third-party implementing agencies, such as the planning institute, have the identity attributes of fairness and independence, industry authority, etc. In the process of system implementation, more consideration is given to the strategic goals of the enterprises, the output of advanced concepts, and the sharing of advanced technologies. Therefore, it is easier to promote the replication and promotion of the smart manufacturing.

5. Unified Planning, Continuous Promotion, and Deep Participation

The smart manufacturing of iron and steel enterprises is a process of continuous advancement. It is necessary to iteratively update the smart manufacturing system for evolution of enterprise strategy, continuous optimization of management philosophy and production organization mode, and accumulation of experience and promotion of knowledge transformation. This requires enterprises to make a unified planning, remain deeply involved in the process of intelligent transformation, establish teams by themselves, and keep a continuous advancement.

## 11.5 Industrial Practices of Intelligentization

In order to respond positively to the call of *Made in China 2025*, China Metallurgical Industry Planning and Research Institute (hereinafter referred to as MPI) proposed to establish a development strategy of "Big Platform of One Body with Two Wings" and "One Industry at Multiple Places, and Simultaneous Development of Multiple Industries". Through planning, consultation, and deep analysis of the problems and bottlenecks encountered in the development of the enterprises, scientific and rigorous overall solutions are developed, which takes standards as the lead to promote the improvement of enterprises' products and technology and the smart manufacturing as the foothold for enterprises to reduce costs and increase efficiency.

MPI creatively proposed the smart manufacturing management and execution system (iMES) and decision-making optimization and support system (DOSS) for the iron and steel industry; established strategic cooperation with Rockcheck Group to create an industry demonstration of "Internet + Smart Manufacturing", which attracted extensive attention of the industry; carried out in-depth research with Benxi Steel and Kunming Steel on smart manufacturing and "Internet+" collaborative manufacturing; has long been paying close attention to the steel e-commerce field; and regularly issued the *Report on the Development of China's Iron and Steel E-Commerce Industry*, which has aroused strong repercussions, and the research results have been widely reported and cited (Table 11.2).

**Table 11.2** Practices of MPI in promoting intelligentization of the industry

| No. | Type | Main content | Typical cases |
|---|---|---|---|
| 1 | Government subjects | Undertake government research subjects, grasp the key points of the subjects, analyze the development trend of the iron and steel industry, and combine the smart manufacturing, "Internet+", and other new business formats to provide valuable reference for government decision-making | The research subject of the Ministry of Industry and Information Technology: *Implementation Path and Key Measures for Intelligent Upgrading Strategy of Iron and Steel Industry* |
| 2 | Special planning | Based on the theoretical research on smart manufacturing of iron and steel, combined with the development status and trend of smart manufacturing in the iron and steel industry, and through analyzing the development status of enterprises' informatization, to provide tailor-made smart manufacturing special plans for enterprises and facilitate the transformation and upgrading of enterprises | *Three-Year Action Plan of Smart Manufacturing of Benxi Steel Group*, *Feasibility Study Report on Internet + Collaborative Manufacturing Intelligent Service Major Project of Kunming Steel*, *Information Industry Plan of Kunming Steel*, and *the "13th Five-Year" Informatization Plan of Rockcheck Group* |
| 3 | System implementation | With the vision of creating an intelligent and comprehensive solution for the iron and steel industry that meets China's national conditions and reflects the wisdom of the Chinese people, on the basis of profoundly grasping the features of the industrial enterprises, closely following the advanced smart manufacturing theories, methods and technologies, focusing on the production process, business management, enterprise decision-making, and other core links, to strive to create a model of iron and steel smart manufacturing with economy, advancement, and applicability as principles | *Smart Manufacturing Management and Execution System (iMES) of Rockcheck Steel*, *Rockcheck Decision-Making Optimization and Support System (RDOSS)*, *Ironmaking Plant Real-Time Database Project of Delong Steel* |

(continued)

**Table 11.2** (continued)

| No. | Type | Main content | Typical cases |
|---|---|---|---|
| 4 | Standard establishment | China Metallurgical Industry Planning and Research Institute is qualified for the standard certification consulting service in respect of the management system for integration of informatization and industrialization, has established close cooperative relationship with the service alliance for integration of informatization and industrialization of China, and actively carried out the standard establishment for integration of informatization and industrialization in the iron and steel industry | *Specification for Advanced Planning and Scheduling (APS) System of Iron and Steel Industry* |
| 5 | Research report | The iron and steel e-commerce research report aims to comprehensively reflect the current situation and development trend of China's iron and steel e-commerce industry, and strive to explore the problems and difficulties encountered in the development of the industry, so as to provide decision-making reference to relevant parties, especially government regulatory authorities, and create a favorable environment for healthy and orderly development of the industry | Development Report on Iron and Steel E-commerce Industry of China (Year/Quarterly) |

## References

1. Ministry of Industry and Information Technology of the People's Republic of China. Development plan of smart manufacturing (2016–2020) [2016-12-08]. http://www.miit.gov.cn/n1146295/N1652858/n1652930/n3757018/c5406111/content.html.
2. Li, X.C., Shi, C.T., & Zhao, F. (2015). Industry 4.0 and China's iron and steel industry. Iron and Steel, (11), 1–13.
3. Ministry of Industry and Information Technology of the People's Republic of China. Adjustment and upgrading plan of iron and steel industry (2016–2020) [2016-11-14]. http://www.miit.gov.cn/n1146295/n1652858/n1652930/n3757016/c5353943/content.html.
4. State Council of the People's Republic of China. Made in China 2025 [2015-05-08]. http://www.gov.cn/zhengce/content/2015-05/19/content_9784.html.
5. Yin, R.R. (2014). Process engineering and manufacturing process. Iron and Steel, (7), 15–22.
6. Gan, Y. (2017). Digitization, networking, intelligentization and industry model of manufacturing industry. China Science and Technology Industry, 1, 18–19.

# Chapter 12
# Diversification

## 12.1 History Review and Status Analysis

### *12.1.1 Historical Stages of Development of Diversified Industries in Iron and Steel Industry*

Reviewing the history of the development of diversified industries in China's iron and steel industry, we can see that it has experienced three important historical stages which can be summarized as the Enterprise-Run Society Stage, the Main Business-focused Development Stage, and the New Leap-Forward Development Stage in chronological order.

1. The Enterprise-Run Society Stage (The Beginning of the Founding of the People's Republic of China-1990)

This stage is the prototype of the development of diversified industries in China's iron and steel industry and can be subdivided into four stages:

(1) The stage of initial restoration (1951–1955). Since the founding of the People's Republic of China, the steel companies left over from 1949, represented by Ansteel, Baotou Steel, Shanghai Steel, Masteel, etc., and their supporting social service facilities were restored and after 3–5 years their own mines, hospitals, etc., all had returned to regular work. The scale of the diversified industry was very small, and the output value created was also very low at that time.
(2) The stage of scale development (1956–1965). After the restoration of the "1st Five-Year Plan", China's iron and steel industry ushered in a great development in the "2nd Five-Year Plan" period. From 1957 to 1958, China began to newly build a large number of local backbone steel enterprises, such as Hangzhou Steel, Xingang Steel (Xinyu), Nansteel (Nanchang), Jinan Steel, Qingdao Steel, Lianyuan Steel, Tong Steel, and Lanzhou Steel, and meanwhile to newly build the corresponding supporting enterprise hospitals, primary and secondary schools, guest houses, etc., as an effort to complement some social



X. C. Li, *The Road Map of China's Steel Industry*,
https://doi.org/10.1007/978-981-15-2074-7_12

functions. Though the scale of diversified industries had been expanded, only mines could create production value while most of the others playing only social functions, and the value-added business was relatively simple.

(3) Business expansion stage (1965–1975). At the end of the 1960s, Chairman Mao raised the "May 7th Instruction" requiring that all walks of life across the country must be "a big school" where "the people can learn politics, military, culture, and engage in agricultural and sideline production while running some small and medium-sized factories to produce products to satisfy their own needs and exchange with the country in equal values". Accordingly, China's iron and steel enterprises had successively built new "May 7th Cadre Schools" and "May 7th Factories" to further expand the scale of industrial diversification and create a large number of new businesses to generate output value such as many new iron and steel construction teams, machine repair plants, deep steel-processing plants, guest houses, and so on.

(4) Rapid development stage (1976–1990). Since the reform and opening-up in 1978, in order to correct the mistake of the "Cultural Revolution", a large number of educated youth returned to the city and new graduates increased in large numbers, so employment problems were prominent. According to the policies at that time, the society provided part of them with a job, and the enterprises arranged employment for most of the employees' children. For this reason, iron and steel enterprises focused on offering jobs for unemployed children of the employees and production of self-help. Following the trend, a large number of iron and steel business-oriented industries had emerged, such as equipment manufacturing, construction, transportation, deep processing of steel products, guest houses, and comprehensive utilization of resources. Different from the state-owned enterprises in the past, such enterprises were basically owned by the subordinate collective (or labor service company) of iron and steel enterprise and the employees' identity also belongs to collective ownership. It was the newly created economic form of socialist collective ownership at that time.

According to the statistics of the *China Steel Yearbook*, as of the end of 1990, there were 4238 factory-running collectively owned enterprises (not including state-owned mines) belonging to the subordinated industry to iron and steel enterprises, which were later converted into diversified industries, with a total number of employees of 870,000. In 1990, it had fixed assets of about 3 billion yuan and realized an operation revenue (sales) of 10.75 billion yuan, and profits and taxes of 1.02 billion yuan. There were 4488 varieties of products and 28,091 product specifications. Between 1979 and 1990, the overall sales revenue of the subordinated industry to iron and steel enterprises has been increased by more than 20% per year, and more than 990,000 young people were employed, with an annual average of more than 80,000 people [1].

2. The Main Business-Focused Development Stage (1991–2000)

In the early 1990s, in order to implement the reform of state-owned enterprises and divest social supporting functions, China's iron and steel enterprises began to accelerate diversified operations and proposed reform ideas of "streamlining the body and separating the subsidiary". Many affiliated industries of steel enterprises began to break away from the main body and gradually transit to socialization or withdraw in accordance with the market positioning of the development, maintenance, and shrinking.

In 1992, WISCO took the lead in implementing the reform of "streamlining the main business and separating the subsidiary" in the state-owned large-scale enterprises across China and vigorously cultivated and developed the non-steel industry. Since 1995, Ansteel had implemented the overall reform policy of "streamlining the main business and separating the subsidiary", established Anshan Iron and Steel Industry Corporation, and carried out the reform of main-supplement separation for 22 supporting units such as mine companies, construction companies, and machinery manufacturing companies.

The policy of "streamlining the main business and separating the subsidiary" had changed the organizational structure of "enterprises burdened with social responsibilities" and "large and all-inclusive enterprises". The personnel separated from the main business actively developed diversified operations and gradually embarked on the road of self-financing and self-development. With the improvement of the modern enterprise system and the competitive demand under the market-oriented economy conditions, the reform of "streamlining the main business and separating the subsidiary" had progressed rapidly. By 1996, 56 out of the state key metallurgical enterprises had implemented the reform of "streamlining the main business and separating the subsidiary" in a different extent. In the same year, the former Ministry of Metallurgical Industry held the first working conference on "streamlining the main business and separating the subsidiary" of iron and steel enterprises, and proposed a new strategy of "one main business, diversified operations, and comprehensive development" for large and medium-sized iron and steel enterprises. In 1997, the main-supplement separation work of iron and steel enterprises achieved remarkable results. According to the statistics of 51 steel enterprises, there were 786 separated supporting units and 835,000 employees, accounting for 68% and 76% of the total supporting units and their total employees, respectively. Fifty-three percentage of the separated supporting units achieved self-financing, the supporting units separated from Ansteel reduced losses by 560 million yuan, and subsidiaries separated from Baosteel made a profit of 400 million yuan. The main-supplement separation reform had effectively promoted the development of diversified operations of iron and steel enterprises. The development of diversified industries in the iron and steel industry had turned from the initial service for main business and staff living to a new stage of self-sustaining and self-development.

At this stage, steel business staff had been reduced and labor productivity had been greatly improved after the reform of "streamlining the main business and separating the supplement business from the main". By the end of 2001, the number of employees of state-owned key large and medium-sized iron and steel enterprises fell to 1.023 million, and the per capita annual output of steel is 134 tons [2]. The iron and steel enterprises had increased their investment in non-steel industry, which significantly boosted their sales revenue. The non-steel industry had been healthily developed and profitable and became a new economic growth point. In 1999, there were 19 large and medium-sized iron and steel enterprises with non-steel industry sales revenue accounting for more than 20% of total sales revenue. Nearly 70% of non-steel industry entities realized self-management, self-financing, and self-development.

Although development of diversified industries in iron and steel enterprises had achieved new results at this stage, however, from the perspective of the overall level of the industry, diversified operation was still very unbalanced and still in its initial stage. Most iron and steel enterprises had merely separated subordinate non-steel units from the main business, namely the iron and steel business, based on the form of production, operation, and service to fulfill the main responsibility of ensuring production, reduce losses, and stabilize employment. The main problems of diversified industries are: more internal services than external development, over-reliance on the iron and steel business as the main and weak self-management; complex industry crossover, lack of pillar industries and leading products, and weak competitiveness; and lack of high-quality talents, few capital and technology investment, which were restricting the development of the industry.

3. The New Leap-Forward Development Stage (2001–Present)

After entering the twenty-first century, in order to keep close pace with the times, iron and steel enterprises vigorously developed diversified industries and achieved leap-forward development in this domain. Some enterprises' diversified industries had achieved equal success to that of the main industry. In 2002, in order to adapt to the new development trend of the industry, China Iron and Steel Association established a Diversified Operation Working Committee, which further promoted the rapid development of the diversified industries in the iron and steel industry. After that, the diversified industries entered a golden decade in which they had experienced the fastest development. The business areas gathered in high-tech and emerging industries such as electronic information, industrial automation, and real estate. The industry continued to expand, and the profitability and development level continued to improve; thus, a well-developed diversified industrial system was basically built. In 2013, China Metallurgical Industry Planning and Research Institute and the Diversified Operation Working Committee of China Iron and Steel Association organized the first conference on non-steel industries in iron and steel industry, reviewed and summarized the past development experience, and jointly discussed the future development of the diversified industries. The successful convening of this conference promoted the diversified industries to the forefront of corporate development strategy. Looking back at the rapid development in more than a decade, we can see that the diversified industries in iron and steel industry had experienced three major historical turning points.

The first turning point was the transition from exploration and development to the formation of an industrial base during the 10th Five-Year Plan period (2000–2005). During the "10th Five-Year Plan" period, the investment in diversified industries was booming. In order to disperse business risks, iron and steel enterprises had launched a development model of "develop one main business and diversified subsidiaries", and boldly explored and tried to enter new fields and new businesses to an extensive and comprehensive extent. The main industries involved were: equipment manufacturing, logistics trade, deep processing of steel products, information, construction, chemical, tourism services, and other industries. Some enterprises such as Baosteel, WISCO, Shougang, Tianjin Tiantie, Panzhihua Steel, etc., had built competitive diversified industry bases and formed a certain industrial scale.

The second turning point was the transition from industrial scale to quality efficiency during the "11th Five-Year Plan" period (2006–2010). During the "11th Five-Year Plan" period, iron and steel enterprises repositioned the development of diversified industries, then focused on optimizing its structure, continuously improving its competitiveness and profitability, and thus significantly improved the development level. The industrial focus of this period was mainly on the high-tech industries and emerging industries such as the development and utilization of information technology, steel structure, waste resources and secondary energy utilization, as well as real estate. A number of new economic growth points in diversified industries were gradually formed. Steel enterprises such as Baosteel, WISCO, and Ansteel had already obtained successful experience. By the end of the "11th Five-Year Plan" period, the revenue, realized profits, and proportion in total revenue of the diversified industries had all increased substantially. Especially, the profits of Shougang, WISCO, and other enterprises accounted for more than 50% of the Group's total profits, and the diversified industries achieved overall profitability. In addition, the revenue and profits achieved by Baosteel, TISCO, and other enterprises accounted for more than 25% of the Group's total.

The third turning point was the transition from an economic growth point to a development strategy during the "12th Five-Year Plan" period (2011–2015). Entering the "12th Five-Year Plan", iron and steel enterprises had integrated the development of diversified industries as an important part into the overall development plan of the enterprises. The development of diversified industries had risen to the same strategic position as the main business. The development of diversified industries was boosted across the board. During the period, the quality of economic operation of diversified industries in the whole industry had been significantly improved, the investment of diversified industries had been further increased, the project structure had become more scientific and reasonable, and the industrial development ideas were more in line with relevant national policies. The revenue of diversified industries had climbed to new heights as non-steel revenue of super-large iron and steel enterprises such as Shougang, Baosteel, and WISCO had exceeded or approached the level of 100 billion yuan.

The historical stage of development of diversified industries in iron and steel industry is shown in Fig. 12.1.

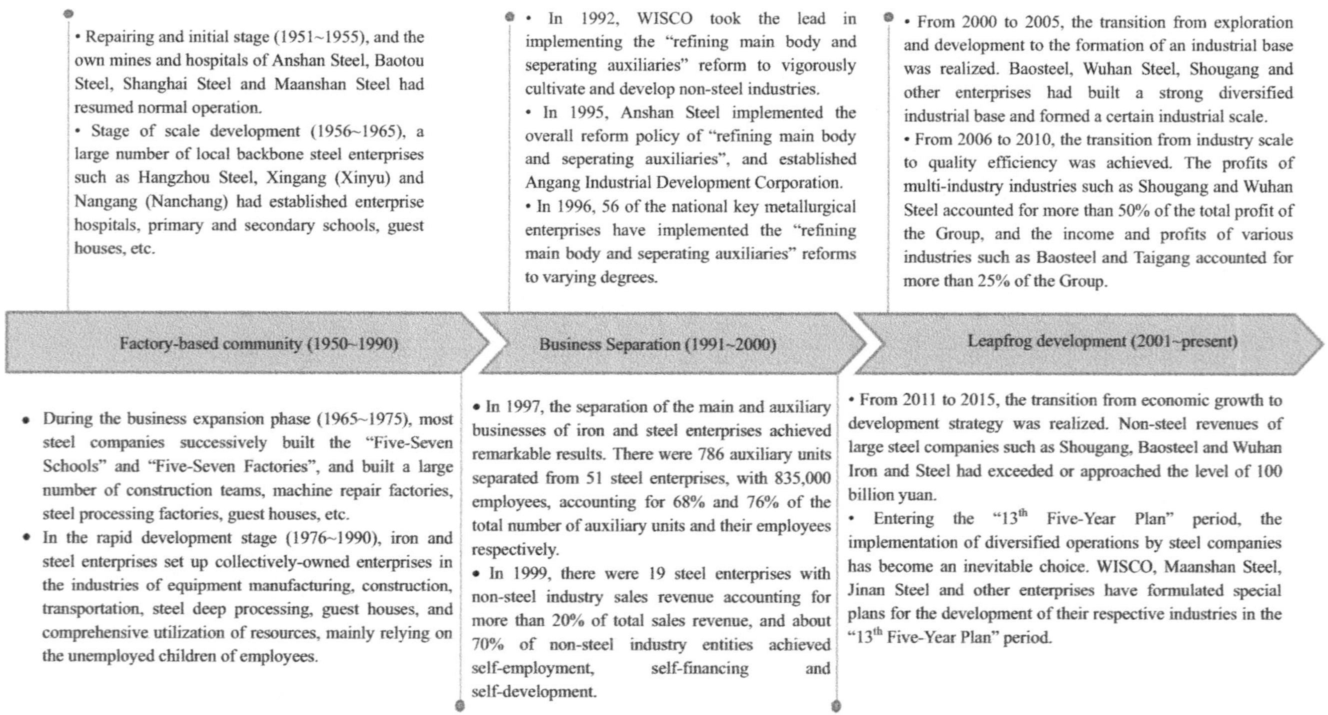


**Fig. 12.1** Historical stage of diversified business in iron and steel industry

### *12.1.2 Development Status of Diversified Industries in Iron and Steel Industry*

1. Industry Scale and Characteristics of the Times

At present, the development of diversified industries in iron and steel enterprises is no longer just a concept of "the subordinated industries", "service companies", and "relocation and re-employment". The development of specialization, scale, and industrialization is gradually taking shape, and many new industrial development modes and sub-industries have emerged. The diversified industries involved in iron and steel enterprises are divided into steel-related industries, emerging industries, and traditional industries according to the correlation with the main business. Among them, the steel-related industries mainly include resource industry, deep steel-processing industry, logistics industry, engineering technology industry, trade and business industry; the emerging industries are mainly based on strategic emerging industries that the country has strongly encouraged in recent years, including equipment manufacturing industry, energy conservation and environmental protection industry, information technology industry, new materials industry, biopharmaceutical industry, 3D printing industry, etc.; the traditional industries include financial investment, food, real estate, education, health care, tourism, and other supporting service industries.

Diversified industries have become more and more supportive and synergistic for the main business through scale and specialization development. It can not only greatly reduce the operating costs of iron and steel business, but also increase the added value of the main business products and even create value and benefits directly. According to the relevant statistics of China Iron and Steel Association, in 2014, large and medium-sized iron and steel enterprises had a profit loss of 4.4 billion yuan in steel business, diversified industries had a profit of 27 billion yuan, and thus an annual profit is 22.6 billion yuan. It was the profits created by diversified industry that made up the loss of iron and steel business; in 2015, the profit loss of steel business was 112.7 billion yuan, the profit of diversified industries was 48.1 billion yuan, so the profit loss for the whole year was 64.6 billion yuan. It can be seen that in recent years, the profit contribution of the diversified industries in iron and steel enterprises is far greater than that of the steel business. In the context of the severe overcapacity of the steel industry and the loss of iron and steel business, the diversified industries have become the pillar industry to restore the advantages of the iron and steel business, improve market competitiveness and profitability of the steel enterprises, and create value for steel enterprises.

2. Overview of Development of Typical Diversified Industries

   (1) Resource Industry

      1) Industry status. The iron and steel industry is closely related to the upstream iron ore, coking coal, auxiliary minerals, and other industries, with high industrial relevance and great influence. After years of development, steel enterprises have become more and more aware of

the importance of the upstream raw material and fuel industrial chain of steel to the development of the iron and steel industry itself. Since the new century, iron and steel enterprises have strengthened development of the upstream-related mining industries in order to ensure a stable supply of resources. The main measures are reflected in two aspects: First, vigorously develop their own mines and increase the production capacity and supply ratio of domestic mines; second, actively "going global" to participate in overseas mineral resources investment and increase the amount of overseas equity mineral resources.

① Iron ore. By the end of 2015, China's iron ore-identified resource reserves were 85.077 billion tons, of which the basic reserves were 20.763 billion tons [3]. The explored reserve of iron ore resource was distributed in 31 provinces (autonomous regions) and municipalities directly under the central government. The top three provinces were Liaoning, Sichuan, and Hebei. The total reserves of the three provinces were 40.18 billion tons, accounting for 47.2% of the country's total reserves.

Iron ores are the main raw material for the iron and steel industry, and almost all of them are used to produce pig iron. According to the output of pig iron, we can speculate that China's iron ore consumption in 2016 was about 1.11 billion tons.

In 2016, China's iron ore output was 1.281 billion tons (raw ore) [4], distributed in 26 provinces (autonomous regions) and municipalities directly under the central government. The top five provinces are Hebei Province, Sichuan Province, Liaoning Province, Inner Mongolia Autonomous Region, and Shanxi Province. The five provinces produced a total of 960 million tons of iron ore, accounting for 75.2% of the country's total output.

Domestic mines cannot meet the demand, so a large amount needs to be imported each year. In 2016, China imported a total of 1.024 billion tons of iron ores [5], mainly from Australia, Brazil, and South Africa. In order to cope with the huge import demand for iron ore, domestic iron and steel enterprises have increased their efforts to obtain iron ore resources from overseas. Ansteel, WISCO, Shougang, and other enterprises have invested in mining in Australia, Canada, Peru, and other places, and achieved certain results. However, in general, the iron ore resources that Chinese enterprises have obtained in recent years have not been rapidly transformed into domestic supply capacity, and the effects and profits are not optimistic. There are only a small number of projects that can form production capacity and effectively supply domestic iron and steel enterprises.

② Coal mines. In 2015, China's explored reserve of coal resource was 1566.3 billion tons [3], of which the proven recoverable reserves were about 169 billion tons. The top three were Inner Mongolia Autonomous Region, Xinjiang Uygur Autonomous Region, and Shanxi Province. The static guarantee period of explored recoverable coal reserves is about 45 years.

In 2015, China's coal output was 3.68 billion tons, imported coal was 204 million tons, exported coal was 5.3 million tons, net imported coal was 200 million tons, and apparent coal consumption was about 3.88 billion tons. In 2016, China's coal output was 3.41 billion tons [6], imported coal was 256 million tons, exported coal was 8.78 million tons [7], net imported coal was 247 million tons, and apparent coal consumption was 3.657 billion tons.

From the perspective of coal consumption in China, fuel (mainly including power generation, boiler fuel, blast furnace injection coal, etc.) occupies the largest proportion, accounting for 74–75% of total coal consumption, followed by coking coal for coke accounting for 15–16%, coal for chemical gasification (synthesis gas, synthesis ammonia, urea, methanol, hydrocarbons, etc.) accounting for about 9.4%, and coal for gasification or liquefaction (natural gas, liquefied oil, coal water slurry, etc.) accounting for about 0.5%.

From the perspective of China's coal sub-sector consumption, the proportion of power generation and heating industry is the largest, accounting for 49–51% of total coal consumption, followed by steel industry accounting for 16–18%, chemical industry accounting for 11–13%, and other industries accounting for 20–22%.

At present, there are more than 70 state-owned key large coal enterprises (groups) in China that have mastered most of the domestic coal resources, and their coal production accounts for more than 60%. The main enterprises are Shenhua, China Coal, Yimin, Longmei, Kailuan, Jizhong, Coking Coal, Datong, Pingdingshan Coal, Panjiang, and so on.

In order to meet the demand of iron and steel enterprises for their own coal consumption, only a small number of steel enterprises in China are involved in or are preparing to enter the domestic coal industry. Typical enterprises include Kunming Steel who masters 460 million tons of coking coal equity resources through investment and stock holding, with an annual output of washed coal for coking of 4 million tons, Shougang and Baosteel who, respectively, invested in Shanxi Coking Coal Group and Yima Coal Industry to form long-term strategic cooperation. However, in general, China's iron and steel industry controls a small amount of coal resources which has not yet formed a scale and their amount is very small compared with the total consumption of the whole industry. In the long run, there are fewer newly explored coal fields of coking coal, which is the largest consumption in the iron and steel industry, because the domestic measured resources have already been divided. Therefore, if the iron and steel enterprises want to control domestic coking coal resources, they can only procure them from the enterprises that already have coal resources or directly purchase shares of coal enterprises to control coking coal resources, but a huge investment is needed.

③ Other mineral resources. Manganese ore: As of the end of 2015, China's explored reserve of manganese ore resources was 1.38 billion tons, and there were 481 manganese ore mines nationwide, mainly distributed in Guangxi, Hebei, Hunan, Guizhou, Yunnan, Chongqing, etc. China's manganese ore production is mainly distributed in provinces (autonomous regions) like Guizhou, Guangxi, Chongqing, Hunan, Yunnan, Liaoning, Hubei, and Shaanxi. Domestic manganese ore is mainly produced from local privately operated mines, and there is no official output statistics. It was estimated that the output of manganese ore in China in 2016 was about 28 million tons (finished product ore). In 2016, China imported a total of 17.05 million tons of manganese ore from 29 countries or regions [8], mainly from South Africa, Australia, Gabon, and so on.

Chrome ore: By the end of 2015, the explored reserve of chromite ore resource in China was 12.458 million tons and there were 64 chromium ore mines nationwide which were mainly distributed in Xinjiang and Tibet. China's chrome ore resources are relatively scarce, and they are scarce resources according to the degree of demand. The main source of domestic chrome ore is from local privately operated mines. There is no complete statistics yet, but it was estimated that the domestic output of chrome ore in 2016 was about 300,000 tons. Since domestic chrome ore cannot meet domestic demand, a large amount of chrome ores need to be imported as a supplement. In 2016, China imported 10.579 million tons of chrome ore from 23 countries or regions [9], mainly from South Africa, Turkey, and other countries. In addition, some steel enterprises in China actively "go out" to invest in overseas chrome ore resources. For example, Sinosteel established Sinosteel South Africa Chromium Co., Ltd. in South Africa in 1996 to engage in the development and production of chrome ore. In addition, TISCO was also investing in chrome ore projects in Turkey.

2) Problems

① Low output of domestic iron ore cannot meet the needs of the domestic iron and steel industry, and external dependence has increased year by year, exceeding 85% in 2016.

② Domestic iron ore resources are in poor condition as they are "poor, fine, miscellaneous, and scattered", and the heavy burden of corporate taxes and fees results in high production costs and weak corporate competitiveness.
③ Iron and steel enterprises control an insufficiently small amount of coal resources, especially the amount of coking coal resources, which is not conducive to the stable production of coking in steel enterprises.
④ The distribution of domestic manganese resources is unbalanced, the scale of the deposit is small, the ore grade is low, and the mining conditions are poor. Besides, more than 80% of the manganese ores comes from local small and medium-sized mines, which determines the difficulty in boosting the productivity.
⑤ Domestic chrome ore resources are in short supply, and the deposit is small and scattered, so imported ore is heavily depended on.

(2) Resources Comprehensive Utilization Industry

1) Status of Comprehensive Utilization of Resources. The resources comprehensive utiliation industry in iron and steel industry mainly refers to the comprehensive utilization of resources such as solid wastes in the iron and steel industry. The key points include: resource utilization of industrial solid wastes such as smelting slags, coal ash, and desulfurization gypsum, efficient utilization by classification of iron-containing dusts such as iron oxide scales and zinc-containing dust, and the regeneration and recycling of waste oil, waste acid, social wastes, and so on [10].

① Output Value of Comprehensive Utilization of Resources. In 2015, the comprehensive utilization output value of "three wastes" of the members of China Steel Association was 33.95 billion yuan and the profit was 6.45 billion yuan, equivalent to 60.4 yuan and 11.5 yuan per ton of steel, respectively, which decreased, respectively, 9.6% and 24.2% compared with the comprehensive utilization output value and profit per ton of steel in 2011. From 2011 to 2015, the comprehensive utilization output value and profit of the "three wastes" of the China Steel Association members are shown in Fig. 12.2 [11–15].

As can be seen from Fig. 12.2, the overall utilization output value of resources and the profit rate of products generally fluctuate downward, mainly because of the continuous downturn in China's iron and steel and building material industries in recent years, and the product homogenization and low-value competition of comprehensive utilization of resources.

② Industrial Structure and Technical Equipment of Comprehensive Utilization of Resources. In recent years, with the promotion of ecological civilization construction and the introduction of relevant

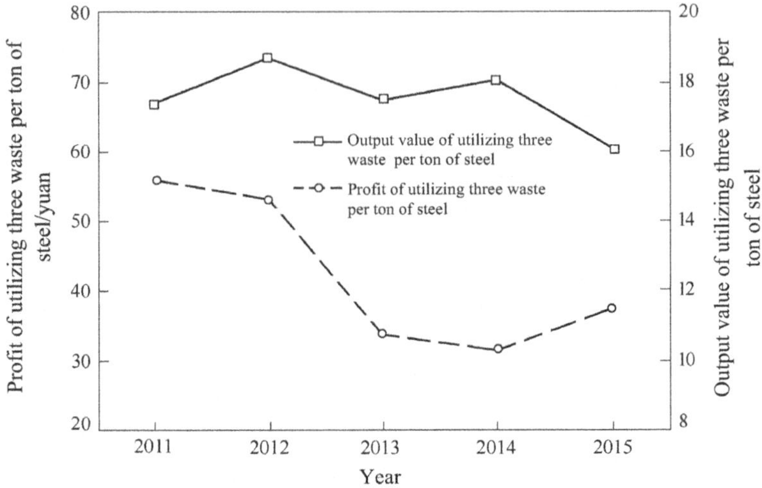


**Fig. 12.2** Output value and profit per ton of steel of the "three wastes" of members of China Iron and Steel Association

incentive policies, industrial structure of comprehensive utilization of resources in China's iron and steel industry has been continuously improved, the technical level, process equipment, and value-added products of comprehensive utilization of resources in key areas have been persistently enhanced, and the overall level of comprehensive utilization of resources has been improved significantly.

At present, the enterprises engaged in comprehensive utilization of resources in China's iron and steel industry have formed an enterprise structure that combines wholly owned subsidiaries of iron and steel enterprises, joint ventures of iron and steel enterprises and specialized companies for comprehensive utilization of resources. The technical level, process equipment, and utilized added value of products of comprehensive utilization of large quantity of industrial solid wastes like iron and steel slags have been significantly improved. According to statistics, in 2016, China's blast furnace slag output was about 260 million tons. Besides, there were about 360 BF slag micro-powder production lines, with a production capacity of about 220 million tons. In 2016, China's steel slag output was about 100 million tons, and steel slag tailings were about 80 million tons. The application and popularization of technology and equipment such as steel slag hot stewing technology and steel slag rolling grinding technology have greatly improved China's overall level of comprehensive utilization of steel slags. In addition, the mature application of separation and extraction technologies of dedusting ash containing zinc and potassium, the regeneration and recycling of waste materials such as waste refractories, waste oil, and waste acid, the development of urban mineral resources, etc., have made a great contribution to expand the comprehensive utilization of resources industry and improve the comprehensive utilization level of resources.

③ Status of Enterprises Engaged in Resources Comprehensive Utilization. In recent years, the scale, professional management level, and scientific research strength of enterprises for comprehensive resources utilization in China's iron and steel industry have been significantly enhanced. For example, Baosteel Development Co., Ltd. mainly undertakes Baosteel's resources regeneration and recycling, steel production services, and other businesses. After nearly ten years of development, its total assets and annual sales revenue have reached 10 billion yuan. The main products of its resources regeneration and recycling include: new-type construction material made from metallurgical waste slags, magnetic materials made from iron oxide, regenerated refractories, waste oil and other regenerated products made from wastes; Baosteel Development Co., Ltd. has thus been crowned the industrial leader for its scale, technological strength, and management level in the resources comprehensive utilization domain. WISCO Metal Resources Co., Ltd. is responsible for the processing and distribution of scrap steel, metallurgical slag treatment, and deep processing and utilization of oxidized scales and other businesses for WISCO, with an annual operating revenue of nearly 10 billion yuan. In addition, the specialized resources comprehensive utilization enterprises such as Huaxia Shixing Co., Ltd. and GreenNovo Co., Ltd. have continuously strengthened their influence in the field of steel slag treatment and zinc-bearing dust and sludge treatment, and the scale of enterprises and market share have been continuously increased.

2) Problems in Comprehensive Utilization of Resources. At present, although the comprehensive utilization of resources in China's iron and steel industry has achieved remarkable results, it still has the following main problems:

① There is Still Room for Improvement of Comprehensive Utilization of Resources. At present, the added value of comprehensive utilization of resources in China's iron and steel industry is relatively low, and the output value of comprehensive utilization of resources is still far behind the advanced iron and steel enterprises at home and abroad. In 2016, the members of the China Iron and Steel Association calculated that the output value of comprehensive utilization of steel resources per ton of steel in iron and steel enterprises was 49.7 yuan, while those of Liugang Group and Shagang Group have reached 139 yuan and 227 yuan, respectively. At present, China's industrial solid waste resources such as blast furnace slags, steel slags, coal ash, and desulfurization gypsum are mainly used for construction materials in road construction, brick making, cement production, and gypsum board. The comprehensive utilization of products has low added value and poor product diversity. In some areas, the markets of applied products are even saturated, resulting in poor economic benefits and low willingness to use solid wastes of enterprises. The levels of iron recovery from steel slags of enterprises are quite different, resulting in the loss of ferrite resources; waste heat from a large amount of high-temperature steel slags has not been effectively utilized, resulting in a huge waste of resources and energy and loss of economic benefits.

② The Concentration and the Technical Level of Resources Comprehensive Utilization Industry Urgently Need to be Improved. At present, the overall level of comprehensive resources utilization industry in China's iron and steel industry has improved significantly, but there are still problems of small-scale resources utilization enterprises, scattered operations, and low level of industrialization. In particular, the research, development, and promotion of resources comprehensive utilization technology and the lack of industrialization of equipment have become important factors restricting the standardization and large-scale development of comprehensive utilization of resources. On the one hand, the products of comprehensive resources utilization are not high end in science and technology, and the entry threshold of the industry is low. That means some resource comprehensive utilization enterprises only carry out simple processing with the incoming materials. On the other hand, the industrialization capacity of equipment for resources comprehensive utilization is insufficient, some advanced technologies and equipment rely on imports, China lacks its own intellectual property rights, the degree of industrialization of equipment for comprehensive resources utilization is low, and the quality of comprehensive resources utilization products is poor.

③ The Industrial Policy and Standard System of Resources Comprehensive Utilization Are Not Well Developed. The management system is not well developed, the incentive mechanism for technology research and development and promotion is unsound, industry-university-research-application cooperation is not close, the development of related technologies, equipment standards, and product standards is relatively lagging, and the degree of industrial standardization is low. These factors are difficult to support the promotion and application of advanced technologies and equipment.

(3) Deep Steel-Processing Industry

1) Development Status of Deep Steel-Processing Industry. In 2017, China's crude steel output was 832 million tons, pig iron output was

710 million tons, and capacity of steel products (including recycled materials) was 1.048 billion tons, increasing by 0.8% on a year-on-year basis. In the same year, the import of steel was 13.297 million tons, the export of steel was 75.431 million tons, and the actual consumption of steel was 728 million tons.

In 2017, the steel industry intensified the supply-side structural reforms and achieved remarkable results in cutting overcapacity, thereby significantly improving the benefits of the enterprises and stabilizing the operation of the industry. However, the pressure of optimization and adjustment of industrial structure became increasingly prominent, and the industry was still facing many challenges.

China's iron and steel industry has experienced rapid development for more than 30 years and has achieved remarkable achievements. However, at present, serious overcapacity, overall losses, resources and environmental protection and many other problems in iron and steel industry are prominent. Most iron and steel enterprises have difficulty in survival. Finding new profit points, broadening the business field, extending the industrial chain, and developing deep steel processing have become the choice of many enterprises [16].

Deep processing of steel in China is generally classified as follows:

① Deep processing products can be classified into wire rod, plate, and pipe by classification.

② Deep processing products can be classified into semifinished products for users and the products for direct market demand by the end-user's requirements on product shapes and performance.

③ Deep processing products can be classified into production-type deep processing, marketing-type deep processing, and industrial deep processing by the division of labor in the industrial chain.

The classification of deep steel processing in China is shown in Fig. 12.3.

2) Main Problems in Deep Steel-Processing Industry. Although China is the largest steel producer, most of the steel is of primary and secondary materials. Compared with industrial developed countries, China has a smaller proportion of deep steel processing and a lower added value. Iron and steel enterprises and downstream industries have not built a mutual beneficial cooperation. Besides, most of the enterprises' sales strategy still follows the line of small profits but quick turnover, thereby leading to small benefits and even loss. Compared with the developed countries, main problems in China's deep steel-processing industry include:

① Low proportion of deep steel processing. At present, the proportion of comprehensive deep steel processing in the developed countries worldwide has reached more than 50%, while the proportion in China is only about 25%, and some high-tech deep processing products still need to be imported.

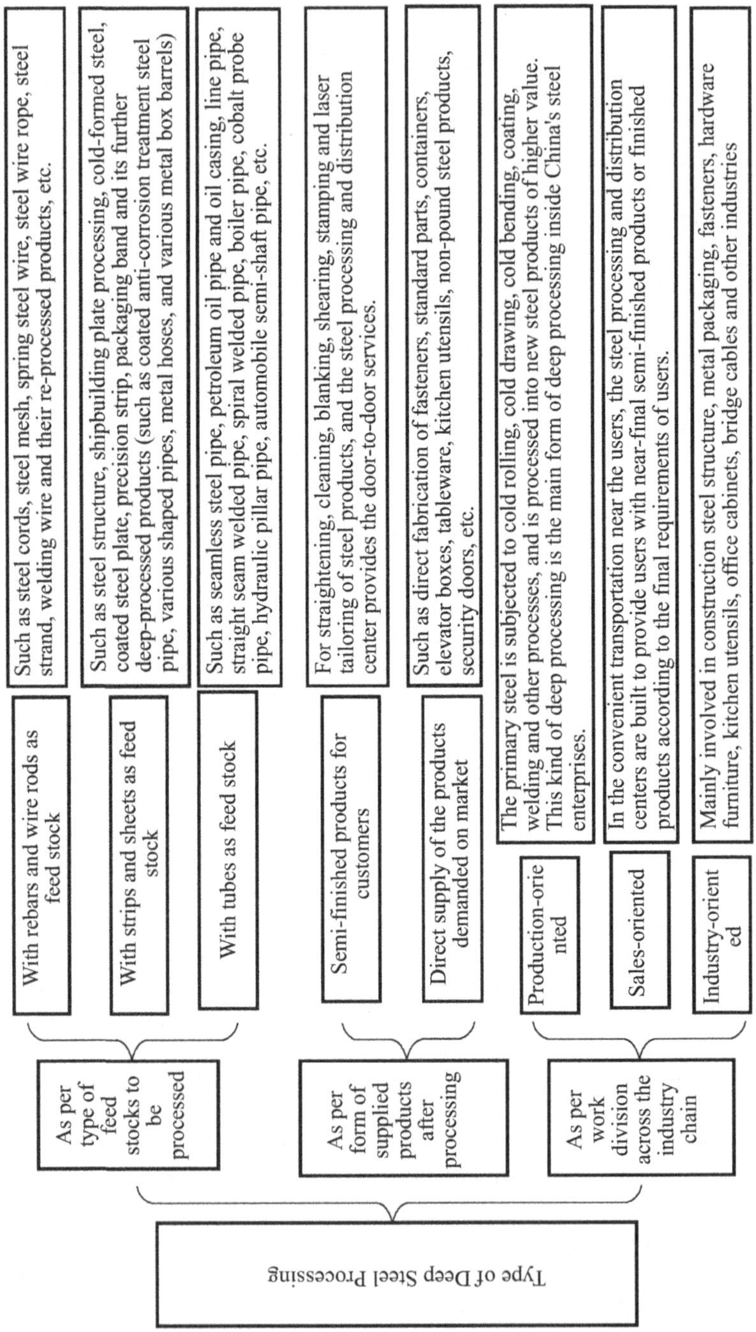


**Fig. 12.3** Classification of deep steel processing

② Overall overcapacity in production-type deep processing within enterprises. Due to historical reasons, almost every enterprise's rolling mills are equipped with various types of finishing shear machine units. At present, these finishing capacities scattered within the enterprises are completely oversupplied.

③ Overlapping network of marketing-oriented deep processing has been formed, and competition has been intensified. Since the beginning of the twenty-first century, domestic steel enterprises, steel traders, and foreign-funded enterprises have continuously built steel processing and distribution centers in cities where the downstream industries are concentrated, which have been spreading everywhere and formed homogenization competition in the gathering of major downstream users.

④ A mutually beneficial and win-win marketing system has not been established between industrial deep processing and downstream market. Industrial deep processing enterprises lack cooperation with downstream industries. Without forming a win-win cooperation, their business revenue is unstable.

3) The Focus of Deep Steel-Processing Industry Concerned in the Future. Extending steel industrial chain and service chain in iron and steel industry is one of the effective ways to solve the difficulties currently faced by the iron steel enterprises. In the future, the downstream demand for deep steel processing has great potential and broad development prospects. However, on the whole, most steel enterprises in China still have not completed the transition from steel producers and suppliers to service providers and cannot meet the needs of steel production and steel trade development.

In order to adapt to this role change, enterprises must find the right market positioning. Before deciding to be engaged in deep processing, enterprises should fully conduct investigation and proofing of the project, and pinpoint the target market and customer base based on the characteristics and grades of their products; otherwise, it will be difficult to survive and develop.

In order to cope with the external environment of overcapacity and homogenization competition, enterprises should focus on refinement, specialization, individualization, and high efficiency in deep steel-processing projects. The users have to be allied in research and development of advanced material technology to ensure satisfactory services and products so as to enhance the core competitiveness of the enterprises.

(4) Mechanical Maintenance and Manufacturing Industry

1) Development Status. The mechanical maintenance and manufacturing industry are closely related to iron and steel industry. It has a stable internal market and obvious advantages in raw materials as one of the pillar industries for the diversified development of iron and steel enterprises and one of the potential industries for key development. Throughout the history of the entire mechanical maintenance and manufacturing industry, it was developed from machine repair and equipment manufacturing business such as the initial equipment repair, simple spare parts processing, and low-end complete equipment manufacturing internally for iron and steel enterprises.

At the beginning of the founding of the People's Republic of China, China's machinery industry followed the Soviet machinery and equipment management and maintenance system. Almost every iron and steel enterprise has its own "big and all-covered" or "small but all-covered" affiliated machine repair shop or machinery shop along with a large number of professional maintenance staff and equipment. By the end of 1984, the iron and steel enterprises had a total of 71 affiliated machinery shops responsible for spare parts processing, equipment repair, and self-made simple equipment for the enterprises internally. However, with a low degree of specialization and cooperation and backward process, equipment, and products, they showed obvious times characteristics of strong self-sufficiency and self-service.

With the development of China's iron and steel industry and the division of production functions, in the 1990s, the machinery maintenance and manufacturing business internally for iron steel enterprises began to be independent from the frontline production departments. Some machine repair departments still focus on new projects, equipment overhaul, and equipment technical transformation within the enterprises; some machine repair departments are independent legal entities, which while serving the enterprise were also facing the society. They implemented independent accounting and self-financing and sought survival and development in the market.

In the twenty-first century, the equipment management system and maintenance methods of iron and steel enterprises are facing new changes, and steel enterprises are beginning to move toward socialization and specialization. In 2002, the eight ministries and commissions of the State Council jointly requested the state-owned enterprises to accelerate main-supplement separation. At earlier stage, the iron and steel industry was focusing on "streamlining the body and separating the supplement business from the main" which increased the proportion of non-steel revenue, achieving initial main-supplement separation. On this basis, the steel industry striped out the machinery maintenance and manufacturing business mainly for internal service, and started external development on the basis of fully ensuring the internal maintenance.

After more than a decade of development in the twenty-first century, the mechanical maintenance and manufacturing industry, separated from steel business, have continuously strengthened management and self-reforms to increase its market competition. Thus, the production scale and equipment technology have reached a higher level. When serving for the iron and steel business, it aims to expand the external market share, develop specialized products, and adapt to the requirements of market-oriented operations.

The machinery maintenance and manufacturing industry in the iron and steel industry follow the development philosophy of "focusing on the main business market, expanding the domestic market, and advancing into the overseas market". After experiencing three stages of self-sufficiency, market restructuring, and in-depth development, it has gradually become a modern machinery maintenance and manufacturing industry that meets the needs of market development. At this stage, most steel enterprises have formed a mechanical maintenance and manufacturing industry with capabilities including equipment maintenance and repair, spare parts processing and manufacturing, repair and remanufacturing, complete equipment manufacturing, technological innovation, and new product development, which plays an important role in the development of iron and steel business.

2) Problems. Affected by unfavorable factors such as severe overcapacity in the iron and steel industry, increased energy and environmental constraints, and slowing growth of market demand, the mechanical maintenance and manufacturing industry, which are closely related to the iron and steel industry, have shown decreased market demand, overcapacity, declining product price, rising price of raw material, and tighter supply, so the industry is clearly at a disadvantage. The main problems are as follows: over-reliance on steel business, lag in response to external markets, poor ability to develop the market; insufficient ability to conduct research and development of products, fixed business philosophy, backward management and control modes; sharp glide-down of market share, simultaneous shrink of internal and external markets; increasing threats from overseas enterprises with core technology, core equipment manufacturing capacity, and general contracting capacity; shift of the cost pressure from iron and steel enterprises to downstream suppliers and the significant increasing survival crisis; increasing competition among equipment manufacturing industries and a low profitability of mechanical products; lack of competitive products with high-tech content, high added value and strong

market influence, poor market competitiveness resulted from zero differentiation, unremarkable brand benefits, and lack of competitiveness; and restriction of key technologies and core technologies which can curb the healthy development of the machinery maintenance and manufacturing industry.

3) Characteristics of Typical Enterprises. Due to different enterprise systems and equipment levels as well as different business scopes and product structures, the competitiveness of the machinery maintenance and manufacturing industry in China's iron and steel industry is quite different. Most large iron and steel enterprises have a relatively long operating history, so their technological strength, brand, and production capacity are in a leading position in China. They lead the market because of their good background and the highest technical level in the domestic industry. The characteristics of typical enterprises are shown in Table 12.1 [17, 18].

(5) Refractory and Auxiliary Materials Industry. Refractory materials are used in various fields of the national economy such as iron and steel, nonferrous metals, glass, cement, electric power, and military industry. They are essential materials to ensure production, operation as well as technological development of the above-mentioned industries. They are irreplaceable in the production and development of high-temperature industries.

1) Industry Status and Existing Problems. From 2001 to 2010, supported by the strong driving of rapid development of high-temperature industries such as iron and steel, nonferrous metals, petrochemicals and building materials, the refractory industry maintained a good growth momentum. China has become a major producer and exporter of refractory materials in the world. During the "12th Five-Year Plan" period, China's refractory output showed a steady and declining trend. In

**Table 12.1** Typical enterprises in mechanical maintenance and manufacturing industry of iron and steel industry

| No. | Enterprise name | Main characteristics | Key products or technology |
|---|---|---|---|
| 1 | Baosteel | Supported by technological integration and innovation capabilities, Baosteel focused on high-end metallurgical equipment and high-precision spare parts, as well as the development and manufacture of the complete equipment and core components in environmental protection, chemical engineering, engineering machinery, plastic molding machinery, etc. | Professional equipment manufacturing technologies such as continuous casting machine, bearing, housing, rolling mill, payoff tension reel, gear, blast furnace cooling stave are world-class |

(continued)

**Table 12.1** (continued)

| No. | Enterprise name | Main characteristics | Key products or technology |
|---|---|---|---|
| 2 | WISCO | Relying on Construction Engineering Group and Heavy Industry Group, WISCO is committed to technology-intensive and management information construction, cultivating and developing engineering contracting and advanced equipment manufacturing capabilities. Now it has formed business such as metallurgical engineering construction, municipal engineering construction, repair for maintaining production and repair, manufacturing and remanufacturing of metallurgical equipment and lifting equipment | Advantageous products include remanufacturing of blast furnace cooling stave, metallurgical slag tank, and payoff (tension) reel |
| 3 | Shougang | Shougang established a base for equipment repair, domestication of spare parts, and maintenance for the iron and steel business which was its main business, and extended such business to other iron and steel plants other than Shougang itself. Multi-approach and multi-channel cooperation platform were set up with the help of foreign high-end patent technology | Leading products include large medium and high-end metallurgical equipment such as continuous casting machine and blast furnace top equipment |
| 4 | Benxi Steel | The products of Benxi Steel fully meet the needs of the transformation and operation of the main equipment, and are sold to Baosteel, QDIS, Tonggang, and other enterprises. Moreover, the products are exported to Russia and other countries and regions. Benxi Steel promoted engineering contracting for iron and steel production lines, and implements "one-stop" service system covering manufacturing, repair, and maintenance. Besides, it undertakes projects extended from the iron and steel industrial chain | The leading products cover the manufacturing and repair of equipment and spare parts required for mining, smelting, coke, cement, petrochemical, and other industries. It is one of the bases producing spare parts for hot air stoves in China |

(continued)

**Table 12.1** (continued)

| No. | Enterprise name | Main characteristics | Key products or technology |
|---|---|---|---|
| 5 | Kungang | It integrates mechanical manufacturing, repair, and installation, and has the capability of research, development, design, and manufacturing for large-scale complete equipment. Besides, its services are extended to metallurgical construction and maintenance and service to serve both the iron and steel industry and the non-steel industry | Its large cast and forged parts fill the gap in Yunnan Province. The wet overflow ball mill is the ball mill with the maximum daily processing capacity developed by the heavy equipment manufacturers in Yunnan |

2016, according to the statistics of China Refractory Industry Association, the national refractory output was 23.9124 million tons, going down by 8.56% on a year-on-year basis. Among them, compact shaped refractory products were 13.5852 million tons, going down by 11.08% on a year-on-year basis; thermal insulation refractory products were 467,200 tons, going down by 1.32% on a year-on-year basis; unshaped refractory products were 9.8599 million tons, going down by 5.19% on a year-on-year basis. At present, China has become a major producer, consumer, and exporter of refractory materials, and its production and sales rank first in the world.

The structure of China's refractory products has been further adjusted to better meet the market demand and to meet the requirements of energy conservation and environmental protection. Although China's refractory industry has achieved remarkable results after years of development, there are still many problems such as severe overcapacity and disorderly market competition which should never be overlooked.

The first problem is severe overcapacity. The refractory industry is currently in a state of complete overcapacity, with a capacity utilization rate of less than 70% or even lower. The capacity in steel, cement, glass, and other refractory consuming industries is also seriously oversupplied. In the next few years, fixed asset investment, especially new construction projects, will decrease rapidly, and the demand for refractory materials will gradually decline. In addition, as refractory materials are developed and the quality is steadily improved, the refractory consumption per unit of product will be gradually reduced. This means that the overcapacity faced by the refractory industry will become increasingly serious.

The second problem is disorderly market competition. Overcapacity has led to intensifying competition in the refractory market. In order to win the market, some refractory enterprises have adopted harmful competitive means such as low-price marketing and delivery before payment. That will directly lead to a lower market price of refractory materials, more arrears in loans, which will seriously damage the industry's business order and overall interests. At present, the refractory industry has lower production concentration and many small enterprises, which has intensified disorderly competition in the industry.

The third problem is environmental protection and resource conservation. Although China's refractory mineral resources are abundant, mineral resources have long been unreasonably allocated, the development has been paid more attention than the protection, and excessive mining, destruction, and waste are serious. As an important part of the high-temperature industry, the refractory industry not only needs to continuously reduce energy consumption, but also bears the important responsibility of providing "green refractory materials" for the downstream industries. The environmental protection, resource conservation, and national ecological construction strategy set higher requirements for chrome-free refractory materials and recycling of refractory materials.

The fourth problem is the insufficient ability of scientific research and innovation. As the profits of the refractory industry continue to decline, the R&D investment of the whole industry grows slowly, and the R&D platform construction lags behind requirements of situation; besides, the insufficiency is also represented by the lack of public welfare research institutions, the unreasonable allocation of R&D resources, the relatively weak basic research, and the insufficient innovation capability. The R&D project is not forward-looking, the connection between production and application is not smooth enough, and the problem solving or application process is not timely enough to guide and support the future development of the refractory industry.

In summary, the production and operation of the refractory industry urgently need to be transformed and upgraded to enhance the industrial competitiveness.

2) Experience in industry transformation and upgrading. The Ministry of Industry and Information Technology officially published *Several Opinions on Promoting the Healthy and Sustainable Development of the Refractory Industry* in March 2013, which has received a huge response in the industry. Local governments at all levels in Henan, Liaoning, Shandong, Shanxi, Zhejiang, and other provinces where refractory industries are relatively concentrated also attach great importance to transformation and have introduced measures to optimize product mix and spatial layout, promote intensive use of

resources, and initiate the construction of new refractory industrialization demonstration bases focusing on eliminating backward capacity, promoting joint restructuring and accelerating transformation and upgrading.

Zhejiang Province regards Changxing County where refractory enterprises are most concentrated as the focus of energy conservation and emission reduction, and puts forward the concept of "no polluted industry, only polluted enterprises", to comprehensively promote the transformation and upgrading of refractory industries. In line with industrial integration reform to promote regional economic development in Changxing County, the refractory industry introduced industry integration policies of "reorganizing, upgrading, and eliminating" and "priorities to guaranteed work items, to policy implementation, and to service and approval". Many local government departments, together with local associations, have taken active measures to put forward higher requirements for enterprises in terms of energy conservation, environmental protection, and safe production, thereby promoting the transformation, upgrading, and healthy development of the refractory industry.

Beijing Lirr is actively transforming and facing difficulties. While striving for policy support and market guarantee, Beijing Lirr focuses on cost reduction and efficiency improvement and increases efforts to contrast the standard, tap the potential, improve the management, protect the main business, expand the market, strengthen the management, and train the team. The production and operation assessment indicators are improving day by day. In addition, the enterprise has established Liaoning Lirr, Baogang Lirr, and Qingdao Sidier New Materials Co., Ltd. in recent years, acquired Ma'anshan Kaiyuan New Material Technology Co., Ltd. and Shanghai Xintaishan High Temperature Engineering Materials Co., Ltd, and reorganized Liaoning Zhongxing Group and Liaoning Jinhong Mining Co., Ltd. by issuing shares and purchasing assets. Through a series of mergers and acquisitions and restructuring, Beijing Lirr has formed three major business segments: magnesium raw material synthesis, magnesium product production, and international refractory trade. Thus, the resource allocation has been optimized, the operation and management costs have been greatly reduced, and the company's profitability has been improved significantly.

(6) New Material Industry

1) Development status. China's new material industry system mainly includes six major fields which are special metal functional materials, high-end metal structural materials, advanced polymer materials, new inorganic nonmetallic materials, high-performance fibers and composite materials, and cutting-edge new materials. The application of new materials in various fields of the national economy has

been continuously expanded, and an industrial system with a complete variety including R&D, design, production, and application has been formed. Since the beginning of the new century, some key technologies in China's new material industry have made major breakthroughs: China's self-developed technologies to produce tantalum niobium beryllium alloys, amorphous alloys, high magnetic orientation silicon steel, super-hard materials, and superconducting materials have reached or are close to international standards. The variety of new materials is also increasing; the high-end metal structural materials, new inorganic nonmetallic materials, and high-performance composite materials are significantly enhanced; the self-sufficiency of special metal functional materials is gradually improved [19].

In recent years, China's new material industry has gradually grown and expanded, with continuously improved industrialization degree and boosted technical level. According to incomplete statistics, the total production value of China's new material industry reached 2.6 trillion yuan in 2016, and the new material industry has made remarkable achievements in system construction, industrial scale, and technological progress. Among them, rare earth functional materials, advanced energy storage materials, super-hard materials, special stainless steel, and other production capacity rank forefront of the world, making significant contributions to the China's economy and national defense construction, and have a good development foundation.

2) Problems. In today's world, the scientific and technological revolution is developing rapidly, new materials and products are changing with each passing day, and industrial upgrading and material replacement are accelerating. Developed countries attach great importance to the cultivation and development of new material industry. With sound technology development and risk investment mechanism, the large multinational companies take the dominant position in high-tech and high value-added new materials relying on advantages such as their technological research and development, capital, talents, and patents, which has placed greater pressure on the development of China's new materials industry.

The overall development of China's new material industry still has a big gap with the developed countries. The industrial development faces some problems that need to be solved urgently, mainly in the following aspects: first, the weak ability of independent development of new materials, less innovative impetus of large-scale materials enterprises, and insufficient ability to guarantee the key and new materials; second, disconnection among industry, university, research institute and application, short industrial chain, difficult promotion and application of the new materials, and defective industrial development model; and third, lack of overall planning and policy guidance of the new material industry, small and scattered R&D investment, and weak basic management [20].

3) Development opportunities. China's economic development has entered the "new normal" period. The transformation and upgrading of traditional industries and the cultivation of strategic emerging industries are the main tone of industrial transformation during the "13th Five-Year Plan" period in China. Strategic emerging industries such as new materials have a promising future as they are driven by dual power from both the industrial environment and the policy environment. The strategic adjustment of economic structure has provided important development opportunities for the new material industry: On the one hand, to accelerate the cultivation and development of strategic emerging industries such as energy conservation and environmental protection, new generation information technology, high-end equipment manufacturing, new energy and new energy vehicles, and to implement the national economy and the national defense construction projects require the support and guarantee of the new material industry, which provides a broad market space for the development of new materials industry. On the other hand, since China's raw material industry is huge and some industries are suffering from overcapacity, the limitation from resources, energy, and environment is pressing. Therefore, there is an urgent need to vigorously develop new material industries, accelerate the transformation and upgrading of the material industry, and foster new growth points.
4) Typical case analysis. With the development of high and new technology, new material and basic material industry are increasingly tied to each other, and the industrial structure is characterized by horizontal diffusion. Basic material enterprises such as iron and steel enterprises are extending businesses into new material industries. The new steel materials in high-end metal structural materials are the hot spot of R&D and innovation of China's iron and steel enterprises and scientific research units, representing the main direction of research and development of new steel materials.

   For example, pure iron materials are high-purity metallic furnace materials for modern high-end equipment manufacturing, EAF cast steel, and EAF smelting high-quality special steel, which is essential for ensuring the quality and economic efficiency of high-end and high-quality special steel. The demand for pure metallic materials is growing year by year, and even just the domestic market has a demand of as much as 30 million tons; at present, many domestic enterprises produce industrial pure iron, but not in large scale due to its complicated process and high smelting cost. Take industrial pure iron as an example: According to incomplete statistics, the annual demand for industrial pure iron materials is more than 800,000 tons, and at present, there is still an insufficiency of about 300,000 tons in China, especially the high-quality industrial pure iron, which needs to be imported from Europe, America, Japan, and other countries every year.

TISCO is the birthplace of pure iron in China and the base for its research and development. It can produce all kinds of products except pipes. The main production equipment and lines include an 80 t and an 180 t steelmaking—external refining—continuous casting lines as well as many electroslag re-melting and vacuum smelting processes for production of special pure iron varieties. In addition to conventional products, it can also undertake the production of pure iron materials for cutting-edge engineering and scientific research, such as pure iron specially for accelerators. TISCO's pure iron has a market share of over 60% in the domestic market, and it has reached more than 95% at the highest level. The pure iron used for China's major projects, aerospace, and military purposes has always been supplied by TISCO. Some enterprises such as Angang, Great Wall Steel, and Fushun Special Steel have started to produce pure iron new materials. In addition, due to the higher cost for treating and smelting industrial pure iron with a converter and external refining process, the market competitiveness is reduced; Benxi Steel uses high-purity concentrate powder to produce sponge iron as it is cheap, and the obtained products made from pure iron material are with high purity and strong competitiveness.

In short, as the most important part of the strategic emerging industry, the basis of the transformation and upgrading of traditional industries, and the important support of the development of other strategic emerging industries, new materials will be one of the most important and most promising areas in the future. It is estimated that by 2020, the new materials industry will have become the leading industry of the national economy, and its main products can meet the needs of the national economy and national defense development. In the future, the rapid development of emerging industries supported by new materials, such as computers, communications, green energy, and nano-industries, will further increase the demand for new materials in both the variety and the quantity.

(7) Coal Chemical Industry

1) Overview. Since the coal chemical industry involved in this section refers to the diversified part of iron and steel industry, it specifically means the traditional coal chemical industry formed by the deep processing of coking by-products in the coking industry of the iron and steel industry, namely coke oven gas, crude benzene, and coal tar.

2) Main products and output

① Coke oven gas. During the process of high-temperature carbonization in a coke oven, coking coals produce a large amount of waste gas, which is converted into coke oven gas after being purified by blast condensate, desulfurization, de-amination, debenzolization, and other processes. The yield and composition vary depending on the quality of coking coal and coking process conditions. Usually, one ton of dry coal can produce 290–380 $m^3$ of coke oven gas (under standard state).

Coke oven gas is a medium heat value gas with a calorific value of 16.3–18.8 MJ per nominal cubic meter. It is suitable fuel for high-temperature industrial furnaces and city gas. The coke oven gas can also be utilized in deep process to produce chemical products such as methanol and natural gas taking advantage of its high hydrogen content.

② Crude benzene. Crude benzene is one of the products produced by pyrolysis of coking coal. It is a benzene-based compound recovered after de-amination of raw gas. The output accounts for about 0.7–1.2% of the coal (dry coal) charged into furnace.

Crude benzene is an intermediate product and can be directly used as a solvent, a fuel, etc., but is limited in the scope of application. For this reason, the crude benzene is usually hydro-refined and separated into pure benzene (also known as refined benzene, accounting for 65–75%), toluene (12–18%), xylene (3–6%), heavy benzene (5–8%), non-aromatics (2–4%), $C^{9+}$ (1.5–2.5%), and so on, and then applied separately to produce various derivatives.

③ Coal tar. Coal tar is also one of the crude gas products produced from coking coal in pyrolysis. It is a black or brownish black viscous liquid with irritating odor. The output accounts for about 3–4% of the coal (dry coal) charged into furnace. It is an organic mixture mainly composed of aromatic hydrocarbons with tens of thousands of components, and there are more than 500 single compounds that have been separated from it and identified.

The component of coal tar is very complex and can only be a fuel for direct use. In order to make full use of coal tar resources and maximize their value, there are two main directions for deep processing of coal tar at home and abroad.

The first is hydrogenation of coal tar which is divided into the light component hydrogenation to produce naphtha, light diesel oil, etc., and the heavy component to produce asphalt; all components are hydrogenated to produce naphtha, light diesel oil and other solvent oils, fuel oil, etc.; selective hydrogenation is to separate high-priced components for further processing or direct sale, and to hydrogenate low-priced components is to produce naphtha, light diesel oil, and other solvent oils. Hydrogenation of coal tar is a new process that has just been developed in recent years. The main advantage is that the domestic market demand for products such as naphtha and light diesel is high and the sales price is only related to the international oil market price, so the fluctuation is relatively small and the revenue is relatively stable; the shortcoming is that the technology is not particularly mature, so the surplus profits cannot be obtained and sales revenue is not high; therefore, an industrial chain cannot be formed.

The second is the traditional separation process which separates the various components by distillation and then performs deep processing. The mainstream process for coal tar treatment in China is the traditional one whose main advantages are mature technology, more deep processes of various components, more options, long industrial chain, high sales revenue, more values and jobs, and excess profits from the right products. However, the shortcoming is that the sales price may experience a sharp market fluctuation and a frequently changing market; therefore, it is hard to predict the market condition and even harder to grasp it.

3) Industrial chain of coking by-products. At present, there are mainly three industrial chains for deep processing of coking by-products at home and abroad. One is the deep processing products produced by comprehensively utilizing coke oven gas as the raw material, the second is the chemical products produced through deep processing from crude benzene as the raw material, and the third is the chemical products produced through deep processing from coal tar as the raw material.
4) Development status of coking by-product deep processing industry. After years of development, the coal chemical industry in coking industry has reached a larger scale and a higher level. As of the end of 2015, according to preliminary statistics, the coking industry has formed a coke oven gas industry chain to produce 12.2 million tons of methanol, 3.6 billion cubic meters of natural gas, and about 2 million tons of ammonia.

   The crude benzene industry chain has a plant capacity of 5.84 million tons of benzene refining (benzene hydrogenation), of which the largest processing capacity of a single plant in China is 200,000 tons/year. In addition, there are also many subsequent deep processing devices for products such as caprolactam, adipic acid, and styrene.

   The coal tar processing industry chain has about 77 sets of traditional coal tar production facilities, with a processing capacity of 1.9 million tons. The coal tar hydrogenation units have a processing capacity of about 1 million tons and many subsequent devices capable of deep processing of products such as carbon black, scrubbing oil, anthracene oil, crude naphthalene, and medium temperature asphalt.

   According to preliminary estimates, in 2016, China's coking by-product deep processing industry achieved a sales revenue of about 161.5 billion yuan while employed about 73,000 people.
5) Typical Enterprise—Baosteel Chemical Co., Ltd.

   At present, Baosteel uses all its coke oven gas for deep steel processing and power generation, instead of deeply processing the gas itself. It has 6 sets of coal tar processing unit with a processing capacity of 1.1 million tons/year, a unit producing carbon black through deep processing of coal tar with a capacity of 250,000 tons/year, a unit producing washing oil with a capacity of 50,000 tons/year, and a unit producing needle coke with a capacity of 100,000 tons/year. Besides, it has 3 sets of benzene refining units (benzene hydrogenation), with a processing capacity of 250,000 tons/year.

At present, Baosteel Chemical's products of naphthalene series, washing oil series, high-purity benzene, carbon black, and other products have been the leader of China and even the forerunner internationally. Among them, the tar processing capacity ranks in the forefront globally, and it has the scale advantage of developing into a world-class coal chemical enterprise.

6) Problems

First, the industrial concentration is low, and the scale economy is not reflected. According to statistics of China's coking by-product processing enterprises, the production scale of most methanol produced from coke oven gas production is 100,000 to 200,000 tons/year, that of benzene hydrogenation is 50,000 tons/year mostly, and that of most coal tar processing unit is 150,000 to 300,000 tons/year; the unit scale is generally small, and the production process and technical equipment are backward.

Second, the industrial chain is insufficiently extended, and the resources are not fully utilized. At present, China's industrial chain of coking by-product is generally insufficiently extended, and the added value of the products is low. Most of them are primary processed products, such as coke oven gas processing to methanol without deep processing; only 20% of pure benzene after benzene hydrogenation has been further processed, and there are even fewer deep processes of toluene and xylene. Besides, there are only 50 kinds of products from coal tar deep processing, which has quite a long way to go to catch up with Germany which has more than 200 kinds of industrial refined products and Japan which has more than 70 kinds.

(8) Financial Industry

1) Background. In the context of overcapacity in the iron and steel industry and increasingly fierce competition among enterprises, low benefit and even loss have become the new normal of the industry. In 2015, China's key large and medium-sized iron and steel enterprises realized a total profit of −64.534 billion yuan, which is the first loss of large and medium-sized iron and steel enterprises since 1980. Driven by profits, the capital in iron and steel industry has gradually penetrated into the finance industry which has higher profit margins and better industry development. Baosteel, Baotou Steel, and other enterprises have clearly announced to regard the finance industry as the main industry

for the future development of the enterprise, making it a new profit growth point for the group.

2) Status. At present, most of the steel enterprises entering the financial industry began with the financial companies. Since the iron and steel industry is a capital-intensive industry, iron and steel enterprises need to implement effective management and control over capital resources to enhance their competitiveness. As a fund management and control platform, the financial companies have played an important role in improving capital efficiency, liquidizing stock assets, and achieving the organic integration of the capital flow, information flow, and material flow of iron and steel enterprises to the utmost extent. As of the foreign investment in the financial sector, the equity investment of China's iron and steel enterprises in financial institutions is mainly concentrated in the financial sub-industry such as banks, securities, insurance, trusts, and funds, but most of them are in shareholding mode [21].

   In the typical enterprises' development of the financial industry, Baosteel established Hwabao Investment Co., Ltd. in 2007, a wholly owned subsidiary, as the flagship of the Group's financial investment business, and managed and controlled the existing securities, funds, and trusts of Baosteel. Through effective integration and optimization of financial resources, Hwabao Investment Co., Ltd. provides new profit points for the Group in market research, investment and financing, mergers and acquisitions, etc., and gradually is built into a financial holding company to achieve listing. In 2014, Hwabao Investment achieved an operating revenue of 1.992 billion yuan, a profit of 1.348 billion yuan, and management assets of 560 billion yuan.

   Baotou Steel is engaged in the financial industry mostly through its finance company and Huachen Trust Co., Ltd. Baotou Steel established the finance company to unify settlement, monitoring, dispatching, and operation of capital resources, and increase capital benefits and efficiency. Baotou Steel invested the trust company to carry out financing business, loan business, security investment business, bond underwriting business, equity investment business, financial leasing business, asset management, and so on, which is beneficial to the Group to achieve cost reduction, debt optimization, etc., through the financing functions of capital market and monetary market. In 2015, Baotou Steel Finance Company expected to achieve a total operating revenue of more than 200 million yuan and an operating profit of more than 150 million yuan. Huachen Trust Co., Ltd. expected to achieve a total operating revenue and a net revenue of more than 200 million yuan, respectively.

3) Problems

   ① Lack of unified management platform in financial industry and weak competitiveness. At present, the overseas financial institutions that iron and steel enterprises invest are often small in scale, their performance is worse than their peers, and their management is relatively scattered. Therefore, they are characterized by "small, weak, and scattered". Besides, the affiliated financial institutions are not competitive, developing slowly, and cannot build a well-known brand.

The development of the financial industry lacks a unified management platform. Although some iron and steel enterprises have gradually formed the prototype of kind of financial holding companies, the layout and development still lack a unified planning. Therefore, the financial institutions under the holding company cannot bond together to fully display the synergy and implement the optimized configuration.

② Inadequate control and lack of scientific equity management. At present, the financial institutions invested and controlled by some iron steel enterprises in China are the largest shareholder, but they have little control over these financial institutions. They have a low voice in their management decisions, and the role of the largest shareholder has not been fully played. Their equity investment in these financial institutions only targets at financial investment, lacking scientific equity management. The economic benefits of enterprises are increased only by obtaining equity revenue, and support of financial services in the iron steel business and other diversified industries is weak.

③ Scarce financial talent and small talent reserve. At present, some iron and steel enterprises in China are developing financial industries, but they are generally faced with a serious shortage of talents which can restrict the development of the financial industry.

(9) Logistics Industry

1) Development Status of Logistics Industry in Chinese Iron and Steel Industry. The logistics industry is a complex service industry that integrates transportation, warehousing, freight forwarding, information services, and other industries, which is a basic and strategic industry that supports the development of the national economy. The iron and steel logistics industry is a branch of the logistics industry and is closely linked to the iron and steel industry.

According to the statistics of 2017, China's crude steel output is 832 million tons, the steel output is 1.048 billion tons, and the apparent consumption of domestic crude steel is 768 million tons. At the same time, the vast geographical distribution of iron and steel production enterprises and sales areas provide the iron and steel logistics industry a huge development space. Over the past decade or so, with the rapid economic growth, the output of crude steel and steel has been hitting new high record, and trade volume of steel has always been rising. On this basis, the demand for iron and steel logistics continues to increase, China's iron and steel logistics industry is also rapidly developing, and the market size is expanding and its growth rate is increasing quickly. In China's steel logistics industry market, only steel sales logistics scale has rapidly been increased from 20.472 billion yuan in 2000 to 246.418 billion yuan in 2014, with a compound annual growth rate of 19.4%, and continued to maintain steady growth. Despite the decline in 2016, it is still large in scale.

As of 2016, the efficiency of iron and steel logistics industry has been steadily improved, and the logistics specialization has continued to be improved. However, in general, the cost of iron and steel logistics in China is still relatively high. According to the calculation, the total cost of the entire supply chain of iron and steel enterprises above the designated size is 790 billion yuan. The cost reduction of iron and steel logistics and the industry has great potential. The general situation of development of the iron and steel logistics industry has the following characteristics: ① improved efficiency of iron and steel logistics and ② effective control by iron and steel enterprises over logistics cost [22, 23].

2) Existing Problems in Logistics Links of Chinese Iron and Steel Industry. Based on current output, the total logistics amount of raw materials and fuel, auxiliary materials, finished steel products, and solid wastes in China's iron and steel enterprises is about 4.1 billion tons per year. In recent years, with the rapid expansion of iron and steel production capacity, China's steel logistics industry has been developed rapidly. But compared with the iron and steel powers, China still has a long way to go in modern logistics, mainly in: ① low marketization of China's iron and steel logistics industry, ② backward logistics technology and equipment, and logistics information resource management being required urgent improvement, ③ lower level of processing and distribution links in the iron and steel logistics, which cannot meet the diversified needs of customers, ④ shortage of logistics talents familiar with iron and steel industry, ⑤ various problems in the development of iron and steel logistics enterprises.

3) Development Trend of Logistics in Chinese Iron and Steel Industry. ① A special logistics management department should be set up to achieve centralized and specialized management of logistics, and gradually enter into the logistics industry. ② It has been a consensus to extend industrial chain and pursue diversified development. ③ The modern logistics management should be built with the support of information technology and network technology. ④ Third-party and fourth-party logistics have been developed rapidly.

(10) Engineering and Technology Industry

1) Status of Engineering and Technology Industry. After experiencing explosive growth, at present, the focus of construction of China's iron and steel industry has shifted from new projects to upgrading

of the existing engineering technology, energy conservation and environmental protection and structural adjustment. The iron and steel engineering industry has shifted from the traditional low-tech, low value-added industries to the emerging high-tech, high value-added and energy-saving and environmental-friendly industries, from traditional engineering design, equipment installation, and commissioning to engineering general contracting and complete equipment supply and transfer, and from the project construction to the project operation management and so on.

① Status of engineering and technology industry. In 2015, there were 20,480 engineering survey and design enterprises participating in statistics, including 1822 engineering survey enterprises, 14,982 engineering design enterprises, and 3676 engineering design and construction integration enterprises. In 2015, the operating revenue of national engineering survey and design enterprises totaled 2708.90 billion yuan, with an increase of 8.6% over the previous year. Among them, engineering surveys contributed 74.34 billion yuan, accounting for 2.7% of operating revenue; engineering design revenue contributed 336.53 billion yuan, accounting for 12.4% of operating revenue; engineering technology management services contributed 37.75 billion yuan, accounting for 1.4%; engineering general contracting contributed 1282.67 billion yuan. The total annual profit of engineering survey and design enterprises was 162.39 billion yuan, with an increase of 9.4% over the previous year; the net profit was 132.05 billion yuan, with an increase of 8.6% over the previous year. In 2015, the total expenditure on scientific and technological activities in the engineering survey and design industry was 52.64 billion yuan, with a decrease of 22.3% over the previous year; the enterprises accumulatively had a total of 93,885 patents, with an increase of 33.2% over the previous year; the enterprises accumulatively had 26,798 proprietary technologies, with a decrease 18.2% over the previous year.

There are hundreds of metallurgical engineering and technology enterprises in China, which are mainly divided into: metallurgical survey and design research units directly under the Ministry of Metallurgy after reform and transformation; metallurgical survey and design research units directly under the provincial and municipal departments and bureaus; and metallurgical survey and design research organized by various large metallurgical enterprises. The annual output value of China's metallurgical engineering design industry reaches hundreds of billions. The scale and technical strength of the enterprises are higher than the average level of the whole national survey and design industry, but it still has a long way to go compared with engineering design units with comprehensive qualifications.

② Overview of key engineering technology enterprises. In recent years, the implementation of a large number of new iron and steel projects and technological transformation projects has brought opportunities for the development of metallurgical engineering and technology industry. In this round of development, the engineering design institutes under the major iron and steel companies have significantly enhanced their comprehensive strength. Their business scope has been continuously expanded, their staff quality and technical level have been continuously improved, and their operating revenue and profitability have continued to grow. For example, Beijing Shougang International Engineering Co., Ltd., founded in 1973, is an international engineering company, with a registered capital of 150 million yuan and more than 1200 employees, established by the former Beijing Shougang Design Institute and controlled by Shougang Group as its relative shareholder. The company mainly provides technical services such as planning consultation, engineering design, complete equipment, project management, and project general contracting in metallurgy, municipal, building, energy conservation, and environmental protection industries. At present, more than 6500 projects have been completed, including over one hundred large-scale general contracting projects. The company has consistently ranked top the list as for business revenues of survey and design enterprises all over China. The undertaken project covers more than 60 iron and steel enterprises including WISCO, TISCO, Baotou Steel, Jigang, Tangsteel, CISC, Xinsteel, Xuansteel, Chenggang, and Xiangtan Steel, as well as more than 20 countries such as Brazil, Peru, India, Malaysia, Vietnam, Bangladesh, the Philippines, South Korea, Saudi Arabia, Oman, Zimbabwe, and Angola. It has undertaken the overall design of the Shougang Jingtang Iron and Steel Plant, which is the national "11th Five-Year Plan" key project and represents the development level of China's iron and steel industry in the twenty-first century. Baosteel Engineering and Technology Group Co., Ltd., established in 2010, was built as a wholly owned subsidiary of Baosteel Group and by the original Baosteel Engineering Technology Co., Ltd. which integrated Baosteel's related businesses such as engineering and steel structure, with a registered capital of 2.82 billion yuan. Baosteel Engineering and Technology Group Co., Ltd. has a complete engineering and technology industry chain and is focusing on developing energy-saving and environmental protection as well as urban building strategic businesses; consolidating and upgrading specialized services such as metallurgical engineering technology, equipment manufacturing, construction management consulting; and guaranteeing to increase competitiveness of iron and steel business as the main business by means of industrial technical services as a regenerative business. The corporate business covers project general contracting, project management and consulting, engineering design, bidding, project supervision, equipment design and manufacturing, complete equipment supply, construction management, operation and maintenance, product inspection, etc. The service extends from metallurgy to environmental protection, energy, finance, coal chemical and municipal and other industries. In 2015, Baosteel Engineering and Technology Group Co., Ltd. achieved an operating revenue of 10.6 billion yuan. In addition, Jigang International Engineering and Technology Co., Ltd., Tang Steel International Engineering Technology Co., Ltd., and other enterprises have achieved leap-forward development as their main business has been developed from the single metallurgical field to metallurgy, construction, municipal, energy saving, and environmental protection, and transformed from engineering design to engineering design, consulting, project general contracting, and technology research and development.

2) Problems in Engineering and Technology Industry. At present, the comprehensive strength of the engineering and technology industry in China's iron and steel industry has been significantly enhanced, and the enterprise scale, business scope, and technological level have been continuously improved. However, the following problems still exist:
   ① Traditional metallurgical engineering business volume decreases. Complete overcapacity and accelerating elimination of backward capacity in China's iron and steel industry declared end of the era when a large number of large-scale projects were constructed in the iron and steel industry. The cumulative investment of the national ferrous metal smelting and rolling processing industry in 2016 was 417.74 billion yuan, decreasing by 1.1% on a year-on-year basis; a decline in the fixed investment also indicated that the volume of engineering design and project construction in China's iron and steel industry has shrunk. The decline in the business volume of traditional metallurgical engineering will increase the pressure of competition in the market of engineering and technology industry, forcing the engineering and technology enterprises to implement transformation and upgrading.
   ② High competitive pressures in new business sector. At present, the key engineering and technology enterprises in China have been continuously expanding their business scope to gradually realize the development from traditional metallurgical engineering design to new business areas such as building, municipal, energy conservation, and environmental protection; however, in the new business

fields, these enterprises still have a long way to go in market competitiveness and scientific and technological research and development strength, etc., when compared with professional design institutes and scientific research institutions, thereby facing greater pressures of competition.

(11) Other Industries. In recent years, many iron and steel enterprises have achieved some success in real estate and urban service industries such as education, health care, and catering. However, facing increasingly fierce market competition, most industries are burdening heavier operational pressures and there is room for further improvement.

1) Practice of Special Enterprises. Shougang entered the urban service industry in accordance with the requirements of adapting to the urban function of capital city. Its business scope covers traditional urban service industries such as real estate, education, tourism, and pension as well as cultural and creative industries. By promoting the development and construction of Beijing Shougang Industrial Park, Shougang closely integrates urban infrastructure construction, land development and industrial layout, park operation, and so on to create a world-class, harmonious, and livable demonstration zone, becoming the innovation-driven hosting platform for the capital and one of the most dynamic areas. Through the establishment of a special operating company to integrate Shougang TV, publishing, art performance, industrial tourism and other resources to develop cultural and creative industries, it has successfully held the two World Animation Conferences and Beijing International Animation Week, the 6th to 10th Shougang Chinese Rose Garden Flower Show, three Light and Shadow Culture Season and Shougang Lighting Festivals, the "Steel Rhythm" Steel Sculpture Art Exhibition, the Live Concert of the Three Chinese Tenors, the 5th Beijing International Electronic Music Festival, and other large-scale events. Shougang Industrial Park has become China's first national-level industrial and cultural tourism AAA scenic spot.

Operation of WISCO's urban service industry mainly relies on its wholly owned subsidiary—the City Service Group. Its core businesses can be divided into four major parts: property services, public utilities, urban services, and manufacturing and transportation, with businesses covering more than 20 varieties. Its market covers all production logistics within WISCO and expands to many social organizations such as universities, financial institutions, hospitals, enterprises, and public institutions. Among them, the City Service Group is the executive member unit of China Property Management Institution and the first enterprise in Hubei Province that has the first first-grade national property management qualification. It has the second-grade qualification for construction of building and municipal engineering projects, and the grade C qualification for design of water supply and fuel gas engineering projects; besides, the City Service Group is the Vice President Unit of Hubei Gas Association and has established WISCO China Resources Gas Company which is a joint venture company with China Resources, serving 136,000 gas users in Qingshan District, occupying gas supply pipelines with a length of more than 600 km, and has obtained the gas franchise right of Port of Fangcheng. The City Service Group's business covers a wide range of scope such as group meals, supermarket chains, beverages, elderly care, early childhood education, hotel and tourism, landscaping, as well as food manufacturing, furniture manufacturing, kitchen equipment, automobile transportation, and ecological agriculture, thereby having a great influence in Wuhan City and even in Hubei Province.

Kunming Steel has extensive experience in real estate, health care and elderly care, hotel, tourism, and so on. The real estate developed by Kunming Steel every year has reached 500,000 to 1 million square meters. Combined with the characteristics of the earthquake-prone area in Yunnan, it promotes the buildings of steel structure which save materials and energy and are environmentally friendly and with short construction period. Based on development of the internal staff residential community of Kunming Steel, it also actively expands the business to the external market covering Kunming, Honghe, Chuxiong, Lijiang, Dali, Shilin, and other areas. The Lijiang Minority National Middle School and the Yulong Middle School projects reconstructed with all steel structures have been well appraised by all sectors of society. Kunming Steel began to explore the healthcare and elderly care service industry in 2009. First, it established the first home-based elderly care service head station in Xincun Community which has the most concentrated population of Kunming Steel. In 2013, Yunnan Kunming Steel Health and Elderly Care Co., Ltd. was established, becoming the first state-owned healthcare and elderly care enterprise in Yunnan Province. Combining Yunnan's rich and colorful tourism resources, Kunming Steel invested hot spring hotels in Yanbian of Panzhihua City which was known as the "Sunshine Flower City" of Sichuan Province. Besides, it invested to build a featured hotel in Shangri-La, Diqing Tibetan Autonomous Prefecture, Yunnan Province, and invested in the Sofitel Hotel, the tallest building in Kunming. Therefore, Kunming Steel has gained some achievements in its exploration into hotel and tourism business.

2) Major Existing Problems. First, natural deficiencies are exposed during passive transformation of industry. The initial functions of such industries are generally for the logistics support of employees and their families. As the pressure in market environment increases, the Group's development philosophy changes, and the organizational structure has

undergone many reforms. Today, such industries are mostly operating as independent units and participate in severe market competition. Because the scale of the industry is generally small, experience in external market expansion is insufficient; thus, the industrial influence is limited and the overall competitiveness is weak.

Second, the lack of human resources becomes the bottleneck of industrial development. There are many problems in real estate, health care, catering, and other traditional urban service industries such as unreasonable personnel structures, aging staff with lower education, resistance to industry standardization and marketization, lack of professional, practical, and all-round technical talents, insufficient comprehensive technical strength, and large staff mobility.

Third, insufficient investment causes equipment aging. Most of the equipment in the property management, health care, catering, and other industries has high usage rates for a long time without repairing, which gradually brought them into the phase-out period. However, as in recent years most iron and steel enterprises are suffering from losses in the iron and steel business as their main business which affects the groups' operating revenue, the investment in equipment renewal lags behind, especially for those without external business qualification who depend entirely on the internal business of the groups. These factors affect the improvement of industrial competitiveness to a certain extent.

## 12.2 Development Environment and Policy Orientation

### *12.2.1 Macroeconomic Environment for Developing Diversified Businesses*

In 2017, by adhering to the general tone of making steady progress, unswervingly implemented the new development concept, upholding the principle of centering on improving quality and efficiency and taking the supply-side structural reform as the main line, economic performance has achieved stable and positive progress which is better than expectation; thus, the economy and society have maintained a steady and sound development. In 2017, the gross domestic product (GDP) for the whole year was 82.71 trillion yuan, with a year-on-year increase of 6.9%, and the growth rate increased by 0.2%. The added value of the primary industry was 6.55 trillion yuan, with an increase of 3.9%; the added value of the secondary industry was 33.46 trillion yuan, with an increase of 6.1%; the added value of the tertiary industry was 42.70 trillion yuan, with an increase of 8.0%. The proportion of the tertiary industry's added value was 51.6%, which is the same as that of 2016. The tertiary industry constitutes the main driving force for economic growth. Investment in fixed assets with high steel consumption increased by 7.2% year on year, of which real estate development investment increased by 7.0%, with the growth rate increased by 0.1%. For the traditional steel-consuming industries like the mining industry, the automobile manufacturing industry, and the metal product industry, the added value growth rate decreased, while that of the general equipment manufacturing industry and the special equipment manufacturing industry increased. The added value of the mining industry decreased by 1.5%, with its growth rate declined by 0.5%; the added value of the automobile manufacturing industry increased by 12.2%, with its growth rate declined by 3.3%; the added value of the metal products industry increased by 6.6%, with its growth rate declined by 1.6%; the added value of the general equipment manufacturing industry increased by 10.5% and that of the special equipment manufacturing industry grew by 11.8%, with their growth rate increased by 4.6 and 5.1%, respectively, compared with 2016. Such growth has promoted the recovery of the demand in the steel product market.

As China's economic development has entered a new normal, the development environment of the iron and steel industry has been undergoing profound changes. In 2015, China's steel consumption and output both entered the peak arc and presented a downward trend. Oversupply contradiction of domestic steel products has become more prominent, and the main iron and steel industry experienced the period of overall loss from that of low profit. In addition, a number of diversified businesses suitable for the development of the iron and steel enterprises represented by the Internet, finance, logistics, deep processing, and emerging industries are maintaining a high-speed growth. The profits realized by the development of diversified businesses in iron and steel enterprises have made up for some of the losses of their main businesses, greatly improved their competitiveness, and promoted their transformation in surmounting setbacks. In 2016, the iron and steel industry deepened the supply-side structural reform and made great efforts in resolving the overcapacity. As a result, the market showed positive changes and corporate benefits got improved. However, the overall condition of overcapacity has not been changed. The foundation for rising prices and recovering returns is still weak. Steel product exports have been declining, and investment in the iron and steel industry has shrunk significantly. 2017 is a crucial year for the supply-side structural reform of the iron and steel industry. Under the premise of striving to improve the efficiency of the main business, the iron and steel enterprises need to tap new profit growth poles so that it is imperative to develop diversified businesses.

### *12.2.2 Policy Environment for Developing Diversified Businesses*

*12th Five-Year Plan for the Iron and Steel Industry* ([2011] No. 480 of the Ministry of Industry and Information Technology) pointed out "transforming service concept, enhancing service awareness, establishing a strategic cooperation mechanism between iron and steel enterprises and their downstream customers, developing steel product deep processing, and improving the logistics distribution system so as to enhance product value and corporate service function, thus promoting the transformation of steel producers to service providers".

*Guiding Opinions of the State Council on Resolving the Severe Overcapacity* (No. 41 [2013] by the State Council) pointed out "deepen the reform of state-owned enterprises and guide the state-owned capital to transfer from the industries with severe overcapacity to strategic emerging industries and public utilities field".

*Adjustment Policies for Iron and Steel Industry* (Revised in 2015) (Draft for Comment) pointed out "encourage qualified iron and steel enterprise to open up new service areas like e-commerce, Internet finance, futures, logistics, and others so as to promote deep integration of production and sales supply chains, accelerate the transformation from manufactures to service providers, maximize customers' value, and create and share the value of the industrial chain".

*The State Council's Opinions on Resolving Overcapacity of the Iron and Steel Industry for Development* (No. 6 [2016] by the State Council) clearly pointed out encouraging enterprises to withdraw part of the steel production capacity via transformation and other means. Commercial banks are encouraged to increase credit support for the transformation of industries with severe overcapacity in accordance with the principles of controllable risks and commercial sustainability.

In the past five years, China has successively introduced relevant policies to encourage iron and steel enterprise to promote their transformation and enhance their competitiveness by advancing the development of diversified businesses with the purpose that the iron and steel industry can resolve its overcapacity, the capital can be transferred to strategic emerging industries and public utilities field, and the process of structural adjustment can be pressed ahead. Therefore, it is in line with the national policy orientation for iron and steel enterprises to develop diversified businesses reasonably and appropriately based on their own situations.

### 12.2.3 Necessity of Developing Diversified Businesses

Developing diversified businesses is an inevitable choice for iron and steel enterprises to guard against systemic risks. The market economy is cyclical, while the iron and steel industry is a typical cyclical industry. As the price fluctuation of the steel products is sharp, iron and steel enterprises must have plans to cope with risks and the cycles. The world's large-scale iron and steel enterprises, such as Nippon Steel, ThyssenKrupp, ArcelorMittal, and Baosteel, all own diverse businesses related to the main iron and steel business. At present, China's economic development has entered a new normal facing greater downward pressure. Thus, rational development of diversified businesses has become a common and inevitable choice for large-scale iron and steel enterprises to avoid cycles and prevent risks. This direction is in line with the development law of the market economy and meets the needs of iron and steel enterprises to enhance their viability, adaptability, and capability of sustainable development under the market economy.

Developing diversified businesses is a necessary choice to promote the transformation of iron and steel enterprises. Developing diversified businesses has become an unstoppable trend for iron and steel enterprises to shift from production-oriented manufacturing to service-oriented manufacturing and from product competition to industrial chain competition. On the one hand, developing diversified businesses of the iron and steel enterprises is a certain requirement for China to achieve steady growth in industrial restructuring. On the other hand, creating diversified businesses that extend the iron and steel industry chain and expand the value chain will realize synergy between diversified businesses and the main iron and steel business, which has strategic significance for reducing the additional cost of the main iron and steel business and restoring the advanced nature of the same.

Developing diversified businesses is an inevitable choice for revitalizing existing assets and improving the competitiveness of iron and steel enterprises. During the 12th Five-Year Plan, some iron and steel enterprises pursued the "big-and-complete" or "small-but-complete" model in the development process, forming a large number of non-operating assets and idle assets. By developing diversified businesses and carrying out diversified business, such stock assets will be revitalized and economic efficiency will be improved. In addition, compared to tapping more profits from the main iron and steel business through technology accumulation and innovation, the development of diversified industries such as deep processing, related finance, and new emerging industries may bring greater benefits. Meanwhile, improvement of the service level and profitability in diversified businesses is conducive to promoting the core competitive advantage of the iron and steel business [24, 25].

Developing diversified businesses is an inevitable choice for the iron and steel industry to resolve excess capacity and breach the bottleneck. *The State Council's Opinions on Resolving Overcapacity of the Iron and Steel Industry for Development* (No. 6 [2016] by the State Council) clearly states that starting from 2016, crude steel production capacity would be cut by 100 million to 150 million tons within five years. According to the rough calculation of the current per capita steel production capacity of iron and steel enterprises in China, cutting the overcapacity of iron and steel industry means that about 500,000 employees in the industry will face adjustment or re-selection of jobs. Developing diversified businesses by iron and steel enterprises can provide more job posts, divert the personnel in the main business as well as prevent large-scale laid off, and meanwhile, reducing the main business personnel is of great significance in improving productivity and cutting the production costs of the main business.

## 12.3 Case Analysis

### *12.3.1 Pohang Iron & Steel Co. Ltd. (POSCO) in South Korea*

1. Overview of Developing Diversified Businesses

Pohang Iron and Steel Co., Ltd. (hereinafter referred to as "POSCO") was founded in 1968. It has established a future-oriented industrial system featuring the transformation from iron and steel to non-steel, from production to service, and from the steel-centric business structure to a balanced one between traditional and new businesses.

Crude steel output of POSCO registers more than 40 million tons. It has the Pohang Steelworks and Gwangyang Steelworks with the largest scale in the world producing hot-rolled steel products, thick plates, wire rods, titanium products, cold-rolled steel products, hot-dip galvanized products, electro-galvanized products, silicon steel sheets, materials for automobiles, stainless steel, etc. Flat products are the main part in the product structure with the cold-rolling ones constituting a large proportion. Automobile and home appliance sheets are its cutting-edge products. From 2010 to 2017, POSCO ranked the first for 8 consecutive years in the world's most competitive iron and steel enterprises released by the World Steel Dynamics (WSD). Industrial characteristics is to construct a diversified industrial structure with the iron and steel industry as the core and the industries of trade, engineering and architecture, information technology, energy and environmental protection, material and chemical engineering as the support [26–28].

2. Briefs and Reviews of Diversified Businesses

POSCO's unique strategy of development of diversified businesses is of significant characteristics. Its diversified businesses focus on five fields: trade, engineering and architecture, information technology, energy and material and chemical engineering. Profit growth in the future is attributed to diversified businesses, and its subsidiaries have shifted from the dependent type to independent development model. Expanding business scope through diversified businesses will increase revenue sources and customer base as well as reduce operational risks.

(1) Trade. In 2010, POSCO acquired Daewoo International, which is specialized in global trade and investment, which is composed by the trade departments in ferrous metal, chemical engineering, petroleum products, machinery, transportation equipment, and agricultural and livestock products, the overseas project departments responsible for overseas plants and infrastructure development, and the resource development departments responsible for the development of petroleum, fuel gas, minerals, grain, and so on. It is promoting new core businesses such as manufacturing and distribution business and the real estate development business at home and abroad. Exploration of copper, uranium, and tin mines in Africa, the Americas, Southeast Asia, and Australia is ongoing.

Expanding new business spaces such as infrastructure construction, compound real estate development, and ship leasing is also being facilitated. The development of the trade industry brings synergies, which provides the main iron and steel business with a stable supply chain.

(2) Engineering and Construction (E&C). As a representative in the construction field, POSCO E&C has been continuing to expand its fields, namely the civil engineering, architecture, energy, urban development, and others based on the construction experience of POSCO steel plants. It has also expanded its business scope to the areas of low-carbon and green growth field through new renewable energy and urban center recycling business.

It conducts the construction and development of industrial plants, civil engineering, commercial and residential projects in ROK and foreign countries and is expected to become a specialized equipment manufacturer centering on high value-added industries like offshore platforms and power generation equipment. Engineering and construction mainly serve for steel plant construction and equipment manufacturing.

(3) Energy. The energy industry includes electricity, renewable energy (waste gas, solar energy, wind energy, and energy recycling), and fuel cells. Iron and steel/environmental-friendly/energy, light rail, skyscrapers, new urban development, and new/renewable energy facilities are also involved.

POSCO is at the leading status in fuel cells. It has participated in the industries concerning waste gas power generation, photovoltaic power generation, wind power generation, fuel cell, and others domestically and globally, and created an energy value chain based on LNG combined cycle power generation. The development of the energy industry plays a role in reducing the volatility of corporate profits and building a value chain to stabilize the profit of the main business.

(4) Material and Chemical Engineering Industry. POSCO CHEMTECH was established in 1994 by the merger of Samhwa Hwasung and POSCO Shipping. The former one was mainly engaged in the production and sales of alkaline refractories, while the latter one was mainly engaged in the maintenance and installation of various industrial furnaces. As a company of comprehensive raw materials that is independently carrying out the manufacturing and construction of refractory materials, it is commissioned to conduct operation of lime sintering equipment, chemical processing plants, and sales of finished chemical products. The company is also making great efforts to leap to the world's top manufacturer of coal chemistry and carbon raw materials. POSCO CHEMTECH has also diversified its business areas by getting involved in the secondary cell cathode materials, needle coke, and other chemical fields.

(5) Information and Communication Technology (ICT). This part is mainly responsible for production automation and IT equipment. By integrating IT and engineering technologies to create synergies and achieve green growth, the concept of "creating a green ICT future" is considered as its development vision. POSCO has been working with many companies around the world to develop a solution

platform and trying to consolidate its status in nuclear energy, cloud computing, green growth, and renewable energy businesses.

It provides ICT services that lead the integration era, by taking the three major areas of engineering, process automation, and IT services as the core businesses, and it also enters the future-oriented business areas such as LED lighting, smart grid, and cloud computing. At the same time, POSCO has been continuing to strengthen the core strength of green businesses such as environmental-friendly solutions, railways and transportation, renewable energy, and nuclear power plants.

3. Summary

Extend the upstream and downstream industrial chains; develop the resource industry and product service industry that are highly related to the main iron and steel business; integrate the comprehensive utilization of steel resources and the power generation by waste energy and waste heat to develop green environmental protection industry; assist the support for the main iron and steel processes to develop the productive service industry and cross-industry and cross-area combined operation to maximize the coupling effect and synergies.

The newly involved diversified businesses are far from the main iron and steel business with large scale of investment and long investment payback period, which leads to a tight budget, worsening financial conditions and declining credit rating. Entering the diversified sector by means of acquisition and reorganizationof the steel sector is helpful to avoid the risks of higher threshold. Diversified businesses are all effective extensions and supplements to the main iron and steel business. They have the coupling effects in stabilizing the profit of the main business and building the advantage of supply chain. Integration, transformation, and separation of the various diversified industrial sectors and establishing professional joint ventures can maximize the coupling effect of the entire industrial system via building the market competitiveness of each sector.

### *12.3.2 ThyssenKrupp Group (ThyssenKrupp) in Germany*

1. Overview of Developing Diversified Businesses

The ThyssenKrupp Group's diversified businesses are almost all oriented to the development of businesses relating to the iron and steel industry. Based on the deep processing capability of steel products, it has developed basic businesses such as the raw material processing, steel cutting and distribution, laser welding of tailored blanks, mechanical parts and components. Automobile system, shipbuilding, elevators, and other manufacturing businesses are actually the industrial extensions of the above basic businesses; it has developed the businesses like raw material service and industrial service by relying on resource integration, logistics, and warehousing management capabilities; it has developed businesses like plant technologies and project management by relying on the experience and technologies in steel plant construction. In ThyssenKrupp's diversified system, each business requires market leadership, usually with the positioning of top three in global or regional markets.

2. Briefs and Reviews of Diversified Businesses

    (1) Technological Industry of Components and Parts. The ThyssenKrupp is a major manufacturer of large-scale pivotal bearing in the world whose components and parts have been applied in the general mechanical engineering equipment, conveyors, mining and mineral beneficiation systems, port equipment, shipbuilding, cranes, earthmoving machinery, etc. This industry also includes the production and supply of track and track assemblies for mining, forestry and agriculture, excavators, bulldozers, and tractors. In addition, the industry has developed and manufactured the steering shafts, steering columns, and steering gears that have been applied in millions of vehicles around the world for safe driving. The ThyssenKrupp Group's component technology industry is synonym of the high-tech and innovative automotive chassis products in the world. It has the ability to produce and supply shock absorbers and suspension components for all passenger cars and trucks as well as state-of-the-art solutions.
    (2) Elevator Industry. The elevator industry brings together the Group's worldwide activities in the passenger conveying system. Its range includes passenger and cargo elevators, escalators, pedestrian walkways, passenger boarding bridges, stair, and platform lifts. In addition to systems that supply products to the mass market, its portfolio products include customized solutions and services, maintenance and modern packaging tailored to customers' requirements. Thanks to the innovative, safe, and reliable technologies, as well as the energy-saving, environmentally friendly, and high-quality products, the ThyssenKrupp Group's elevator technology industry has been always able to set new market benchmarks.
    (3) Industry of Industrial Solutions. This industry offers a full range of professional engineering and construction services, and its history can be dated back to several centuries ago in shipbuilding. It is also the core strength of the ThyssenKrupp Group. High-quality engineering is a key to a company's success, and this industry provides chemical plants, refineries and other industrial facilities, industrial equipment for cement, minerals, and others, machinery and systems for ore mining, processing and treatment and raw materials and mineral transportation, as well as bodywork and final assembly equipment, including A–Z engineering services from the design to construction of related controlling and experimental equipment for automotive manufacturing plants and suppliers. New fields of the industry include solutions for automotive alternative battery and driving system, innovative automotive lightweight design solutions, and equipment and experimental systems for the aerospace industry. Built on its innovation and technical expertise, its industrial solution industry is one of the world's leading players in the shipping industry, which can provide customers with high-quality military products and services.

(4) Material Service Industry. The materials service industry has 500 branches in 40 countries. The industry is committed to the global distribution of materials and providing comprehensive technical services to the productive and manufacturing businesses. Forty percent of the Group's products are sold through these material service centers in Europe. Products of ThyssenKrupp Group's materials services industry include carbon steel, stainless steel, steel pipes, nonferrous metals, and plastics. It also offers a wide range of customized materials services extending from processing and logistics to warehousing and inventory management. Service centers and branch offices of this industry have extensive processing equipment, which are able to conduct cutting, clipping, plasma or laser cutting, sawing, drilling, milling, and coating for materials. The development of the material service industry has enabled the ThyssenKrupp Group to transform from a material manufacturer to a material service provider, thus greatly enhancing the Group's core competitiveness.

3. Summary

(1) Some diversified businesses have strong correlation with the iron and steel business. The development of diversified businesses is inseparable from the basic support by iron and steel materials.
(2) Long-term accumulation and strong technology R&D foundation and ability lead to the current high-level and strong technical innovation ability to ensure its global leadership in the diversified businesses.
(3) A clear understanding of the development of diversified businesses is acquired. It resolutely withdraws from the diversified sectors with bleak prospects or saturated market, such as the sector of selling laser tailor-welded blank. Diversified businesses with promising prospects are getting more investment to seek higher returns.

### *12.3.3 China Baowu Steel Group Corporation Ltd. (China Baowu)*

On December 1st, 2016, the former Baosteel Group Co., Ltd. and Wuhan Iron and Steel (Group) Co., Ltd. formally got merged to form China Baowu Steel Group Co. Ltd. (hereinafter referred to as China Baowu). This case focuses on the analysis of the Baosteel Group's diversified businesses.

1. Overview of Developing Diversified Businesses

Based on practical exploration, development of Baosteel Group's diversified businesses is achieved in stages gradually via analyzing and drawing on the experiences on diversified businesses of the world's steel giants. The process can be roughly divided into five stages. The first stage (before 1995) is a start-up stage of diversified business during which the decision-making operation concept of "self-supporting and self-digesting" was introduced; the second stage (1995–2004) is the booming stage of diversified business during which a strategic orientation of building a transnational, cross-industry, and large-scale enterprise group with one main industry and several supporting businesses was pursued; the third phase (2005–August 2007) is the adjustment stage of diversified business during which an overall strategic framework with one cutting-edge industry and appropriately developed related diversified businesses was formulated; the fourth stage (August 2007–December 2012) is the optimization and convergence stage of diversified business during which the new strategic concept featuring innovative development mode and dynamic optimization of diversified businesses' structure was determined; the fifth stage (2013–2016) is Baosteel's new round of planning period with the focus on development, transformation, and building a healthy industrial portfolio [29].

2. Briefs and Reviews of Diversified Businesses

By the end of November 2016, Baosteel Group has formed "1 + 6" development model with iron and steel business as its score business and resource development and logistics, steel extension processing, engineering and technical services, coal chemical engineering, financial investment, production services, and so on as its strategic diversified businesses. The following are the introduction for each diversified business sector:

(1) Resource development and logistics industry. The resource development and logistics industry is the upstream industry of the iron and steel industry's main supply chain. It is mainly engaged in the development and operation of iron ore, coal, and other resources to ensure the resource supply for steel production. As the main unit of this industry sector, Baosteel Resources Co., Ltd. is mainly engaged in the investment, trade, and logistics services of mineral resources. This company also strives to build resource investment, trade, and integrated logistics service platform for iron and steel industry and the relevant industrial fields. In 2015, Baosteel Resources realized operating revenue of 29.4 billion yuan.
(2) Steel Extension Processing Industry. The steel extension processing industry is the downstream industry of the iron and steel industry's main supply chain. It mainly utilized Baosteel's advantage in steel production to develop the steel extension processing industry. As the main unit of this industry sector, Baosteel Metal mainly engaged in metal packaging, industrial gases, metal products, automobile trade, etc. In 2015, Baosteel Metal achieved an operating revenue of 10.14 billion yuan with a profit of 310 million yuan.
(3) Engineering and Technical Service Industry. The engineering and technical service industry shoulders the missions of enhancing Baosteel's independent capabilities in integrated innovation, supporting the iron and steel business to keep streamlined and efficient, and promoting the industrialization of engineering technology. As the main unit of this industry sector, Baosteel Engineering Technology Group Co., Ltd. covers a business scope of EPC project, project management and engineering consulting, engineering design, engineering bidding, engineering supervision, equipment design and manufacturing, supply of

complete set of equipment, construction management, operation and maintenance, product inspection, etc. Its service range extends from metallurgy to environmental protection, energy, finance, coal chemical, municipal industries, and others. In 2015, Baosteel's Engineering Technology Company achieved an operating revenue of 10.67 billion yuan.

(4) Coal Chemical Industry. The coal chemical industry is a resource utilization industry, which is mainly engaged in the gas refining after coke oven in iron and steel enterprises and the production and sales of coal chemical products. It has achieved recycled usage and product conversion based on coking by-products. By the end of 2015, Baosteel's Chemical had an original fixed assets of 9.11 billion yuan, total assets of 5.34 billion yuan, the treatment capacity of 4.2 billion cubic meters of coke oven gas, 950,000 tons of tar and 250,000 tons of crude benzene treatment, and production capacity of 240,000 tons of carbon black. The quality of needle coke products has further improved, and it has already occupied a place in the high-end market. Its tar processing capacity has ranked among the international top level with the scale advantage of developing into a world-class coal chemical enterprise. In 2015, Baosteel Chemical achieved a total operating revenue of 7.63 billion yuan with a profit of 40 million yuan.

(5) Financial Investment Industry. The combination of the financial investment business and the main iron and steel business can effectively promote the development of the iron and steel business and enhance its comprehensive competitiveness. Hwabao Investment Co., Ltd. is committed to the equity investment business and selects outstanding leading enterprises to make direct investment. While actively seeking investment opportunities in the upstream and downstream industrial chains related to the mainstream financial industry and the iron and steel business, it has also increased its investment in anti-cyclical industries such as public facilities, large-scale consumption, and medicine and health care. Focus investment at right timing was also made in energy saving and environmental protection, new materials, high-end equipment, and some high-tech industries; in addition, with market-oriented operational advantages, it has indirectly participated in the equity investment business. In 2015, Hwabao Investment realized an operating revenue of 5.67 billion yuan with a profit of 2.83 billion yuan and management assets of 763 billion yuan.

(6) Production Service Industry. The production service industry has two strategic tasks, namely providing highly efficient services for the iron and steel business and realizing the industrialization of renewable resources. As the main undertaker of this business sector, Baosteel Development Co., Ltd.'s main business includes the service covering new building materials, magnetic materials, environmental improvement, quality life, auxiliary production, logistics and distribution, factory property, and others, which can provide systematic and integrated solutions for industrial enterprises and urban systems. By based in Baosteel and extending to the whole society, it is committed to becoming a comprehensive utilization service platform for solid waste resources, a quality life integration service platform, and an industrial collaborative service platform.

In 2015, Baosteel Development achieved an operating income of 7.25 billion yuan and a profit of 590 million yuan.

3. Summary

After development of over 30 years, China Baowu's diversification has witnessed its success and twists. Through summarizing the past and facing the future, several points are worth our pondering:

(1) Developing diversified businesses by centering on the value chain of main iron and steel business. It must be ensured that the development direction of diversified business should be insisted around the main value chain of iron and steel business, so as to select appropriate businesses to make moderate development and form the industrial development policy of the joint industrial structure. Meanwhile, improving the core competitiveness of main iron and steel business should be always taken as the first element the Group's development.
(2) Core competence is the only way for related diversified businesses to become stronger. To make the diversified businesses stronger, it is necessary to find its own position in a certain link of the iron and steel production industry chain, share the resources among the similar main businesses, and pursue the coordinated development among them. Cultivating its own core competence constitutes the key to the success of the diversified businesses.
(3) Uphold the principle of marketization in developing diversified businesses. Separation of main and supporting business and the deconstruction of the value chain releases part of the resources of iron and steel enterprises. However, to truly cultivate a strong diversified industry, the enterprise must go out of the internal market and adhere to the principle of market competition.
(4) Effectively guard against the risks of diversified operations. Entering new industries is not a simple process of "buying". After entering a new industry, enterprises need to continuously inject follow-up resources, foster their own workforce, and establish corporate brands. Meanwhile, enterprises should try not to enter areas that they are not familiar with, which means "no involvement in unfamiliar area".
(5) Improving the market exit mechanism. Due to historical reasons, the diversified businesses involved in iron and steel enterprises are scattered in many fields with large extent of overlapping business. Some of the industries have serious losses, and some do not meet the company's overall development strategy. There must be innovation, new ideas, and an effective exit mechanism.

### *12.3.4 HBIS Group Co. Ltd. (HBIS)*

1. Overview of Developing Diversified Businesses

HBIS Group Co., Ltd. (hereinafter referred to as "HBIS") is the second largest steel group in China. It now has 20 sub-branches such as Tangshan Iron and Steel Co.,

Ltd. and Handan Iron and Steel Co., Ltd. with a total of 121,400 employees and an asset size of 324.287 billion yuan. In recent years, HBIS has been actively taking the path of diversified development, extending the industrial chain, and broadening the industrial scope. By putting equal emphasis on "creating brand-new businesses" and "making innovations in existing businesses", it has scored remarkable fruits.

The development of diversified businesses in the HBIS presents the "double-layered" pattern between the Group and its subsidiaries: The Group fully leverages its integration advantages and has established three major operating companies for procurement, sales, and its international business as well as several specialized companies in mining, international logistics, finance, and other areas. It has initially formed five industry sectors in mining resources, modern logistics, financial securities, steel trade, and equipment manufacturing; based on revitalizing stock assets and optimizing resource allocation, Tangshan Steel, Handan Steel, Xuanhua Steel, and other steel plants have promoted the market-oriented transformation to the diversified businesses and cultivated business sectors such as comprehensive resource utilization, energy conservation and environmental protection, steel extension processing, engineering technologies, and social services. Diversified businesses have become an important point for HBIS's efficiency improvement and profit supplement, and have contributed 30% of the Group's revenue as well as 50% of its employment [30].

2. Briefs and Reviews of Diversified Businesses

    (1) Mineral resources. During the 12th Five-Year Plan period, resource control of HBIS Mining Company has increased significantly from 889 million tons to 3.9 billion tons, with an increase of 338.7%; the output of iron ore concentrate increased from 5.13 million tons to 10 million tons. During the 13th Five-Year Plan period, the seven key mine projects currently under construction will also be put into operation.
    (2) Modern logistics. HBIS owns 8 companies in logistics industry. The total logistics throughput of main iron and steel business stands at 180 million tons/year, and the annual value of their logistics resources is over 20 billion. However, the logistics entities at the Group level face "limited resources and lack of strength", while the logistics business at the level of each steel plants is "separate and uneven". There is no overall integration, so that the value is scattered with great losses.
    (3) Iron and Steel trade. HBIS International Company is mainly responsible for the import of raw materials and fuels, the export of products, equipment import management and agency, overseas investment, investment introduction, and other businesses. In recent years, its ore imports accounted for about 5.0% of the national total imports, and its steel exports accounted for 7.5% of the national total exports. The main value-added services cover two major aspects: One is the implementation of a large-scale logistics strategy to carry out freight forwarding business on their own; the other

is the expansion of financing channels to optimize financing structure and realize financing value adding.

(4) Finance and securities. Currently, HBIS's institutions involving in financial business mainly include the HBIS Finance Company and Caida Securities Company. In 2014, operating revenue of HBIS's financial and securities business registered 1.910 billion yuan with a profit of 1.232 billion yuan and total assets of 28.895 billion yuan.

(5) Deep processing of steel products. The steel product deep processing of HBIS is currently carried out by the Tangshan Huaye (Tianjin) Steel Marketing Co., Ltd., Handan Iron and Steel (Linzhang) Industrial Park, and Wugang Sanhesheng Machinery Manufacturing Construction Co., Ltd. As of the end of 2014, HBIS's steel product deep processing capacity has reached about 3.6 million tons, which needs to be further strengthened.

(6) Equipment manufacturing. HBIS has 7 equipment manufacturing enterprises. Tangshan Steel Machinery (Heavy Machinery) Equipment Company, Handan Steel Equipment Manufacturing and Installation Co., Ltd., and Shijiazhuang Steel and Jingcheng Equipment Technology Co., Ltd. perform well, but the overall development of the industry is relatively weak.

(7) Comprehensive utilization of resources. The comprehensive utilization of HBIS mainly involves in industrialized gas, solid waste, and other resource utilization businesses. Its Industrial Gas Company has realized the development and utilization of high-purity gases, rare gases, and special gases. The added value and profit margin of the products have been greatly improved, and the external market has been effectively exploited with an annual output of about 1.6 billion yuan and the total profit of 300 million yuan. The overall level of solid waste comprehensive utilization needs to be improved. Most of the solid waste resources are directly sold out, and the added value of solid waste utilization is relatively low. The annual output of solid waste resources is about 30 million tons with a total value of 650 million yuan by comprehensive utilization.

(8) Engineering technologies. The engineering technology industry of HBIS is relatively scattered, which is currently mainly carried out by the Tangshan Steel International Engineering Technology Co., Ltd. and the Design Institute of Handan Iron and Steel Group. The annual revenue of this sector has achieved 300 million yuan.

(9) Medical and health. The medical and health industry of HBIS mainly includes five Grade II A affiliated staff hospitals. After years of development, it has cultivated an employee team with good comprehensive quality and strong ability, which has made prominent contribution for protecting the health and safety of the employees in steel plants and keeps the personnel stability of the same. Under the impetus of the reform of the enterprise mechanism and system in recent years, the hospitals have

been gradually facing the society. But due to the lack of flexible market-oriented operational mechanisms, most hospitals are confronted with many difficulties.

(10) Social services. The social service industry of HBIS Group mainly includes property management, vocational education, catering and accommodation, landscaping, industrial tourism and real estate, etc. During the 12th Five-Year Plan period, the Group actively promoted the market-oriented transformation of its subsidiaries' life logistics service system. By relying on internal related markets, this system fully participated in social competition, with the help of innovated management system and mechanism to revitalize the existing resources and assets.

3. Summary

Through years of development, HBIS has provided a good resource guarantee for the development of the Group's diversified businesses, including capital, land, talents, technology, and many other aspects based on its strong scale advantages, regional influence, and resource integration capabilities. The Group has initially formed five major industrial sectors, namely mining resources, modern logistics, financial securities, iron and steel trade, and equipment manufacturing. Subsidiaries like Tangshan Steel, Handan Steel, and Xuanhua Steel have focused on cultivating comprehensive resource utilization, energy conservation and environmental protection, steel product extension processing and engineering technology, social services, etc. Various subsidiaries in diversified businesses have basically formed their own leading products and have acquired certain market competitiveness in related fields. A solid foundation for the rapid rise of diversified businesses has been laid.

In addition, although HBIS has achieved outstanding results in developing its diversified businesses in recent years, such development still does not meet its goal of being a world-class steel complex. Compared to those advanced steel groups at home and abroad, HBIS lags behind in the development of diversified businesses. For example, the development of its financial industry lacks impetus, the combination of industry and finance is weak, and its contribution to the overall development of the Group is insufficient; its management mechanism needs to be further streamlined, the synergy between the various subsidiaries has not been realized, and the ability to cope with the external market is lacking; restrictions posed by talent shortage have become increasingly obvious so that the institutional mechanism guarantee for the introduction and training of talents in the diversified businesses needs to be further improved; high labor costs have seriously weakened the market competitiveness of enterprises; leading enterprises with strong influence are insufficient; problems such as insufficient exploration of external market need to be solved step by step.

### 12.3.5 *Kunming Iron & Steel Co. Ltd.*

1. Overview of Developing Diversified Businesses

Kunming Iron & Steel Co. Ltd. (hereinafter referred to as Kunming Steel) was founded in 1939 and was formerly known as China Electric Steel Plant and Yunnan Steel Plant. After years of development, Kunming Steel has been built into a huge group with the coordinated development of iron and steel business and diversified business. It is the largest state-owned enterprise in Yunnan Province and has been among the top 500 Chinese enterprises for many years. In 2007, Kunming Steel's main business and WISCO Group made a strategic restructuring and formed the WISCO Group Kunming Iron and Steel Co., Ltd., which is controlled by WISCO Group. After the reorganization, work focus of Kunming Steel has gradually been shifted to the development of diversified businesses. The related diversified businesses have seen rapid development, and a multi-industrial system with synchronous development of coal–coke chemical industry, mining exploitation, equipment manufacturing, new materials, cement building materials, and modern service industry has been established. In 2015, the revenue of Kunming Steel's diversified businesses has exceeded 50% of total operating revenue, making diversified businesses an important pillar for its development (Fig. 12.4). The characteristics of its diversified business development are: basing on the main iron and steel business, starting from the steel-related businesses, developing the resource industry necessary for steel production, enriching the steel-consuming industry and the steel service industry, and expanding the new materials, modern service industry, and others with certain relations to its iron and steel business.

2. Briefs and Reviews of Diversified Businesses

   (1) Coal–coke chemical business. Through various measures such as capacity replacement, mergers and acquisitions, and technological transformation, Kunming Steel has basically established a complete industrial chain of coal–coke chemical industry, owning coal mines and coal washing plants. Its coke output capacity stands at 4.5 million tons/year, and its deep processing capacity of coal-based products registers 1 million tons/year. The product varieties in this sector have reached 24. Based on the "backdoor" restructuring of Malong Company, it has successfully been listed in the name of "Yunmei Energy". As a result, the coal–coke chemical industry has become an important one of the diversified businesses of Kunming Steel. Kunming Steel's coal–coke chemical industry is in line with the development model of being economical, clean, and recycling. It has the technical and economic conditions in coal resource control and industry integration. While satisfying the demand for coke in the main iron and steel business, it also creates new profit growth points for the enterprise.
   (2) Mining exploitation business. Via strengthening the ore prospecting and resource integration, promoting the comprehensive utilization of tailings,

**Fig. 12.4** Map of Kunming Steel's diversified businesses

focusing on the R&D and promotion of new technologies in beneficiation and smelting, and implementing such strategies and measures as going global, Kunming Steel has been continuously improving the overall level of its mining exploitation industry with nine iron ore mines being built successively. It now owns an iron ore reserve of 800 million tons and an iron ore concentrate output capacity of 6 million tons per year. The mining exploitation industry of Kunming Steel has effectively provided low-cost raw materials for its iron and steel business, broadened the resource channels, and consolidated the foundation for the company's sustainable development.

(3) Equipment manufacturing business. In 2008, the utility and auxiliary facility maintenance works and their personnel of Kunming Steel had been separated from Kunming Steel to set up a heavy equipment manufacturing group

so as to systematically upgrade its research, development, design, and manufacturing capabilities on large-scale complete sets of equipment. It gives priority to the development of heavy lifting equipment, wear-resistant materials, heavy castings and forgings to provide the equipment required for the manufacturing of large-scale equipment in Yunnan Province and promote the differential development of the equipment manufacturing business. The shield machine, ball mill, large steel castings, and other advanced products can meet the various customized needs of customers. While supporting the rapid development of the local equipment manufacturing industry, Kunming Steel's equipment manufacturing business has also created a number of regional leading technical records.

(4) New materials business. It is mainly aimed at developing high-end materials consumed by the manufacturing industry, dedicated to the R&D and production of new materials like stainless steel composite materials, iron-based powder for 3D printing, cold-rolled titanium coil, stone paper, etc. A number of its products have filled the domestic blank, and the industry has basically formed a certain scale. New materials business of Kunming Steel is aiming at the world's advanced level. As it has been continuously improving the innovative R&D capabilities and core competitiveness, this sector enjoys a promising prospect.

(5) Building materials business. Through direct investment, restructuring and merger, and joint venture holding, Kunming Steel has promoted the rapid development of the cement building materials business. It has completed 16 new dry process cement clinker production lines with a production capacity of 20 million tons/year. By rapidly developing the building materials business through multiple measures at the same time, Kunming Steel has become a leading enterprise in the cement building materials industry in Yunnan and ranked as top 500 Chinese building materials company.

(6) Modern service business. On the basis of integrating the original service business, the company has been continuously expanding the fields of modern service business and formed a modern service business system with modern logistics, real estate, health care, and social welfare services as its main business segments. Sales revenue of Kunming Steel's modern service business accounted for nearly 30% of its total revenue, which effectively accelerated its transformation and development.

3. Summary

Kunming Steel takes the extension of the industrial chain as a breakthrough, prioritizes the development of familiar businesses, integrates relevant resources and elements, and continuously expands the business system. It has soon created a sound situation where diversified businesses have generated profits and efficiency. A good result has been achieved in helping the company make transformation and surmounting difficulties. As the development model of the diversified businesses in Kunming Steel is easy to push forward and can realize scale benefits rapidly, it is worthy of reference by other iron and steel enterprises. Kunming Steel's diversified businesses cover a wide range of fields and can achieve significant results in various areas. However, the development core is not clear enough so that it is necessary to further integrate the elements, concentrate the advantageous resources to build the core competitiveness, and form a well-known brand so as to develop the diversified businesses into a new engine in driving the transformation and development of enterprises.

### 12.3.6 *Shagang Group*

1. Overview of Developing Diversified Businesses

Under the guidance of "strengthening the main iron and steel business, building greater modern logistics, and operating well the non-steel industries", Shagang Group takes active exploration, steady advancement, and whole process risk control as the basic principles. While developing the main iron and steel business, it also actively accelerates the development of the non-steel business, effectively utilizes existing resources, and gradually implements a diversified business strategy in a moderate way. At present, the main investment areas of Shagang's non-steel business include resources and energy, trade logistics, financial investment, industrial chain extension, venture capital, and real estate with an accumulated investment amounting to tens of billions of yuan. It has formed 69 non-steel companies successively. Among them, the resource and energy sector has an annual output of nearly 40 million tons of metallurgical raw and auxiliary materials such as iron ore, coal, coke, alloy, and refractories and a processing capability of 10 million tons of solid waste resources such as steel slag and Fe-containing dust and mud. The logistics sector mainly includes trade in steel products, metallurgical raw materials and equipment as well as the business of Jiulong Steel Logistics Park. In 2016, raw materials trade volume of Shagang International Trade Co., Ltd. registered 56.117 million tons with a sales revenue of 37.09 billion yuan, while Jiulong Logistics Park realized an operating revenue of 120 billion yuan; in extension of the industrial chain, it has realized an annual output of 160,000 tons of steel strands, 300,000 tons of welded steel pipes, and 400,000 tons of coal and chemical products; by the end of July 2015, 4.1 billion yuan of petty loans and a total loans of 21.6 billion yuan have been granted in the financial sector; by the end of 2014, the venture capital investment in 23 venture capital enterprises and one private equity investment fund with a total investment of over 6 billion yuan have been made; a total construction area of 800,000 $m^2$ has been developed in the real estate sector. In 2015, Shagang's sales revenue was 203.8 billion yuan with a profit of 1.897 billion yuan. Among them, the non-steel businesses' profit accounted for about 20% of the total profit from January to August. In the next five years, Shagang aims to increase the profit proportion of the non-steel businesses to 50%.

2. Briefs and Reviews of Diversified Businesses

   (1) Resource and energy sector. In order to ensure the stable supply of raw materials for the main iron and steel business, Shagang Group has invested

in iron ore, coal, coke, alloy, limestone, metallurgical auxiliary materials, and other supporting resources via mergers and acquisitions, joint ventures, and other methods. Now it has formed a scale with an annual output of 14.38 million tons of iron ore, 15.5 million tons of coal, 6.3 million tons of coke, and 150,000 tons of silicomanganese. At the same time, it has the capacity to process 6 million tons of BF slag micro-powder, 3.2 million tons of steelmaking slag, and 450,000 tons of Fe-containing dust and mud annually. Apart from meeting its own demand as the metallurgical auxiliary materials, the extra BF slag micro-powder, steelmaking slag and tailings, steel slag baking-free bricks, and other products are sold to the market, which has made important contributions to the normal production and economic growth of the main iron and steel business.

(2) Trade logistics sector. The trading business mainly covers steel products, bulk raw materials (ores, coking coal, and scrap steel), import and export metallurgical equipment and others, scrap steel recycling, as well as the procurement and supply of limestone and other metallurgical auxiliary materials. In 2016, the trade volume of raw materials amounted to 56.17 million tons with a sales revenue of 37.09 billion yuan. Among them, the total export volume of steel products was 4.31 million tons, ranking the first in the country's individual steel plants for three consecutive years. The Qianlong Logistics Park focuses on becoming the "information center, trading center, settlement center, price center, processing center" for steel products logistics and establishing six supporting platforms in "warehousing and distribution, extended processing, e-commerce, bonded logistics, financing guarantee, and integrated services". At present, more than 1600 enterprises in total have settled in the park, creating an annual business revenue of over 160 billion yuan.

(3) Financial investment sector. In order to realize the effective allocation of the funds in the Group, maximize the value of funds, effectively promote the virtuous growth of the Group, and form a beneficial supplement to the main iron and steel business, Shagang Group has established several companies such as Shagang Finance Co., Ltd., which formed a certain scale in the financial investment field. As of the end of July 2015, Shagang's petty loans totaled 4.1 billion yuan; its cumulative issuance of MTN reached 10 billion yuan, short-term financing 9.6 billion yuan, and ultra-short-term financing 2 billion yuan; the bank's credit granted to Shagang totaled more than 150 billion yuan; in addition, its financial institutions like Shagang Investment Bank and Shagang Securities have all achieved a certain return on investment.

(4) Industrial chain extension. In order to extend the industrial chain of steel products, Shagang Group fully leverages its resource edge steel products and coking by-products and others as well as cooperates with the leading enterprises in the industry to jointly invest in steel product deep processing and manufacturing enterprises and fine coal chemical enterprises, which has further enhanced its product added value and economic profits. Therefore,

it has achieved resource sharing, advantages complementing, and improvement of the Group's competitiveness. Currently, it has an annual output of 160,000 tons of steel strands, 300,000 tons of welded steel pipes, and 400,000 tons of coal chemical products.

(5) Venture capital sector. In order to effectively utilize the capital advantages of the Group and to support the main iron and steel business through capital investment, Shagang Group has cooperated with several well-known domestic investment institutions such as China Science and Merchants Investment Management Group (CSC Group) and GP Capital Co., Ltd. to gradually develop the venture capital business. On April 26, 2016, Shagang Group Investment Co., Ltd., a wholly owned subsidiary of Jiangsu Shagang Group, was formally established with a registered capital of 2 billion yuan. On this basis, the Group founded a number of equity investment companies and investment management companies in Shanghai, Zhangjiagang, Shenzhen, and Hong Kong, which constitutes an investment platform for the secondary industries the Group's priority in the future. While performing well in the main iron and business, Shagang has been actively trying new investment so as to improve the profit margin of the company.

(6) Real estate sector. Adhering to the development concept of "building quality projects and creating value", Hongrun Real Estate Co., Ltd., a subsidiary of Shagang Group, has operated a number of real estate projects with a total construction area of 800,000 $m^2$. The company owns rich experience in real estate development. At present, projects under the company's operation include Jiyang Lake No. 1 and Jiyang Lake Crown with a total planned area of nearly 260,000 $m^2$, a construction area of nearly 360,000 $m^2$, and a saleable area of nearly 220,000 $m^2$. It is one of the top-level high-end communities in the Yangtze River Delta region and has won high praise for its advanced planning concept and exquisite architecture quality.

3. Summary

While maintaining the major status of the main iron and steel business, Shagang has chosen to moderately develop non-steel business which shares a relevance to its main iron and steel business. Through the developing of resources, energy, and trade logistics, it provides a stable supply of raw materials and high-quality supporting services for steel production, laying a foundation for the green, stable, and high-efficiency development. In particular, the logistics business is the strategic focus in developing the non-steel businesses by Shagang. In response to the broad development space of steel logistics, Shagang proposes the goal of "creating another world's top 500" in modern logistics. At the same time, with the core of improving economic benefits, the Group has been actively promoting the integration of industry and finance, giving full play to the respective advantages of the main iron and steel business and the financial business so as to enhance the market response rate and efficiency. In addition, cooperation around steel products and secondary resources has been carried out to extend the industrial chain and further improve the economic performance and risk resistance. The development of Shagang's non-steel businesses has cultivated and formed a new growth point for the Group, thus increasing the overall strength and comprehensive competitiveness of the Group. However, the non-steel business' share and profit contribution rate still have a certain gap compared with those advanced iron and steel enterprises at home and abroad.

## 12.4 Prospects and Path Analysis of Diversification Trend

During the 12th Five-Year Plan period, the main iron and steel industry experienced the new normal of development featuring low growth, low price, low efficiency, and high pressure. The whole industry entered a state of loss. Faced with huge pressure for survival, operation with diversification of iron and steel enterprises has become a new trend for development. In the future, three major development tasks of China's large and medium-sized iron and steel enterprises are: to strengthen the main iron and steel business, enhance the extension to the upstream and downstream of the iron and steel industrial chain, and moderately develop diversified businesses. At this stage, developing diversified businesses in iron and steel enterprises will pay more attention to the transformation from related diversification to non-related diversification, and to the strategic synergy between the diversified businesses.

To this end, the following suggestions and countermeasures are proposed to help the iron and steel enterprises develop their diversified businesses.

### *12.4.1 Scientifically Chart the Development Road Map of Diversified Businesses and Highlight the Guiding Role of Planning*

The strategy of diversified business constitutes an important part of the development strategies of most iron and steel enterprises in the future. It has a bearing on the successful realization of the overall development goals and the effective improvement of the overall competitiveness of enterprises. The development of diversified businesses requires overall planning, systematic arrangements, key highlights, and scientific implementation. Some iron and steel enterprises lack strategic coordination. They chose to develop the diversified businesses as a contingency plan when the main business was in serious loss. Such rigorous decision-making would bring about unsettled payments and new debts. Enterprises should have clear understanding of their own status and find out the weakness and advantages to scientifically chart the road map for the related sectors of diversified businesses and decide reasonable scale, specific measures, and key projects by combining with the characteristics and development trend of the target businesses. It is suggested that iron and steel enterprises should put the planning work of diversified businesses on the equal footing as that of the main business with early arrangement and scientific preparation.

### 12.4.2 Shape the Main Body for Market Competition and Accelerate Innovations in Institutions, Mechanisms, and Management

The main iron and steel business is the cornerstone of the development of iron and steel enterprises. The key to the development of diversified businesses lies in the system and mechanism. There are two main types of management modes for diversified businesses by iron and steel enterprises currently. Most of them adopt the parent–subsidiary operation mode where the group takes control and the subsidiaries carry out independent management, while a small number of enterprises set up specialized companies for the management of their diversified businesses to conduct centralized management and operation of all related businesses. For the time being, iron and steel enterprises face various problems in the management system and mechanism of diversified businesses. In the future, they must accelerate the innovation of management systems and mechanisms to truly establish a business management and control system that is compatible with its own strategies in developing diversified businesses and is in line with relevant development laws. At the same time, they should accelerate the establishment of a modern enterprise system, actively carry out the reform of the property rights system, promote the diversification of property rights and investment subjects in a view to stimulate the development vitality and strive to achieve the transformation from single ownership to multiple forms of ownership. Iron and steel enterprises should also speed up to shift from the "internal-oriented" mode to "market-oriented" mode and make a structural transition from an "industrial group with diversified businesses" to the "non-steel group industry". Relevant enterprises should truly become the legal entity and the body for market competition with clear property rights, clear power and responsibilities, separation of government and enterprises, and scientific management by improving corporate governance structure and standardizing enterprise operation mechanism.

### 12.4.3 Allocate Resources Reasonably and Highlight Development Priorities

During the 13th Five-Year Plan period, main iron and steel business faced multiple pressures on structural adjustment, and the adjustment in the structure of diversified business was also an imperative. In accordance with the principle of "doing something, and not doing something", enterprises should take market competitiveness as the criterion for judging the survival of their diversified businesses. Through comprehensive assessment of the environment for enterprise's survival and development and the trend and conditions for the development of target business, iron and steel enterprises should establish an internal elimination mechanism to optimize integration and allocate the limited resources to those businesses with their promising prospects. In particular, we must focus on seizing the historical opportunities and actively participating in the labor division of social specialization and the economic restructuring. The large-scale enterprises, especially, should have the ability to participate in the exploration and development of emerging industries. For example, by seizing the major opportunities in which the country attaches great importance to the development of strategic emerging industries, iron and steel enterprises can accelerate the development of energy-saving and environmental protection industries. Development projects of urban mineral resources are an industry with broad prospects as another example. With the deepening of China's industrialization and urbanization, cities will become the largest mines in the future. The development of urban mineral resources should be based on scrap recycling and processing status to make further steps gradually.

### *12.4.4 Emphasize the Cultivation of Innovative Ability and Strengthen the Building of the Talent Team*

In developing diversified businesses, iron and steel enterprises in China have their own successful experiences and profound lessons. It is an important consensus to pay attention to the cultivation of innovative ability and the building of a talent team. Insufficient innovation and talent shortage constitute a "bottleneck" restricting the rapid development of the diversified businesses in most iron and steel enterprises currently. For the time being, requirements for the rapid development of the non-steel business cannot be fully met by the technological innovation system and human resource structure in the iron and steel enterprises, especially in the cultivation and introduction of high-level talents. In formulating the development strategy for diversified businesses, it must be focused on those nurturing high-end talents with capabilities in capital operation, international management, international investment and trade, and business innovation to create a group of leading talents for diversified business development. Meanwhile, independent innovation should be put at the height of the enterprise's survival and development. It is of great significance for enterprises to constantly improve the independent innovation development strategy according to their own needs and continuously increase the independent innovation capability. This also serves as the basis for survival in the brutal market competition.

## 12.5 Industrial Practices of Diversification

China Metallurgical Industry Planning and Research Institute (hereinafter referred to as MPI) has a research center for the non-steel business development, which is dedicated to provide guidance and suggestions for iron and steel enterprises to improve the quality and level of development of diversified businesses. The non-steel business development research center provides services for governments at all levels, industrial associations, various enterprises and public institutions, financial institutions, etc. It has long been engaged in studying on the development of non-steel industries at home and abroad as well as carrying out consulting services for iron and steel enterprises in areas such as diversified business planning and strategic research, comprehensive utilization of resources and special planning for circular economy, logistics park planning, e-commerce, and intelligent manufacturing.

Over the years, MPI has made a large quantity of explorations and practices in helping the iron and steel enterprises to achieve diversified development. It has provided consulting services in nearly 30 projects for Anshan Steel, Wuhan Steel, Shandong Steel, Ma Steel, and others, which has scored fruitful results. See details in Table 12.2.

**Table 12.2** Practices of MPI in promoting industrial diversification

| No. | Business sector | Introduction | Typical cases |
|---|---|---|---|
| 1 | Planning for enterprises' diversification strategy | Focusing on the development of diversified business such as industrial chain extension, recycling, energy conservation and environmental protection, financial services, logistics and e-commerce as well as functional manufacturing around the main iron and steel business, formulating the overall strategic plan and diversified development plan in *13th Five-Year Plan* period for many enterprises | *Development Plan of Guangxi Liuzhou Iron and Steel Group during 13th Five-Year Plan* |
| 2 | | | *Development Plan of Wuhan Iron and Steel (Group) Company during 13th Five-Year Plan* |
| 3 | | | *Development Plan of Baogang Group during 13th Five-Year Plan* |
| 4 | | | *Related Diversified Business Development Plan of Shandong Iron and Steel Group during 13th Five-Year Plan* |
| 5 | | | *Entity Industry Development Plan from 2015 to 2020 of Guizhou Hongbo Company* |
| 6 | | | *Development Plan of Masteel Group during 12th Five-Year Plan* |
| 7 | Comprehensive utilization of resources and circular economy | It undertakes researches on circular economy issues in government and industry and carries out circular economy planning of metallurgical enterprises, industrial park circular economy planning, circular economy planning for mineral development and deep processing, etc. | *Researches on the Development of Circular Economy in the Iron and Steel Industry during the "12th Five-Year Plan Period"* |
| 8 | | | *Planning on Building Demonstration Enterprises in Circular Economy of Anshan Iron and Steel Group* |

(continued)

**Table 12.2** (continued)

| No. | Business sector | Introduction | Typical cases |
|---|---|---|---|
| 9 | | | *Planning on Comprehensive Utilization of Vanadium and Titanium and Industrial Restructuring of Panzhihua Iron and Steel (Group) Co., Ltd.* |
| 10 | | | *Circular Economy Industrial Park Planning of Mazongshan in Gansu Province* |
| 11 | | | *Planning on Industrial Demonstration Zone of Circular Economy in Hebei and Tianjin (Tiantie Metallurgy Group Co., Ltd. in Shexian County, Hebei)* |
| 12 | | | *Planning on Circular Economy in Shenyang Metal Deep Processing Industrial Park* |
| 13 | Logistics and e-commerce planning | Focusing on the cost reduction and efficiency improvement of iron and steel enterprises, researches and planning on integration and system optimization for metallurgical logistics management and logistics cost optimization have been carried out. It conducts studies on the development trend of steel-related e-commerce, industrial policies to help enterprises in building e-commerce, operation management, system development, and other e-commerce packaged designs | *Development Plan and Logistics Cost Optimization of Anhui Changjiang Steel* |
| 14 | | | *Diagnosis, Optimization, and Planning Scheme for the Logistics System of HBIS Tangshan Steel* |
| 15 | | | *Strategic Adjustment and System Optimization Plan for Masteel's Efficient Logistics* |
| 16 | | | *Planning for Shagang Group's Jiulong Steel Logistics Park* |
| 17 | | | *Overall Planning for Huaigang Logistics Industrial Park* |
| 18 | | | *E-commerce Planning of Nanjing Iron and Steel Group Corp.* |

(continued)

**Table 12.2** (continued)

| No. | Business sector | Introduction | Typical cases |
|---|---|---|---|
| 19 | | | *Special Planning for Logistics Optimization of Shaanxi Iron and Steel (Group) Co., Ltd.* |
| 20 | Industrial intelligence consulting service | By closely tracking the advanced intelligent manufacturing theories, methods, and technologies, focusing on the core processes like production process, business operation, and corporate decision-making, MPI aims to create a unique platform for industry, universities, and research so as to accurately interface with the actual needs of enterprises and provide related services for enterprises | Business Intelligence System Planning of Tianjin Rockcheck Steel (Group) Co., Ltd. |
| 21 | | | *Feasibility Study Report of Kunming Steel's Internet Plus Collaborative Manufacturing Project* |
| 22 | | | *Informatization Special Plan of Guangxi Liuzhou Iron and Steel Group Company Ltd. during the 13th Five-Year Plan Period* |
| 23 | | | *Three-Year Action Plan on Intelligent Manufacturing of Benxi Steel Group Corporation* |
| 24 | | | Special plans for WISCO, Baotou Steel, CITIC Pacific, Shaanxi Steel, etc. |
| 25 | Other special businesses | Special planning services in finance, management, human resource optimization, and others have been provided | *13th Five-Year Plan and Financial Plan on Capital Operation of Shaanxi Iron and Steel (Group) Co., Ltd.* |
| 26 | | | *Strategic Planning and Financial Leasing of Tianjin Rockcheck Zhonghe Energy Co., Ltd.* |
| 27 | | | *Special Research Report (Capital Operation, Management and Control Model, Human Resource Optimization, etc.) of Ma Steel during 13th Five-Year Plan* |
| 28 | | | *Corporate Culture Consulting of Anhui Yuntian Metallurgical Technology Co., Ltd.* |

## References


1. China Steel Development and Research Institute. (1980–1999). China steel yearbook. Beijing: Metallurgical Industry Press.
2. Wu, X.C. (2002). Thoughts on the development of non-steel industries under the new circumstances. China Steel Focus, 12, 20–23.
3. Department of Mineral Resources Reserves of Ministry of Land and Resources, & Information Center of Ministry of Land and Resources. (2015). Bulletin on national mineral resources reserves.
4. National Bureau of Statistics of the PRC. Raw ore output of iron ore [2017-02-18]. http://data.stats.gov.cn/easyquery.htmcn=A01.
5. www.mmi.gov.cn. (2016, December). Statistics of iron ore imports by countries [2017-02-20]. http://www.mmi.gov.cn/content-117-6457193-1.html.
6. National Bureau of Statistics. (2017). Statistical Communiqué of the People's Republic of China on the 2016 national economic and social development. Beijing: Information Center of the National Bureau of Statistics.
7. www.haiguan.info. (2017). Customs data. Beijing: Beijing Zhonghaitong Technology Co., Ltd.
8. www.mmi.gov.cn. (2016, December). Partial manganese ore import statistics by country [2017-02-12]. http://www.mmi.gov.cn/content-151-6447630-1.html.
9. www.mmi.gov.cn. (2016, December). Partial chrome ore import statistics by country [2017-02-12]. http://www.mmi.gov.cn/content-155-6447632-1.html.
10. GB50405-2007. (2007). Design code for comprehensive resources utilization of iron and steel industry. Beijing: China Planning Press.
11. Information and Statistics Department of China Iron and Steel Association, & Metallurgical Statistics Branch of China Statistics Society of China. Environmental protection statistics of China's Iron and Steel Industry in 2011.
12. Information and Statistics Department of China Iron and Steel Association, & Metallurgical Statistics Branch of China Statistics Society of China. Environmental protection statistics of China's Iron and Steel Industry in 2012.
13. Information and Statistics Department of China Iron and Steel Association, & Metallurgical Statistics Branch of China Statistics Society of China. Environmental protection statistics of China's Iron and Steel Industry in 2013.
14. Information and Statistics Department of China Iron and Steel Association, & Metallurgical Statistics Branch of China Statistics Society of China. Environmental protection statistics of China's Iron and Steel Industry in 2014.
15. Information and Statistics Department of China Iron and Steel Association, & Metallurgical Statistics Branch of China Statistics Society of China. Environmental protection statistics of China's Iron and Steel Industry in 2015.
16. www.chyxx.com. Analysis on the status of deep processing of china's iron and steel industry and the output of the major industrial segmentation products [2016-02-18]. http://www.chyxx.com/industry/201602/386826.html.
17. China Metallurgical Industry Planning and Research Institute. (2013). Materials of 2013 (First) conference on the non-steel industries for the iron and steel industry (internal materials). Beijing.
18. China Metallurgical Industry Planning and Research Institute. (2014). Materials of 2014 (Second) conference on the non-steel industries for the iron and steel industry (internal materials). Beijing.
19. Ministry of Industry and Information Technology. 12th five-year plan development plan for new materials issued by Ministry of Industry and Information Technology [2012-02-22]. http://www.gov.cn/gzdt/2012-02/22/content_2073383.html.
20. Nie, Z.Y., Wen, D.B., Zhao, A.Z., et al. (2017). Research on the status and trend of the new material industry at home and abroad. China Science and Technology, (6).
21. Ma, C. (2011). Research on financial application of the supply chain in iron and steel industry. Qingdao: Ocean University of China.

22. Wang, J.L. (2017). Brief discussion on the status and development of steel logistics. Marine Equipment Materials & Marketing, (05).
23. Li, B.Q. (2013). Road of breakthroughs for steel logistics. China Storage and Transportation, (12).
24. Wan, X.M. (2014). Promoting the coupling development of diversified industries to boost the transformation and upgrading of the iron and steel industry. Metallurgical Economics and Management, (6), 12–14, 15.
25. Zhang, X.F. (2014). Thinking on the future development direction of diversified industries for iron and steel industry. Metallurgical Economics and Management, 06, 4–6.
26. Introduction to POSCO. http://www.posco-china.com.
27. Luo, Y., Yuan, Y.F., Luo, X.G., et al. (2014). Development status of diversified industries in POSCO. China Steel Focus, (8), 41–46.
28. Zhan, H.S. (2015). Analysis on the diversified independent industrial structure model of POSCO. China Steel Focus, 2, 14–16.
29. Dai, M.Y. (2012). Practices of diversified businesses building of baosteel in recent years. China Steel Focus, 06, 35–38.
30. Xie, L. HBIS group going to absorb 50,000 employees by non-steel industry in the next two years (2016-03-25) [2016-03-25]. http://www.ccstock.cn/gscy/qiyexinxi/2016-03-25/A1458851281122.html.

# Chapter 13
# Internationalization

## 13.1 History Review and Status Analysis

### *13.1.1 Product Internationalization*

1. Increase in Total Exports

In recent years, as the global economy recovers, the demand for steel products in many countries has risen. The international competitiveness of China's steel companies has improved, and therefore, China's steel exports have been greatly increased. All these make China gradually transform from the world's largest steel importer to the largest steel exporter. In 2005, China's steel export volume of steel products exceeded 20 million tons for the first time; China's yearly export volume until 2014 all exceeded 40 million tons, except for 2009; in 2015, steel exports continued growing, reaching an all-time high of 112.4 million tons. In 2016, the steel exports reached 108.49 million tons, declining by 3.5% year-on-year.

The changes of China's steel exports from 2005 to 2016 are shown in Fig. 13.1.

2. Export Coverage Area Expanded, and the Concentration Ratio of Major Exporting Destinations Reduced

China's iron and steel enterprises are actively expanding overseas markets and the number of export destination countries is increasing year by year. From 2005 to 2016, China's steel export destination area continued expanding. In 2005, China exported steel products to 193 countries and regions and the number had increased to 234 by 2016, reaching 96% coverage, increasing by ten percentage points if compared with that of 2005.

With China's rapid increase in steel exports, anti-dumping cases initiated by the developed countries take place frequently. In addition, influenced by the decline in demand for steel products due to the slowing economic growth in developed economies, China has shifted the focus of China's steel exports to Southeast Asia and emerging economies. In 2016, the major destination countries for steel exports



X. C. Li, *The Road Map of China's Steel Industry*,
https://doi.org/10.1007/978-981-15-2074-7_13

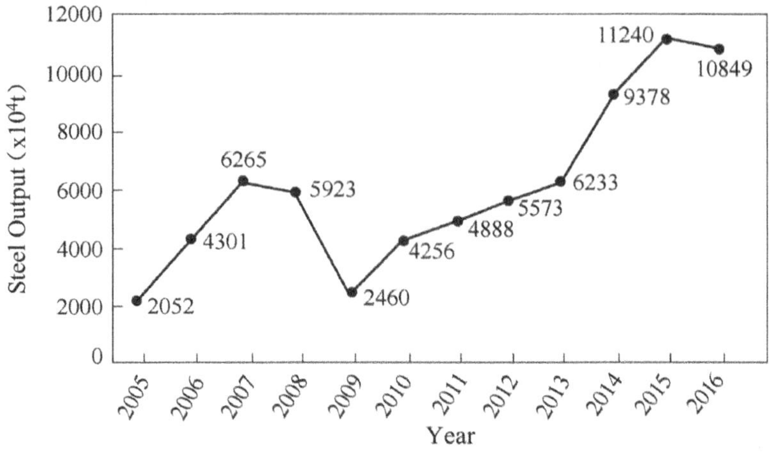


**Fig. 13.1** Changes of China's steel exports from 2005 to 2016

were South Korea accounting for 13.2%, Vietnam 10.8%, Philippines 6.0%, Thailand 5.7%, Indonesia 5.4%, Malaysia 3.1%, India 3.1%, Saudi Arabia 2.9%, Singapore 2.7%, and Pakistan 2.7%.

Judging from the changes in the major export destination countries, in 2016, the top 10 destination countries and regions accounted for 55.6% of steel exports, declining 16.9 percentage points from 2005 [1]. The proportions of exports of China's steel export destination countries in 2005 and 2016 are shown in Figs. 13.2 and 13.3, respectively.

3. Structural Optimization of Export Varieties

In 2016, China's cumulative export of steel products was 108.49 million tons, of which sheet and plate exports were 48.02 million tons, accounting for 44.3% that was 10.3 percentage points lower than the highest proportion of 58% in 2010; rod

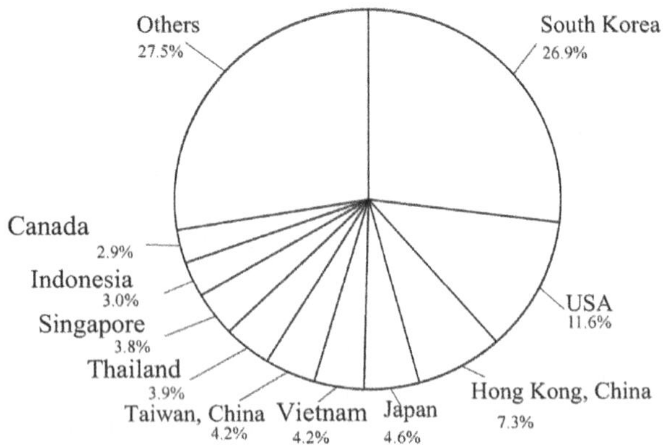


**Fig. 13.2** Proportion of exports of China's steel export destination countries in 2005

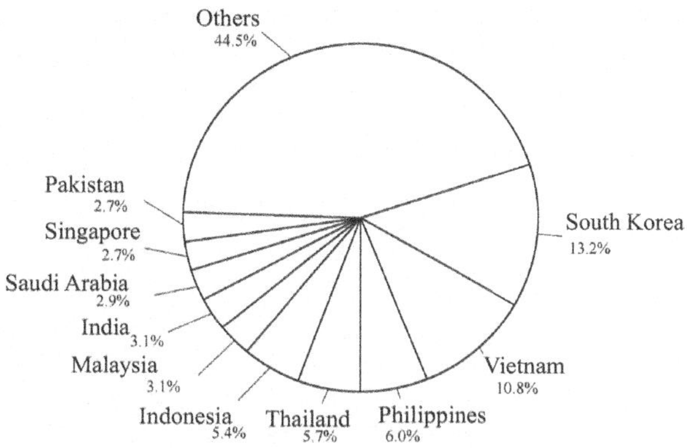


**Fig. 13.3** Proportion of exports of China's steel export destinations in 2016

and wire exports were 41.25 million tons, accounting for 38%, a big increase than the previous peak 10% in 2009; the exports of pipes were 9.67 million tons, accounting for 8.9% that was 17 percentage points lower than the highest proportion of 26% in 2009. The proportion of bar and wire exports increased sharply.

Thanks to China's export policy orientation and the structural adjustment of China's steel products, China's steel export structure continues to optimize itself, and the proportion of high value-added export products has increased. In 2016, the average export prices of China's sheets and plates, rods and wires, pipes, angles and sections were 514 US dollars/ton, 336 US dollars/ton, 811 US dollars/ton, and 408 US dollars/ton, and their export volumes accounted for 45.3%, 25.5%, 14.4%, and 3.8%, respectively. Judging from the unit prices of steel products exported in 2016, the unit prices of plates and strips and pipes are higher, and the unit prices of rods and wires are the lowest.

The rapid increase in the exports of high value-added products in China has created a greater profit margin for enterprises. At the same time, it has created favorable conditions for enterprises to build a good image of international brands and enhance their international competitiveness.

4. The Analysis of Steel Export Competitiveness

The competitiveness of high-end products needs to be improved. The international market share of China in exporting plates has surpassed Japan in only a few years. If considering the low price advantage, we can say that China's competitiveness in the high-end plate industry is actually not as good as Japan. From the perspective of steel imports, China's competitiveness of high-end steel products is nowhere near that of Japan and other countries. The labor productivity of China's manufacturing industry is much lower than that of Japan, Germany, and South Korea.

The technological innovation does not match the scale of the iron and steel industry. The output of China's crude steel production is the world's largest, but the patents in steel industry are only 54% of Japan's. It can be concluded that the innovation power of China's steel industry does not match the industry scale. At the beginning of the twenty-first century, when the world steel production center moved from

Japan and South Korea to China, there was no revolutionary production technology emerged. China's innovation capacity in iron and steel industry still has room for improvement.

5. Major Problems in Steel Exports
   (1) Frequent trade conflicts increase exporting difficulties. In recent years, the USA, the European Union, and ASEAN have frequently initiated trade remedy investigations on Chinese steel products. In the first half of 2017, China's products encountered 37 trade remedy investigations initiated by 15 countries and regions, including 28 anti-dumping cases, 4 anti-subsidies, and 5 safeguard investigations; the total amount of money involved was 5.3 billion US dollars. In the same period of 2016, the number of trade remedy investigations in China reached an all-time high, totaling 65 cases, involving an amount of 8.5 billion US dollars. In the first half of 2017, the number and amount of trade remedy investigations against China dropped significantly, but compared with the same period of the past five years, the overall change was small and was still at a high level.
   (2) Global steel production capacity is in excess and steel export prices are falling. According to the annual statistics of the International Steel Association, in 2016, the global steel overcapacity further intensified. The global crude steel output reached 1.63 billion tons, increasing 0.8% year-on-year; the annual crude steel capacity utilization was about 70%, which was in a downward trend; the per capita apparent consumption of steels in the world continued to decline, the oversupply of the global steel market intensified, and the average price of steel continued to drop, greatly reducing the profit margin of China's export steel products. In 2016, the average export price of steel products from China was 502 US dollars/ton, declining 10% year-on-year. In the next few years, many countries in the world will continue to invest in building steel production capacity. At the same time, the steel consumption power brought down by economic growth is insufficient, and the oversupply of steel will continue to intensify. Therefore, China's steel export prospect is not optimistic.
   (3) The steel export trade mode lacks diversity and the export added value is low. In recent years, China's export of plates and strips has continued to rise, but importers are still dominated by foreign steel traders and its export model is single. Compared with the world well-known plates producers building overseas trade mode of steel products such as POSCO in Korea and Nippon Steel in Japan, China's iron and steel enterprises have not made significant progress in optimizing services and increasing product added value due to the lack of a vertical sales system established with foreign users, which has weakened the competitiveness of China's plates and strips in overseas market. The vertical sales system for plates and strips established for foreign users is mainly to build a steel processing and distribution center, a steel processing center in the production area where the main customers are located, to process the products based on the customer's needs, to provide

customers with timely and accurate steel cutting, processing, and logistics services, and also to provide customers with a higher-level quality services and convenient and diversified services for steel processing and distribution centers through EVI activities in the relevant areas to collect steel demand dynamics and technical information. The establishment of overseas steel processing and distribution centers, on the one hand, can improve the industrial chain of plates and strips of the production enterprises and increase the added value of products; on the other hand, it can provide quality services for downstream users and enhance the competitiveness of China's plates and strips production enterprises in overseas markets.

Furthermore, except HBIS, the overseas outlets of China's iron and steel enterprises mostly adopt overseas representative offices which lack control over overseas markets. China's iron and steel enterprises should learn from the experiences of HBIS which are to adopt joint stock, acquisitions, and other methods to work together with large-scale steel sales companies abroad in expanding overseas sales channels.

(4) The dependence on steel exports and on major exporting destination countries is high. In 2016, China's apparent dependence on steel exports was 9.5%, an increase of 2.3 percentage points over 2005. As China's export dependence increases, the risk of steel exports increases. In 2016, the steel products exports to China's top ten importers accounted for 55.6%. Among them, exports of steel products to South Korea, Vietnam, the Philippines, Thailand, and Indonesia accounted for 41.1%. Considering the political factors such as territorial disputes and anti-Chinese sentiment between China and Southeast Asian countries in recent years, the export environment of the top ten destination countries is not optimistic and the export risks are increasing.

### *13.1.2 Capacity Internationalization*

1. Status Analysis

Since 2006, China's crude steel capacity utilization rate has shown a significant downward. It is especially true for the years since 2013 when the indicator has continued going below the reasonable level, reflecting the fact that the overcapacity problems are prominent, as shown in Fig. 13.4. "*Opinions of the State Council on the Development of the Steel Industry to Resolve Excess Capacity to Eliminate Poverty*" (National Issue [2016] No. 6) clearly states that "encourage enterprises, if conditions permit, to participate in the 'Belt and Road' initiative and carry out international capacity cooperation to transfer part of the production capacity in the principle of mutual benefits". At the same time, the local governments have also introduced policies to encourage iron and steel enterprises to combine the overseas

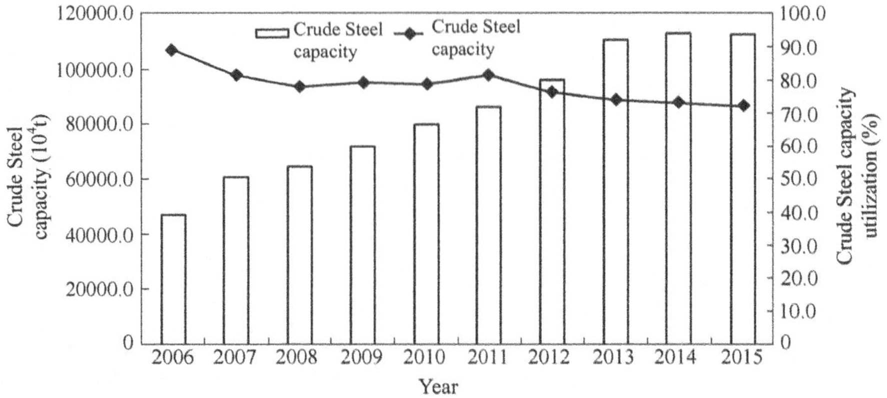


**Fig. 13.4** Crude steel production capacity and its utilization rate in China from 2006 to 2015

mineral resources development with deep processing of overseas resources, developing mineral resources in areas rich in mineral resources and investing in building steel production and processing bases, and to extend the industrial chain of smelting and processing. In the 2017 Report on the Work of the Government by the State Council, it reiterated that "International capacity cooperation should be expanded". Adhere to the enterprise-oriented, government-driven, and market-oriented operation and implement a number of major demonstration projects. Implement and improve fiscal, taxation, and financial support policies, establish RMB overseas cooperation funds, and make good use of bilateral capacity cooperation funds. Promote equipment, technology, standards, and services to go abroad and make the "Made in China" shine.

China's iron and steel enterprises have established overseas plants since the 1990s. In recent years, with the slowdown in domestic steel demand growth and the stimulation of the "Belt and Road" strategy, investment mania to the Belt and Road countries has gained momentum and iron and steel enterprises are making investments actively in setting up overseas factories. In January 2014, NISCO, together with GGS (PT Gunung Gahapi Sakti, a subsidiary of the Indonesian Gunung Steel Group), jointly built a steel plant in Medan, Indonesia, and it will reach a production scale of 1 million tons of steel within five years. In May 2014, the Vietnam-China Iron and Steel Plant jointly established by KISC and VN Steel was officially put into operation. The annual output of the first phase project was 500,000 tons of pig iron and 500,000 tons of steel billets. After the completion of the second phase of the project, the production capacity will double. In September 2014, HBIS concluded a cooperation intention with South African Industrial Development Corporation and China-Africa Development Fund. And now, they have started the construction of a 5 million tons steel project. In March 2015, China Metallurgical Group Corporation, Ma Steel, and Kazakhstan Ferrum Corp. jointly signed a memorandum of joint ventures for building a 1 million tons/year integrated steel plant project. In September 2015, WISCO and West Africa signed an agreement to build and operate a joint venture steel plant with an annual output of 500,000 tons of steel in Liberia, West Africa. In April 2016, HBIS acquired the Smeder Revo Steel Plant in Serbia by means of asset packaging and acquisition [2], which is a major achievement since the promotion of the "Belt and Road" initiative and the international capacity cooperation with Central and Eastern European countries.

In addition, a number of large private iron and steel enterprises in Hebei Province are also actively planning to "go global". The projects currently under construction include Bazhou Xinya Metal Products Co., Ltd. acquiring 30% equity of Indonesian Java Pacific Co., Ltd. to expand production of galvanized steel strip and furniture pipes, and three enterprises including Xingtai Delong Iron and Steel Co., Ltd. and Thailand's Permsin Steel Company have jointly established a 600,000-ton hot-rolled narrow strip line in Thailand. Projects to be finished include a 300,000 tons/a steel plant project jointly established by Qinhuangdao Tonglian Group and First Pacific Mining (Lao) Co. Ltd., and a 2 million tons/a steel plant project in the first phase in Indonesia jointly built by Wu'an Yongcheng Casting Co., Ltd. and Indonesia Lippo Group. Handan Yuhua Iron and Steel Co., Ltd. has signed a cooperation intention [3] to build a 2 million to 3 million tons/a steel project in the Indonesian Chinese Park. In addition, a number of iron and steel enterprises such as Huarui Casting Pipe, Tianzhu Iron and Steel, Guofeng Iron and Steel, Ganglu Steel and Donghai Iron and Steel have also stepped out and conducted field visits and cooperation negotiations with some countries along the "Belt and Road".

2. Pending Problems

At present, China's steel production capacity "going out" is mainly focused on countries of Southeast Asia, Africa, and West Asia. With the help of policies, the "going out" of China's steel production capacity will continue to accelerate in the future. Compared with exports, it is easier to avoid trade conflicts by using joint-stock enterprises as platforms or building factories abroad, but at the same time, political factors, overseas regulations, local customs, infrastructure, environmental assessment, land acquisition, employment systems, etc., may hinder domestic steel enterprises to "going out". The main challenges faced by Chinese iron and steel enterprises in their current "going out" efforts include the following points.

(1) Market capacity is small and environment could be hostile. At present, the steel market in Southeast Asia, Africa, and other countries is generally small in capacity, and a business with a local production capacity of more than 3 million tons is deemed super-large. One million ton production capacity, if transferred to these countries, is big enough to change the local supply landscape and exert competitive pressure on local iron and steel enterprises. Therefore, it is easy to be regarded as an enemy by local competitors who will use their power to force the government to introduce new restriction policies.
(2) High environmental protection costs and strict inspection. Different from the development path of China's past "treatment after pollution", people of Southeast Asian countries are deeply influenced by the Western countries, so they put forward the high-demanding environmental protection requirements while

building and developing the country. Therefore, Chinese equipment transported to ASEAN countries will face more strict environmental protection review and incur greater environmental protection expenditures. This also results in many projects denied or, although approved by government, put on hold or suspended due to public opposition.

(3) Imperfect legal system and disordered competitions. The tax system in Southeast Asian countries is still not perfect, officials are corrupt, and many local enterprises do not pay taxes in accordance with the tax law. Faced with such disorderly competition in Southeast Asia, compliant Chinese enterprises can only be treated unfairly and bear corresponding losses.

(4) Lack of talents and high cost of employment. Although there are many labor force in Southeast Asia and Africa, many of them have basically zero or have just started the steel industry. It is very difficult to find a large number of talents needed in the steel industry. If talents are introduced from the Chinese market, the cost will rise and the local labor cost advantage will not be exerted. Training for locals also costs a lot of time.

### *13.1.3 Resource Supply Internationalization*

1. Iron Ore

   (1) Supply Status China's iron ore supply develops in three stages: the first stage see mainly domestic ore supply; domestic mines could basically meet the demand for iron ores in China's steel industry. In the second phase, ore supply is based on the long-term ore supply contracts (negotiated pricing); as China's iron and steel industry's demand for iron ore increases, the scale of imported ore gradually increases. Thanks to the booming development and trade globalization, the international trade of iron ore also gradually matured and formed its own specific trading practices and price mechanism in the early 1980s, namely the annual pricing long-term mechanism. In the third phase, long-term ore and spot ore coexisted (index pricing, swaps, futures); the traditional iron ore long-term agreement pricing mechanism was shifted to a more flexible pricing mechanism. In addition to pricing indexation, iron ore financial derivatives have emerged in recent years, such as the iron ore swap transactions of the SGX, the iron ore futures of India, and the Chinese iron ore futures of the Dalian Commodity Exchange of China.

   In recent years, due to the sharp increase in demand for iron ore in China, and because the growth rate of China's finished ore production is not as fast as demand growth, China's imported iron ore and external dependence have increased year by year. The internationalization of iron ore supply has been continuously enhanced. The increased import volume from 275 million tons in 2005 to 1.024 billion tons in 2016 [4] made the external dependence increase from 50.2% in 2005 to 87.3% in 2016, as shown in Fig. 13.5.

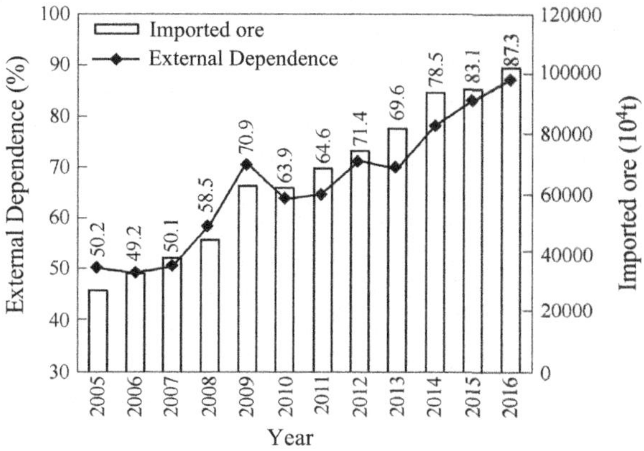


**Fig. 13.5** Changes in the scale and external dependence on imported iron ore in China in recent years

At present, the main sources of iron ore used in China's iron and steel industry are mainly domestic and imported. In 2016, China's domestic iron raw ore production was 1.281 billion tons [5], and after the mineral processing process, the iron ore was provided as finished ore for the iron and steel industry. In 2016, China imported 1.024 billion tons of iron ore [4]. From the perspective of variety, nearly all of domestic ores are iron ore concentrates after undergoing the mineral processing process (including magnetite and hematite, of which magnetite accounts for 3/4), while imported ore includes iron concentrate, ore fines, lump ore, pellet, and burnt pyrites, among which ore fines are majority, reaching 739 million tons, accounting for 72.2% of the imported ore, as shown in Fig. 13.6.

Among the imported iron ore, the supply sources are mainly from three countries. In 2016, the top three sources of iron ore were Australia, Brazil, and South Africa, and their supply volumes were 639.87 million tons,

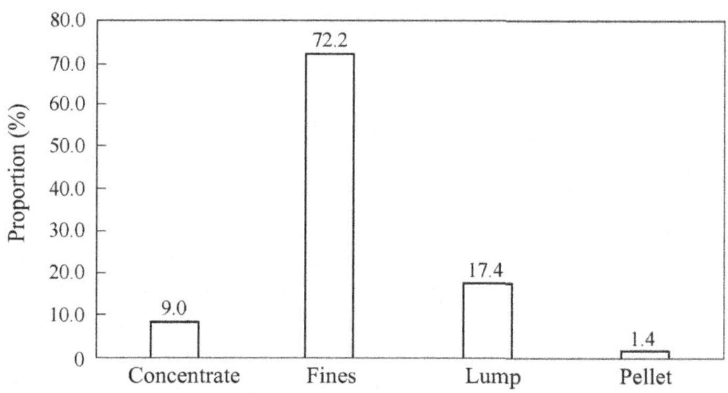


**Fig. 13.6** Proportions of iron ore imports in 2016 by variety

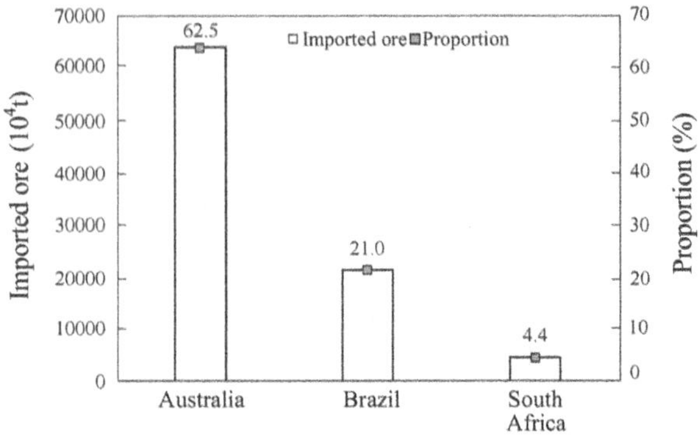


**Fig. 13.7** Proportions of the top three countries in iron ore supply in 2016

214.69 million tons, and 44.85 million tons respectively, and the ratios were 62.5%, 21.0%, and 4.4%, respectively. The three countries together accounted for 87.9% of the total imported ores. However, the relatively dependent and too concentrated procuring channels will bring price and supply risks. See Fig. 13.7 for details.

(2) The International Layout of Iron Ore Supply. China's overseas iron ore development began in the 1980s. Since the twenty-first century, especially after 2005, the price of iron ore sharply has been increasing sharply and Chinese enterprises has begun to develop iron ore overseas on a large scale and the investment in iron ore abroad is accelerating. At present, China needs to import a large amount of iron ore every year, but the proportion of equity ore is low, which has caused huge losses to the iron and steel industry; although China has made a lot of efforts in upgrading its equity ores, it still faces many difficulties in carrying on projects due to insufficient experience in overseas ore investment. The development of Chinese overseas investment in iron ore can be divided into the following four stages.

The first stage (1980–2000): China's steel production was not large, the price of iron ore was low, and the impact of resources on the iron and steel industry was not particularly prominent. The overseas resources investment was inactive, and only two overseas iron ore joint ventures on the development and acquisitions were successfully carried out, namely Sinosteel-Australia Channar and Shougang Peru Iron Ore.

The second stage (2000–2008): China's steel output expanded rapidly, resource demand increased sharply, and iron ore prices continued to rise at a high speed, which brought certain impact to the safe operation of the iron and steel industry. Enterprises gradually attached importance to the establishment of the resource security system and a fever of investing on overseas mines through joint ventures and acquisitions swept the country.

The third stage (2008–2012): China's steel production further increased. The iron and steel industry was paying more and more attention to the stable supply of raw materials and market risks. It came at the time when the country was gradually implementing a basket of policy measures to deal with the economic crisis and, as a result, many enterprises were "going out" to develop overseas resources in that period.

The fourth stage (2012–present): Due to the sluggish recovery of the world economy and the fact that few overseas Chinese investments were successful, Chinese enterprises have taken cautious in overseas investment and there were only two cases, as shown in Table 13.1.

Expansion of international investment on iron ore resources shows that Chinese enterprises have accelerated steps in "going out" in recent years. According to statistics, from 2006 to 2016, the overseas iron ore equity investment of various types by Chinese enterprises totaled more than 24.6 billion US dollars, and Chinese investors participated in the exploration, design and construction of 31 large overseas iron ore projects. As of 2016, the proven and controlled reserves of the mine-related projects by Chinese investors were about 98 billion tons. The planned equity ore production capacity was about 270 million tons, accounting for 54% of the total production capacity of finished ores of 500 million tons in the projects involved.

China's overseas investment in iron ore is mainly concentrated in Western Australia, Quebec, Canada, and West Africa. Among them, there are 18 cooperative projects in Australia possessing high iron ore grades and abundant resources, accounting for half of all overseas projects; there are mainly five projects in Canada including WISCO Group; there are mainly three projects including Guinea Simandou Iron Ore invested by Aluminum Corporation of China and other enterprises in West Africa.

(3) Current Problems

1) Despite a great amount of resources, and a large planning and development area, the rich ore resources and the actual output are small. As of 2014, only 15 mine projects have been put into operation, and the actual supply of equity finished ore is only 83 million tons/year, accounting for 10% of total imports. Compared with countries such as Japan and Europe with more than 50% equity ore proportion, China still has a long way to go.
2) The supporting infrastructure is weak, which restricts the development of ore exploration. Geographically, China's overseas iron ore projects are mostly located in remote places, lacking supporting infrastructure such as railways, ports, and power plants. Some of these problems require the Chinese investors to solve through huge investment, and some are subject to third parties.
3) Insufficient research in the early stage. Preliminary researches on projects were mostly defective, such as insufficient exploration, unclear geological conditions, and failure in conducting ore dressing test. This not only increases the risk and uncertainty of subsequent resource development, but also adversely affects the progress of the project.

**Table 13.1** List of development progresses of overseas iron ore resources

| No. | Country | Name of overseas ore | Name of enterprises | China | China proportion of shares/% | Joint venture time | Putting into operation time |
|---|---|---|---|---|---|---|---|
| I | 1980–2000 | | | | | | |
| 1 | Australia | Channar | Channar Joint Venture | Sinosteel | 40 | 1987 | 1990 |
| 2 | Peru | Peru Iron Ore | Shougang Peru Iron Ore Co., Ltd. | Shougang | 98.4 | 1992 | 1992 |
| II | 2000–2008 | | | | | | |
| 3 | Brazil | Agua Limpa Iron Ore | Baohuarui Mining Co., Ltd. | Baosteel | 50 | 2001 | 2001 |
| 4 | Australia | Paraburdoo Eastern Ranges Mine | Baoruiji Ore Mine Co., Ltd. | Baosteel | 46 | 2002 | 2002 |
| 5 | Australia | Wheelarra Iron Ore Mine | Wheelarra Joint Venture | Tang Gang, Wuhan Iron and Steel, Ma Steel, Shagang | 40 | 2004 | 2004 |
| 6 | Russia | Berezov Iron Ore Mine | | The Western Group | 100 | 2005 | Early stage |
| 7 | Vietnam | Guisha Iron Ore | Vietnam-China Minerals and Metallurgy Co., Ltd. | Kunming Iron and Steel | 45 | 2005 | 2006 |
| 8 | Australia | Sino Iron Ore | Mineralogy | CITIC Pacific MCC Group | 100 | 2006 | 2012 |
| 9 | Australia | Karara Iron Mine | Gindalbie Metals | Anshan Steel | 69.3 | 2007 | 2012 |

(continued)

**Table 13.1** (continued)

| No. | Country | Name of overseas ore | Name of enterprises | China | China proportion of shares/% | Joint venture time | Putting into operation time |
|---|---|---|---|---|---|---|---|
| 10 | Argentina | Sierra Grande | | MCC Group | 70 | 2007 | Production |
| III | 2008–2012 | | | | | | |
| 11 | Australia | Cape Lambert Iron Ore | | MCC Group | 100 | 2008 | Feasibility study |
| 12 | Australia | Weld Range Iron Ore | Midwest | Sinosteel Group | 100 | 2008 | Feasibility study |
| 13 | Australia | Mount Gibson Iron Ore | Mount Gibson Mining Company | Shougang | 40 | 2008 | Production |
| 14 | Australia | Bungalow Magnetite Iron Ore | | Baosteel | 50 | 2008 | Early stage |
| 15 | Madagascar | Solala Iron Ore | Hong Kong WISCO Guangxin Jinhua Resources Co., Ltd. | Wuhan Iron and Steel Guangxin | 80 | 2008 | Early stage |
| 16 | Cameroon | Lobi Iron Ore | Sinosteel Cameroon Co., Ltd. | Sinosteel | 97.5 | 2008 | Exploration |
| 17 | Australia | Eyre Iron Ore | CXM | Wuhan Iron and Steel | 60 | 2009 | Feasibility study |
| 18 | Australia | | FMG | Hunan Hualing | 17.4 | 2009 | Production |
| 19 | Australia | Extension Hill | Extension Hill Co., Ltd. | Chongqing Iron and Steel Company (CISC) | 60 | 2009 | Feasibility study |

(continued)

**Table 13.1** (continued)

| No. | Country | Name of overseas ore | Name of enterprises | China | China proportion of shares/% | Joint venture time | Putting into operation time |
|---|---|---|---|---|---|---|---|
| 20 | Brazil | | MMX of EBX Group | Wuhan Iron and Steel | 21.52 | 2010 | Production |
| 21 | Sierra Leone | Tonkolili Iron Ore | | Shandong Iron and Steel Group | 100 | 2010 | Shutdown |
| 22 | Liberia | Liberia State Mine | | Wuhan Iron and Steel China-Africa Development Fund | 100 | 2010 | Production |
| 23 | Guinea | Simandou Iron Ore | SIMFER | Chinese Consortium | 41.3 | 2010 | Feasibility study |
| 24 | Canada | Century Iron Mines Corporation (Sunny Lake, Duncan, Attikamagen Projects) | Century Iron Mine Co., Ltd. | Wuhan Iron and Steel | 40 | 2011 | Early stage |
| 25 | Canada | Lac Otelnuk Project\December Lake Project | ADI | Wuhan Iron and Steel | 60 | 2011 | Early stage |
| 26 | Canada | Kami Iron Ore Project | Alderon Iron Ore Company | HBIS Group | 25 | 2012 | Pre-feasibility research |
| 27 | Canada | Tuktu Iron Ore Project | Advanced Explorations Inc | Xinxing Cathay International Group Co., Ltd. | | 2012 | Exploration |
| 28 | Canada | Astray-X | Northern Star Minerals Ltd. | Xinxing Ductile Iron Pipes Co. Ltd. | | 2012 | Exploration |

(continued)

**Table 13.1** (continued)

| No. | Country | Name of overseas ore | Name of enterprises | China | China proportion of shares/% | Joint venture time | Putting into operation time |
|---|---|---|---|---|---|---|---|
| 29 | South Africa | PMC Mining Industry | | HBIS Group | 74.5 | 2012 | Production |
| IV | Since 2012 | | | | | | |
| 30 | Mexico | ZANIZA Iron Ore Project | Mexico MINOSA Company | Xinxing Hanfang | | 2013 | |
| 31 | Australia | Aquila Mining Company | | Baosteel | 85 | 2014 | Early stage |

4) High production cost and weak anti-risk ability. At present, most of the overseas iron ore mines invested by China produce low-grade ore with a grade of about 30%. It is necessary to increase investment in construction of ore dressing, power plants, water intake facilities, etc. Most of the projects require large investment in supporting infrastructure and suffer from long transportation distance, high construction cost, and high cost of mining and dressing. Therefore, the projects generally cost a lot. It is estimated that the average FOB cost of overseas ore is 58 US dollars/ton, and the average CIF cost is as high as 79 US dollars/ton. Compared with the average cost of 20 US dollars of the major international iron ore producers, the competitiveness is weak and the cost is higher than that of the domestic mines of 75–85 US dollars/ton.
5) Investments are generally high. The overall average investment on the projects is above the medium level among international iron ore investors, and 20% of them is at the high level in terms of international iron ore investment.
6) Lack of unified management leads to vicious competition. In recent years, many iron and steel enterprises and related enterprises in China have recognized the necessity and importance of overseas resource development and have "gone out" to develop iron ore projects in flocks. However, due to the lack of national unified planning, enterprises are fighting their own battles and vicious competition among enterprises and financial institutions is not in rare cases, which has led to failure of some overseas projects or made it difficult to survive for some overseas mining projects.

2. Coal

(1) Supply Status In 2014, the world's proven reserves of coal resources were 891.5 billion tons [6], in keeping with that in 2013, with 403.2 billion tons anthracite and bituminous coals and 488.3 billion tons of sub-bituminous coal and lignite. The top three were Europe and Eurasia, Asia Pacific and North America. According to the 2014 mining level, the static guaranteed period of proven recoverable reserves of coal was about 110 years.

In 2015, the world's coal production was 7.861 billion tons [4], which was 4% lower than the 8.165 billion tons in 2014. The decrease was mainly attributable to China. It is estimated that the world's coal production in 2016 is about 7.45 billion tons, a further decrease of about 5% from 2015. The decrease is mainly attributable to China, the USA, and Indonesia.

The regional pattern of coal consumption can be roughly divided into 70.0% in Asia Pacific, 12.4% in North America, 13.9% in Europe and Eurasia, 2.6% in Africa, 0.8% in Central and South America, and 0.3% in the Middle East.

In 2015, China's iron and steel industry consumed 650.22 million tons of coal [7], of which the physical consumption of coal was 140.6 million tons, and the actual consumption of coke was 387.25 million tons, equivalent to the consumption of washed coal of 509.62 million tons. By types of coals, coking coal consumption was about 509.6 million tons, injection coal consumption was 97 million tons, and anthracite and thermal coal for sintering were 43.6 million tons. In the same year, China's iron and steel industry imported 47.84 million tons of coking coal and did not import other coals. The external dependence on coking coal was 9.38%.

In 2016, China's iron and steel industry consumed 645.94 million tons of coal, of which the physical consumption of coal was 152.84 million tons, and the actual consumption of coke was 374.70 million tons, equivalent to the consumption of washed coal of 493.10 million tons. By types of coals, coking coal consumption was about 493 million tons, injection coal consumption was 105.4 million tons, and anthracite and thermal coal for sintering were 47.4 million tons. In the same year, China's iron and steel industry imported 59.23 million tons of coking coal and did not import other coals. The external dependence on coking coal was 12.01%.

From 2015 to 2016, China's imported coking coals by countries are shown in Table 13.2 [8].

(2) The International Layout of Iron Ore Supply. In the period from the reform and opening up to the end of the 1990s, China's coking coal was mainly exported in exchange for foreign exchange, and the import volume was less than 200,000 tons in most of the time, which is very small. With the rapid development of China's iron and steel industry, the demand for coking coal increased sharply. The turning point happened in 2004 when the coking coal imports exceeded the exports for the first time. Since then, the import volume has been increasing and hit staged record high in 2013 with imported coking coal being 75.39 million tons.

**Table 13.2** China imported coking coals list by countries from 2015 to 2016 [5]

| No. | Name of country | 2015 quantity/$\times 10^6$ t | 2016 quantity/$\times 10^4$ t | Remarks |
|---|---|---|---|---|
| 1 | Russia | 323 | 260 | |
| 2 | Canada | 571 | 518 | |
| 3 | The United States | 12 | 0 | |
| 4 | New Zealand | 0 | 54 | |
| 5 | Indonesia | 23 | 57 | |
| 6 | Malaysia | 1 | | |
| 7 | Mongolia | 1272 | 2356 | The second in import volume |
| 8 | Kazakh | 0 | 1 | |
| 9 | Australia | 2555 | 2677 | The largest in import volume |
| | Total | 4784 | 5923 | |

In order to meet the domestic needs on coking coal, since the beginning of the twenty-first century, some large state-owned enterprises such as Shenhua Group have invested in the mining and production of coal mines through joint stock or holdings in Mongolia, Russia, Australia, the USA, Canada, etc., and some private enterprises have also made their own investments and hold shares in some coal mines in Mongolia, Russia, Australia, and other countries.

China's iron and steel enterprises have started late on investment of overseas coal mines, and the numbers are few. And their focus is on the investment of exploration and production of coking coal through joint stock. For example, Baosteel invested and held shares in Australian coal mines, and HBIS Group and Shougang held shares in Canadian coal mines. But the results of investment are not as expected and the main problems lie in the poor prediction on the long-term trend of coal prices and the lack of all-round talents who familiarize laws of different countries possesses good value judgments on coal mines and investment estimates.

## 13.2 Development Environment and Policy Orientation

### *13.2.1 The Necessity of International Development for Iron and Steel Enterprises*

At present, the domestic environment and international environment faced by the Chinese economy are far more complicated. On the one hand, the international economy is going through profound adjustment, international trade is growing at a low rate, and disparity between developed economies and emerging economies is further enlarged; the FED raises interest rates and the prices of bulk commodities such as oil and iron ore are on the downside, which causes global financial and commodity market fluctuations; geopolitical conflicts are endless, and uncertainty and instability have further increased. On the other hand, the downward pressure on the domestic economy has increased, the challenges faced by the real economy have multiplied, and the risk of economic inertia drop and the systemic risks fueled by the superposition of fluctuations in the real economy and financial markets have continued building up. Under such circumstances, it is of great strategic significance to further promote the going out drive, accelerate the construction of the "Belt and Road" initiative, and enhance international capacity cooperation. The "Belt and Road" initiative is an overall strategy for China's opening-up and international economic cooperation in a long period of time to come. To implement the "going out" strategy in the new era, we must take the "Belt and Road" strategy as a guide, strengthen overall planning and guidance, and take all measures to establish all ties with the Belt and Road countries and regions so that policies are compatible, facilities are connected, trades are smooth, transactions are circulated, and people-to-people ties are strengthened. International capacity cooperation is the main focus and platform for promoting the "Belt and Road" initiative and is also a new feature and a new task in a stage for implementing the strategy of going out. Therefore, it plays a vital role.

First, it is conducive to expanding China's economic development space. In order to maintain rapid economic growth during the 13th Five-Year Plan period, we must speed up the pace of going out, expand external development, accelerate the export of advantageous production capacity and equipment such as steel, shift focus from exporting products to outputting industries and capitals, and maintain the balance of international payments by compensating deficit under capital item using the surplus under trade item. Second, it blazes a new path for promoting economic transformation and upgrading. On the one hand, China's iron and steel industry has developed a large amount of high-quality production capacity, advanced and practical equipment, mature and reliable technology and enjoys outstanding cost performance and unique external advantages; on the other hand, domestic resources and environmental constraints have intensified, labor costs continue to rise, and steel overcapacity is increasingly prominent. Therefore, we must actively promote the transformation of the development mode, shifting from quantitative growth to quality benefits, effectively cut the overcapacity in domestic steel production through international capacity cooperation, transfer excessive and high-quality production capacity, boost structural reform, and enhance supply-side quality and efficiency. Third, it helps develop a high-level open economy. China's foreign economic structure is undergoing profound changes. Innovations must be made to create new ways of foreign investment. We must vigorously promote international capacity cooperation and shift focus from the export of consumer goods to the output of investment goods, and from focusing on imports to laying equal stress on imports and exports. Fourth, it helps implement a mutually beneficial and win-win strategy at a higher level. At present, developing countries and emerging economies are continuing to promote industrialization and urbanization. Developed countries are accelerating industrialization, transforming existing infrastructure. Promoting international capacity cooperation in industries such as steel industry provides opportunities of linking the supply and demand of countries at different development stages and boosting the organic integration of global industrial chains, which is conducive to improving China's discourse power in global economic governance.

### *13.2.2 The Government's Policies Supporting the International Development of Iron and Steel Enterprises and the Macroeconomic Environments*

Internationally, the world economy is undergoing profound adjustment, the recovery is sluggish, international trade growth remains slow, financial and bulk commodity markets are volatile, geopolitical risks are on the rise, and uncertainties and instabilities in the external environment are increasing, the impact on China's development should never be underestimated. Domestically, the problems and risks escalated in a period of time are further manifested, economic growth gears down, structural adjustment pains and replacement of old growth drivers with new ones are intertwined, and economic downward pressure is increasing. Now, it is an important opportunity for Chinese industries such as iron and steel industry to carry out international capacity cooperation. The National Development and Reform Commission and the Ministry of Foreign Affairs jointly compiled the "*Country Wise Planning for International Capacity and Equipment Manufacturing Cooperation*". In the near future, a regional production cooperation pattern consisting of "one axis and two wings" and including 45 key countries will take shape with the China's neighboring countries on the main axis and Africa, Middle East, and Central and Eastern Europe on the west wing, and Latin American countries on the east wing, aiming for steady development of international capacity cooperation.

According to the instructions of the State Council, relevant departments compiled the *"13th Five-Year Plan" for International Capacity Cooperation.* Twelve key industries including steel industry, nonferrous metals industry, and building materials industry will play core role. The global layout of industries must be designed as whole system with key projects leading the way; the countries near and along the "Belt and Road" will be given the priority and strategies on each country shall be systematically designed in order to deepen multi-bilateral pragmatic cooperation. The government will release the *"Regulations on Overseas Investment"*, further enhancing the autonomy of enterprises in investment, and leaving focus on strengthening the in-process and after-event supervision and supporting services. We will increase efforts in coordinating major projects for overseas investment cooperation and standardizing market competitive order. We will strengthen statistical analysis and operational monitoring of capacity cooperation in key countries.

The fiscal and taxation financial support policy will favor strategic projects in international capacity cooperation. By expanding the special funds for foreign trade and economic development and scaling up "concessional loans and preferential export buyer's credit", we will flexibly adjust sovereign loans or sovereign guarantee policies to further improve tax policies. We will increase policy-based and open financial supports, accelerate the overseas establishment of financial institutions, and encourage the use of domestic and foreign capital markets and bond markets to finance for going out projects. We shall effectively give play to the role of foreign exchange reserves, steadily expand the scale and coverage of entrusted loans, and give full play to the investment promotion role of investment and financing platforms such as CIC, Silk Road Fund, and CNIC Co. Ltd. We shall put emphasis on strategic pivotal landmark projects in the fields of steel, cement, nonferrous metals, electric power, infrastructure, etc., accelerate the construction of the "Belt and Road" initiative and international capacity cooperation, vigorously implement the action plan on construction of high-speed railway network, express roads, airline network, and industrialization in African, and participate in construction of infrastructure of surrounding countries for connectivity and overseas projects of economic and trade cooperation zones, and accelerate the formation of relevant production capacity overseas.

## 13.3 Case Analysis

### *13.3.1 Pohang Iron & Steel Co., Ltd.*

1. Internationalization Overview

The development history of Pohang Iron and Steel Co. Ltd. (hereinafter referred to as POSCO) can be roughly divided into three stages, namely the start-up period from 1968 to 1997, the rapid development period from 1998 to 2007, and the strategic adjustment period from 2008 to this day. At present, its industrial scale measured in production of crude steel has registered over 40 million tons. In the 1990s, POSCO began to implement the strategy of "Make Expansion Overseas with Epoch-Making Technology". With high-end technology and products in its possession, it has pursued the strategy of globalized steel industry and expanded its international market share through overseas projects. The main focus of overseas expansion was China and Japan and gradually shifted to emerging economies such as India, Vietnam, and Indonesia, and then reached out to Brazil, Mexico, the USA, Turkey, and other European and American countries. After years of meticulous cultivation, it has basically established its existence in the world as an international steel industry player.

POSCO's internationalization strategy adopts a gradual development model, and its development stages are shown in Table 13.3 [9].

2. Analysis on Internationalization Experiences

POSCO invests in and builds bases and technical service centers for steel production and steel processing in emerging economies to seize markets in Southeast Asia and the Middle East and consistently strengthens precision marketing targeting at high-yield regions and industries on a global scale. At the same time, it increases the supporting exports of high value-added technologies. With a global vision of building steel capacity, its businesses, branches, and manufacturing plants spread across Asia, the Americas, and Europe.

POSCO entered the Chinese market in 1991 and established POSCP (China) Investment Co., Ltd. in July 2003 to provide business support for its investment corporations in China in conducting business activities and made investment in expansion by planning and expanding the steel processing centers. POSCO established a

**Table 13.3** Globalization history of POSCO

| 1972–1993 | 1993–2001 | 2002–2010 | 2010–now |
|---|---|---|---|
| Exploration period before market entry | Market entry period | Market expansion period | Internationalization stage |

strategic existence by building steel processing bases in China's coastal areas and entered the Chinese market with the output of "high-end" products and technologies. Thanks to these efforts, it has expanded its scale, improved service quality, and expanded diversified sales channels in a continuous way [10].

It sets up steel production enterprises. From 1991 to 2011, POSCO established five steel production enterprises including Pujin Company in Dalian, Pohang in Zhangjiagang, Pohang in Shunde, Guangdong, Pohang in Qingdao, and Benxi Steel-Pohang Rolling Company. The enterprises boasted over 2 million tons of cold rolling annual production capacity, 1 million tons of galvanizing annual production capacity, and stainless annual output reached up to 1 million tons.

Prepared and expanded steel processing centers. In Tianjin, Qingdao, Foshan, Suzhou, Chongqing, and other cities, more than 30 integrated steel plate processing centers for producing automobile steel plates, stainless steel plates, electrical steel, and color-coated plates with an annual processing capacity of over 3 million tons have been built, which has increased its products' market share.

Built model steel mill and exported FINEX technology. POSCO and Chongqing Iron and Steel Co., Ltd. built a FINEX integrated demonstration steel plant by applying the FINEX-CEM technology to produce high-quality steel plates and wires with a focus on serving the industries like automobiles, home appliances, and architectural ornament. The export of FINEX technology in China marks the beginning of forming a business model featuring the export of steel technology. POSCO can also get the royalty in such model. The building of integrated demonstration steel mills will help POSCO to make expansion to India, Southeast Asia and the Middle East.

(1) Global Steel Projects Steel projects and steel processing centers mainly produce automotive sheets, stainless steel cold-rolled coils, and other products [11]. See POSCO's major overseas projects of flat steel products in Table 13.4 for details.

    In order to meet the rapid development of the automotive and home appliance industries in all major regions around the world, POSCO has also established coil processing centers in foreign countries, which is easier than building a large steel plant in the local area. POSCO built six coil processing centers in Japan, followed by coil processing centers in Mexico, Poland, China, Thailand, and Vietnam, etc. Currently, POSCO has more than 50 coil processing centers around the world with an annual processing capacity of more than 5 million tons.

(2) Steel Products Processing Centers POSCO built its first steel processing center in Thailand as early as 1998. Afterward, processing centers were built in Malaysia, Indonesia, Vietnam, and the Philippines. At present, POSCO has 10 steel processing centers in the Southeast Asian market with an annual processing capacity of about 1.15 million tons. This provides a sound pivot for the development of POSCO in the ASEAN countries. The distribution of processing centers is shown in Table 13.5.

3. Conclusions

POSCO's industrial expansion model is to establish legal entities in overseas regions, build integrated steel plants and steel processing bases through joint venture or M&A,

**Table 13.4** POSCO's major overseas projects of flat products

| Description | Annual capacity/$\times 10^4$ t | Place | Time of completion |
|---|---|---|---|
| Processing Center | 2.4 | India | February, 2012 |
| Galvanizing Line | 45 | India | May, 2012 |
| Non-oriented Electrical Steel Plant | 30 | India | October, 2013 |
| Cold Rolled Coil Plant | 180 | India | January, 2015 |
| Equipment Relocation of No. 1 FINEX | 150 | India | April, 2019 |
| Stainless Steel Cold Rolled Coil Plant | 8.5 | Vietnam | Started from 2009 |
| Stainless Steel Cold Rolled Coil Plant | 15 | Vietnam | Started from March, 2012 |
| Cold Rolling Mill | 120 | Vietnam | September, 2009 |
| Integrated Steel Plant | 300 | Indonesia | December, 2013 |
| Cold Rolling Mill | 120 | Thailand | 2009 |
| Stainless Steel Plant | 30 | Thailand | Production expansion completed in 2015 |
| Electrogalvanizing Line | 18 | Malaysia | 2007 |
| Galvanizing Line | 3 | Myanmar | 1999 |
| Galvanizing Line | 40 | Mexico | June, 2013 |
| Stainless Steel Cold Rolled Coil Plant | 20 | Turkey | April, 2013 |
| Integrated Steel Plant | 300 | Brazil | Middle of 2015 |
| Steel Pipe Plant | 27 | The United States | 2009 |
| Integrated Steel Plant (FINEX-CEM) | 160 | Iran | 2019 |

and set up technology service centers to increase the export of high value-added technologies (such as FINEX-CEM). Extending the upstream and downstream industry chain, making continuous innovations, understanding and integrating into foreign cultures are POSCO's way "to make overseas expansion with the epoch-making technologies" [12].

(1) In its global expansion, POSCO sets eyes on the local steel demand in energy, home appliances, construction, electronics, and other industries, makes targeted overseas strategic investment, and gives priority to downstream processes and the automotive industry.
(2) By closely cooperating with the downstream industries and building steel processing centers in emerging markets, POSCO has been actively carrying out overseas expansion as well as expanding its deep processing capacity of high value-added steel products such as automobile sheets, electrical steel, and stainless steel. It aims to increase products' added value, provide more customer-oriented services, and expand the market share.

**Table 13.5** Distribution of POSCO's steel products processing centers

| Name | Country | Address | Time of put into operation | Annual processing capacity/$\times 10^4$ t |
|---|---|---|---|---|
| POSCO-TBPC | Thailand | Chonburi | 1998 | 12 |
| | | Rayong | 2006 | 12 |
| | | Wellgrow Industrial Zone | 2009 | 12 |
| POSCO-MKPC | Malaysia | Rawang | 2006 | 14 |
| | | Port Klang | 2009 | 12 |
| POSCO-VNPC | Vietnam | Hai Duong Province | 2007 | 12 |
| POSCO-VHPC | | Ho Chi Minh City | 2008 | 15 |
| POSMI | Indonesia | Bekasi | 2002 | 8.5 |
| POSCO-IJPC | | Karawang | 2006 | 12 |
| POSCO-PMPC | The Philippines | First Philippine Industrial Park (FPIP) | 2009 | 5 |

(3) The expansion of overseas production capacity by means of joint ventures cannot only make up for the lack of global management capabilities via local partners, but also enable it to achieve cultural compatibility.

Overseas investment is full of risks and challenges. Political ecology, economic foundation, environmental protection policies, and other factors will affect the success of the projects. Meanwhile, enterprises are confronted with risks like being highly dependent on the market, debts, rating, laws, and project delay. But in the long run, it is right decision to stick to the strategy of "focusing on emerging markets and meeting the local growing demand for steel by procuring local raw materials". Due to the limited capacity in the domestic market, future capacity expansion will basically resort to building plants overseas. The long-range plan is to achieve a crude steel capacity of 71 million tons in 2020, of which overseas production capacity will reach 26 million tons, accounting for 36.6%.

### 13.3.2 ArcelorMittal

1. Internationalization Overview

The ArcelorMittal Group is currently the world's largest steel group and the largest steel multinational enterprise. It is known for its largest "cross-sector" and cross-border M&A, the fastest growth speed after M&A, and the highest success rate of M&A in the world. Lakshmi Mittal, the founder and chairman, established his own steel company (LNM) in 1980; from 1986 to 1995, LNM Group acquired a series of steel mills that were in trouble at that time in different countries (including Canada, Mexico, Ireland, Trinidad, Tobago, Britain, Germany, and Indonesia); in 2004, after the acquisition of the US International Steel Group, a blockbuster around the world, the Mittal Steel Group (LNM Group) was incorporated. In 2006, it acquired the Arcelor Steel Company in Europe and formed the ArcelorMittal Group. The headquarters was therefore moved to Luxembourg from the Netherlands.

Through 130 M&As of poorly managed small steel plants and 2 M&As of world-class large enterprises in just 20 years, ArcelorMittal Group has emerged as the largest steel enterprise in the world. Its business covers both the emerging markets and mature markets and has occupied a leading position on a global scale in the fields like automobiles, construction, household appliances, and packaging. According to incomplete statistics, ArcelorMittal Group possesses more than 60 steel plants in 27 countries and regions in Europe, Asia, Africa, and the Americas. Its annual production capacity of crude steel is about 130 million tons, and that of pig iron is about 72.1 million tons. With nearly 330,000 employees in more than 60 countries across the world, the group has been described as "developing at the speed of the Internet in the steel industry".

2. Analysis on Internationalization Experiences

The prerequisite for the success of ArcelorMittal Group is: clear positioning—global strategy. Prior to the Mittal Group, there were no real global steel enterprises in the world. At that time, the so-called world-class steel companies were mostly formed by acquiring the steel companies in their own countries. The US International Steel Group and Arcelor Group in Europe were among them. Some even thought that "steel industry can never be globalized and will always be a regional industry". Mittal aimed to become the "Henry Ford" in the steel kingdom—not only to rule but also bring changes to the operating model in the entire steel industry.

The first step of ArcelorMittal Group toward success was the formulating of a development path of making mergers and acquisitions and transforming loss-making state-owned enterprises. It acquired endangered small steel mills for expansion at low cost and turned losses into gains through reducing the operation cost and utilizing the modern processing technologies and management concepts. After the oil crisis, the steel industry fell into a long-term downturn. Global steel demand remained stagnant for a long time and steel industry was once considered as a sunset industry. However, during this period, what Mittal did was to acquire the "garbage enterprises" in others' eyes. For example, in 1992, the third largest steel company in Mexico, SIBAISA Steel, was caught in great difficulty because it had no orders with an operation rate of only 25%. While the Mexican government wanted to privatize it, Mittal spent only $220 million on acquisition of this highly modernized steel conglomerate with $2.2 billion government investment. The company's production line with an annual output of 330,000 tons of welded pipes, port equipment, and mines with 3 million tons of iron ore were among the acquired. After Mittal's transformation, the company's output tripled in less than five years, becoming the largest steel manufacturer and exporter in Mexico. Following the initial success, Mittal acquired many small steel companies suffering from surplus labor, backward technologies, and poor operation at low prices in such low-cost regions as Poland, Romania, Kazakhstan, and the Czech Republic.

The key to the grand success of ArcelorMittal Group is the real-time adjustment of its global M&A strategic focus, which means to move from the strategy of "low-cost M&A" to "mastering the high-end ones". Thus, it has achieved a leap-frog transition from large enterprise to a strong one. When it came to the twenty-first century, the steel industry witnessed a cyclical peak. The growth of steel demand finally brought the international steel industry to a fast-growing track. With its growing strength, Mittal made a timely adjustment of the strategic focus in the transnational operation, aiming at becoming the globe master in steel industry, based on the trend of globalization and multinational enterprises' rapid development. It sets eyes on mature markets and targeted at the world's top steel enterprises. By launching a series of M&A battles against the world's top companies, it formed a high-end multinational business network system with monopoly and a global coverage in the steel industry. In 2004, Mittal Steel was listed in Amsterdam and New York through 16% assets of its subsidiary "Ispat International". This opened the window for the group's capitalization process, and then it merged the assets of Mittal and Ispat. It then acquired US International Steel Group with approximately $4.5 billion in "cash and stock", therefore forming the Mittal Steel Group. After the acquisition of such world-class enterprises, Mittal Group's annual steel output reached 70 million tons with an operating revenue of $32 billion. Mittal became the world's largest steel producer surpassing the then Arcelor Group. In 2006, Mittal made another effort to acquire Arcelor Steel spending huge sums, which took up a dominant position in the high-end steel market. The ArcelorMittal after merger boasted an annual production capacity of 120 million tons, which was three times of that of the world's third steel enterprise, Nippon Steel, and accounted for 10% of the global steel output. It is the stock market value amounted to 46 billion US dollars, becoming the undisputed emperor in the global steel industry.

3. Conclusions

The secret of ArcelorMittal's success is to pursue a global M&A strategy, make M&A at the most appropriate time, and grow the acquired companies with the smartest business practices so that a leap-forward development from "small to large" and from "large to strong" has been realized. The expansion model of Mittal Group's global M&A strategy is worth learning by China's large steel enterprises:

First is to obtain the largest assets with the least investment. Mittal acquired Karaganda Steel in a scale of 4 million tons/year at a cost of $240/ton steel and acquired Inland Steel at a cost of $315/ton. However, a newly built large steel complex needs $1200/ton in terms of investment.

Second is not only to purchase the hardware of steel plants, but also acquire the software added up over years of production practices, namely production know-how and management experience. In particular, companies like the time-honored Inland Steel possessed great knacks in many fields, such as making steel plates for the automobiles and large household appliances, which also became property of Mittal Group after acquisition. In addition, Mittal Group were able to take over the original market and blaze new markets in other parts for the group in a rapid way.

Third is to acquire the existing corporate talents apart from obtaining the hardware and software from the steel plant. This allows Mittal Group to quickly take full possession of the talents of operation and management. By virtue of its advanced management and production model, it helps the acquired companies to quickly get rid of the predicament and make contributions to the group.

### *13.3.3 HBIS Group Co. Ltd.*

1. Internationalization Overview

HBIS Group Co. Ltd. (hereinafter referred to as HBIS) is one of the largest steel material producers and integrated service providers in China. It has more than 30 first-level subsidiaries and more than 120,000 registered employees. The Group is mainly engaged in steel materials businesses and sets foot in many sectors like mining resources, financial securities and modern logistics, steel trade, equipment manufacturing, and others. In 2016, its operating revenue stood at 290.8 billion yuan, and the total assets at the end of the year reached 360.4 billion yuan. HBIS was listed among the *Fortune* Top 500 for 9 consecutive years, ranking No. 221 once. It also earned the highest rating, namely "owning extremely strong competitiveness", among the "MPI China Steel Enterprise Competitiveness Ranking" issued by China Metallurgical Industry Planning and Research Institute. HBIS is a member of the Executive Committee of the World Steel Association and the rotating president of the China Iron and Steel Association.

While enhancing strengths and striving for better performance in steel industry, the Group also seeks the strategic transformation in the light of "deepening and broadening industry chain" with a global version across the whole industry chain. By giving full play to the advantages of the entire industrial chain and the resource chain of iron and steel enterprises, it pursues the deepened extension of the industrial chain and the broadened use of the entire resource chain and vigorously expands overseas businesses. Based on the vision of "having resource, market and customers around the globe", it has accelerated the strategic expansion of building global marketing service platform, global technology R&D platform, and global steel manufacturing platform. In recent years, HBIS Group has completed the acquisition of PMC, the largest copper smelting enterprise in South Africa, and the Duferco in Swiss, world's largest steel marketing service provider, thus becoming a multinational group with overseas mature smelting companies and globalized marketing service platforms. HBIS is moving toward the European region which has the high-end manufacturing capability via the acquisition of Serbia's only state-owned large-scale pillar enterprise, Smeder Revo. This has laid a solid foundation for expanding overseas industrial bases and building a global industrial manufacturing platform.

2. Analysis on Internationalization Experiences

As the largest steel conglomerate in China and the second largest in the world, HBIS is committed to building "an international HBIS" and has made remarkable achievements in actively planning overseas manufacturing bases and promoting close cooperation with the mature foreign strategic partners.

In December 2012, HBIS led several domestic enterprises to form a Chinese consortium, and then jointly with South Africa's IDC contributed to subscribing the PMC company in South Africa. HBIS held 35% of its share at a price of $234 million, becoming the biggest shareholder of PMC and dominating the operation management. South Africa's IDC is the second largest shareholder and uses the capital ties to form a community of shared interests. Share acquisition of PMC company in South Africa is the first large-scale overseas M&A project led by HBIS and has set a successful example for Chinese enterprises' investment projects in Africa. In September 2014, HBIS signed an agreement with South African Industrial Development Corporation and China-Africa Development Fund Co., Ltd. to launch the 5 million ton steel project in South Africa, setting a record as the largest whole-process steel project by China in foreign countries.

In November 2014, HBIS announced an equity cooperation project with Duferco Holding Group. HBIS increased its stake in Duferco International Trading Company (world's largest steel trading and integrated service provider) to 51%. At the same time, the comprehensive strategic partnership between Tangshan Steel (a former subsidiary) and Duferco was expanded to the level between the two groups. This equity cooperation is the first time that a China's steel enterprise has acquired an international mature business network, creating a new model for the integrated cooperation between China's largest steel enterprise and the world's largest steel trade and integrated service provider.

In April 2016, building on the strength of the national "Belt and Road" Initiative, HBIS continued to deepen international capacity cooperation and acquired the Smeder Revo Steel Plant in Serbia. After the acquisition, HBIS sent 11 batches of technical teams to Serbia, with a total of nearly 200 people. These teams helped Smeder Revo diagnose and solve its problems in equipment, technology, information application, and process. It took only half a year to reverse the loss lasting seven consecutive years, demonstrating the advantages and capabilities of the Chinese steel industry and HBIS. HBIS's steel plant in Serbia has become a model project in the international capacity cooperation between China and Central and Eastern Europe.

In recent years, HBIS has been continuously increasing its pace of international development via carrying out a series of practices in export channel expansion, financing, talent training, project cooperation, etc. It also carried out close cooperation with Duferco in Switzerland, POSCO in South Korea, South African Industrial Development Corporation, China-Africa Fund, and other companies. At present, HBIS Group has more than 20 subsidiaries or joint ventures in Canada, Australia, Singapore, South Africa, and other countries with more than 5000 overseas employees.

3. Conclusions

The root of the successful experience of HBIS internationalization strategy lies in liberating the mind, adhering to the international development concept and business philosophy. All its practices, like leading equity acquisition of South Africa PMC company, increasing shareholding of Duferco International Trade Co., acquiring the Serbian Smeder Revo Steel Plant, encouraging "brainstorming", inviting well-known experts at home and abroad to give lectures, as well as comprehensively bringing itself in line with the world-class steel enterprises in terms of profitability, marketing model and others, have shown that HBIS is making great endeavor to guide its development with an international version and market-oriented perspective.

The strategy of setting up HBIS overseas manufacturing bases is characterized by high positioning and aims to build model projects for international capacity cooperation. Focusing on the sustainable development of overseas manufacturing bases, HBIS attaches great importance to establishing effective dialog and communication mechanism with local trade unions, associations, and other organizations and actively fulfills social responsibilities while supporting and promoting project construction and enhancing business competitive advantages, which has brought a good reputation for HBIS. The business philosophy, product capability, technical team, and marketing network have laid a solid foundation for building model projects in steel industry's international capacity cooperation.

### *13.3.4 Tianjin Seamless Pipe Corporation*

1. Internationalization Overview

(1) Historical Origin Tianjin Seamless Pipe Corporation (hereinafter referred to as TPCO) was put into operation in 1992. It is a large-scale modern producer of seamless steel pipes founded in the "Eighth Five-Year Plan" period. Its main processes are steelmaking, tube rolling and pipe processing with imported technical equipment from Britain, Germany, Italy, and the USA, which was at the advanced level in the 1990s.

Currently, TPCO has a blast furnace of 1,000 $m^3$ with an annual production capacity of 900,000 tons; two 150-ton ultra-high-power electric furnaces and one 90-ton ultra-high-power electric furnace with an annual steelmaking capacity of 3.3 million tons; six units of $\phi$250MPM, $\phi$168PQF, $\phi$219ASSEL, $\phi$258PQF and $\phi$460PQF and $\phi$720 rotary expander.

For now, TPCO has developed from a producer of seamless steel pipes into a group corporation that integrates various metallurgical products such as seamless steel pipes, stainless steel plates, color-coated sheets, sponge iron, iron alloy, and copper wire rods.

(2) Capacity Status In 1992, a 563,000 tons/a, as originally designed, tube line was constructed to produce 500,000 tons seamless steel pipes mainly for

oil casing, among which 350,000 tons were oil casings. At present, TPCO annual production capacity of seamless steel pipes exceeds 3 million tons.

(3) Industry Position Compared with the products of world-famous Sumitomo, TPCO's products enjoy quality indicators comparable in terms of chemical composition, mechanical properties, geometric dimensions, and thread processing accuracy and are very much the same to the physical quality boasted by world-class companies. TPCO is now able to compete with those powerful steel pipe producers around the world.

TPCO is currently the largest production base for oil well pipes and seamless steel pipes in China. It is also the largest producer of seamless steel in terms of single plant's scale. Its process, technology, and equipment levels represent the highest level of seamless steel pipe rolling in the world today, surpassing that of the Sumitomo Metal Co., Ltd. TPCO has already been ranked among the three major groups of seamless steel pipes in the world.

(4) Internationalization Strategy and Achievements TPCO has been actively implementing the international development strategy and has embarked on a characteristic development path from technology introduction, digestion, and absorption to technology export and overseas investment and construction. In 2007, the joint construction of the seamless steel pipe project in the Republic of Belarus marked its shift to technology export from technology introduction. In 2009, the Indonesian oil well pipe processing project was completed and put into production. The first phase of the United States Seamless Steel Pipe Project, China's largest investment project on the steel pipe industry in the USA has been completed and put into operation, and the second phase is in smooth progress.

2. Analysis on Internationalization Experiences

After years of struggles, TPCO's seamless steel pipes are widely sold in foreign countries. TPCO's main product, oil casing, passed the certification of American Petroleum Institute (API) in 1994 and obtained the use rights of API monogram. In early 1996, it obtained the ISO9001 quality management system certification. TPCO's oil castings have been used in major oil fields in China and ended the country's dependence on imported casings. TPCO's products have been exported to nearly 100 countries and regions, and TPCO has become the main force in China's international competition.

TPCO always adheres to the principle of high starting point and high standards. In the earlier stage of construction, TPCO introduced the world's most advanced technologies and equipment in steelmaking, rolling, pipe processing and direct reduced iron from Germany, Italy, the USA, the UK, and other countries. With a complete automation system, it became a professional manufacturer of oil tubes with the most advanced technologies and the largest scale in China in the 1990s. Since its production, it has maintained a leading edge in technologies through continuous transformation. It ranks among the Top 3 steel pipe producers globally in terms of the total output of seamless steel tubes, core equipment and technologies. In March

2005, TPCO and German SMS Meer jointly won the bid of the Belarusian steel pipe project, marking that TPCO realized the transformation from technology import to technology export.

TPCO also has experience in investing in foreign projects and has built a tube processing plant in Indonesia. For establishment of the plant, TPCO and a Singaporean company, Tubulars International Pte Ltd. (TIPL) established a joint venture in Singapore, namely TPCO Pan Asia Pte Ltd., of which TPCO holds 51% of the shares and TIPL holds 49%. Afterward, TPCO Pan Asia Pte Ltd. invested in Indonesia and established PT. TPCO Pan Asia as its production plant. This plant was mainly designed for pipe processing with an annual output of 60,000 tons as designed. The plant produces oil casings, and the business scope covers the production and sales of steel pipes and the related products. The equipment and facilities employed for this project were mainly China-made with a total investment of $9.2 million.

TPCO has established a national-level technical center, possessed an advanced steel pipe processing test line, and enjoyed strong research and development power. TPCO's products have been developed from three steel grades in the original design to 25 steel grades with 235 varieties and tens of thousands of specifications, of which 62 have filled the domestic gap. With 33 national patents, a series of TP products with proprietary intellectual property rights came into being. Its oil casing is awarded as "Chinese Top Brand Products", and the seamless steel pipe is awarded as "National Inspection-Free Product" and "China Brand-Name Export Commodity".

3. Conclusions

TPCO's strategy for international development can be seen as a process starting from product export to technology export and then to capacity export. It first established a marketing network through product export, got to know the local market and the characteristics, and further sought opportunities to promote mature and reliable technologies overseas, gaining more overseas engineering experience. Starting from pipe processing which has relatively easier requirements on process and equipment, it chosen the joint venture method to directly build plants in foreign countries, thus accumulating experience in overseas construction and production management. TPCO has been making solid efforts step by step in a few years and embarked on a road of international development with its own characteristics. All its experiences and lessons are worth learning by Chinese steel enterprises.

### *13.3.5 Tsingshan Holding Group*

1. Internationalization Overview

Founded in 1992, Tsingshan Holding Group is a multinational enterprise that started its career in Wenzhou City, grew bigger to cover Zhejiang Province and China and finally went abroad. It now holds more than 30 enterprises like the Ruipu Technology Group, Qingtuo Industrial Group and Guangdong Jirui Technology Group. In 2016, the Group's output of stainless steel crude steel stood at 5.8 million tons with a sales value of 102.8 billion yuan, and its total employees reached more than 20,000. As the largest private enterprise in domestic stainless steel industry, Tsingshan Holding Group has been listed among China's Top 500 Enterprises, China's Top 500 Manufacturing Enterprises, China's Top 500 Private Enterprises, and China's Top 500 Private Manufacturing Enterprises.

Tsingshan Holding Group has established three production bases for nickel-chromium alloy smelting, stainless steel smelting, and steel rolling in Lishui (Zhejiang), Fuyang (Fujian), and Yangjiang and Qingyuan (Guangdong). It also owns a 47,000-ha nickel mine base in Sulawesi, Indonesia, and a 5000-ha chrome ore base in Zimbabwe. Building on this strength, Tsingshan Group has formed a complete industrial chain from the exploitation of raw materials of stainless steel such as nickel-chromium ore, nickel-chromium-iron smelting, and stainless steel smelting, to processing of bars, wires and sheets, steel pipe manufacturing, precision wire processing, terminal transportation, and international trade. The companies that the Group holds shares in mainly produces stainless steel ingots, steel bars, plates, wires, seamless pipes, and other products, which are widely used in petroleum, chemical engineering, machinery, electric power, automobile, shipbuilding, aerospace, food, pharmaceutical, decoration, etc. The Group has a well-established marketing network at home and abroad, and its sales channels covers the entire country and has extended to Southeast Asia, Europe, and America. It has established direct sales outlets in Wenzhou, Shanghai, Wuxi, and Foshan.

2. Analysis on Internationalization Experiences

Facing the grave situation of highly fluctuating international nickel prices, overcapacity of the domestic stainless steel industry, market downturn, and shrinking profit margins, Tsingshan Group has greatly leveraged its advantage and potential by acquiring and utilizing the high-quality mineral resources overseas, integrating mining companies and stainless steel production enterprises, and securing pricing discretion. By now, Tsingshan Group has obtained the mining rights of the nickel ore in Indonesia and chrome ore in Zimbabwe and has successively built plants in the mining area, making it a true multinational enterprise. In addition, the "RKEF + AOD" process for stainless steel production adopted by Tsingshan Group can save more than 20% of the cost, thus setting the lowest and most internationally competitive deadline for stainless steel production in the world.

The early action in deploying resources in the mining and export businesses of nickel ore in Indonesia as well as the ferronickel smelting industry chain constitutes a key link in the global layout of Tsingshan Group. In 2009, Tsingshan Group and Indonesia Eight Star Investment Co., Ltd. jointly established the Sulawesi Mining Investment Co., Ltd. and obtained the mining rights of the 47,000-ha laterite-nickel ore in Indonesia. In February 2010, the company's first batch of nickel ore from this mining area in Indonesia was successfully shipped to China. In October 2013, President Xi Jinping and former Indonesian President Susilo held a signing ceremony for the China-Indonesia Business Agreement in Jakarta. Listed as one of the projects under the framework of this agreement, Qingshan Industrial Park in China-Indonesia Economic and Trade Cooperation Zone in Sulawesi, Indonesia was signed successfully.

In 2014, the Indonesian government began to implement export restrictions on raw ore, which completely banned the export of nickel ore. Tsingshan Industrial Park had become the fastest-constructing and most successful foreign-invested mining project with the fastest capital injection in Indonesia after the prohibition policy was launched. The planned land area of the park is over 1300 ha. It is mainly equipped with a thermal power plant with a total capacity of more than 1,000 MW, a 100,000-ton-level wharf, a simple airport and a living area with a total construction area of about 200,000 $m^2$. The total planned investment currently is $5 billion. The ferronickel smelting and the supporting thermal power plant project in Phase-I occupied an area of 95 ha and was put into trial operation in January 2015. In April 2015, the first ship laden with 22,000 tons of ferronickel left the port, marking the official operation of Phase-I ferronickel smelting project of Tsingshan Industrial Park. In May 2015, Indonesian President Joko attended the ribbon-cutting ceremony of Tsingshan Industrial Park in the China-Indonesia Economic and Trade Cooperation Zone. The Phase-II ferronickel smelting and supporting thermal power plant project of the industrial park was completed and put into operation in March 2016. In June of the same year, a one million ton/a stainless steel line was completed. In April 2017, the Phase-III ferronickel smelting project was officially put into operation. The project constructs a total of 20 RKEF production lines to produce 1.5 million tons ferronickel a year. In addition, the Phase-IV 600,000-ton ferrochrome and 700,000-ton cold rolling projects are under construction. Tsingshan Holding Group signed a 3.5-million ton steel works investment agreement in June 2017 with Delong Co., Ltd. and Eight Star Group.

3. Conclusions

There is experience in three perspectives worth learning from Tsingshan Group's global layout: the first is that enterprises should define their own industrial positioning, accumulate funds, and experience for production and operation, timely track the market trends in domestic and international market, and seize the opportunity to occupy international resources; second is that enterprises should pay attention to the state-to-state relations between China and foreign countries, adapt to the changes of local political environment and economic trends, and strive to raise its business to the attention of strategic cooperation among countries so that there will be favorable policies in resource allocation at home and abroad; third is that enterprises should strengthen the research and development of advanced technologies and process, integrate and extend the industrial chain, keep the technology, quality and cost controllable to enhance the core competitiveness.

### 13.3.6 Magang (Group) Holding Co. Ltd.

1. Internationalization Overview

Magang (Group) Holding Co. Ltd. (hereinafter referred to as Masteel) is a large-scale iron and steel complex and an important steel production base in China with a production scale of 20 million tons. In 2016, its crude steel output reached 18.63 million tons with a total asset of 85 billion yuan. The operating income stood at 55.1 billion yuan with a total profit of 976 million yuan. For a long time, Masteel has been paying great attention to internationalized development and has scored remarkable achievements. On the one hand, Masteel uses joint ventures with overseas consortiums and famous enterprises to sell its products to more than 50 countries and regions in Europe, the Americas, Southeast Asia, etc. On the other hand, the Group establishes strategic partnership with many internationally renowned companies like the United StatesGE, GermanySiemens, Danieli, DB, and KoreaPOSCO to enhance brand influence in the international market. In addition, with the successful development of a number of high-end products such as corrosion-resistant steel, weather-resistant H-beam, low-noise wheels, and 40-ton high-power locomotive wheels by Masteel, its competitiveness in the international market has been increasing and it has won recognition from many well-known overseas manufacturers. By now, Masteel's wheels have obtained the AAR Certification for American railway products, EU Railway Industry Standard Certification, German IRIS International Railway Industry Standard Certification, and its H-beam has obtained the certification of American standard, Japanese standard, German standard, and European standard.

2. Analysis on Internationalization Experiences

Against the backdrop that the global steel industry has entered an age of meager profit or even loss, Masteel's lucrative wheel business has made an important contribution to the profit of its main steel business. Upgrading the wheel products, expanding the wheels' market and extending the industrial chain have now become a significant strategy for its development [1]. On May 30, 2014, Masteel acquired the century-old world-renowned brand, SAS VALDUNES in France, at a price of 13 million Euro. SAS VALDUNES, NSSC (Japan), BVV (Germany), and Lucini (Italy) are recognized as the world's four major manufacturers of high-speed axles. VALDUNES was once known as the "Industrial Flower of France" with a history of over 100 years, and it was the only specialized enterprise in producing wheels, axles, and wheel sets. VALDUNES possesses core technologies in high-speed rail wheels, axles, and wheel sets. Its high-speed rail wheels have created a world record of 574.8 km per hour. Suffering from the industrial downturn and other factors, Valdunes' profitability was greatly affected and it then was mired in financial distress. It entered the insolvency proceedings on October 11, 2013, and went into bankruptcy reorganization on March 31, 2014. Through this acquisition, Masteel acquired all the assets of Valdunes, including the real estates, tangible assets, intangible assets and all of Valdunes' inventories and work-in-progress, Valdunes' lease contracts, contracts with the customers, and contracts with suppliers of various types, maintenance contracts, etc., and kept all of the employees. After this acquisition, Masteel took charge of its high-quality talents who master the core technology and its R&D systems, as well as a series of technical patents of Valdunes high-speed axles. This enables Masteel to make a quick integration of self-owned technologies and the world-leading technologies and create the core technology with independent intellectual property rights, thus promoting the industrialization and high-end process of China's high-speed axle products. Meanwhile, through the acquisition of the Valdunes, Masteel also took over its sales network across more than 40 countries and regions around the world and its seats in a few international organizations like the International Union of Railways and International Railway Industry Standard. This has played an active role in supporting the internationalization process of China's high-speed axle products and increasing the international market share. Gao Haijian, chairman of Masteel, said that, by taking the acquisition of the Valdunes, the Group will build itself into a leading enterprise globally in the full range of high-speed axle products in about 3 years [13].

Expanding overseas markets through reorganization and M&A of foreign superior enterprises are important ways for the "internationalization" of Chinese iron and steel enterprises. Masteel's acquisition of the Valdunes has completely broken the long-term monopoly and blockade of products and technologies in high-speed rail axles to China by foreign countries. By quickly taking possession of the internationally advanced high-speed axle technologies, the Group will help to upgrade China's high-speed axle technologies, accelerate the localization of high-speed axles, reduce the cost of manufacturing, maintenance and use of high-speed railways. Furthermore, it will fundamentally guarantee the safe operation of China's high-speed railways, support the future development of China's high-speed railway with self-owned axles, and support the national strategy of high-speed railway "going out". Masteel is a successful example of the "internationalized development" of steel enterprises.

3. Conclusions

At present, some foreign superior enterprises are affected by the financial crisis. Due to problems such as financial matters and debts, they will or preparing to sell or spin-off assets or businesses in whole or in part. This is a rare opportunity for Chinese companies to implement cross-border mergers and acquisitions and accelerate the "internationalized" development. It is also a strategic pivot in taking the high ground of international advanced technologies, expanding international marketing channels, and enhancing its brand image in the Post-Financial Crisis Era. Through integration, the two parties involved take coordinated steps in system organization, management ideas, production management, and marketing strategies, which can produce "1 + 1 > 2" economic benefits and M&A synergy.

## 13.4 Prospects and Path Analysis of Internationalization Trend

China's iron and steel enterprises have not achieved obvious results under the simple international development model of "going out" for either markets or resources. They have not obtained the expected results in terms of holding resources, expanding markets and increasing profits. In the future, the steel industry will not "go global" simply for internationalization. The international development will pay more attention to efficiency and promotion of the competitiveness of enterprises. The development model will transform from the following five aspects: (1) focus changes from product trade to equal emphasis on product trade and service trade; (2) transform from product's "going global" to capital's "going global;" (3) transform from self-motivated "going global" to organized "going global;" (4) transform from investing in resource sector to investing in manufacturing and service sector; and (5) transform from "going global" of a single enterprise to "going global" of the whole industrial chain by combining both the upstream and downstream enterprises and the related enterprises.

As one of the most competitive industries for "going global", China's steel industry should lay emphasis on the new trend of international development in implementing the internationalization strategy, especially actively participating in the "Belt and Road" Initiative to achieve positive results. Efforts should be focused on the following five aspects: (1) to carry out project cooperation and provide supporting services riding on major development opportunities like the international capacity cooperation; (2) to explore and establish strategic cooperation and restructuring that covers the entire industry chain, and unite the international and domestic strategic partners to deepen industrial cooperation and improve the level and ability of international cooperation; (3) to actively integrate and utilize global innovation resources to make breakthroughs in overseas talent training, international project cooperation, and introduction of overseas high-end talents; (4) to provide value-added supply chain services to global metallurgical enterprises and industrial enterprises, and vigorously develop and cultivate overseas strategic terminal customers depending on the self-owned trading companies or the domestic strategic alliance of trading companies; and (5) to carefully prepare for and seriously deal with international trade frictions so as to safeguard the rights and interests of enterprises and promote sound development of international trade.

## 13.5 Industrial Practices of Internationalized Development

Responding to the national "Belt and Road" Initiative, China Metallurgical Industry Planning and Research Institute (hereinafter referred to as MPI) has been actively carrying out relevant researches on the production capacity and investment cooperation between China and foreign countries and providing guidance for the "going global" enterprises of the steel industry and other advantageous ones; entrusted by the National Development Bank to provide consultation for national planning, MPI has sent the researchers for several times to Australia, Africa, South America, the Commonwealth of Independent States, Southeast Asia, and other dozens of countries to conduct field visits. They have made in-depth studies on the exploitation and utilization of mineral resources, put forward planning suggestions to provide decision-making basis for the better cooperation and win-win results between China and foreign countries; MPI developed the consulting business for steel industry with a global perspective and has provided services in market research, feasibility study, equipment procurement, project evaluation, and proposals for domestic and foreign enterprises and institutions; with its long-term commitment to introducing advanced metallurgical technologies abroad, MPI has been keeping close cooperation with many foreign well-reputed universities and academic institutions like the National Metallurgic Academy of Ukraine and the National Titanium Design Institute of Ukraine. MPI has achieved outstanding results in organizing the introduction of RKEF ferronickel production process and GOR converter smelting of stainless steel. In addition, in recent years, MPI has conducted various exchanges and cooperation with relevant enterprises in Australia, Japan, the USA, Austria, Germany, the UK, Italy, Zimbabwe, and others, as shown in Table 13.6.

**Table 13.6** Practices of MPI in promoting internationalization of iron and steel industry

| No. | Business segments | Segment introduction | Typical cases |
|---|---|---|---|
| 1 | Researches on Related Issues of the "Belt and Road" Initiative | Entrusted by the National Development and Reform Commission, the Ministry of Industry and Information Technology, and the National Development Bank, the Institute has carried out researches on related issues of the "Belt and Road" Initiative | *Study on the "Going Global" Thinking of the Steel Industry under the "Belt and Road" Strategy*<br>*Planning Research on Capacity Cooperation with African Countries*<br>*Special Report on China-Saudi Arabia Joint Research on Capacity Cooperation in Bilateral Strategic Planning*<br>*Planning Study of China-Jamaica Capacity and Investment Cooperation*<br>*Promotion of China-Pakistan Joint Industrial Zone Project of "Two Parks in Two Countries"*<br>and others |

(continued)

**Table 13.6** (continued)

| No. | Business segments | Segment introduction | Typical cases |
|---|---|---|---|
| 2 | National Industrial Development Planning Research | Entrusted by the National Development Bank, the Institute carried out consultation on the development of mining and steel industry in different countries | Development planing of the mining and steel industry in Zambia, Sweden, Cambodia, Guinea, Venezuela, Tajikistan, Myanmar, South Africa, Mozambique, Bolivia, Western Africa, etc. |
| 3 | Consulting and Information Service for Foreign Companies | Provide research reports and technical advices to foreign companies | It has offered the cost analysis report on producing magnesium alloy to the Boeing Company, bank-level feasibility study report and evaluation report to Rio Tinto Group's Simandou project, research report on the major strategic issues of China's steel industry from 2013 to 2018 to Australia's Arrium Company, and the data information services on "*Blue Paper of China's Steel Industry*" to Oldendorff |
| 4 | Consulting Service for Domestic Enterprise's "Going Global" | Provide consulting service for domestic companies' "Going Global" | *Feasibility Study Report on Fuhai Steel Project in Indonesia of Dongfang Mineral Resources Co., Ltd.*<br>*Feasibility Review Report on the Integrated Steel Plant of Bolivia ElMutún Iron and Steel Company*<br>*Evaluation Report of Anshan Steel's Project in Sulawesi, Indonesia*<br>*Project Proposal of China-Egypt 5-Million Ton Steel Base (Phase-I) and others* |
| 5 | Introduction of Foreign Advanced Technologies | It has organized the introduction of Ukraine RKEF ferronickel process and GOR converter smelting of stainless steel | *RKEF Ferronickel Project of Baosteel Desheng Stainless Steel Co., Ltd.*<br>*GOR Converter Smelting of Stainless Steel Project of Hebei Taigang Group*<br>*GOR Converter Smelting Project of Southwest Stainless Steel and others* |

# References


1. China Iron and Steel Association. (2017). China steel statistics 2017. Beijing: Statistics Department of China Iron and Steel Industry Association Information.
2. Jia, C.N., & Zhang, L.H. (2017). 'Belt and Road' is not only a dream of China but also a dream of Serbia. Beijing: China Metallurgical News.
3. Zhu, Z.X., & Ding, G.F. (2016). Legal risk analysis of Chinese enterprises' going global. Journal of Hebei Tangshan Teachers College, 38(1).

4. General Administration of Customs of the People's Republic of China. http://www.customs.gov.cn/customs/.
5. National Bureau of Statistics of the People's Republic of China. http://www.stats.gov.cn/.
6. Information Center of Ministry of Land and Resources. (2015). Annual review of world mineral resources. Beijing: Geological Publishing House.
7. Energy Statistics Department, National Bureau of Statistics. (2016). China energy statistics yearbook. Beijing: China Statistics Press.
8. Customs Information Network. (2017). Customs data. Beijing: Beijing Zhonghaitong Technology Co., Ltd.
9. Introduction to POSCO Korea. http://www.posco-china.com.
10. China Steel News. (2017). Analysis on the layout of Posco in overseas emerging markets. Beijing: China Steel News.
11. Zhang, W.W. (2012). Analysis on overseas layout strategy of Pohang. Beijing: World Metals (A06 version).
12. Luo, Y., Ke, H.F., Yuan, Y.F., et al. (2014). Operation and development direction of Pohang Steel Corporation. World Steel, (3), 67–72.
13. Ma Steel Co., Ltd. (2015). Ma steel: Main practice and experience on acquiring French Walton Company. Metallurgical Finance, (08), 18–20.

The manufacturer's authorised representative in the EU is Springer Nature Customer Service Centre GmbH, Europaplatz 3, 69115 Heidelberg, Germany. If you have any concerns regarding our products, please contact ProductSafety@springernature.com

Printed and bound by CPI Group (UK) Ltd, Croydon, CR0 4YY
30/09/2026
02225333-0014